Student Solutions Manual

Intermediate Algebra
An Applied Approach

NINTH EDITION

Richard N. Aufmann
Palomar College

Joanne S. Lockwood
Nashua Community College

Prepared by

Ellena Reda
Dutchess Community College

BROOKS/COLE
CENGAGE Learning

Australia • Brazil • Japan • Korea • Mexico • Singapore • Spain • United Kingdom • United States

For product information and technology assistance, contact us at **Cengage Learning Customer & Sales Support, 1-800-354-9706**

For permission to use material from this text or product, submit all requests online at **www.cengage.com/permissions** Further permissions questions can be emailed to **permissionrequest@cengage.com**

ISBN-13: 978-1-285-41745-5
ISBN-10: 1-285-41745-3

Brooks/Cole
20 Davis Drive
Belmont, CA 94002-3098
USA

Cengage Learning is a leading provider of customized learning solutions with office locations around the globe, including Singapore, the United Kingdom, Australia, Mexico, Brazil, and Japan. Locate your local office at: **www.cengage.com/global**

Cengage Learning products are represented in Canada by Nelson Education, Ltd.

To learn more about Brooks/Cole, visit **www.cengage.com/brookscole**

Purchase any of our products at your local college store or at our preferred online store **www.cengagebrain.com**

Printed in the United States of America
1 2 3 4 5 6 7 17 16 15 14 13

Contents

Chapter 1: Review of Real Numbers

Prep Test

1. $\dfrac{5}{12}+\dfrac{7}{30}=\dfrac{25}{60}+\dfrac{14}{60}=\dfrac{39}{60}=\dfrac{3\cdot13}{3\cdot30}=\dfrac{13}{20}$

2. $\dfrac{8}{15}-\dfrac{7}{20}=\dfrac{32}{60}-\dfrac{21}{60}=\dfrac{11}{60}$

3. $\dfrac{5}{6}\cdot\dfrac{4}{15}=\dfrac{5\cdot2\cdot2}{3\cdot2\cdot5\cdot3}=\dfrac{2}{9}$

4. $\dfrac{4}{15}\div\dfrac{2}{5}=\dfrac{4}{15}\cdot\dfrac{5}{2}=\dfrac{2\cdot2\cdot5}{5\cdot3\cdot2}=\dfrac{2}{3}$

5. $8.000+29.340+7.065=44.405$

6. $92.00-18.37=73.63$

7. $2.19(3.4)=7.446$

8. $32.436\div0.6=324.36\div6=54.06$

9.
a) $-6>-8$ Yes
b) $-10<-8$ No
c) $0>-8$ Yes
d) $8>-8$ Yes

10.
$\dfrac{1}{2}=0.5$ C

$\dfrac{7}{10}=0.7$ D

$\dfrac{3}{4}=0.75$ A

$\dfrac{89}{100}=0.89$ B

Section 1.1

Concept Check

1. a) integers: $0, -3$

b) rational numbers: $-\dfrac{15}{2}, 0, -3, 2.\overline{33}$

c) irrational numbers:
$$\pi, 4.232232223..., \dfrac{\sqrt5}{4}, \sqrt7$$

d) real numbers: all

3. $\dfrac{2}{3}$

5. $-1.\overline{45}$

7. 5.6

9. 0

11. 54

Objective A Exercises

13. Replace y with each element in the set to determine if the inequality is true or false.
$-6>-4$ False
$-4>-4$ False
$7>-4$ True

15. Replace w with each element in the set to determine if the inequality is true or false.
$-2\le-1$ True
$-1\le-1$ True
$0\le-1$ False
$1\le-1$ False

17. Replace b with each element in the set then evaluate the expression.
$-(-9)=9$
$-(0)=0$
$-(9)=-9$

19. Replace c with each element in the set then evaluate the expression.
$|-4|=4$
$|0|=0$
$|4|=4$

21. $-x>0$ whenever the value of x is a negative real number.

Objective B Exercises

23. $\{-2, -1, 0, 1, 2, 3, 4\}$

25. $\{2, 4, 6, 8, 10, 12\}$

27. $\{3, 6, 9, 12, 15, 18, 21, 24, 27, 30\}$

29. $\{x|\, x > 4, x \in \text{integers}\}$

31. $\{x|\, x \geq -2\}$

33. $\{x|\, 0 < x < 1\}$

35. $\{x|\, 1 \leq x \leq 4\}$

37. Yes

39. No

41. No

43. $\{x|\ x < 2\}$

45. $\{x|\ x \geq 1\}$

47. $\{x|\ -1 < x < 5\}$

49. $\{x|\ 0 \leq x \leq 3\}$

51. $(-2, 4)$

53. $[-1, 5]$

55. $(-\infty, 1)$

57. $[-2, \infty)$

59. $\{x|\, 0 < x < 8\}$

61. $\{x|-5 \leq x \leq 7\}$

63. $\{x|\,-3 \leq x < 6\}$

65. $\{x|\, x \leq 4\}$

67. $\{x|\, x > 5\}$

69. $(-2, 5)$

71. $[-1, 2]$

73. $(-\infty, 3]$

75. $[3, \infty)$

Objective C Exercises

77. $A \cup B = \{1, 2, 4, 6, 9\}$

79. $A \cup B = \{2, 3, 5, 8, 9, 10\}$

81. $A \cup B = \{-4, -2, 0, 2, 4, 8\}$

83. $A \cup B = \{1, 2, 3, 4, 5\}$

85. $A \cap B = \{6\}$

87. $A \cap B = \{5, 10, 20\}$

89. $A \cap B = \emptyset$

91. $A \cap B = \{4, 6\}$

93. $[5, \infty) \cap (0, 5)$

95. $\{x|\ x > 1\} \cup \{x|\ x < -1\}$

97. $\{x|\ x \leq 2\} \cap \{x|\ x \geq 0\}$

99. $\{x|\ x > 1\} \cap \{x|\ x \geq -2\}$

-5 -4 -3 -2 -1 0 1 2 3 4 5

101. $\{x|\ x > -3\} \cup \{x|\ x < 1\}$

-5 -4 -3 -2 -1 0 1 2 3 4 5

103. $[-5, 0) \cup (1,4]$

-5 -4 -3 -2 -1 0 1 2 3 4 5

105. $[-3, 3] \cap [0,5]$

-5 -4 -3 -2 -1 0 1 2 3 4 5

Critical Thinking

107. $|x| < 4$

109. $|x-1| \leq 5$

111. $|x+2| < 2$

113. $|x-a| > b$

Projects or Group Activities

115. $A \cup B = A$

117. $B \cap B = B$

119. $A \cap R = A$

121. $B \cup R = R$

123. $R \cup R = R$

Section 1.2

Concept Check

1. a) To add two numbers with the same sign add the absolute values of the numbers. Then attach the sign of the addends.

b) To add two numbers with different signs find the absolute value of each number. Subtract the smaller of these two numbers from the larger. Then attach the sign of the larger absolute value.

3. No. For example, $-5 + (-3) = -8$

5. One number is positive and one number is negative

7. At least one of the numbers is zero.

9. $(-5)^6$

Objective A Exercises

11. Negative

13. Positive

15. $-15(-11) = 165$

17. $-39 - 5$
$= -39 + (-5)$
$= -44$

19. $87 \div (-3) = -29$

21. $39 - (-6)$
$= 39 + (6)$
$= 45$

23. $(-12)(15) = -180$

25. $238 \div 17 = 14$

27. $-5 + 19 = 14$

29. $44 \div (-4) = -11$

31. $-14 - 8$
$= -14 + (-8)$
$= -22$

33. $9 - (-9)$
$= 9 + (9)$
$= 18$

35. $27 + (-8) = 19$

37. $2(-26) = -52$

39. The total of -6 and 24 is 18.
$-6 + 24 = 18$

41. 80 divided by 80 is 1.
$80 \div 80 = 1$

43. $-56 \div (-7) = 8$

45. $-18 + (-10) = -28$

47. $(-17)(-12) = 204$

49. $26 + (-6) = 20$

51. $24 - (-2)$
$= 24 + (2)$
$= 26$

53. $-21 - (-10)$
$= -21 + (10)$
$= -11$

55. $107 - (-153)$
$= 107 + (153)$
260
The difference between the average temperatures during the day and at night on the moon is 260° C.

57. $-282 + 223 = -59$
The new elevation of the hiker is -59 ft.

Objective B Exercises

59. Negative

61. $-2^4 = -(2 \cdot 2 \cdot 2 \cdot 2) = -16$

63. $(-5)^3 = (-5) \cdot (-5) \cdot (-5) = -125$

65. $(-2)^{10} = (-2)(-2)(-2)(-2)(-2)(-2)(-2)(-2)(-2)(-2)$
$= 1024$

67. $(-1)^{50} = 1$

69. $5 - 8 - 12 + 3$
$= 5 + (-8) + (-12) + 3$
$= -3 + (-12) + 3$
$= -15 + 3$
$= -12$

71. $24 \div 2 \cdot 6 = 12 \cdot 6 = 72$

73. $-28 - 12 \div 4 = -28 - 3 = -28 + (-3) = -31$

75. $-9 + 4(8 - 15) = -9 + 4(-7) = -9 + (-28)$
$= -37$

77. $27 \div (-3) + 5(2 - 8)^2$
$= 27 \div (-3) + 5(-6)^2$
$= 27 \div (-3) + 5(36)$
$= 27 \div (-3) + 180$
$= -9 + 180$
$= 171$

79. $3 \cdot 8 - 6(4-9) + \dfrac{3-15}{4}$

$= 3 \cdot 8 - 6(-5) + \dfrac{-12}{4} = 24 - (-30) + (-3)$

$= 24 + (30) + (-3)$
$= 51$

81. $-3(5-8)^3 + (19-7) \div (1-3)$
$= -3(-3)^3 + (19-7) \div (1-3)$
$= -3(-27) + (12) \div (-2)$
$= 81 + (-6)$
$= 75$

83. $\dfrac{15-19}{2^2} \cdot \dfrac{3(2-7)^2}{3 \cdot 2 - 1} \div \dfrac{3 \cdot 2 \cdot 2^4}{10 - 4 \cdot 3}$

$= \dfrac{-4}{4} \cdot \dfrac{3(25)}{5} \div \dfrac{96}{-2} = -1 \cdot 15 \div (-48)$

$= -15 \div -48$

$= \dfrac{5}{16}$

Critical Thinking

85. There is a 7 in the ones place in 7^{25}.

87. $\left(2^3\right)^4 < 2^{(3^4)}$

Section 1.3

Concept Check

1. The least common multiple of two numbers is the smallest number that is a multiple of each of the two numbers. When adding fractions with different denominators we need to first rewrite the fractions as equivalent fractions with a common denominator. The lowest common denominator is the LCM of the denominators.

3. All integers are rational numbers.

5. There is a smallest possible positive integer. However, there is not a smallest positive rational number.

7. a. The reciprocal of $-\dfrac{8}{27}$ is $-\dfrac{27}{8}$.

   b. $\dfrac{2}{3} \cdot \left(-\dfrac{27}{8}\right) = -\dfrac{9}{4}$

Objective A Exercises

9. Find the prime factorization of 83 and 2.
   $83 = 83 \cdot 1$
   $2 = 2 \cdot 1$
   The LCM $= 83 \cdot 2 = 166$
   The GCF is 1.

11. Find the prime factorization of 15 and 75.
   $15 = 5 \cdot 3$
   $75 = 5 \cdot 5 \cdot 3$
   The LCM is $5 \cdot 5 \cdot 3$.
   The GCF is $5 \cdot 3$

13. Find the prime factorization of 6 and 2.
   $6 = 3 \cdot 2$
   $2 = 2 \cdot 1$
   The LCM is 6.
   The GCF is 2.

15. Find the prime factorization of 10 and 55.
   $10 = 5 \cdot 2$
   $55 = 11 \cdot 5$
   The LCM is 110.
   The GCF is 5.

17. Find the prime factorization of 34 and 85.
   $34 = 17 \cdot 2$
   $85 = 17 \cdot 5$
   The LCM is 170.
   The GCF is 17.

19. Find the prime factorization of 140 and 7.
   $140 = 7 \cdot 5 \cdot 2 \cdot 2$
   $7 = 7 \cdot 1$
   The LCM is 140.
   The GCF is 7.

21. Find the prime factorization of 14, 42 and 18.
   $14 = 7 \cdot 2$
   $42 = 7 \cdot 3 \cdot 2$
   $18 = 3 \cdot 3 \cdot 2$
   The LCM is 126.
   The GCF is 2.

23. Find the prime factorization of 30, 5 and 10.
   $30 = 5 \cdot 3 \cdot 2$
   $5 = 5 \cdot 1$
   $10 = 5 \cdot 2$
   The LCM is 30.
   The GCF is 5.

25. Find the prime factorization of 15, 45 and 10.
   $15 = 5 \cdot 3$
   $45 = 5 \cdot 3 \cdot 3$
   $10 = 5 \cdot 2$
   The LCM is 90.
   The GCF is 5.

27. Find the prime factorization of 12, 42 and 14.
   $12 = 3 \cdot 2 \cdot 2$
   $42 = 7 \cdot 3 \cdot 2$
   $14 = 7 \cdot 2$
   The LCM is 84.
   The GCF is 2.

Objective B Exercises

29. $\left(-\dfrac{35}{12}\right)\left(\dfrac{4}{3}\right) = -\dfrac{35 \cdot 4}{12 \cdot 3}$

   $= -\dfrac{7 \cdot 5 \cdot 2 \cdot 2}{3 \cdot 2 \cdot 2 \cdot 3} = -\dfrac{35}{9}$

31. $\left(-\dfrac{1}{3}\right)\left(\dfrac{9}{29}\right) = -\dfrac{1 \cdot 9}{3 \cdot 29}$

$= -\dfrac{1 \cdot 3 \cdot 3}{3 \cdot 29} = -\dfrac{3}{29}$

33. $-\dfrac{3}{2} - \left(-\dfrac{1}{9}\right) = -\dfrac{3}{2} + \dfrac{1}{9} = -\dfrac{27}{18} + \dfrac{2}{18}$

$= \dfrac{-27 + 2}{18} = \dfrac{25}{18}$

35. $-\dfrac{5}{4} + \dfrac{19}{22} = \dfrac{-55}{44} + \dfrac{38}{44}$

$= \dfrac{-55 + 38}{44} = -\dfrac{17}{44}$

37. $-\dfrac{11}{4} \div \left(-\dfrac{1}{8}\right) = -\dfrac{11}{4} \cdot \left(-\dfrac{8}{1}\right) = \dfrac{11 \cdot 8}{4 \cdot 1}$

$= \dfrac{11 \cdot 2 \cdot 2 \cdot 2}{2 \cdot 2} = \dfrac{22}{1} = 22$

39. $\dfrac{32}{39} + \left(-\dfrac{14}{3}\right) = \dfrac{32}{39} + \dfrac{-182}{39}$

$= \dfrac{32 + (-182)}{39} = -\dfrac{150}{39} = -\dfrac{50}{13}$

41. $\left(\dfrac{3}{4}\right)\left(-\dfrac{17}{27}\right) = -\dfrac{3 \cdot 17}{4 \cdot 27}$

$= -\dfrac{3 \cdot 17}{2 \cdot 2 \cdot 3 \cdot 3 \cdot 3} = -\dfrac{17}{36}$

43. $\dfrac{3}{4} \div \left(-\dfrac{39}{16}\right) = \dfrac{3}{4} \cdot \left(-\dfrac{16}{39}\right) = -\dfrac{3 \cdot 16}{4 \cdot 39}$

$= -\dfrac{3 \cdot 2 \cdot 2 \cdot 2 \cdot 2}{2 \cdot 2 \cdot 3 \cdot 13} = -\dfrac{4}{13}$

45. $-\dfrac{10}{37} \div \left(-\dfrac{5}{2}\right) = -\dfrac{10}{37} \cdot \left(-\dfrac{2}{5}\right) = \dfrac{10 \cdot 2}{37 \cdot 5}$

$= \dfrac{5 \cdot 2 \cdot 2}{37 \cdot 5} = \dfrac{4}{37}$

47. $-\dfrac{11}{24}\left(\dfrac{32}{9}\right) = -\dfrac{11 \cdot 32}{24 \cdot 9}$

$= -\dfrac{11 \cdot 2 \cdot 2 \cdot 2 \cdot 2 \cdot 2}{3 \cdot 2 \cdot 2 \cdot 2 \cdot 3 \cdot 3} = -\dfrac{44}{27}$

49. $\dfrac{12}{11} \div \left(-\dfrac{1}{3}\right) = \dfrac{12}{11} \cdot \left(-\dfrac{3}{1}\right) = -\dfrac{12 \cdot 3}{11 \cdot 1} = -\dfrac{36}{11}$

51. $\dfrac{5}{9} + \left(-\dfrac{34}{9}\right) = -\dfrac{29}{9}$

53. $\dfrac{14}{9} + \left(-\dfrac{11}{18}\right) = \dfrac{28}{18} - \dfrac{11}{18} = \dfrac{17}{18}$

55. $-\dfrac{2}{5}\left(-\dfrac{8}{7}\right) = \dfrac{2 \cdot 8}{5 \cdot 7} = \dfrac{16}{35}$

57. $\dfrac{5}{39} \div \left(-\dfrac{40}{13}\right) = \dfrac{5}{39} \cdot \left(-\dfrac{13}{40}\right) = -\dfrac{5 \cdot 13}{39 \cdot 40}$

$= -\dfrac{5 \cdot 13}{3 \cdot 13 \cdot 2 \cdot 2 \cdot 2 \cdot 5} = -\dfrac{1}{24}$

59. $-\dfrac{11}{4} - \left(-\dfrac{16}{9}\right) = \dfrac{-99}{36} + \dfrac{64}{36} = \dfrac{-99 + 64}{36}$

$= -\dfrac{35}{36}$

61. $-\dfrac{3}{4} + \dfrac{5}{6} = \dfrac{-9}{12} + \dfrac{10}{12} = \dfrac{1}{12}$

63. $-\dfrac{11}{12} - \left(\dfrac{13}{18}\right) = \dfrac{-33}{36} - \dfrac{26}{36} = \dfrac{-33 - 26}{36} = -\dfrac{59}{36}$

65. $-\dfrac{4}{5} - \left(\dfrac{2}{3}\right) = \dfrac{-12}{15} - \dfrac{10}{15} = \dfrac{-12 - 10}{15} = -\dfrac{22}{15}$

67. $\dfrac{5}{8} \div \left(-\dfrac{7}{5}\right) = \dfrac{5}{8} \cdot \left(-\dfrac{5}{7}\right) = -\dfrac{5 \cdot 5}{8 \cdot 7} = -\dfrac{25}{56}$

69. $\dfrac{5}{8} + \left(\dfrac{13}{12}\right) = \dfrac{15}{24} + \dfrac{26}{24} = \dfrac{15 + 26}{24} = \dfrac{41}{24}$

71. No. For instance, there is no integer between 2 and 3.

Objective C Exercises

73. $\dfrac{17}{20} = 0.85$

75. $\dfrac{7}{22} = 0.3\overline{18}$

77. $\dfrac{5}{12} = 0.41\overline{6}$

79. $\dfrac{6}{13} = 0.\overline{461538}46$

81. $0.0986 \div 0.29 = 0.34$

83. $3.4 + (-0.09) = 3.31$

85. $0.08 + (-3.6) = 0.08 - 3.6 = -3.52$

87. $-60 - (-1.3) = -60 + 1.3 = -58.7$

89. $-1.8 + (-3.4) = -5.2$

91. $0.2 + (-90) = -89.8$

93. $-2.7(-0.28) = 0.756$

95. $-0.033 - (-3.6) = -0.033 + 3.6 = 3.567$

97. $0.13 - (-0.017) = 0.13 + 0.017 = 0.147$

99. $-\dfrac{3}{4} \div \dfrac{5}{8} \cdot \left(-\dfrac{10}{11}\right) = -\dfrac{3}{4} \cdot \dfrac{8}{5} \cdot \left(-\dfrac{10}{11}\right)$

$= -\dfrac{24}{20} \cdot \left(-\dfrac{10}{11}\right) = \dfrac{12}{11}$

101. $\left(\dfrac{2}{3} - \dfrac{5}{6}\right)^2 - \dfrac{5}{12} = \left(-\dfrac{1}{6}\right)^2 - \dfrac{5}{12}$

$= \dfrac{1}{36} - \dfrac{5}{12} = -\dfrac{14}{36} = -\dfrac{7}{18}$

103. $-\dfrac{5}{16} \cdot \dfrac{8}{9} - \dfrac{5}{6} \div \dfrac{3}{4} = -\dfrac{5}{16} \cdot \dfrac{8}{9} - \dfrac{5}{6} \cdot \dfrac{4}{3}$

$= -\dfrac{5}{18} - \dfrac{20}{18} = -\dfrac{25}{18}$

105. $\dfrac{1}{2} - \left(\dfrac{2}{3} \div \dfrac{5}{9}\right) + \dfrac{5}{6} = \dfrac{1}{2} - \left(\dfrac{2}{3} \cdot \dfrac{9}{5}\right) + \dfrac{5}{6}$

$= \dfrac{1}{2} - \dfrac{18}{15} + \dfrac{5}{6} = \dfrac{15 - 36 + 25}{30} = \dfrac{4}{30} = \dfrac{2}{15}$

107. $\dfrac{1}{6} - \dfrac{5}{4}\left(-\dfrac{7}{12} + \dfrac{1}{24}\right) = \dfrac{1}{6} - \dfrac{5}{4}\left(-\dfrac{13}{24}\right)$

$= \dfrac{1}{6} + \dfrac{65}{96} = \dfrac{81}{96} = \dfrac{27}{32}$

109. $\left(-\dfrac{1}{2}\right)^3 \div \left(-\dfrac{3}{2} - \dfrac{1}{4}\right) - \dfrac{2}{3}$

$= \left(-\dfrac{1}{2}\right)^3 \div \left(-\dfrac{7}{4}\right) - \dfrac{2}{3} = -\dfrac{1}{8} \div -\dfrac{7}{4} - \dfrac{2}{3}$

$= -\dfrac{1}{8} \cdot \dfrac{4}{7} - \dfrac{2}{3} = \dfrac{1}{14} - \dfrac{2}{3} = -\dfrac{25}{42}$

111. $0.4(1.2 - 2.3)^2 + 5.8$
$= 0.4(-1.1)^2 + 5.8$
$= 0.4(1.21) + 5.8$
$= 6.284$

113. $1.75 \div 0.25 - (1.25)^2$
$= 1.75 \div 0.25 - 1.5625$
$= 7 - 1.5625$
$= 5.4375$

115. $25.76 \div (6.54 \div 3.27)^2$
$= 25.76 \div (2)^2$
$= 25.76 \div 4$
$= 6.44$

117.
$$\frac{\dfrac{2}{3}}{\dfrac{4}{5}} = \frac{2}{3} \div \frac{4}{5} = \frac{2}{3} \cdot \frac{5}{4} = \frac{5}{6}$$

119.
$$\frac{\dfrac{2}{3} - \dfrac{5}{6}}{\dfrac{3}{4} - \dfrac{1}{2}} = \frac{-\dfrac{1}{6}}{\dfrac{1}{4}} = -\frac{1}{6} \div \frac{1}{4} = -\frac{1}{6} \cdot \frac{4}{1} = -\frac{2}{3}$$

121.
$$\frac{\dfrac{2}{3} - \left(\dfrac{1}{2}\right)^2}{\dfrac{5}{4}\left(\dfrac{1}{2} - \dfrac{3}{4}\right)} = \frac{\dfrac{2}{3} - \dfrac{1}{4}}{\dfrac{5}{4}\left(-\dfrac{1}{4}\right)} = \frac{\dfrac{5}{12}}{-\dfrac{5}{16}}$$

$$= \frac{5}{12} \div -\frac{5}{16} = \frac{5}{12} \cdot -\frac{16}{5} = -\frac{4}{3}$$

123.
$$\frac{5}{8} - \frac{\dfrac{2}{3} + \dfrac{1}{4}}{\dfrac{5}{6} - \dfrac{7}{8}} = \frac{5}{8} - \frac{\dfrac{11}{12}}{-\dfrac{1}{24}} = \frac{5}{8} - \frac{11}{12} \div -\frac{1}{24}$$

$$= \frac{5}{8} - \frac{11}{12} \cdot -\frac{24}{1} = \frac{5}{8} + \frac{22}{1} = \frac{181}{8}$$

125.
$$\frac{1}{2} - \frac{\dfrac{17}{25}}{4 - \dfrac{3}{5}} + \frac{1}{5} = \frac{1}{2} - \frac{\dfrac{17}{25}}{\dfrac{17}{5}} + \frac{1}{5}$$

$$= \frac{1}{2} - \frac{17}{25} \div \frac{17}{5} + \frac{1}{5} = \frac{1}{2} - \frac{17}{25} \cdot \frac{5}{17} + \frac{1}{5}$$

$$= \frac{1}{2} - \frac{1}{5} + \frac{1}{5} = \frac{1}{2}$$

127.
$$\frac{\dfrac{1 - 2 \cdot 3}{4(5-4)}}{\dfrac{3 - 5 \cdot 2}{3 \cdot 5 - 1}} = \frac{\dfrac{-5}{4}}{\dfrac{-7}{14}} = \frac{-5}{4} \div \frac{-7}{14}$$

$$= \frac{-5}{4} \cdot -\frac{14}{7} = \frac{5}{2}$$

Critical Thinking

129. The decimal representation of $\dfrac{5}{23}$ is a rational number since it is the ratio of two integers. Therefore it cannot be a non-terminating, non-repeating decimal.

131.
$$2 + \frac{2}{2 - \dfrac{2}{2+1}} = 2 + \frac{2}{2 - \dfrac{2}{3}} = 2 + \frac{2}{\dfrac{4}{3}}$$

$$= 2 + 2 \div \frac{4}{3} = 2 + 2 \cdot \frac{3}{4} = 2 + \frac{3}{2} = \frac{7}{2}$$

133.
$$3 - \frac{1}{3 - \dfrac{1}{3 - \dfrac{1}{3}}} = 3 - \frac{1}{3 - \dfrac{1}{\dfrac{8}{3}}}$$

$$= 3 - \frac{1}{3 - 1 \div \dfrac{8}{3}} = 3 - \frac{1}{\dfrac{21}{8}} = 3 - 1 \div \frac{21}{8}$$

$$= 3 - 1 \cdot \frac{8}{21} = 3 - \frac{8}{21} = \frac{55}{21}$$

Projects or Group Activities

135. The denominator is the product of a power of 2 and a power of 5.

Check Your Progress: Chapter 1

1. $\{x - 4 < x \le 5\}$

2. $\{x \mid x > 2\}$

3. $[4, 8]$

4. $(-\infty, -3)$

5. $\{x \mid x \le -2\} \cup \{x \mid 3 \le x \le 5\}$

$$-5 \ -4 \ -3 \ -2 \ -1 \ 0 \ 1 \ 2 \ 3 \ 4 \ 5$$

6. $\{x|\ x < 0\} \cap \{x|\ x \ge -3\}$

$$-5\ -4\ -3\ -2\ -1\ \ 0\ \ 1\ \ 2\ \ 3\ \ 4\ \ 5$$

7. $-495 \div (-33) = 15$

8. $-13 - 6 = -19$

9. $9(-15) = -135$

10. $-6 + (-30) = -36$

11. $-28 + (-10) = -38$

12. $31 + (-7) = 24$

13. $-13 - (-17) = -13 + 17 = 4$

14. $-18 + (-4) = -22$

15. $-\dfrac{3}{17}\left(-\dfrac{17}{39}\right) = \dfrac{3 \cdot 17}{17 \cdot 3 \cdot 13} = \dfrac{1}{13}$

16. $-\dfrac{7}{39} - \dfrac{8}{3} = -\dfrac{7}{39} + \dfrac{-104}{39} = -\dfrac{111}{39} = -\dfrac{37}{13}$

17. $\dfrac{5}{16} + \left(-\dfrac{5}{2}\right) = \dfrac{5}{16} + \dfrac{-40}{16} = -\dfrac{35}{16}$

18. $\dfrac{37}{18} + \left(-\dfrac{1}{18}\right) = \dfrac{36}{18} = 2$

19. $\dfrac{35}{4} + \left(-\dfrac{35}{12}\right) = \dfrac{105}{12} + \dfrac{-35}{12} = \dfrac{70}{12} = \dfrac{35}{6}$

20. $-\dfrac{13}{9} - \dfrac{10}{9} = -\dfrac{13}{9} + \dfrac{-10}{9} = -\dfrac{23}{9}$

21. $\dfrac{3}{8} - \left(-\dfrac{1}{6}\right) = \dfrac{9}{24} + \dfrac{4}{24} = \dfrac{13}{24}$

22. $\dfrac{15}{6} \div \left(-\dfrac{3}{2}\right) = \dfrac{15}{6} \cdot \left(-\dfrac{2}{3}\right) = -\dfrac{5 \cdot 3 \cdot 2}{3 \cdot 2 \cdot 3} = -\dfrac{5}{3}$

23. $-5 - (-32) = -5 + 32 = 27$

24. $-2 - 3 = -2 + (-3) = -5$

25. $-4 + 28 = 24$

26. $-84 \div (-4) = 21$

27. $\dfrac{7}{8} \div \left(\dfrac{5}{16}\right) = \dfrac{7}{8} \cdot \left(\dfrac{16}{5}\right) = -\dfrac{7 \cdot 2 \cdot 2 \cdot 2 \cdot 2}{2 \cdot 2 \cdot 2 \cdot 5} = \dfrac{14}{5}$

28. $\dfrac{7}{15} + \left(-\dfrac{11}{20}\right) = \dfrac{28}{60} - \dfrac{33}{60} = -\dfrac{5}{60} = -\dfrac{1}{12}$

29. $\dfrac{11}{12} = 0.91\overline{6}$

30. $\dfrac{27}{32} = 0.84375$

31. $48 - 36 \div 2^2 = 48 - 36 \div 4 = 48 - 9 = 39$

32. $3(4-7)^2 - 6 \div 2 \cdot 3$
$= 3(-3)^2 - 3 \cdot 3$
$= 3(9) - 9$
$= 27 - 9$
$= 18$

33. $\dfrac{3}{4} - \left(\dfrac{1}{3} - \dfrac{1}{2}\right)^2 \div \dfrac{4}{9} = \dfrac{3}{4} - \left(-\dfrac{1}{6}\right)^2 \div \dfrac{4}{9}$

$= \dfrac{3}{4} - \left(\dfrac{1}{36}\right) \div \dfrac{4}{9} = \dfrac{3}{4} - \left(\dfrac{1}{36}\right) \cdot \dfrac{9}{4} = \dfrac{3}{4} - \dfrac{1}{16}$

$= \dfrac{12}{16} - \dfrac{1}{16} = \dfrac{11}{16}$

34. $\dfrac{5}{6} \div \dfrac{2}{3} - \dfrac{1}{6}\left(\dfrac{4}{5} - \dfrac{7}{15}\right) = \dfrac{5}{6} \div \dfrac{2}{3} - \dfrac{1}{6}\left(\dfrac{5}{15}\right)$

$= \dfrac{5}{6} \cdot \dfrac{3}{2} - \dfrac{1}{6}\left(\dfrac{5}{15}\right) = \dfrac{5}{4} - \dfrac{1}{18}$

$= \dfrac{45}{36} - \dfrac{2}{36} = \dfrac{43}{36}$

35. $\dfrac{1}{2} - \dfrac{\frac{3}{4} - \frac{5}{6}}{\left(\frac{1}{2} - \frac{3}{4}\right)^2} = \dfrac{1}{2} - \dfrac{-\frac{1}{12}}{\frac{1}{16}} = \dfrac{1}{2} - \dfrac{-1}{12} \div \dfrac{1}{16}$

$= \dfrac{1}{2} - \dfrac{-1}{12} \cdot \dfrac{16}{1} = \dfrac{1}{2} + \dfrac{4}{3} = \dfrac{11}{6}$

Section 1.4

Concept Check

1. Addition and multiplication have a commutative property.

3. The inverse of $-a$ is a.

5. Distributive Property

7. No. The variable parts are not the same.

Objective A Exercises

9. $3 \cdot 4 = 4 \cdot 3$

11. $(3 + 4) + 5 = 3 + (4 + 5)$

13. $\dfrac{5}{0}$ is undefined.

15. $3(x + 2) = 3x + 6$

17. $\dfrac{0}{-6} = 0$

19. $\dfrac{1}{mn} \cdot mn = 1$

21. $2(3x) = (2 \cdot 3)x$

23. The Division Property of Zero

25. The Inverse Property of Multiplication

27. The Addition Property of Zero

29. The Division Property of Zero

31. The Distributive Property

33. The Associative Property of Multiplication

35. When the sum of a positive number n and its additive inverse are multiplied by the reciprocal of the number n the result is 0.

Objective B Exercises

37. $ab + dc$
$2(3) + (-4)(-1) = 6 + 4 = 10$

39. $4cd \div a^2$
$4(-1)(-4) \div (2)^2 = 4(-1)(-4) \div 4$
$= (-4)(-4) \div 4 = 16 \div 4 = 4$

41. $(b - 2a)^2 + c$
$(3 - 2(2))^2 + (-1)$
$= (3 - 4)^2 + (-1)$
$= (-1)^2 + (-1) = 1 + (-1) = 0$

43. $(bc + a)^2 \div (d - b)$
$(3(-1) + 2)^2 \div (-4 - 3) = (-3 + 2)^2 \div (-4 - 3)$
$= (-1)^2 \div (-7) = 1 \div (-7) = -\dfrac{1}{7}$

45. $\dfrac{1}{4}a^4 - \dfrac{1}{6}bc$
$\dfrac{1}{4}(2)^4 - \dfrac{1}{6}(3)(-1) = \dfrac{1}{4}(16) - \dfrac{1}{6}(-3)$
$= 4 + \dfrac{1}{2} = \dfrac{9}{2}$

47. $\dfrac{3ac}{-4} - c^2$
$\dfrac{3(2)(-1)}{-4} - (-1)^2 = \dfrac{3(2)(-1)}{-4} - 1$
$= \dfrac{6(-1)}{-4} - 1 = \dfrac{-6}{-4} - 1$
$= \dfrac{6}{4} - \dfrac{4}{4} = \dfrac{2}{4} = \dfrac{1}{2}$

49.
$$\frac{3b-5c}{3a-c}$$
$$\frac{3(3)-5(-1)}{3(2)-(-1)} = \frac{9+5}{6-(-1)}$$
$$= \frac{14}{6+1} = \frac{14}{7} = 2$$

51.
$$\frac{a-d}{b+c}$$
$$\frac{2-(-4)}{3+(-1)} = \frac{2+4}{2} = \frac{6}{2} = 3$$

53. $-a|a+2d|$
$$-2|2+2(-4)| = -2|2+(-8)|$$
$$= -2|-6| = -2(6) = -12$$

55.
$$\frac{2a-4d}{3b-c}$$
$$\frac{2(2)-4(-4)}{3(3)-(-1)} = \frac{4-(-16)}{9+1}$$
$$= \frac{4+16}{10} = \frac{20}{10} = 2$$

57. $-3d \div \left|\dfrac{ab-4c}{2b+c}\right|$
$$-3(-4) \div \left|\frac{2(3)-4(-1)}{2(3)+(-1)}\right| = 12 \div \left|\frac{6+4}{6-1}\right|$$
$$= 12 \div \left|\frac{10}{5}\right| = 12 \div |2| = 12 \div 2 = 6$$

59.
$$2(d-b) \div (3a-c)$$
$$2(-4-3) \div (3(2)-(-1))$$
$$= 2(-7) \div (6-(-1))$$
$$= 2(-7) \div (6+1) = 2(-7) \div 7$$
$$= -14 \div 7 = -2$$

61. $-d^2 - c^3a$
$$-(-4)^2 - (-1)^3(2) = -16 - (-1)(2)$$
$$= -16 - (-2) = -16 + 2 = -14$$

63. $-d^3 + 4ac$
$$-(-4)^3 + 4(2)(-1) = -(-64) + 4(2)(-1)$$
$$= -(-64) + (8)(-1) = -(-64) + (-8)$$
$$= 64 + (-8) = 56$$

65. $4^{(a^2)}$
$$4^{(2^2)} = 4^4 = 256$$

67. If $a = -38$, $b = -52$ and $c > 0$, the numerator is positive and the denominator is negative therefore, the expression is negative.

Objective C Exercises

69. $5x + 7x = 12x$

71. $3x - 5x + 9x = 7x$

73. $5b - 8a - 12b = -8a - 7b$

75. $\dfrac{1}{3}(3y) = y$

77. $-\dfrac{2}{5}\left(-\dfrac{5}{2}z\right) = z$

79. $3(a-5) = 3a - 15$

81. $-5(x-9) = -5x + 45$

83. $-(x+y) = -x-y$

85. $4x - 3(2y-5) = 4x - 6y + 15$

87. $25x + 10(9-x) = 25x + 90 - 10x$
$$= 15x + 90$$

89. $3[x - 2(x+2y)] = 3[x - 2x - 4y]$
$$= 3[-x - 4y] = -3x - 12y$$

91. $7 - 3(4a-5) = 7 - 12a + 15 = -12a + 22$

93. $-3m - 2(4m+3) = -3m - 8m - 6$
$$= -11m - 6$$

95. $-5 - 6(2y-3) = -5 - 12y + 18 = -12y + 13$

97. $3[a - 5(5 - 3a)] = 3[a - 25 + 15a]$
$= 3[16a - 25] = 48a - 75$

99. $-2(x - 3y) + 2(3y - 5x) = -2x + 6y + 6y - 10x$
$= -12x + 12y$

101. $5(3a - 2b) - 3(-6a + 5b)$
$= 15a - 10b + 18a - 15b$
$= 33a - 25b$

103. $3x - 2[y - 2(x + 3[2x + 3y])]$
$= 3x - 2[y - 2(x + 6x + 9y)]$
$= 3x - 2[y - 2(7x + 9y)]$
$= 3x - 2[y - 14x - 18y]$
$= 3x - 2[-14x - 17y] = 3x + 28x + 34y$
$= 31x + 34y$

105. $4 - 2(7x - 2y) - 3(-2x + 3y)$
$= 4 - 14x + 4y + 6x - 9y$
$= 4 - 8x - 5y$

107. Simplify $31a - 102b + 73 - 88a + 256b - 73$ to $-57a + 154b$
a) The coefficient of a is negative
b) The coefficient of b is positive
c) The constant term is 0

Critical Thinking

109. $0.052x + 0.072(x + 1000)$
$= 0.052x + 0.072x + 72 = 0.124x + 72$

111. $\dfrac{t}{20} + \dfrac{t}{30} = \dfrac{3t}{60} + \dfrac{2t}{60} = \dfrac{5t}{60} = \dfrac{t}{12}$

Projects or Group Activities

113. a. $4 \otimes 5 = 6$
b. $6 \otimes 3 = 4$
c. $3^2 = 3 \otimes 3 = 2$

115. Yes, because $2 \otimes (3 \otimes 5) = 2$ and
$(2 \otimes 3) \otimes 5 = 2$.

117. The multiplicative inverse of for the $\otimes$ operation is 3 because $5 \otimes 3 = 1$.

Section 1.5

Concept Check

1. a. "The sum of a and b" means to <u>add</u> a and b.
b. "The product of a and b" means to <u>multiply</u> a and b.
c. "The quotient of a and b" means to <u>divide</u> a and b.
d. "The difference between a and b" means to <u>subtract</u> a and b.

3. No. The first is $m - 10$; the second is $10 - m$.

Objective A Exercises

5. Let n represent the unknown number.
Eight <u>less than</u> a number: $n - 8$

7. Let n represent the unknown number.
Four-fifths <u>of</u> a number : $\dfrac{4}{5}n$

9. Let n represent the unknown number.
The <u>quotient</u> of a number and fourteen: $\dfrac{n}{14}$

11. Let n represent the unknown number.
Five <u>subtracted</u> from the <u>product</u> of the cube of eight and a number: $8^3 n - 5$
No. This is not the same as the given expression $8n^3 - 5$.

13. Let n represent the unknown number.
A number <u>minus</u> the <u>sum</u> of the number and two: $n - (n + 2)$
$n - (n + 2) = n - n - 2 = -2$

15. Let n represent the unknown number.
Five <u>times</u> the <u>product</u> of eight and a number:
$5(8n) = 40n$

17. Let n represent the unknown number.
The <u>difference</u> <u>between</u> seventeen <u>times</u> a number and <u>twice</u> the number:
$17n - 2n = 15n$

19. Let n represent the unknown number.
The <u>difference</u> between the <u>square</u> of a number and the <u>total</u> of twelve and the <u>square</u> of the number:
The square of the number: n^2
The total of twelve and the square of the number: $(12 + n^2)$
$n^2 - (12 + n^2) = n^2 - 12 - n^2 = -12$

21. Let n represent the unknown number.
The <u>sum</u> of five <u>times</u> a number and twelve is <u>added</u> to the <u>product</u> of fifteen and the number.
Five times the number and twelve: $5n + 12$
The product of fifteen and the number: $15n$
$(5n + 12) + 15n = 20n + 12$

23. The <u>sum</u> of two numbers is fifteen. Let x represent the smaller of the two numbers.
The larger number is $15 - x$.
The <u>sum</u> of two <u>more than</u> the larger number and <u>twice</u> the smaller number:
$(15 - x + 2) + 2x = 17 + x$

25. The <u>sum</u> of two numbers is thirty-four. Let x represent the larger of the two numbers.
The smaller number is $34 - x$.
The <u>quotient</u> of five <u>times</u> the smaller number and the <u>difference</u> <u>between</u> the larger number and three:
$\dfrac{5(34 - x)}{x - 3}$

Objective B Exercises

27. a) Let j represent the number of jobs in November.
December jobs number: $j + 200{,}000$
b) Let r represent the unemployment rate in November.
December unemployment rate: $r - 0$.

29. Let d represent the distance from Earth to the moon.
Distance from the Earth to the sun is 390 times the distance from the Earth to the moon: $390d$

31. Let x represent the amount in the first account.
The total amount in both accounts is $10,000.
The amount in the second account: $10{,}000 - x$

33. Let x represent the measure of angle B.
The measure of angle A is twice the measure of angle B: $2x$
The measure of angle C is twice the measure of angle A: $2(2x) = 4x$

35. In 2013, a house sold for $30,000 less than the same house in sold for in 2010.
s represents the selling price of the house in 2013.

Critical Thinking

37. a. The product of mass m and acceleration a:
ma
b. The product of area A and the square of the velocity v: Av^2
c. the sum of the principal P and the interest I: $P + I$
d. the sum of twice the length L and twice the width W: $2L + 2W$
e. the product of 16 and the time t squared: $16t^2$
f. the product of four-thirds π and the radius r cubed: $\dfrac{4}{3}\pi r^3$.
g. the quotient of the product of mass 1 m and mass 2 M and the distance r squared: $\dfrac{mM}{r^2}$

Projects or Group Activities

39. Answers will vary. For instance, the quotient of a and b; the ratio of a to b.

41. The sum of twice x and 3.

43. Twice the sum of x and 3.

Chapter 1 Review Exercises

1. $\{-2, -1, 0, 1, 2, 3\}$

2. $A \cap B = \{2, 3\}$

3. $(-2, 4]$

$$-5 \ -4 \ -3 \ -2 \ -1 \ 0 \ 1 \ 2 \ 3 \ 4 \ 5$$

4. The Associative Property of Multiplication

5. $-4.07 + 2.3 - 1.07 = -1.77 - 1.07 = -2.84$

6. $(a - 2b^2) \div (ab)$

$(4 - 2(-3)^2) \div (4(-3))$

$= (4 - 2(9)) \div (4(-3))$

$= (4 - 18) \div (-12) = -14 \div (-12)$

$= \dfrac{-14}{-12} = \dfrac{7}{6}$

7. $-2 \cdot (4^2) \cdot (-3)^2 = -2 \cdot 16 \cdot 9 = -288$

8. $4y - 3[x - 2(3 - 2x) - 4y]$

$\quad = 4y - 3[x - 6 + 4x - 4y]$

$\quad = 4y - 3[5x - 4y - 6]$

$\quad = 4y - 15x + 12y + 18 = 16y - 15x + 18$

9. The additive inverse of $-\dfrac{3}{4}$ is $\dfrac{3}{4}$.

10. $\{x \,|\, x < -3\}$

11. $\{x \,|\, x < 1\}$

$$-5 \ -4 \ -3 \ -2 \ -1 \ 0 \ 1 \ 2 \ 3 \ 4 \ 5$$

12. $-10 - (-3) - 8 = -10 + 3 - 8 = -15$

13. $-\dfrac{2}{3} + \dfrac{3}{5} - \dfrac{1}{6} = -\dfrac{20}{30} + \dfrac{18}{30} - \dfrac{5}{30}$

$\quad = \dfrac{-20 + 18 - 5}{30} = \dfrac{-7}{30} = -\dfrac{7}{30}$

14. $3 + (4 + y) = (3 + 4) + y$

15. $-\dfrac{3}{8} \div \dfrac{3}{5} = -\dfrac{3}{8} \cdot \dfrac{5}{3} = -\dfrac{3 \cdot 5}{8 \cdot 3} = -\dfrac{5}{8}$

16. Replace x with each element in the set to determine if the inequality is true or false.
a) $-4 > -1$ False
b) $-2 > -1$ False
c) $0 > -1$ True
d) $2 > -1$ True

17. $2a^2 - \dfrac{3b}{a}$

$2(-3)^2 - \dfrac{3(2)}{-3} = 2(9) - \dfrac{6}{-3}$

$= 18 - (-2) = 18 + 2 = 20$

18. $18 - |-12 + 8| = 18 - |-4| = 18 - 4 = 14$

19. $20 \div \dfrac{3^2 - 2^2}{3^2 + 2^2} = 20 \div \dfrac{9 - 4}{9 + 4} = 20 \div \dfrac{5}{13}$

$\quad = 20 \cdot \dfrac{13}{5} = 52$

20. $[-3, \infty)$

$$-5 \ -4 \ -3 \ -2 \ -1 \ 0 \ 1 \ 2 \ 3 \ 4 \ 5$$

21. $A \cup B = \{1, 2, 3, 4, 5, 6, 7, 8\}$

22. $-204 \div (-17) = 12$

23. $[x \,|\, {-2} \le x \le 3\}$

24. $\dfrac{\dfrac{2}{3} - \dfrac{5}{6}}{\dfrac{1}{2} - \dfrac{3}{4}} - \dfrac{2}{3} \div \dfrac{4}{9} = \dfrac{-\dfrac{1}{6}}{-\dfrac{1}{4}} - \dfrac{2}{3} \div \dfrac{4}{9}$

$= -\dfrac{1}{6} \div \left(-\dfrac{1}{4}\right) - \dfrac{2}{3} \div \dfrac{4}{9}$

$= -\dfrac{1}{6} \cdot \left(-\dfrac{4}{1}\right) - \dfrac{2}{3} \cdot \dfrac{9}{4}$

$= \dfrac{2}{3} - \dfrac{3}{2} = -\dfrac{5}{6}$

25. $6x - 21y = 3(2x - 7y)$

26. $\{x| \ x \le -3\} \cup \{x| \ x > 0\}$

-5 -4 -3 -2 -1 0 1 2 3 4 5

27. $-2(x - 3) + 4(2 - x) = -2x + 6 + 8 - 4x$
$= -6x + 14$

28. Replace p with each element in the set to determine the value of $|-p|$
a) $-|-4| = -4$
b) $-|0| = 0$
c) $-|7| = -7$

29. The Inverse Property of Addition

30. $-3.286 \div (-1.06) = 3.1$

31. $\dfrac{7}{12} = 0.58\overline{3}$

32. Replace y with each element in the set to determine if the inequality is true or false.
a) $-4 > -2$ False
b) $-1 > -2$ True
c) $4 > -2$ True

33. $\{-3, -2, -1, 0, 1\}$

34. $\{x/ \ x < 7\}$

35. $A \cup B = \{-4, -2, 0, 2, 4, 5, 10\}$

36. $A \cap B = \varnothing$

37. $\{x| \ x \le 3\} \cap \{x| \ x > -2\}$

-5 -4 -3 -2 -1 0 1 2 3 4 5

38. $(-3,4) \cup [-1,5]$

-5 -4 -3 -2 -1 0 1 2 3 4 5

39. $2 - (-3) = 2 + 3 = 5$

40. $-\dfrac{5}{6} + \left(\dfrac{1}{3}\right) - -\dfrac{5}{6} + \left(\dfrac{2}{6}\right) - -\dfrac{3}{6} - -\dfrac{1}{2}$

41. $(-3)^4 - 3(20-1) = (-3)^4 - 3(19)$
$= 81 - 57 = 24$

42. $(-3)^3 - (2 - 6)^2 \cdot 5 = (-3)^3 - (-4)^2 \cdot 5$
$= -27 - 16 \cdot 5 = -27 - 80 = -107$

43. $-8ac \div b^2$
$-8(-1)(-3) \div 2^2 = -8(-1)(-3) \div 4$
$= -24 \div 4 = -6$

44. $-(3a + b) - 2(-4a - 5b) = -3a - b + 8a + 10b$
$= 5a + 9b$

45. Let x represent the unknown number.
Four <u>times</u> the <u>sum</u> of a number and four:
$4(x + 4) = 4x + 16$

46. Let t represent the flying time between San Diego to New York.
Total flying time is 13 hrs.
The flying time between New York and San Diego: $13 - t$

47. Let C represent the number of calories burned by walking at 4 mph for one hour.
The number of calories burned cross country skiing for one hour is 396 more than the number of calories burned walking: $C + 396$

48. Let x represent the unknown number.
Eight <u>more than</u> twice the <u>difference</u> between a number and two: $2(x - 2) + 8$
$2(x - 2) + 8 = 2x - 4 + 8 = 2x + 4$

49. Let x represent the first integer.
The second integer is 5 more than four times the first integer.
The second integer: $4x + 5$

50. Let x represent the unknown number.
Twelve <u>minus</u> the <u>quotient</u> of three <u>more than</u> a number and four. $12 - \dfrac{x+3}{4}$

$12 - \dfrac{x+3}{4} = \dfrac{48}{4} - \dfrac{x+3}{4}$
$= \dfrac{48 - x + 3}{4} = \dfrac{45 - x}{4}$

51. The sum of two numbers is forty.
Let x represent the smaller number.
Let $40 - x$ represent the larger number.
The <u>sum</u> of <u>twice</u> the smaller number and five
<u>more than</u> the larger number: $2x + (40 - x) + 5$
$2x + (40 - x) + 5 = 2x + 40 - x + 5 = x + 45$

52. Let w represent the width of the rectangle.
The length is three feet <u>less than</u> three <u>times</u>
the width: $3w - 3$

Chapter 1 Test

1. $-52(4) = -208$

2. $A \cap B = \{5, 7\}$

3. $(-2)^3(-3)^2 = (-8)(9) = -72$

4. $(-\infty, 1]$

$$\underleftrightarrow{\quad\underset{-5\ -4\ -3\ -2\ -1\ 0\ 1\ 2\ 3\ 4\ 5}{\mid\ \mid\ \mid\ \mid\ \mid\ \mid\ \mid\ \mid\ \mid\ \mid\ \mid}\quad}$$

5. $A \cap B = \{-1, 0, 1\}$

6. $(a - b)^2 \div (2b + 1)$
$(2 - (-3))^2 \div (2(-3) + 1)$
$= (2 + 3)^2 \div (-6 + 1)$
$= (5)^2 \div (-5)$
$= 25 \div (-5) = -5$

7. $-3 - (-5) = -3 + 5 = 2$

8. $2x - 4[2 - 3(x + 4y) - 2]$
$= 2x - 4[2 - 3x - 12y - 2]$
$= 2x - 4[-3x - 12y] = 2x + 12x + 48y$
$= 14x + 48y$

9. The additive inverse of -12 is 12.

10. $-5^2 \cdot 4 = -25 \cdot 4 = -100$

11. $\{x \mid x < 3\} \cap \{x \mid x > -2\}$

$$\underleftrightarrow{\quad\underset{-5\ -4\ -3\ -2\ -1\ 0\ 1\ 2\ 3\ 4\ 5}{\mid\ \mid\ \mid\ (\mid\ \mid\ \mid\ \mid\ \mid\ )\ \mid\ \mid}\quad}$$

12. $8 - 5(3 - 5)^3 \div 10 \cdot 2 = 8 - 5(-8) \div 10 \cdot 2$
$= 8 + 40 \div 10 \cdot 2 = 8 + 4 \cdot 2 = 8 + 8$
$= 16$

13. $\dfrac{3}{4} - \dfrac{\left(\dfrac{2}{3} - \dfrac{5}{6}\right)^2}{\dfrac{2}{3} - \dfrac{3}{4}} = \dfrac{3}{4} - \dfrac{\left(-\dfrac{1}{6}\right)^2}{-\dfrac{1}{12}} = \dfrac{3}{4} - \dfrac{\dfrac{1}{36}}{-\dfrac{1}{12}}$

$= \dfrac{3}{4} - \dfrac{1}{36} \div \left(-\dfrac{1}{12}\right) = \dfrac{3}{4} - \dfrac{1}{36} \cdot \left(-\dfrac{12}{1}\right)$

$= \dfrac{3}{4} + \dfrac{1}{3} = \dfrac{13}{12}$

14. $(3 + 4) + 2 = (4 + 3) + 2$

15. $\left(-\dfrac{2}{3}\right)\left(\dfrac{9}{16}\right) = -\dfrac{2 \cdot 3 \cdot 3}{3 \cdot 2 \cdot 2 \cdot 2 \cdot 2} = -\dfrac{3}{8}$

16. Replace x with each element in the set to
determine if the inequality is true or false.
$-5 < -1 \qquad$ True
$3 < -1 \qquad$ False
$7 < -1 \qquad$ False

17. $\dfrac{b^2 - c^2}{a - 2c}$

$\dfrac{(3)^2 - (-1)^2}{2 - 2(-1)} = \dfrac{9 - 1}{2 + 2} = \dfrac{8}{4} = 2$

18. $-180 \div 12 = -15$

19. $12 - 4\left(\dfrac{5^2 - 1}{3}\right) \div 16 = 12 - 4\left(\dfrac{25 - 1}{3}\right) \div 16$

$= 12 - 4\left(\dfrac{24}{3}\right) \div 16 = 12 - 4(8) \div 16$

$= 12 - 32 \div 16 = 12 - 2 = 10$

20. $(3, \infty)$

$$\underleftrightarrow{\quad\underset{-5\ -4\ -3\ -2\ -1\ 0\ 1\ 2\ 3\ 4\ 5}{\mid\ \mid\ \mid\ \mid\ \mid\ \mid\ \mid\ \mid\ (\mid\ \mid\ \mid}\quad}$$

21. $A \cup B = \{1, 2, 3, 4, 5, 7\}$

22. $3x - 2(x - y) - 3(y - 4x)$
   $= 3x - 2x + 2y - 3y + 12x$
   $= 13x - y$

23. $8 - 4(2 - 3)^2 \div 2 = 8 - 4(-1)^2 \div 2$
   $= 8 - 4(1) \div 2 = 8 - 4 \div 2$
   $= 8 - 2 = 6$

24. $\dfrac{3}{5}\left(-\dfrac{10}{21}\right)\left(-\dfrac{7}{15}\right) = \dfrac{3 \cdot 2 \cdot 5 \cdot 7}{5 \cdot 3 \cdot 7 \cdot 3 \cdot 5} = \dfrac{2}{15}$

25. The Distributive Property

26. $\{x \mid x \le 3\} \cup \{x \mid x < -2\}$

$-5\ -4\ -3\ -2\ -1\ 0\ 1\ 2\ 3\ 4\ 5$

27. $\dfrac{7}{18} = 0.3\overline{8}$

28. $A \cup B = \{-2, -1, 0, 1, 2, 3\}$

29. The sum of two numbers is nine.
   Let x represent the larger of the two numbers.
   Let $9 - x$ represent the smaller number.
   The <u>difference</u> <u>between</u> one <u>more than</u> the
   larger number and <u>twice</u> the smaller number:
   $(x + 1) - 2(9 - x)$
   $(x + 1) - 2(9 - x) = x + 1 - 18 + 2x = 3x - 17$

30. Let x represent the amount of cocoa produced
   in Ghana.
   The Ivory Coast produces three times the
   amount of cocoa produced in Ghana: $3x$

Chapter 2: First Degree Equations and Inequalities

Prep Test

1. $8 - 12 = -4$

2. $-9 + 3 = -6$

3. $\dfrac{-18}{-6} = 3$

4. $-\dfrac{3}{4}\left(-\dfrac{4}{3}\right) = \dfrac{3 \cdot 2 \cdot 2}{2 \cdot 2 \cdot 3} = 1$

5. $-\dfrac{5}{8}\left(\dfrac{4}{5}\right) = -\dfrac{5 \cdot 2 \cdot 2}{2 \cdot 2 \cdot 2 \cdot 5} = -\dfrac{1}{2}$

6. $3x - 5 + 7x = 10x - 5$

7. $6(x - 2) + 3 = 6x - 12 + 3 = 6x - 9$

8. $n + (n + 2) + (n + 4) = 3n + 6$

9. $0.08x + 0.05(400 - x)$
$= 0.08x + 20 - 0.05x$
$= 0.03x + 20$

10. The total weight of a snack mix of nuts and pretzels is twenty ounces.
Let n represent the ounces of nuts in the mix.
Ounces of peanuts in the mix: $20 - n$

Section 2.1

Concept Check

1. An equation contains an equal sign while an expression does not contain an equal sign.

3. The Addition Property of Equations states that the same quantity can be added to each side of an equation without changing the solution of the equation.
This property is used to remove a term from one side of an equation by adding the opposite of that term to each side of the equation.

5. $7 - 3(1) = 4$
$\quad\ 7 - 3 = 4$
$\qquad\quad 4 = 4$ ⠀⠀Yes, 1 is a solution.

7. $6(0) - 1 \neq 7(0) + 1$
$\quad\ 0 - 1 \neq 0 + 1$
$\qquad\ -1 \neq 1$ No, 0 is not a solution.

Objective A Exercises

9. ⠀⠀⠀⠀$x - 2 = 7$
$x - 2 + 2 = 7 + 2$
$\qquad\quad x = 9$
The solution is 9.

11. ⠀⠀⠀$-7 = x + 8$
$-7 - 8 = x + 8 - 8$
$\quad -15 = x$
The solution is -15.

13. $3x = 12$
$\dfrac{3x}{3} = \dfrac{12}{3}$
$x = 4$
The solution is 4.

15. ⠀$-3x = 2$
$\dfrac{-3x}{-3} = \dfrac{2}{-3}$
$x = -\dfrac{2}{3}$
The solution is $-\dfrac{2}{3}$.

17. ⠀⠀$-\dfrac{3}{2} + x = \dfrac{4}{3}$
$-\dfrac{3}{2} + \dfrac{3}{2} + x = \dfrac{4}{3} + \dfrac{3}{2}$
$x = \dfrac{8}{6} + \dfrac{9}{6}$
$x = \dfrac{17}{6}$
The solution is $\dfrac{17}{6}$.

19.

$$x + \frac{2}{3} = \frac{5}{6}$$

$$x + \frac{2}{3} - \frac{2}{3} = \frac{5}{6} - \frac{2}{3}$$

$$x = \frac{5}{6} - \frac{4}{6}$$

$$x = \frac{1}{6}$$

The solution is $\frac{1}{6}$.

21.

$$\frac{2}{3}y = 5$$

$$\frac{3}{2}\left(\frac{2}{3}y\right) = \frac{3}{2}(5)$$

$$y = \frac{15}{2}$$

The solution is $\frac{15}{2}$.

23.

$$\frac{4}{5} = -\frac{5}{8}x$$

$$-\frac{8}{5}\left(-\frac{5}{8}y\right) = -\frac{8}{5}\left(\frac{4}{5}\right)$$

$$y = -\frac{32}{25}$$

The solution is $-\frac{32}{25}$.

25.

$$-12 = \frac{4x}{7}$$

$$\frac{7}{4}\left(\frac{4}{7}x\right) = \frac{7}{4}(-12)$$

$$x = -21$$

The solution is -21.

27.

$$-\frac{5y}{7} = \frac{10}{21}$$

$$-\frac{7}{5}\left(-\frac{5}{7}y\right) = -\frac{7}{5}\left(\frac{10}{21}\right)$$

$$y = -\frac{70}{105} = -\frac{2}{3}$$

The solution is $-\frac{2}{3}$.

29.

$$-\frac{3b}{5} = -\frac{3}{5}$$

$$-\frac{5}{3}\left(-\frac{3}{5}b\right) = -\frac{5}{3}\left(-\frac{3}{5}\right)$$

$$b = 1$$

The solution is 1.

31.

$$-\frac{2}{3}x = -\frac{5}{8}$$

$$-\frac{3}{2}\left(-\frac{2}{3}x\right) = -\frac{3}{2}\left(-\frac{5}{8}\right)$$

$$x = \frac{15}{16}$$

The solution is $\frac{15}{16}$.

33. $1.5x = 27$

$$\frac{1.5x}{1.5} = \frac{27}{1.5}$$

$$x = 18$$

The solution is 18.

35. $-0.015x = -12$

$$\frac{-0.015x}{-0.015} = \frac{-12}{-0.015}$$

$$x = 800$$

The solution is 800.

37.
$$3x + 5x = 12$$
$$8x = 12$$
$$\frac{8x}{8} = \frac{12}{8}$$
$$x = \frac{3}{2}$$

The solution is $\frac{3}{2}$.

39.
$$3y - 5y = 0$$
$$-2y = 0$$
$$\frac{-2y}{-2} = \frac{0}{-2}$$
$$y = 0$$

The solution is 0.

41. If a is a negative number less than -5, the solution to the equation $a = -5b$ is greater than 1.

Objective B Exercises

43.
$$2x - 4 = 12$$
$$2x - 4 + 4 = 12 + 4$$
$$2x = 16$$
$$\frac{2x}{2} = \frac{16}{2}$$
$$x = 8$$

The solution is 8.

45.
$$4x - 6 = 3x$$
$$4x + (-4x) - 6 = 3x + (-4x)$$
$$-6 = -x$$
$$(-1)(-6) = (-1)(-x)$$
$$6 = x$$

The solution is 6.

47.
$$7x + 12 = 9x$$
$$7x + (-7x) + 12 = 9x + (-7x)$$
$$12 = 2x$$
$$\frac{12}{2} = \frac{2x}{2}$$
$$6 = x$$

The solution is 6.

49.
$$4x + 2 = 4x$$
$$4x + (-4x) + 2 = 4x + 4x$$
$$2 = 0$$

There is no solution.

51.
$$2x + 2 = 3x + 5$$
$$2x + (-2x) + 2 = 3x + (-2x) + 5$$
$$2 = x + 5$$
$$2 + (-5) = x + 5 + (-5)$$
$$-3 = x$$

The solution is -3.

53.
$$2 - 3t = 3t - 4$$
$$2 - 3t + 3t = 3t + 3t - 4$$
$$2 = 6t - 4$$
$$2 + 4 = 6t - 4 + 4$$
$$6 = 6t$$
$$\frac{6}{6} = \frac{6t}{6}$$
$$1 = t$$

The solution is 1.

55.
$$3b - 2b = 4 - 2b$$
$$b = 4 - 2b$$
$$b + 2b = 4 - 2b + 2b$$
$$3b = 4$$
$$\frac{3b}{3} = \frac{4}{3}$$
$$b = \frac{4}{3}$$

The solution is $\frac{4}{3}$.

57.
$$3x + 7 = 3 + 7x$$
$$3x + (-3x) + 7 = 3 + 7x + (-3x)$$
$$7 = 3 + 4x$$
$$7 + (-3) = 3 + (-3) + 4x$$
$$4 = 4x$$
$$\frac{4}{4} = \frac{4x}{4}$$
$$1 = x$$

The solution is 1.

59.
$$5x - 3 = -3 + 8x$$
$$5x - 5x - 3 = -3 + 8x - 5x$$
$$-3 = -3 + 3x$$
$$-3 + 3 = -3 + 3 + 3x$$
$$0 = 3x$$
$$\frac{0}{3} = \frac{3x}{3}$$
$$x = 0$$
The solution is 0

61.
$$5.3y + 0.35 = 5.02y$$
$$5.3y + (-5.3y) + 0.35 = 5.02y + (-5.3y)$$
$$0.35 = -0.28y$$
$$-1.25 = y$$
The solution is -1.25.

63.   If A is a negative number, the solution of the equation $Ax - 2 = -5$ is positive.

Objective C Exercises

65. $2x + 3(x - 5) = 15$
$$2x + 3x - 15 = 15$$
$$5x - 15 = 15$$
$$5x = 30$$
$$\frac{5x}{5} = \frac{30}{5}$$
$$x = 6$$
The solution is 6.

67. $3(y - 5) - 5y = 2y + 9$
$$3y - 15 - 5y = 2y + 9$$
$$-2y - 15 = 2y + 9$$
$$-15 = 4y + 9$$
$$-24 = 4y$$
$$\frac{-24}{4} = \frac{4y}{4}$$
$$-6 = y$$
The solution is -6.

69. $4 - 3x = 7x - 2(3 - x)$
$$4 - 3x = 7x - 6 + 2x$$
$$4 - 3x = 9x - 6$$
$$10 - 3x = 9x$$
$$10 = 12x$$
$$\frac{10}{12} = \frac{12x}{12}$$

$$x = \frac{5}{6}$$
The solution is $\frac{5}{6}$.

71. $-3x - 2(4 + 5x) = 14 - 3(2x - 3)$
$$-3x - 8 - 10x = 14 - 6x + 9$$
$$-13x - 8 = 23 - 6x$$
$$-8 = 23 + 7x$$
$$-31 = 7x$$
$$-\frac{31}{7} = \frac{7x}{7}$$
$$-\frac{31}{7} = x$$
The solution is $-\dfrac{31}{7}$.

73. $3y = 2[5 - 3(2 - y)]$
$$3y = 2[5 - 6 + 3y]$$
$$3y = 2[-1 + 3y]$$
$$3y = -2 + 6y$$
$$-3y = -2$$
$$\frac{-3y}{-3} = \frac{-2}{-3}$$
$$y = \frac{2}{3}$$
The solution is $\frac{2}{3}$.

75. $2[3 - 2(z + 4)] = 3(4 - z)$
$$2[3 - 2z - 8] = 12 - 3z$$
$$2[-2z - 5] = 12 - 3z$$
$$-4z - 10 = 12 - 3z$$
$$-4z - 22 = -3z$$
$$-22 = z$$
The solutions is -22.

77.
$$\frac{3}{2} - \frac{10x}{9} = -\frac{1}{6}$$
$$18\left(\frac{3}{2} - \frac{10x}{9}\right) = 18\left(-\frac{1}{6}\right)$$
$$27 - 20x = -3$$
$$27 - 27 - 20x = -3 - 27$$
$$-20x = -30$$
$$\frac{-20y}{-20} = \frac{-30}{-20}$$
$$y = \frac{3}{2}$$

The solution is $\frac{3}{2}$.

79.
$$\frac{2x-5}{12} - \frac{3-x}{6} = \frac{11}{12}$$
$$12\left(\frac{2x-5}{12} - \frac{3-x}{6}\right) = 12 \cdot \frac{11}{12}$$
$$\frac{12(2x-5)}{12} - \frac{12(3-x)}{6} = 11$$
$$(2x-5) - 2(3-x) = 11$$
$$2x - 5 - 6 + 2x = 11$$
$$4x - 11 = 11$$
$$4x = 22$$
$$\frac{4x}{4} = \frac{22}{4}$$
$$x = \frac{11}{2}$$

The solution is $\frac{11}{2}$.

81.
$$\frac{2x-1}{4} + \frac{3x+4}{8} = \frac{1-4x}{12}$$
$$24\left(\frac{2x-1}{4} + \frac{3x+4}{8}\right) = 24 \cdot \frac{1-4x}{12}$$
$$\frac{24(2x-1)}{4} + \frac{24(3x+4)}{8} = \frac{24(1-4x)}{12}$$
$$6(2x-1) + 3(3x+4) = 2(1-4x)$$
$$12x - 6 + 9x + 12 = 2 - 8x$$
$$21x + 6 = 2 - 8x$$
$$29x + 6 = 2$$
$$29x = -4$$

$$\frac{29x}{29} = \frac{-4}{29}$$
$$x = -\frac{4}{29}$$

The solution is $-\frac{4}{29}$.

83. Solve $5 - 2(4x - 1) = 3x + 7$.
$$5 - 2(4x - 1) = 3x + 7$$
$$5 - 8x + 2 = 3x + 7$$
$$7 - 8x = 3x + 7$$
$$7 - 11x = 7$$
$$-11x = 0$$
$$x = 0$$
Now evaluate $x^4 - x^2$.
$$(0)^4 - (0)^2 = 0$$

85. The Distributive Property is used three times to remove grouping symbols from the equation $-3[5 - 4(x - 2)] = 5(x - 5)$.

Objective D Exercises

87. $C = 2\pi r$
$$\frac{C}{2\pi} = \frac{2\pi r}{2\pi}$$
$$r = \frac{C}{2\pi}$$

89. $A = \frac{1}{2}bh$
$$2 \cdot A = 2 \cdot \frac{1}{2}bh$$
$$2A = bh$$
$$\frac{2A}{b} = \frac{bh}{b}$$
$$h = \frac{2A}{b}$$

91.
$$I = \frac{100M}{C}$$
$$C \cdot I = C \cdot \frac{100M}{C}$$
$$IC = 100M$$
$$\frac{IC}{100} = \frac{100M}{100}$$
$$M = \frac{IC}{100}$$

93.
$$A = P + Prt$$
$$A - P = Prt$$
$$\frac{A - P}{Pt} = \frac{Prt}{Pt}$$
$$t = \frac{A - P}{Pt}$$

95.
$$s = \frac{1}{2}(a + b + c)$$
$$2 \cdot s = 2 \cdot \frac{1}{2}(a + b + c)$$
$$2s = a + b + c$$
$$c = 2s - a - b$$

97.
$$S = 2\pi r^2 + 2\pi rh$$
$$S - 2\pi r^2 = 2\pi rh$$
$$\frac{S - 2\pi r^2}{2\pi r} = \frac{2\pi rh}{2\pi r}$$
$$\frac{S - 2\pi r^2}{2\pi r} = h$$

99.
$$P = \frac{R - C}{n}$$
$$n \cdot P = n \cdot \frac{R - C}{n}$$
$$nP = R - C$$
$$R = nP + C$$

101.
$$A = P(1 + i)$$
$$\frac{A}{1 + i} = \frac{P(1 + i)}{1 + i}$$
$$\frac{A}{1 + i} = P$$

103. To solve the formula $P = 2L + 2W$ for L subtract $2W$ from each side of the equation.

Critical Thinking

105.
$$0.05(300 - x) + 0.07x = 45$$
$$15 - 0.05x + 0.07x = 45$$
$$15 + 0.02x = 45$$
$$0.02x = 30$$
$$x = 1500.$$

107.
$$3(2x + 1) = 5 - 2(x - 2)$$
$$6x + 3 = 5 - 2x + 4$$
$$6x + 3 = 9 - 2x$$
$$8x = 6$$
$$x = \frac{6}{8} = \frac{3}{4}$$
Now evaluate $2x^2 + 1$
$$2\left(\frac{3}{4}\right)^2 + 1 = 2\left(\frac{9}{16}\right) + 1 = \frac{17}{8}$$

109. The result is not correct. The correct solution is
$$5x + 15 = 2x + 3(2x + 5)$$
$$5x + 15 = 2x + 6x + 15$$
$$5x + 15 = 8x + 15$$
$$5x + 15 - 15 = 8x + 15 - 15$$
$$5x = 8x$$
$$5x - 5x = 8x - 5x$$
$$0 = 3x$$
$$\frac{0}{3} = \frac{3x}{3}$$
$$0 = x$$
The solution is 0.

Projects or Groups Activities

111.   Strategy:
Let n represent the first integer.
The second integer is $n + 1$.
The third integer is $n + 2$.
The first integer is forty-four less than three times the sum of the second and third integers:
$$n - 3[(n + 1) + (n + 2)] - 44$$

Solution: $n = 3[(n + 1) + (n + 2)] - 44$

$n = 3[2n + 3] - 44$

$n = 6n + 9 - 44$

$n = 6n - 35$

$35 = 5n$

$7 = n$

$n + 1 = 7 + 1 = 8$

$n + 2 = 7 + 2 = 9$

The integers are 7, 8 and 9.

113.   Strategy:

Let n represent the first even integer.
The second even integer is $n + 2$.
The third even integer is $n + 4$.
Three times the sum of the first and third integers is twenty more than four times the middle integer:

$3[n + (n + 4)] = 4(n + 2) + 20$

Solution: $3[n + (n + 4)] = 4(n + 2) + 20$

$3(2n + 4) = 4n + 8 + 20$

$6n + 12 = 4n + 28$

$2n = 16$

$n = 8$

$n + 2 = 8 + 2 = 10$

$n + 4 = 8 + 4 = 12$

The even integers are 8, 10 and 12.

Section 2.2

Concept Check

1.   The amount A of coffee is 10 lb. The total value V is $83.60. The cost per unit is C of the coffee is unknown.

$$AC = V$$

$$10C = 83.60$$

$$C = \$8.36$$

The cost per pound is $8.36

3.   The cost per unit of a blend is between the costs per unit of the two ingredients that make up the blend. Therefore (iii) $11.00 and (iv) $6.50 are not possible answers for the cost per pound of the mixture.

5.   The amount A of mixture is 100 g. The percent concentration r of sugar is 15% or 0.15. The unknown is quantity Q of sugar.

$$Ar = Q$$

$$100(0.15) = 15$$

There are 15 g of sugar.

7.   The amount A of mixture is 200 L. The percent concentration r of fruit juice is 10% or 0.10. The unknown is quantity Q of fruit juice.

$$Ar = Q$$

$$200(0.10) = 20$$

$$200 - 20 = 180$$

There is 180 L of water.

9.   The percent concentration of a mixture is between the percent concentrations of the two solutions or alloys that make up the mixture. Therefore (ii), (iii), (iv) and (v) are not possible answers for the percent concentration of the resulting mixture.

11. The bicycle speed r is unknown. The time t spent biking is 1hr 30min, or 1.5 hr . The distance traveled is 24 mi.

$$rt = d$$

$$r(1.5) = 24$$

$$r = 16$$

His average speed was 16 mph.

13. a. The distance traveled by Lois is greater than the distance traveled by Michael.

   b. The time walked by Lois is equal to the time walked by Michael.

   c. The total distance traveled by Lois and Michael is 2 mi.

Objective A Exercises

15. Strategy: Let x represent the cost of the mixture.

	Amount	Cost	Value
Cashews	40	9.20	40(9.20)
Peanuts	100	3.32	100(3.32)
Mixture	140	x	140x

The sum of the values before mixing equals the value after mixing.

Solution: $40(9.20) + 100(3.32) = 140x$

$$368 + 332 = 140x$$
$$700 = 140x$$
$$5 = x$$

The cost of the mixture is $5.00 per pound.

17. **Strategy**: Let x represent the number of adult tickets.
The number of children's tickets is $460 - x$.

	Amount	Cost	Value
Adult	x	10	$10x$
Children	$460 - x$	4	$4(460 - x)$

The total value of the tickets sold is $3760.

Solution: $10x + 4(460 - x) = 3760$
$$10x + 1840 - 4x = 3760$$
$$6x = 1920$$
$$x = 320$$

There were 320 adult tickets sold.

19. **Strategy**: Let x represent the number of liters of imitation maple syrup.

	Amount	Cost	Value
Imitation	x	4.00	$4x$
Maple	5	9.50	$5(9.50)$
Mixture	$5 + x$	5.00	$5(5 + x)$

The sum of the values before mixing equals the value after mixing.

Solution: $4x + 5(9.50) = 5(5 + x)$
$$4x + 47.5 = 25 + 5x$$
$$47.5 - 25 = x$$
$$22.5 = x$$

The mixture must contain 22.5 L of imitation maple syrup.

21. **Strategy**: Let x represent ounces of pure gold.
The ounces of gold alloy is $50 - x$.

	Amount	Cost	Value
Gold	x	890	$890x$
Alloy	$50 - x$	360	$360(50 - x)$
Mixture	50	519	$519(50)$

The sum of the values before mixing equals

the value after mixing.

Solution: $890x + 360(50 - x) = 519(50)$
$$890x + 18{,}000 - 360x = 25{,}950$$
$$530x + 18{,}000 = 25{,}950$$
$$530x = 7950$$
$$x = 15$$

$50 - x = 50 - 15 = 35$

There were 15 oz of pure gold and 35 oz of gold alloy used in the mixture.

23. **Strategy**: Let x represent the cost per pound of the mixture.

	Amount	Cost	Value
$5.40 tea	40	5.40	$5.4(40)$
$3.25 tea	60	3.25	$3.25(60)$
Mixture	100	x	$100x$

The sum of the values before mixing equals the value after mixing.

Solution: $5.4(40) + 3.25(60) = 100x$
$$216 + 195 = 100x$$
$$411 = 100x$$
$$4.11 = x$$

The cost of the mixture is $4.11 per pound.

25. **Strategy**: Let x represent the number of gallons of cranberry juice.

	Amount	Cost	Value
Cranberry	x	28.50	$28.5(x)$
Apple	20	11.25	$11.25(20)$
Mixture	$20 + x$	17.00	$17(20 + x)$

The sum of the values before mixing equals the value after mixing.

Solution: $28.5x + 11.25(20) = 17(20 + x)$
$$28.5x + 225 = 340 + 17x$$
$$11.5x + 225 = 340$$
$$11.5x = 115$$
$$x = 10$$

The mixture must contain 10 gal of cranberry juice.

Objective B Exercises

27. Strategy: Let x represent the number of pounds of 15% aluminum alloy.

	Amount	Percent	Quantity
15%	x	0.15	$0.15x$
22%	500	0.22	$0.22(500)$
20%	$500 + x$	0.20	$0.20(500 + x)$

The sum of the quantities before mixing is equal to the quantity after mixing.
Solution: $0.15x + 0.22(500) = 0.20(500 + x)$
$$0.15x + 110 = 100 + 0.20x$$
$$110 = 100 + 0.05x$$
$$10 = 0.05x$$
$$200 = x$$
200 lb of the 15% aluminum alloy must be used.

29. Strategy: Let x represent the number of ounces of pure water.
The ounces of 70% alcohol is $3.5 - x$.

	Amount	Percent	Quantity
Water	x	0	$0x$
70% alcohol	$3.5 - x$	0.70	$0.7(3.5 - x)$
45% mixture	3.5	0.45	$0.45(3.5)$

The sum of the quantities before mixing is equal to the quantity after mixing.

Solution: $0x + 0.7(3.5 - x) = 0.45(3.5)$
$$2.45 - 0.7x = 1.575$$
$$-0.7x = -0.875$$
$$x = 1.25$$
$3.5 - x = 3.5 - 1.25 = 2.25$
The solution should contain 2.25 oz of rubbing alcohol and 1.25 oz of water.

31. Strategy: Let x represent the number of ounces of pure water.

	Amount	Percent	Quantity
Pure water	x	0	$0x$
8%	75	0.08	$0.08(75)$
5%	$75 + x$	0.05	$0.05(75 + x)$

The sum of the quantities before mixing is

equal to the quantity after mixing.

Solution: $0x + 0.08(75) = 0.05(75 + x)$
$$6 = 3.75 + 0.05x$$
$$2.25 = 0.05x$$
$$45 = x$$
45 oz of pure water must be added.

33. Strategy: Let x represent the number of milliliters of alcohol.

	Amount	Percent	Quantity
Alcohol	x	0	$0x$
25% iodine	200	0.25	$0.25(200)$
10% iodine	$200 + x$	0.10	$0.1(200 + x)$

The sum of the quantities before mixing is equal to the quantity after mixing.

Solution: $0x + 0.25(200) = 0.1(200 + x)$
$$50 = 20 + 0.1x$$
$$30 = 0.1x$$
$$300 = x$$
300 ml of alcohol must be added.

35. Strategy: Let x represent the percent concentration of the remaining fruit drink.

	Amount	Percent	Quantity
5% fruit	12	0.05	$0.05(12)$
Water	2	0	$0(2)$
Mixture	10	x	$x(10)$

The sum of the quantities before mixing is equal to the quantity after mixing.

Solution: $0.05(12) + 0(2) = x(10)$
$$0.6 = 10x$$
$$0.06 = x$$
The remaining fruit drink is 6% fruit juice.

37. Strategy: Let x represent the number of quarts of 40% antifreeze replaced.
The quarts of pure antifreeze added is x.

	Amount	Percent	Quantity
40% antifreeze	12	0.40	0.40(12)
40% antifreeze	x	0.40	0.40(x)
Replaced by pure antifreeze	x	1.00	1.00(x)
Added 60% antifreeze	12	0.60	0.60(12)

The quantity in the radiator minus the quantity replaced plus the quantity added equals the quantity in the resulting solution.

Solution: $0.4(12) - 0.4(x) + x = 0.6(12)$
$$4.8 + 0.6x = 7.2$$
$$0.6x = 2.4$$
$$x = 4$$

4 qt will have to be replaced with pure antifreeze.

Objective C Exercises

39. Strategy: Let t represent policeman's time, in seconds.
The speeding car's time is $10 + t$ seconds.

	Rate	Time	Distance
Police officer	100	t	100(t)
Speeding car	80	$10 + t$	80($10 + t$)

Police and the speeding car travel the same distance.

Solution $100t = 80(10 + t)$
$$100t = 800 + 80t$$
$$20t = 800$$
$$40 = t$$

The police officer will catch up to the speeding car in 2/3 minute.

41. Strategy: Let r represent the speed of the first car.
The speed of the second car is $r + 8$.

	Rate	Time	Distance
1st car	r	2.5	2.5(r)
2nd car	$r + 8$	2.5	2.5($r + 8$)

The total distance traveled by the two cars is 310 mi.

Solution: $2.5r + 2.5(r + 8) = 310$
$$2.5r + 2.5r + 20 = 310$$
$$5r = 290$$
$$r = 58$$

$r + 8 = 58 + 8 = 66$
The speed of the first car is 58 mph.
The speed of the second car is 66 mph.

43. Strategy: Let t be the time flying to the city.
Time returning to the airport is $4 - t$.

	Rate	Time	Distance
Going	250	t	250(t)
Returning	150	$4 - t$	150($4 - t$)

The distance to the city is the same as the distance returning to the airport.

Solution: $250t = 150(4 - t)$
$$250t = 600 - 150t$$
$$400t = 600$$
$$t = 1.5$$
$d = rt = 250(1.5) = 375$
The distance between the airports is 375 mi.

45. Strategy: Let r represent the speed of the freight train.
The speed of the passenger train is $r + 18$.

	Rate	Time	Distance
Freight	r	4	4(r)
Passenger	$r + 18$	2.5	2.5($r + 18$)

The distance traveled by the freight train is equal to the distance traveled by the passenger train.

Solution: $4r = 2.5(r + 18)$
$$4r = 2.5r + 45$$
$$1.5r = 45$$
$$r = 30$$
$$r + 18 = 30 + 18 = 48$$
The speed of the freight train is 30 mph.
The speed of the passenger train is 48 mph.

47. Strategy: Let x represent the speed of the jogger.
The speed of the cyclist is $4x$.

	Rate	Time	Distance
Jogger	x	2	$2(x)$
Cyclist	$4x$	2	$2(4x)$

In two hours the cyclist is 33 mi ahead of the jogger.

Solution: $2(4x) - 2x = 33$
$$8x - 2x = 33$$
$$6x = 33$$
$$x = 5.5$$
$$4x = 4(5.5) = 22$$
The speed of the cyclist is 22 mph.
$$d = rt = 22(2) = 44$$
The cyclist traveled 44 mi.

Critical Thinking

49. The distance between them 2 min before impact is equal to the sum of the distance each one can travel during 2 min.
$$2 \text{ minutes} \cdot \frac{1 \text{ hour}}{60 \text{ minutes}} = 0.03\overline{3} \text{ hour}$$

Distance between cars = rate of first car $\cdot 0.03\overline{3}$ + rate of second car $\cdot 0.03\overline{3}$.

Distance between cars
$$= 40(0.03\overline{3}) + 60(0.03\overline{3}) = 3.3\overline{3}$$

The cars are $3.3\overline{3}$ mi apart 2 min before impact.

51. Strategy: Let x represent the amount of 12-karat gold.

	Amount	Percent	Quantity
12-karat gold	x	$\dfrac{12}{24}$	$\dfrac{12}{24}x$
24-karat gold	3	$\dfrac{24}{24}$	$\dfrac{24}{24}(3)$
14-karat gold	$x + 3$	$\dfrac{14}{24}$	$\dfrac{14}{24}(x+3)$

The sum of the quantities before mixing is equal to the quantity after mixing.

Solution:
$$\frac{12}{24}x + \frac{24}{24}(3) = \frac{14}{24}(x + 3)$$
$$12x + 24(3) = 14(x + 3)$$
$$12x + 72 = 14x + 42$$
$$72 = 2x + 42$$
$$30 = 2x$$
$$15 = x$$
15 oz of the 12-karat gold should be used.

Projects or Group Activities

53. Let d represent the distance traveled in one direction. Bianca walks x miles on level ground and $(d - x)$ miles uphill while going to Nadia's house. On the return trip she travels $(d - x)$ downhill and x miles on level ground.

Total time going is $\dfrac{x}{4} + \dfrac{d - x}{3}$

Total time returning is $\dfrac{d - x}{6} + \dfrac{x}{4}$

The total trip takes 1 hour.
$$\frac{x}{4} + \frac{d - x}{3} + \frac{d - x}{6} + \frac{x}{4} = 1$$
Multiply both sides of the equation by 12.
$$3x + 4(d - x) + 2(d - x) + 3x = 12$$
$$3x + 4d - 4x + 2d - 2x + 3x = 12$$
$$6d = 12$$
$$d = 2$$
The distance from Bianca's house to Nadia's apartment is 2 mi.

Check Your Progress: Chapter 2

1. $3 - 2x = 9$
$-2x = 6$
$x = -3$

2. $4x + 7 = 9x - 3$
$7 = 9x - 4x - 3$
$7 = 5x - 3$
$10 = 5x$
$2 = x$

3. $6 - 2(3x + 4) = 5(2x - 1)$
$6 - 6x - 8 = 10x - 5$
$-6x - 2 = 10x - 5$
$-2 = 16x - 5$
$3 = 16x$
$\dfrac{3}{16} = x$

4. $3(2x - 5) - 5(4x + 2) = 25$
$6x - 15 - 20x - 10 = 25$
$-14x - 25 = 25$
$-14x = 50$
$x = -\dfrac{50}{14} = -\dfrac{25}{7}$

5. $\dfrac{4x - 1}{3} - \dfrac{x + 1}{2} = \dfrac{3x + 1}{4} - \dfrac{x - 4}{6}$
$12\left(\dfrac{4x - 1}{3} - \dfrac{x + 1}{2}\right) = 12\left(\dfrac{3x + 1}{4} - \dfrac{x - 4}{6}\right)$
$4(4x - 1) - 6(x + 1) = 3(3x + 1) - 2(x - 4)$
$16x - 4 - 6x - 6 = 9x + 3 - 2x + 8$
$10x - 10 = 7x + 11$
$3x - 10 = 11$
$3x = 21$
$x = 7$

6. $P = 2L + 2W$
$P - 2W = 2L$
$L = \dfrac{P - 2W}{2}$

7. **Strategy**: Let x represent the number of pounds of coffee.

	Amount	Cost	Value
Coffee 1	x	7	$7(x)$
Coffee 2	10	5	$5(10)$
Mixture	$10 + x$	5.75	$5.75(10 + x)$

The sum of the values before mixing equals the value after mixing.
Solution: $7x + 5(10) = 5.75(10 + x)$
$7x + 50 = 57.5 + 5.75x$
$1.25x = 7.5$
$x = 6$
The mixture must contain 6 lb of coffee 1.

8. **Strategy:** Let x represent the number of milliliters of 11% acetic acid solution. The milliliters of 6% solution is $600 - x$.

	Amount	Percent	Quantity
11%	x	0.11	$0.11x$
6%	$600 - x$	0.06	$0.06(600 - x)$
8%	600	0.08	$0.08(600)$

The sum of the quantities before mixing is equal to the quantity after mixing.

Solution: $0.11x + 0.06(600 - x) = 0.08(600)$
$0.11x + 36 - 0.06x = 48$
$0.05x = 12$
$x = 240$
The mixture must contain 240 ml of the 11% solution and 360 ml of the 6% solution.

9. **Strategy:** Let t represent joggers time. The cyclist's time is $30 + t$ minutes.

	Rate	Time	Distance
jogger	8	t	$8(t)$
cyclist	18	$t - 30$	$18(t - 30)$

The jogger and the cyclist travel the same distance.

Solution $8t = 18(t - 30)$
$$8t = 18t - 540$$
$$-10t = -540$$
$$t = 54$$

The cyclist overtakes the jogger in 54 min.

Section 2.3

Concept Check

1. The Addition Property of Inequalities states that the same number can be added to each side of an inequality without changing the solution set of the inequality.
    Examples will vary. For instance:

    $8 > 6$ $-5 < -1$
    $8 + 7 > 6 + 7$ and $-5 + (-2) < -1 + (-2)$
    $15 > 13$ $-7 < -3$

3. Replace x with each value to determine if the inequality holds.
    i) $-17 + 7 \leq -3$; $-10 \leq -3$; solution
    ii) $8 + 7 \leq -3$; $15 \leq -3$; not a solution
    iii) $-10 + 7 \leq -3$; $-3 \leq -3$; solution
    iv) $0 + 7 \leq -3$; $7 \leq -3$; not a solution

5. If $-x > 0$, then $x < 0$.

Objective A Exercises

7. $x - 3 < 2$
    $$x < 5$$
    $$\{x | x < 5\}$$

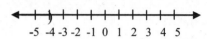

9. $4x \leq 8$
    $$\frac{4x}{4} \leq \frac{8}{4}$$
    $$x \leq 2$$
    $$\{x | x \leq 2\}$$

11. $-2x > 8$
    $$\frac{-2x}{-2} < \frac{8}{-2}$$
    $$x < -4$$
    $$\{x | x < -4\}$$

13. $3x - 1 > 2x + 2$
    $$x - 1 > 2$$
    $$x > 3$$
    $$\{x | x > 3\}$$

15. $2x - 1 > 7$
    $$2x > 8$$
    $$\frac{2x}{2} > \frac{8}{2}$$
    $$x > 4$$
    $$\{x | x > 4\}$$

17. $5x - 2 \leq 8$
    $$5x \leq 10$$
    $$\frac{5x}{5} \leq \frac{10}{5}$$
    $$x \leq 2$$
    $$\{x | x \leq 2\}$$

19. $6x + 3 > 4x - 1$
    $$6x > 4x - 4$$
    $$2x > -4$$
    $$\frac{2x}{2} > \frac{-4}{2}$$
    $$x > -2$$
    $$\{x | x > -2\}$$

21. $8x + 1 \geq 2x + 13$
    $$6x + 1 \geq 13$$
    $$6x \geq 12$$
    $$\frac{6x}{6} \geq \frac{12}{6}$$
    $$x \geq 2$$
    $$\{x | x \geq 2\}$$

23. $4 - 3x < 10$
$-3x < 6$
$\dfrac{-3x}{-3} > \dfrac{6}{-3}$
$x > -2$
$\{x \mid x > -2\}$

25. $7 - 2x \geq 1$
$-2x \geq -6$
$\dfrac{-2x}{-2} \leq \dfrac{-6}{-2}$
$x \leq 3$
$\{x \mid x \leq 3\}$

27. $-3 - 4x > -11$
$-4x > -8$
$\dfrac{-4x}{-4} < \dfrac{-8}{-4}$
$x < 2$
$\{x \mid x < 2\}$

29. $4x - 2 < x - 11$
$3x - 2 < -11$
$3x < -9$
$\dfrac{3x}{3} < \dfrac{-9}{3}$
$x < -3$
$\{x \mid x < -3\}$

31. $x + 7 \geq 4x - 8$
$-3x + 7 \geq -8$
$-3x \geq -15$
$\dfrac{-3x}{-3} \leq \dfrac{-15}{-3}$
$x \leq 5$
$\{x \mid x \leq 5\}$

33. $3x + 2 \leq 7x + 4$
$-4x + 2 \leq 4$
$-4x \leq 2$
$\dfrac{-4x}{-4} \geq \dfrac{2}{-4}$
$x \geq -\dfrac{1}{2}$
$\{x \mid x \geq -\dfrac{1}{2}\}$

35. The solution set of the inequality $nx > a$, where both n and a are negative contains both positive and negative numbers.

37. The solution set of the inequality $x - n > -a$, where both n and a are positive and $n < a$ contains both positive and negative numbers.

39. $7x + 3 < 4x + 1$
$3x + 3 < 1$
$3x < -2$
$\dfrac{3x}{3} < -\dfrac{2}{3}$
$x < -\dfrac{2}{3}$
$\left(-\infty, -\dfrac{2}{3}\right)$

41. $\dfrac{2}{3}x - \dfrac{3}{2} < \dfrac{7}{6} - \dfrac{1}{3}x$
$6\left(\dfrac{2}{3}x - \dfrac{3}{2}\right) < 6\left(\dfrac{7}{6} - \dfrac{1}{3}x\right)$
$4x - 9 < 7 - 2x$
$6x - 9 < 7$
$6x < 16$
$\dfrac{6x}{6} < \dfrac{16}{6}$
$x < \dfrac{8}{3}$
$\left(-\infty, \dfrac{8}{3}\right)$

43. $\dfrac{1}{2}x - \dfrac{3}{4} < \dfrac{7}{4}x - 2$
$4\left(\dfrac{1}{2}x - \dfrac{3}{4}\right) < 4\left(\dfrac{7}{4}x - 2\right)$
$2x - 3 < 7x - 8$
$-5x - 3 < -8$
$-5x < -5$
$\dfrac{-5x}{-5} > \dfrac{-5}{-5}$
$x > 1$
$(1, \infty)$

45. $4(2x - 1) > 3x - 2(3x - 5)$
$8x - 4 > 3x - 6x + 10$
$8x - 4 > -3x + 10$
$11x - 4 > 10$
$11x > 14$
$\dfrac{11x}{11} > \dfrac{14}{11}$
$x > \dfrac{14}{11}$
$\left(\dfrac{14}{11}, \infty \right)$

47. $2 - 5(x + 1) \geq 3(x - 1) - 8$
$2 - 5x - 5 \geq 3x - 3 - 8$
$-3 - 5x \geq 3x - 11$
$-5x \geq 3x - 8$
$-8x \geq -8$
$\dfrac{-8x}{-8} \leq \dfrac{-8}{-8}$
$x \leq 1$
$(-\infty, 1]$

49. $3 + 2(x + 5) \geq x + 5(x + 1) + 1$
$3 + 2x + 10 \geq x + 5x + 5 + 1$
$2x + 13 \geq 6x + 6$
$-4x + 13 \geq 6$
$-4x \geq -7$
$\dfrac{-4x}{-4} \leq \dfrac{-7}{-4}$
$x \leq \dfrac{7}{4}$
$\left(-\infty, \dfrac{7}{4} \right]$

51. $3 - 4(x + 2) \leq 6 + 4(2x + 1)$
$3 - 4x - 8 \leq 6 + 8x + 4$
$-4x - 5 \leq 10 + 8x$
$-12x - 5 \leq 10$
$-12x \leq 15$
$\dfrac{-12x}{-12} \geq \dfrac{15}{-12}$
$x \geq -\dfrac{5}{4}$
$\left[-\dfrac{5}{4}, \infty \right)$

53. $12 - 2(3x - 2) \geq 5x - 2(5 - x)$
$12 - 6x + 4 \geq 5x - 10 + 2x$
$16 - 6x \geq 7x - 10$
$-6x \geq 7x - 26$
$-13x \geq -26$
$\dfrac{-13x}{-13} \leq \dfrac{-26}{-13}$
$x \leq 2$
$(-\infty, 2]$

Objective B Exercises

55. $x - 3 \leq 1$ and $\quad 2x \geq -4$
$\quad x \leq 4 \qquad\qquad x \geq -2$
$\{x \mid x \leq 4\} \qquad \{x \mid x \geq -2\}$
$\{x \mid x \leq 4\} \cap \{x \mid x \geq -2\} = [-2, 4]$

57. $2x < 6 \qquad$ or $\qquad x - 4 > 1$
$\quad x < 3 \qquad\qquad x > 5$
$\{x \mid x < 3\} \qquad \{x \mid x > 5\}$
$\{x \mid x < 3\} \cup \{x \mid x > 5\} = (-\infty, 3) \cup (5, \infty)$

59. $\dfrac{1}{2}x > -2$ and $\quad 5x < 10$
$\quad x > -4 \qquad\qquad x < 2$
$\{x \mid x > -4\} \qquad \{x \mid x < 2\}$
$\{x \mid x > -4\} \cap \{x \mid x < 2\} = (-4, 2)$

61. $\dfrac{2}{3}x > 4 \qquad$ or $\qquad 2x < -8$
$\quad x > 6 \qquad\qquad x < -4$
$\{x \mid x > 6\} \qquad \{x \mid x < -4\}$
$\{x \mid x > 6\} \cup \{x \mid x < -4\} = (-\infty, -4) \cup (6, \infty)$

63. $3x < -9$ and $\quad x - 2 < 2$
$\quad x < -3 \qquad\qquad x < 4$
$\{x \mid x < -3\} \qquad \{x \mid x < 4\}$
$\{x \mid x < -3\} \cap \{x \mid x < 4\} = (-\infty, -3)$

65. $2x - 3 > 1$ and $\quad 3x - 1 < 2$
$\quad 2x > 4 \qquad\qquad 3x < 3$
$\quad x > 2 \qquad\qquad x < 1$
$\{x \mid x > 2\} \qquad \{x \mid x < 1\}$
$\{x \mid x > 2\} \cap \{x \mid x < 1\} = \emptyset$

67. $4x + 1 < 5$ and $4x + 7 > -1$
$\quad\quad 4x < 4 \quad\quad\quad\quad 4x > -8$
$\quad\quad\quad x < 1 \quad\quad\quad\quad\quad x > -2$
$\{x \mid x < 1\} \quad\quad\quad \{x \mid x > -2\}$
$\{x \mid x < 1\} \cap \{x \mid x > -2\} = (-2, 1)$

69. The inequality $x > -3$ or $x < 2$ describes all real numbers.

71. The inequality $x < -3$ or $x > 2$ describes two intervals of real numbers.

73. $6x - 2 < -14$ or $5x + 1 > 11$
$\quad\quad 6x < -12 \quad\quad\quad 5x > 10$
$\quad\quad\quad x < -2 \quad\quad\quad\quad x > 2$
$\{x \mid x < -2\} \quad\quad\quad \{x \mid x > 2\}$
$\{x \mid x < -2\} \cup \{x \mid x > 2\} = \{x \mid x < -2 \text{ or } x > 2\}$

75. $5 < 4x - 3 < 21$
$\quad 5 + 3 < 4x - 3 + 3 < 21 + 3$
$\quad\quad\quad 8 < 4x < 24$
$\quad\quad\quad \dfrac{8}{4} < \dfrac{4x}{4} < \dfrac{24}{4}$
$\quad\quad\quad\quad 2 < x < 6$
$\quad\{x \mid 2 < x < 6\}$

77. $-2 < 3x + 7 < 1$
$\quad -2 + (-7) < 3x + 7 + (-7) < 1 + (-7)$
$\quad\quad\quad -9 < 3x < -6$
$\quad\quad\quad \dfrac{-9}{3} < \dfrac{3x}{3} < \dfrac{-6}{3}$
$\quad\quad\quad\quad -3 < x < -2$
$\quad\{x \mid -3 < x < -2\}$

79. $3x - 5 > 10$ or $3x - 5 < -10$
$\quad\quad 3x > 15 \quad\quad\quad\quad 3x < -5$
$\quad\quad\quad x > 5 \quad\quad\quad\quad\quad x < -\dfrac{5}{3}$
$\{x \mid x > 5\} \quad\quad\quad \{x \mid x < -\dfrac{5}{3}\}$

$\{x \mid x > 5\} \cup \{x \mid x < -\dfrac{5}{3}\} = \{x \mid x > 5 \text{ or } x < -\dfrac{5}{3}\}$

81. $8x + 2 \leq -14$ and $4x - 2 > 10$
$\quad\quad 8x \leq -16 \quad\quad\quad 4x > 12$
$\quad\quad\quad x \leq -2 \quad\quad\quad\quad x > 3$
$\{x \mid x \leq -2\} \quad\quad \{x \mid x > 3\}$
$\{x \mid x \leq -2\} \cap \{x \mid x > 3\} = \emptyset$

83. $5x + 12 \geq 2$ or $7x - 1 \leq 13$
$\quad\quad 5x \geq -10 \quad\quad\quad\quad 7x \leq 14$
$\quad\quad\quad x \geq -2 \quad\quad\quad\quad\quad x \leq 2$
$\{x \mid x \geq -2\} \quad\quad\quad \{x \mid x \leq 2\}$
$\{x \mid x \geq -2\} \cup \{x \mid x \leq 2\} =$ the set of real numbers

85. $3 \leq 7x - 14 \leq 31$
$\quad 3 + 14 \leq 7x - 14 + 14 \leq 31 + 14$
$\quad\quad\quad 17 \leq 7x \leq 45$
$\quad\quad\quad \dfrac{17}{7} \leq \dfrac{7x}{7} \leq \dfrac{45}{7}$
$\quad\quad\quad \dfrac{17}{7} \leq x \leq \dfrac{45}{7}$
$\{x \mid \dfrac{17}{7} \leq x \leq \dfrac{45}{7}\}$

87. $1 - 3x < 16$ and $1 - 3x > -16$
$\quad\quad -3x < 15 \quad\quad\quad\quad -3x > -17$
$\quad\quad\quad x > -5 \quad\quad\quad\quad\quad x < \dfrac{-17}{-3}$
$\{x \mid x > -5\} \quad\quad\quad \{x \mid x < \dfrac{17}{3}\}$

$\{x \mid x > -5\} \cap \{x \mid x < \dfrac{17}{3}\} = \{x \mid -5 < x < \dfrac{17}{3}\}$

89. $6x + 5 < -1$ or $1 - 2x < 7$
$\quad\quad 6x < -6 \quad\quad\quad\quad -2x < 6$
$\quad\quad\quad x < -1 \quad\quad\quad\quad\quad x > -3$
$\{x \mid x < -1\} \quad\quad \{x \mid x > -3\}$
$\{x \mid x < -1\} \cup \{x \mid x > -3\} =$ the set of real numbers.

91. $9 - x \geq 7$ and $9 - 2x < 3$
$\quad\quad -x \geq -2 \quad\quad\quad\quad -2x < -6$
$\quad\quad\quad x \leq 2 \quad\quad\quad\quad\quad x > 3$
$\{x \mid x \leq 2\} \quad\quad\quad \{x \mid x > 3\}$
$\{x \mid x \leq 2\} \cap \{x \mid x > 3\} = \emptyset$

Objective C Exercises

93. "The temperature did not go above 42°F" can be written as $t \leq 42$.

95. "The high temperature was 42°F." can be written as $t < 42$.

97. Strategy: Let x represent the width of the rectangle.
The length of the rectangle is $2x - 5$.
To find the maximum width, solve the inequality $2L + 2W < 60$.

Solution:
$$2L + 2W < 60$$
$$2(2x - 5) + 2x < 60$$
$$4x - 10 + 2x < 60$$
$$6x - 10 < 60$$
$$6x < 70$$
$$x < \frac{70}{6} = 11\frac{2}{3}$$

The smallest width of the rectangle is 11 cm.

99. Strategy: Let d represent the number of days to run the advertisement.
To find the maximum number of days the advertisement can run on the website, solve the inequality $250 + 12d \leq 1500$.

Solution:
$$250 + 12d \leq 1500$$
$$12d \leq 1250$$
$$d \leq \frac{1250}{12}$$
$$d \leq 104\frac{1}{6}$$

You can run the advertisement for 104 days.

101. Strategy: Let x represent the cost of a gallon of paint.
Since a gallon of paint covers 100 ft^2 and the room is 320 ft^2 the homeowner will need to buy 4 gal of paint.
To find the maximum cost per gallon, solve the inequality $24 + 4x \leq 100$.

Solution:
$$24 + 4x \leq 100$$
$$4x \leq 76$$
$$x \leq 19$$

The maximum that the homeowner can pay for a gallon of paint is $19.

103. Strategy: To find the temperature range in Fahrenheit degrees, solve the compound inequality $0 < \frac{5(F - 32)}{9} < 30$.

Solution:
$$0 < \frac{5(F - 32)}{9} < 30$$
$$\frac{9}{5}(0) < \frac{9}{5}\left(\frac{5(F - 32)}{9}\right) < \frac{9}{5}(30)$$
$$0 < F - 32 < 54$$
$$0 + 32 < F - 32 + 32 < 54 + 32$$
$$32° < F < 86°$$

105. Strategy: Let N represent the amount of sales.
To find the minimum amount of sales, solve the inequality $1000 + 0.05N \geq 3200$.

Solution:
$$1000 + 0.05N \geq 3200$$
$$0.05N \geq 2200$$
$$N \geq 44{,}000$$

George's amount of sales must be $44,000 or more per month.

107. Strategy: Let x represent the number of gallons of juice produced in the first month.
To find the minimum number of gallons produced solve the inequality
$x + x + 400 + x + 800 + x + 1200 + x + 1600 \geq 8500$

Solution:
$x + x + 400 + x + 800 + x + 1200 + x + 1600 \geq 8500$
$5x + 4000 \geq 8500$
$5x \geq 4500$
$x \geq 900$
The minimum number of gallons of juice produced is 900 gal.

109. Strategy: Let n represent the score on the last test.
To find the range of scores, solve the inequality
$$70 \leq \frac{56 + 91 + 83 + 62 + n}{5} \leq 79.$$

Solution:
$$70 \leq \frac{56 + 91 + 83 + 62 + n}{5} \leq 79$$
$$70 \leq \frac{292 + n}{5} \leq 79$$
$$5(70) \leq 5 \cdot \frac{292 + n}{5} \leq 5(79)$$
$$350 \leq 292 + n \leq 395$$

$$350 - 292 \le 292 - 292 + n \le 395 - 292$$
$$58 \le n \le 103$$

Since 100 is the maximum score, the range of scores needed to receive a C grade is $58 \le n \le 100$.

Critical Thinking

111. a) $a \le 2x + 1 \le b$
$a - 1 \le 2x \le b - 1$
Since $-2 \le x \le 4$, we have $-4 \le 2x \le 8$,
$a - 1 \le 2x$
$a - 1 \le -4$
$a \le -3$
The largest possible value of a is -3.

b) $2x \le b - 1$
$6 \le b - 1$
$7 \le b$
The smallest possible value of b is 7.

113.   True

115.   True

Section 2.4

Concept Check

1. $|2 - 8| = 6$
$|-6| = 6$
$6 = 6$
Yes, 2 is a solution.

3. $|3(-1) - 4| = 7$
$|-3 - 4| = 7$
$|-7| = 7$
$7 = 7$
Yes, -1 is a solution.

5. $|x| = 7$
$x = 7$ or $x = -7$
The solutions are 7 and -7.

7. $|-y| = 6$
$-y = 6$ or $-y = -6$
$y = -6$ or $y = 6$
The solutions are 6 and -6.

9. $|x| = -4$
There is no solution to this equation because the absolute value of a number must be nonnegative.

11. $|-t| = -3$
There is no solution to this equation because the absolute value of a number must be nonnegative.

13. $|x| > 3$
$x > 3$ or $x < -3$
$\{x \mid x > 3\}$ $\{x \mid x < -3\}$
$\{x \mid x > 3\} \cup \{x \mid x < -3\}$

15. $|x - 2| < 5$

Objective A Exercises

17. $|x + 2| = 3$
$x + 2 = 3$ or $x + 2 = -3$
$x = 1$ $x = -5$
The solutions are 1 and -5.

19. $|y - 5| = 3$
$y - 5 = 3$ or $y - 5 = -3$
$y = 8$ $y = 2$
The solutions are 2 and 8.

21. $|a - 2| = 0$
$a - 2 = 0$
$a = 2$
The solution is 2.

23. $|x - 2| = -4$
There is no solution to this equation because the absolute value of a number must be nonnegative.

25. $|3 - 4x| = 9$
$3 - 4x = 9$ or $3 - 4x = -9$
$-4x = 6$ $-4x = -12$
$x = -\dfrac{3}{2}$ $x = 3$

The solutions are 3 and $-\dfrac{3}{2}$.

27. $|2x - 3| = 0$
$2x - 3 = 0$
$2x = 3$
$x = \dfrac{3}{2}$

The solution is $\dfrac{3}{2}$.

29. $|3x - 2| = -4$
There is no solution to this equation because the absolute value of a number must be nonnegative.

31. $|x - 2| - 2 = 3$
$|x - 2| = 5$
$x - 2 = 5$ or $x - 2 = -5$
$x = 7$ $\qquad x = -3$
The solutions are 7 and -3.

33. $|3a + 2| - 4 = 4$
$|3a + 2| = 8$
$3a + 2 = 8$ or $3a + 2 = -8$
$3a = 6$ $\qquad 3a = -10$
$a = 2$ $\qquad a = -\dfrac{10}{3}$

The solutions are 2 and $-\dfrac{10}{3}$.

35. $|2 - y| + 3 = 4$
$|2 - y| = 1$
$2 - y = 1$ or $2 - y = -1$
$-y = -1$ $\qquad -y = -3$
$y = 1$ $\qquad y = 3$
The solutions are 1 and 3.

37. $|2x - 3| + 3 = 3$
$|2x - 3| = 0$
$2x - 3 = 0$
$2x = 3$
$x = \dfrac{3}{2}$

The solution is $\dfrac{3}{2}$.

39. $|2x - 3| + 4 = -4$
$|2x - 3| = -8$
There is no solution to this equation because the absolute value of a number must be nonnegative.

41. $|6x - 5| - 2 = 4$
$|6x - 5| = 6$
$6x - 5 = 6$ or $6x - 5 = -6$
$6x = 11$ $\qquad 6x = -1$
$x = \dfrac{11}{6}$ $\qquad x = -\dfrac{1}{6}$

The solutions are $\dfrac{11}{6}$ and $-\dfrac{1}{6}$.

43. $|3t + 2| + 3 = 4$
$|3t + 2| = 1$
$3t + 2 = 1$ or $3t + 2 = -1$
$3t = -1$ $\qquad 3t = -3$
$x = -\dfrac{1}{3}$ $\qquad x = -1$

The solutions are $-\dfrac{1}{3}$ and -1.

45. $3 - |x - 4| = 5$
$-|x - 4| = 2$
$|x - 4| = -2$
There is no solution to this equation because the absolute value of a number must be nonnegative.

47. $8 - |2x - 3| = 5$
$-|2x - 3| = -3$
$|2x - 3| = 3$
$2x - 3 = 3$ or $2x - 3 = -3$
$2x = 6$ $\qquad 2x = 0$
$x = 3$ $\qquad x = 0$
The solutions are 3 and 0.

49. $|2 - 3x| + 7 = 2$
$|2 - 3x| = -5$
There is no solution to this equation because the absolute value of a number must be nonnegative.

51. $|8 - 3x| - 3 = 2$
$|8 - 3x| = 5$
$8 - 3x = 5$ or $8 - 3x = -5$
$-3x = -3$ $\qquad -3x = -13$
$x = 1$ $\qquad x = \dfrac{13}{3}$

The solutions are 1 and $\dfrac{13}{3}$.

53. $|2x - 8| + 12 = 2$

$|2x - 8| = -10$

There is no solution to this equation because the absolute value of a number must be nonnegative.

55. $2 + |3x - 4| = 5$

$|3x - 4| = 3$

$3x - 4 = 3$ or $3x - 4 = -3$

$3x = 7$ $3x = 1$

$x = \dfrac{7}{3}$ $x = \dfrac{1}{3}$

The solutions are $\dfrac{7}{3}$ and $\dfrac{1}{3}$.

57. $5 - |2x + 1| = 5$

$-|2x + 1| = 0$

$2x + 1 = 0$

$2x = -1$

$x = -\dfrac{1}{2}$

The solution is $-\dfrac{1}{2}$.

59. $6 - |2x + 4| = 3$

$-|2x + 4| = -3$

$|2x + 4| = 3$

$2x + 4 = 3$ or $2x + 4 = -3$

$2x = -1$ $2x = -7$

$x = -\dfrac{1}{2}$ $x = -\dfrac{7}{2}$

The solutions are $-\dfrac{1}{2}$ and $-\dfrac{7}{2}$.

61. $8 - |1 - 3x| = -1$

$-|1 - 3x| = -9$

$|1 - 3x| = 9$

$1 - 3x = 9$ or $1 - 3x = -9$

$-3x = 8$ $-3x = -10$

$x = -\dfrac{8}{3}$ $x = \dfrac{10}{3}$

The solutions are $-\dfrac{8}{3}$ and $\dfrac{10}{3}$.

63. $5 + |2 - x| = 3$

$|2 - x| = -2$

There is no solution to this equation because the absolute value of a number must be nonnegative.

65. Two positive solutions

67. Two negative solutions

Objective B Exercises

69. $|x + 1| > 2$

$x + 1 > 2$ or $x + 1 < -2$

$x > 1$ $x < -3$

$\{x \mid x > 1\}$ $\{x \mid x < -3\}$

$\{x \mid x > 1\} \cup \{x \mid x < -3\} = \{x \mid x > 1 \text{ or } x < -3\}$

71. $|x - 5| \le 1$

$-1 \le x - 5 \le 1$

$-1 + 5 \le x - 5 + 5 \le 1 + 5$

$4 \le x \le 6$

$\{x \mid 4 \le x \le 6\}$

73. $|2 - x| \ge 3$

$2 - x \ge 3$ or $2 - x \le -3$

$-x \ge 1$ $-x \le -5$

$x \le -1$ $x \ge 5$

$\{x \mid x \le -1\}$ $\{x \mid x \ge 5\}$

$\{x \mid x \le -1\} \cup \{x \mid x \ge 5\} = \{x \mid x \le -1 \text{ or } x \ge 5\}$

75. $|2x + 1| < 5$

$-5 < 2x + 1 < 5$

$-5 - 1 < 2x + 1 - 1 < 5 - 1$

$-6 < 2x < 4$

$-3 < x < 2$

$\{x \mid -3 < x < 2\}$

77. $|5x + 2| > 12$

$5x + 2 > 12$ or $5x + 2 < -12$

$5x > 10$ $5x < -14$

$x > 2$ $x < -\dfrac{14}{5}$

$\{x \mid x > 2\}$ $\{x \mid x < -\dfrac{14}{5}\}$

$\{x \mid x > 2\} \cup \{x \mid x < -\dfrac{14}{5}\} = \{x \mid x > 2 \text{ or } x < -\dfrac{14}{5}\}$

79. $|4x - 3| \le -2$

The absolute value of a number must be nonnegative. The solution set is the empty set $\varnothing$.

81. $|2x + 7| > -5$

$$
\begin{array}{lll}
2x + 7 > -5 & \text{or} & 2x + 7 < 5 \\
2x > -12 & & 2x < -2 \\
x > -6 & & x < -1 \\
\{x \mid x > -6\} & & \{x \mid x < -1\}
\end{array}
$$

$\{x \mid x > -6\} \cup \{x \mid x < -1\} =$ The set of all real numbers.

83. $|4 - 3x| \ge 5$

$$
\begin{array}{lll}
4 - 3x \ge 5 & \text{or} & 4 - 3x \le -5 \\
-3x \ge 1 & & -3x \le -9 \\
x \le -\dfrac{1}{3} & & x \ge 3 \\[2mm]
\left\{x \mid x \le -\dfrac{1}{3}\right\} & & \{x \mid x \ge 3\}
\end{array}
$$

$\left\{x \mid x \le -\dfrac{1}{3}\right\} \cup \{x \mid x \ge 3\} = \left\{x \mid x \le -\dfrac{1}{3} \text{ or } x \ge 3\right\}$

85. $|5 - 4x| \le 13$

$$
\begin{aligned}
-13 &\le 5 - 4x \le 13 \\
-13 + (-5) &\le 5 + (-5) - 4x \le 13 + (-5) \\
-18 &\le -4x \le 8 \\
\frac{18}{4} &\ge x \ge -2
\end{aligned}
$$

$\left\{x \mid -2 \le x \le \dfrac{9}{2}\right\}$

87. $|6 - 3x| \le 0$

$$
\begin{array}{lll}
6 - 3x \le 0 & \text{or} & 6 - 3x \ge 0 \\
-3x \le -6 & & -3x \ge -6 \\
x \ge 2 & & x \le 2 \\
\{x \mid x \ge 2\} & & \{x \mid x \le 2\}
\end{array}
$$

$\{x \mid x \ge 2\} \cup \{x \mid x \le 2\} = \{x \mid x = 2\}$

89. $|2 - 9x| > 20$

$$
\begin{array}{lll}
2 - 9x > 20 & \text{or} & 2 - 9x < -20 \\
-9x > 18 & & -9x < -22 \\
x < -2 & & x > \dfrac{22}{9} \\[2mm]
\{x \mid x < -2\} & & \left\{x \mid x > \dfrac{22}{9}\right\}
\end{array}
$$

$\{x \mid x < -2\} \cup \left\{x \mid x > \dfrac{22}{9}\right\} = \left\{x \mid x < -2 \text{ or } x > \dfrac{22}{9}\right\}$

91. $|2x - 3| + 2 < 8$

$$
\begin{aligned}
|2x - 3| &< 6 \\
-6 &< 2x - 3 < 6 \\
-6 + 3 &< 2x - 3 + 3 < 6 + 3 \\
-3 &< 2x < 9 \\
-\frac{3}{2} &< x < \frac{9}{2}
\end{aligned}
$$

$\left\{x \mid -\dfrac{3}{2} < x < \dfrac{9}{2}\right\}$

93. $|2 - 5x| - 4 > -2$

$$
\begin{array}{lll}
|2 - 5x| > 2 \\
2 - 5x > 2 & \text{or} & 2 - 5x < -2 \\
-5x > 0 & & -5x < -4 \\
x < 0 & & x > \dfrac{4}{5} \\[2mm]
\{x \mid x < 0\} & & \left\{x \mid x > \dfrac{4}{5}\right\}
\end{array}
$$

$\{x \mid x < 0\} \cup \left\{x \mid x > \dfrac{4}{5}\right\} = \left\{x \mid x < 0 \text{ or } x > \dfrac{4}{5}\right\}$

95. $8 - |2x - 5| < 3$

$$
\begin{array}{lll}
-|2x - 5| < -5 \\
|2x - 5| > 5 \\
2x - 5 < -5 & \text{or} & 2x - 5 > 5 \\
2x < 0 & & 2x > 10 \\
x < 0 & & x > 5 \\
\{x \mid x < 0\} & & \{x \mid x > 5\}
\end{array}
$$

$\{x \mid x < 0\} \cup \{x \mid x > 5\} = \{x \mid x < 0 \text{ or } x > 5\}$

97. All negative solutions

Objective C Exercises

99. The desired dosage is 3 ml. The tolerance is 0.2 ml.

101. **Strategy:** Let d represent the diameter of the bushing, T the tolerance and x the lower and upper limits of the diameter.
Solve the absolute value inequality
$|x - d| \le T$.

Solution: $|x - d| \leq T$

$|x - 1.75| \leq 0.008$

$-0.008 \leq x - 1.75 \leq 0.008$

$-0.008 + 1.75 \leq x - 1.75 + 1.75 \leq 0.008 + 1.75$

$1.742 \leq x \leq 1.758$

The lower and upper limits of the diameter of the bushing are 1.742 in. and 1.758 in.

103. **Strategy:** Let L represent the length of the piston.

Solve the absolute value inequality

$|L - 9\frac{5}{8}| \leq \frac{1}{32}$.

Solution: $|L - 9\frac{5}{8}| \leq \frac{1}{32}$

$-\frac{1}{32} \leq L - 9\frac{5}{8} \leq \frac{1}{32}$

$-\frac{1}{32} + 9\frac{5}{8} \leq L - 9\frac{5}{8} + 9\frac{5}{8} \leq \frac{1}{32} + 9\frac{5}{8}$

$9\frac{19}{32} \leq L \leq 9\frac{21}{32}$

The lower and upper limits of the length of the piston are $9\frac{19}{32}$ in. and $9\frac{21}{32}$ in.

105. **Strategy:** Let P represent the range of voters who felt the economy is the most important election issue.

Let E represent the margin of error.

$41\% - E < P < 41\% + E$

Solution: $E = 3\%$

$41\% - 3\% < P < 41\% + 3\%$

$38\% < P < 44\%$

The lower and upper limits of the percentage who felt the economy was the most important issue is 38% and 44%.

107. **Strategy:** Let M represent the range, in ohms, for a resistor.

Let T represent the tolerance of the resistor.

Solve the absolute value inequality

$|M - 29,000| \leq T$.

Solution: $T = (0.02)(29,000)$

$= 580$ ohms

$|M - 29,000| \leq 580$

$-580 \leq |M - 29,000| \leq 580$

$-580 + 29,000 \leq |M - 29,000 + 29,000|$

$\leq 580 + 29,000$

$28,420 \leq M \leq 29,580$

The lower and upper limits of the resistor are 28,420 ohms and 29,580 ohms.

Critical Thinking

109. a) The equation $|x + 3| = x + 3$ is true for all x for which $x + 3 \geq 0$.

$x + 3 \geq 0$

$x \geq -3$

$\{x \mid x \geq -3\}$

b) The equation $|a - 4| = 4 - a$ is true for all a for which $4 - a \geq 0$.

$4 - a \geq 0$

$-a \geq -4$

$a \leq 4$

$\{a \mid a \leq 4\}$

111. $-2 \leq x \leq 2$

$-a \leq 3x - 2 \leq a, \ a \geq 0$

For $x = 2$, we have $3x - 2 = 4$, and $3x - 2 < 4$ for $-2 \leq x \leq 2$.

For $3x - 2 \leq a$ to be true, a must be greater than or equal to 4. The smallest possible value of a is 4.

Projects or Group Activities

113. $|3x - 4| = 2x + 10$ or $|3x - 4| = -(2x + 10)$

$3x - 4 = 2x + 10 \qquad 3x - 4 = -2x - 10$

$x = 14 \qquad\qquad 5x = -6$

$x = -\frac{6}{5}$

The solutions are $x = 14$ and $x = -\frac{6}{5}$.

115. $|3x + 1| = 2x - 5$ or $|3x + 1| = -(2x - 5)$

$3x + 1 = 2x - 5 \qquad 3x + 1 = -2x + 5$

$x = -6 \qquad\qquad 5x = 4$

$x = \frac{4}{5}$

Neither value for x checks.

There is no solution.

Chapter 2 Review Exercises

1. $3t - 3 + 2t = 7t - 15$

$\qquad 5t - 3 = 7t - 15$

$\qquad -3 = 2t - 15$

$\qquad -3 + 15 = 2t - 15 + 15$

$\qquad 12 = 2t$

$\qquad 6 = t$

The solution is 6.

2. $3x - 7 > -2$

$\qquad 3x > 5$

$\qquad \dfrac{3x}{3} > \dfrac{5}{3}$

$\qquad x > \dfrac{5}{3}$

$\qquad \left(\dfrac{5}{3}, \infty \right)$

3. $\qquad P = 2L + 2W$

$\quad P - 2W = 2L$

$\quad \dfrac{P - 2W}{2} = L$

4. $\qquad x + 4 = -5$

$\quad x + 4 - 4 = -5 - 4$

$\qquad x = -9$

The solution is -9.

5. $3x < 4 \quad$ and $\quad x + 2 > -1$

$\quad x < \dfrac{4}{3} \qquad\qquad x > -3$

$\quad \{x \mid x < \dfrac{4}{3}\} \qquad \{x \mid x > -3\}$

$\{x / x < \dfrac{4}{3}\} \cap \{x \mid x > -3\} = \{x \mid -3 < x < \dfrac{4}{3}\}$

6. $\qquad \dfrac{3}{5}x - 3 = 2x + 5$

$\quad 5\left(\dfrac{3}{5}x - 3 \right) = 5(2x + 5)$

$\qquad 3x - 15 = 10x + 25$

$\qquad -15 = 7x + 25$

$\qquad -40 = 7x$

$\qquad -\dfrac{40}{7} = x$

The solution is $-\dfrac{40}{7}$.

7. $\qquad -\dfrac{2}{3}x = \dfrac{4}{9}$

$\quad -\dfrac{3}{2}\left(-\dfrac{2}{3}x \right) = -\dfrac{3}{2}\left(\dfrac{4}{9} \right)$

$\qquad x = -\dfrac{2}{3}$

The solution is $-\dfrac{2}{3}$.

8. $|x - 4| - 8 = -3$

$\qquad |x - 4| = 5$

$\quad x - 4 = 5 \quad$ or $\quad x - 4 = -5$

$\qquad x = 9 \qquad\qquad x = -1$

The solutions are 9 and -1.

9. $|2x - 5| < 3$

$\quad -3 < 2x - 5 < 3$

$-3 + 5 < 2x - 5 + 5 < 3 + 5$

$\qquad 2 < 2x < 8$

$\qquad 1 < x < 4$

$\{x \mid 1 < x < 4\}$

10. $\qquad \dfrac{2x - 3}{3} + 2 = \dfrac{2 - 3x}{5}$

$\quad 15\left(\dfrac{2x - 3}{3} + 2 \right) = 15\left(\dfrac{2 - 3x}{5} \right)$

$\quad 5(2x - 3) + 15(2) = 3(2 - 3x)$

$\qquad 10x - 15 + 30 = 6 - 9x$

$\qquad 10x + 15 = 6 - 9x$

$\qquad 19x + 15 = 6$

$\qquad 19x = -9$

$\qquad x = -\dfrac{9}{19}$

The solution is $-\dfrac{9}{19}$.

11. $2(a - 3) = 5(4 - 3a)$

$\qquad 2a - 6 = 20 - 15a$

$\quad 17a - 6 = 20$

$\qquad 17a = 26$

$\qquad a = \dfrac{26}{17}$

The solution is $\dfrac{26}{17}$.

12. $5x - 2 > 8$ or $3x + 2 < -4$
$\qquad\quad 5x > 10 \qquad\qquad 3x < -6$
$\qquad\quad x > 2 \qquad\qquad\quad x < -2$
$\qquad \{x \mid x > 2\} \qquad\qquad \{x \mid x < -2\}$
$\{x \mid x > 2\} \cup \{x \mid x < -2\} = \{x \mid x > 2 \text{ or } x < -2\}$

13. $|4x - 5| \geq 3$
$\qquad 4x - 5 \geq 3$ or $4x - 5 \leq -3$
$\qquad 4x \geq 8 \qquad\qquad\quad 4x \leq 2$

$\qquad\quad x \geq 2 \qquad\qquad\qquad x \leq \dfrac{1}{2}$

$\qquad \{x / x \geq 2\} \qquad\qquad \{x / x \leq \dfrac{1}{2}\}$

$\{x / x \geq 2\} \cup \{x / x \leq \dfrac{1}{2}\} = \{x \mid x \geq 2 \text{ or } x \leq \dfrac{1}{2}\}$

14. $\qquad P = \dfrac{R - C}{n}$

$\qquad n \cdot P = n\left(\dfrac{R - C}{n}\right)$

$\qquad\quad nP = R - C$
$\quad nP + C = R$
$\qquad\quad C = R - nP$

15. $\qquad \dfrac{1}{2}x - \dfrac{5}{8} = \dfrac{3}{4}x + \dfrac{3}{2}$

$\qquad 8\left(\dfrac{1}{2}x - \dfrac{5}{8}\right) = 8\left(\dfrac{3}{4}x + \dfrac{3}{2}\right)$

$\qquad\qquad 4x - 5 = 6x + 12$
$\qquad\qquad\quad -5 = 2x + 12$
$\qquad\qquad\; -17 = 2x$

$\qquad\qquad -\dfrac{17}{2} = x$

The solution is $-\dfrac{17}{2}$.

16. $6 + |3x - 3| = 2$
$|3x - 3| = -4$
There is no solution to this equation because the absolute value of a number must be nonnegative.

17. $3x - 2 > x - 4$ or $7x - 5 < 3x + 3$
$\qquad 2x - 2 > -4 \qquad\qquad 4x - 5 < 3$
$\qquad 2x > -2 \qquad\qquad\qquad 4x < 8$
$\qquad\; x > -1 \qquad\qquad\qquad\; x < 2$
$\{x \mid x > -1\} \qquad\qquad \{x \mid x < 2\}$
$\{x \mid x > -1\} \cup \{x \mid x < 2\} = \{x \mid x \text{ is any real number}\}$
$(-\infty, \infty)$

18. $2x - (3 - 2x) = 4 - 3(4 - 2x)$
$\qquad 2x - 3 + 2x = 4 - 12 + 6x$
$\qquad\quad 4x - 3 = -8 + 6x$
$\qquad\qquad\; -3 = -8 + 2x$
$\qquad\qquad\quad 5 = 2x$

$\qquad\qquad \dfrac{5}{2} = x$

The solution is $\dfrac{5}{2}$.

19. $\qquad x + 9 = -6$
$\quad x + 9 - 9 = -6 - 9$
$\qquad\qquad x = -15$
The solution is -15.

20. $\qquad \dfrac{2}{3} = x + \dfrac{3}{4}$

$\qquad \dfrac{2}{3} - \dfrac{3}{4} = x + \dfrac{3}{4} - \dfrac{3}{4}$

$\qquad \dfrac{8}{12} - \dfrac{9}{12} = x$

$\qquad\qquad -\dfrac{1}{12} = x$

The solution is $-\dfrac{1}{12}$.

21. $\qquad -3x = -21$

$\qquad \dfrac{-3x}{-3} = \dfrac{-21}{-3}$

$\qquad\qquad x = 7$
The solution is 7.

22.
$$\frac{2}{3}a = \frac{4}{9}$$
$$\frac{3}{2}\left(\frac{2}{3}a\right) = \frac{3}{2}\left(\frac{4}{9}\right)$$
$$a = \frac{2}{3}$$
The solution is $\frac{2}{3}$.

23. $3y - 5 = 3 - 2y$
$$5y - 5 = 3$$
$$5y = 8$$
$$y = \frac{8}{5}$$
The solution is $\frac{8}{5}$.

24. $4x - 5 + x = 6x - 8$
$$5x - 5 = 6x - 8$$
$$-5 = x - 8$$
$$3 = x$$
The solution is 3.

25. $3(x - 4) = -5(6 - x)$
$$3x - 12 = -30 + 5x$$
$$-12 = -30 + 2x$$
$$18 = 2x$$
$$9 = x$$
The solution is 9.

26.
$$\frac{3x - 2}{4} + 1 = \frac{2x - 3}{2}$$
$$8\left(\frac{3x - 2}{4} + 1\right) = 8\left(\frac{2x - 3}{2}\right)$$
$$2(3x - 2) + 8(1) = 4(2x - 3)$$
$$6x - 4 + 8 = 8x - 12$$
$$6x + 4 = 8x - 12$$
$$4 = 2x - 12$$
$$16 = 2x$$
$$8 = x$$
The solution is 8.

27.
$$5x - 8 < -3$$
$$5x - 8 + 8 < -3 + 8$$
$$5x < 5$$
$$x < 1$$
$(-\infty, 1)$

28. $2x - 9 \le 8x + 15$
$$-9 \le 6x + 15$$
$$-24 \le 6x$$
$$-4 \le x$$
$[-4, \infty)$

29.
$$\frac{2}{3}x - \frac{5}{8} \ge \frac{3}{4}x + 1$$
$$24\left(\frac{2}{3}x - \frac{5}{8}\right) \ge 24\left(\frac{3}{4}x + 1\right)$$
$$16x - 15 \ge 18x + 24$$
$$-15 \ge 2x + 24$$
$$-39 \ge 2x$$
$$-\frac{39}{2} \ge x$$
$\{x \mid x \le -\frac{39}{2}\}$

30. $2 - 3(2x - 4) \le 4x - 2(1 - 3x)$
$$2 - 6x + 12 \le 4x - 2 + 6x$$
$$14 - 6x \le 10x - 2$$
$$14 \le 16x - 2$$
$$16 \le 16x$$
$$1 \le x$$
$\{x \mid x \ge 1\}$

31.
$$-5 < 4x - 1 < 7$$
$$-5 + 1 < 4x - 1 + 1 < 7 + 1$$
$$-4 < 4x < 8$$
$$-1 < x < 2$$
$(-1, 2)$

32. $|2x - 3| = 8$
$$2x - 3 = 8 \qquad 2x - 3 = -8$$
$$2x = 11 \qquad\quad 2x = -5$$
$$x = \frac{11}{2} \qquad\quad x = \frac{-5}{2}$$
The solutions are $\frac{11}{2}$ and $\frac{-5}{2}$.

33. $|5x + 8| = 0$
$$5x + 8 = 0$$
$$5x = -8$$
$$x = \frac{-8}{5}$$
The solution is $-\frac{8}{5}$.

34. $|5x - 4| < -2$

$\varnothing$

The solution is the empty set because the absolute value of a number must be nonnegative.

35. Strategy: Let t be the time to travel to the island. Time returning to dock is $2\frac{1}{3} - t$.

	Rate	Time	Distance
To island	16	t	$16(t)$
Back to dock	12	$\frac{7}{3} - t$	$12\left(\frac{7}{3} - t\right)$

The distance to the island is the same as the distance back to the dock.
Determine t and then find the distance.

Solution: $16t = 12\left(\dfrac{7}{3} - t\right)$

$16t = 28 - 12t$
$28t = 28$
$t = 1$
$d = rt = 16(1) = 16$
The island is 16 min from the dock.

36. Strategy: Let x represent the number of gallons of apple juice.

	Amount	Cost	Value
Apple	x	12.50	$12.5(x)$
Cranberry	25	31.50	$25(31.5)$
Mixture	$25 + x$	25.00	$25(25 + x)$

The sum of the values before mixing equals the value after mixing.

Solution: $12.5x + 25(31.5) = 25(25 + x)$
$12.5x + 787.5 = 625 + 25x$
$787.5 = 625 + 12.5x$
$162.5 = 12.5x$
$13 = x$

The mixture must contain 13 gal of apple juice.

37. Strategy: Let N represent the amount of sales.
To find the minimum amount of sales solve the inequality $1200 + 0.08N \geq 5000$.

Solution: $1200 + 0.08N \geq 5000$
$\qquad\qquad 0.08N \geq 3200$
$\qquad\qquad\qquad N \geq 40{,}000$

The executive's amount of sales must be $40,000 or more per month.

38. Strategy: Let d represent the diameter of the bushing, T the tolerance and x the lower and upper limits of the diameter.
Solve the absolute value inequality $|x - d| \leq T$.

Solution: $|x - d| \leq T$
$|x - 2.75| \leq 0.003$
$-0.003 \leq x - 2.75 \leq 0.003$
$-0.003 + 2.75 \leq x - 2.75 + 2.75 \leq 0.003 + 2.75$
$2.747 \leq x \leq 2.753$
The lower and upper limits of the diameter of the bushing are 2.747 in. and 2.753 in.

39. Strategy: Let N represent the score on the last test.
To find the range of scores, solve the inequality
$$80 \leq \frac{92 + 66 + 72 + 88 + N}{5} \leq 90$$

Solution:
$$80 \leq \frac{92 + 66 + 72 + 88 + N}{5} \leq 90$$
$$80 \leq \frac{318 + N}{5} \leq 90$$
$$5(80) \leq 5 \cdot \frac{318 + N}{5} \leq 5(90)$$
$$400 \leq 318 + N \leq 450$$
$$400 - 318 \leq 318 - 318 + N \leq 450 - 318$$
$$82 \leq N \leq 132$$
Since 100 is the maximum score, the range of scores needed to receive a C grade is $82 \leq N \leq 100$.

40. Strategy: Let r be the speed of the first plane.
The speed of the second plane is $r + 80$.

	Rate	Time	Distance
1st plane	r	1.75	$1.75r$
2n plane	$r + 80$	1.75	$1.75(r + 80)$

The total distance traveled by the two planes is 1680 mi.

Solution:
$$1.75r + 1.75(r + 80) = 1680$$
$$1.75r + 1.75r + 140 = 1680$$
$$3.5r + 140 = 1680$$
$$3.5r = 1540$$
$$r = 440$$

$r + 80 = 440 + 80 = 520$
The speed of the first plane is 440 mph and the speed of the second plane is 520 mph.

41. Strategy: Let x represent the number of pounds of 30% tin.
The pounds of 70% tin is $500 - x$.

	Amount	Percent	Quantity
30%	x	0.30	$0.30x$
70%	$500 - x$	0.70	$0.70(500 - x)$
40%	500	0.40	$0.40(500)$

The sum of the quantities before mixing is equal to the quantity after mixing.

Solution:
$$0.3x + 0.7(500 - x) = 0.4(500)$$
$$0.3x + 350 - 0.7x = 200$$
$$-0.4x + 350 = 200$$
$$-0.4x = -150$$
$$x = 375$$

$500 - x = 500 - 375 = 125$
375 lb of the 30% tin alloy and 125 lb of the 70% tin alloy were used.

42. Strategy: Let L represent the range in length of the piston.
Solve the absolute value inequality
$$|L - 10\tfrac{3}{8}| \le \frac{1}{32}.$$

Solution: $|L - 10\tfrac{3}{8}| \le \dfrac{1}{32}$

$$-\frac{1}{32} \le L - 10\frac{3}{8} \le \frac{1}{32}$$

$$-\frac{1}{32} + 10\frac{3}{8} \le L - 10\frac{3}{8} + 10\frac{3}{8} \le \frac{1}{32} + 10\frac{3}{8}$$

$$10\frac{11}{32} \le L \le 10\frac{13}{32}$$

The lower and upper limits of the length of the piston are $10\dfrac{11}{32}$ in. and $10\dfrac{13}{32}$ in.

Chapter 2 Test

1.
$$x - 2 = -4$$
$$x - 2 + 2 = -4 + 2$$
$$x = -2$$
The solution is -2.

2.
$$b + \frac{3}{4} = \frac{5}{8}$$
$$b + \frac{3}{4} - \frac{3}{4} = \frac{5}{8} - \frac{3}{4}$$
$$b = \frac{5}{8} - \frac{6}{8}$$
$$b = -\frac{1}{8}$$

The solution is $-\dfrac{1}{8}$.

3.
$$-\frac{3}{4}y = -\frac{5}{8}$$
$$-\frac{4}{3}\left(-\frac{3}{4}\right)y = -\frac{4}{3}\left(-\frac{5}{8}\right)$$
$$y = \frac{5}{6}$$
The solution is $\dfrac{5}{6}$.

4.
$$3x - 5 = 7$$
$$3x - 5 + 5 = 7 + 5$$
$$3x = 12$$
$$x = 4$$
The solution is 4.

5. $\dfrac{3}{4}y - 2 = 6$

$\dfrac{3}{4}y - 2 + 2 = 6 + 2$

$\dfrac{3}{4}y = 8$

$\dfrac{4}{3}\left(\dfrac{3}{4}y\right) = 8\left(\dfrac{4}{3}\right)$

$y = \dfrac{32}{2}$

The solution is $\dfrac{32}{2}$.

6. $2x - 3 - 5x = 8 + 2x - 10$

$-3x - 3 = -2 + 2x$

$-3x - 1 = 2x$

$-1 = 5x$

$-\dfrac{1}{5} = x$

The solution is $-\dfrac{1}{5}$.

7. $2[a - (2 - 3a) - 4] = a - 5$

$2[a - 2 + 3a - 4] = a - 5$

$2[4a - 6] = a - 5$

$8a - 12 = a - 5$

$7a - 12 = -5$

$7a = 7$

$a = 1$

The solution is 1.

8. $E = IR + Ir$

$E - Ir = IR$

$\dfrac{E - Ir}{I} = R$

9. $\dfrac{2x+1}{3} - \dfrac{3x+4}{6} = \dfrac{5x-9}{9}$

$18\left(\dfrac{2x+1}{3} - \dfrac{3x+4}{6}\right) = 18\left(\dfrac{5x-9}{9}\right)$

$6(2x + 1) - 3(3x + 4) = 2(5x - 9)$

$12x + 6 - 9x - 12 = 10x - 18$

$3x - 6 = 10x - 18$

$-6 = 7x - 18$

$12 = 7x$

$\dfrac{12}{7} = x$

The solution is $\dfrac{12}{7}$.

10. $3x - 2 \geq 6x + 7$

$-3x \geq 9$

$x \leq -3$

$\{x \,/\, x \leq -3\}$

11. $4 - 3(x + 2) < 2(2x + 3) - 1$

$4 - 3x - 6 < 4x + 6 - 1$

$-3x - 2 < 4x + 5$

$-2 < 7x + 5$

$-7 < 7x$

$-1 < x$

$(-1, \infty)$

12. $4x - 1 > 5$ or $2 - 3x < 8$

$4x > 6$ $-3x < 6$

$x > \dfrac{3}{2}$ $x > -2$

$\{x \mid x > \dfrac{3}{2}\} \cup \{x \mid x > -2\} = \{x \mid x > -2\}$

13. $4 - 3x \geq 7$ and $2x + 3 \geq 7$

$-3x \geq 3$ $2x \geq 4$

$x \leq -1$ $x \geq 2$

$\{x \mid x \leq -1\} \cap \{x \mid x \geq 2\} = \emptyset$

14. $|3 - 5x| = 12$

$3 - 5x = 12$ $3 - 5x = -12$

$-5x = 9$ $-5x = -15$

$x = -\dfrac{9}{5}$ $x = 3$

The solutions are $-\dfrac{9}{5}$ and 3.

15. $2 - |2x - 5| = -7$

$-|2x - 5| = -9$

$|2x - 5| = 9$

$2x - 5 = 9$ $2x - 5 = -9$

$2x = 14$ $2x = -4$

$x = 7$ $x = -2$

The solutions are 7 and -2.

16. $|3x - 5| \leq 4$

$$-4 \leq 3x - 5 \leq 4$$
$$-4 + 5 \leq 3x - 5 + 5 \leq 4 + 5$$
$$1 \leq 3x \leq 9$$
$$\frac{1}{3} < x < 3$$

$$\{x \mid \frac{1}{3} < x < 3\}$$

17. $|4x - 3| > 5$

$4x - 3 < -5$ or $4x - 3 > 5$
$\qquad 4x < -2 \qquad\qquad 4x > 8$

$$x < -\frac{1}{2} \qquad\qquad x > 2$$

$$\{x \mid x < -\frac{1}{2}\} \qquad \{x \mid x > 2\}$$

$$\{x \mid x < -\frac{1}{2}\} \cup \{x \mid x > 2\} = \{x \mid x < -\frac{1}{2} \text{ or } x > 2\}$$

18. Strategy: Let N represent the number of miles.
The cost of the Gambelli car < the cost of the McDougal car.

Solution: $40 + 0.25N < 58$
$$0.25N < 18$$
$$N < 72$$
It costs less to rent from Gambelli if the car is driven less than 72 mi.

19. Strategy: Let d represent the diameter of the bushing, T the tolerance and x the lower and upper limits of the diameter.

Solution: $|x - d| \leq T$
$|x - 2.65| \leq 0.002$
$$-0.002 \leq x - 2.65 \leq 0.002$$
$$-0.002 + 2.65 \leq x - 2.65 + 2.65 \leq 0.002 + 2.65$$
$$2.648 \leq x \leq 2.652$$
The lower and upper limits of the diameter of the bushing are 2.648 in. and 2.652 in.

20. Strategy: Let x represent the number of ounces of silver alloy at $11 per oz..

	Amount	Cost	Value
Alloy2	9	8	9(8)
Alloy1	x	11	11x
Mixture	$9 + x$	9	$9(9 + x)$

The sum of the values before mixing is equal to the quantity after mixing.

Solution:
$$9(8) + 11x = 9(9 + x)$$
$$72 + 11x = 81 + 9x$$
$$2x = 9$$
$$x = 4.5$$
4.5 oz of silver alloy must be used in the mixture..

21. Strategy: Let x represent the number of ounces of 8% solution.

	Amount	Percent	Quantity
mixture	50	0.104	50(0.104)
8%	x	0.08	0.08(x)
12%	$50 - x$	0.12	0.12$(50 - x)$

The sum of the quantities before mixing is equal to the quantity after mixing.

Solution:
$$0.08x + 0.12(50 - x) = 50(0.104)$$
$$0.08x + 6 - 0.12x = 5.2$$
$$-0.04x = -0.8$$
$$x = 20$$
20 oz of 8% solution ; 30 oz of 12% solution.

22. Strategy: Let x represent the price of the hamburger mixture.

	Amount	Cost	Value
$3.10 hamburger	100	3.10	3.10(100)
$4.38 hamburger	60	4.38	4.38(60)
Mixture	160	x	x(160)

The sum of the values before mixing is equal to the quantity after mixing.

Solution: $3.1(100) + 4.38(60) = 160x$
$$310 + 262.8 = 160x$$
$$572.8 = 160x$$
$$3.58 = x$$

The price of the hamburger mixture is $3.58 per pound.

23. Strategy: Let t represent the time a jogger runs a distance.

Total running time is $1\dfrac{45}{60} = 1\dfrac{3}{4} = \dfrac{7}{4}$

	Rate	Time	Distance
Jogger runs a distance	8	t	$8t$
Jogger returns same distance	6	$\dfrac{7}{4} - t$	$6\left(\dfrac{7}{4} - t\right)$

A jogger runs a distance and returns the same distance.

Solution: $8t = 6\left(\dfrac{7}{4} - t\right)$
$$8t = \dfrac{21}{2} - 6t$$
$$14t = \dfrac{21}{2}$$
$$t = \dfrac{3}{4}$$

$8t = 8 \cdot \dfrac{3}{4} = 6$

The jogger ran 6 mi one way or a total distance of 12 mi.

24. Strategy: Let r represent the speed of the slower train.

	Rate	Time	Distance
Slower train	r	2	$2r$
Faster train	$r + 5$	2	$2(r + 5)$

The total distance traveled by the two trains is 250 mi.

Solution: $2r + 2(r + 5) = 250$
$$2r + 2r + 10 = 250$$
$$4r + 10 = 250$$
$$4r = 240$$

$$r = 60$$
$$r + 5 = 60 + 5 = 65$$
The rate of the slower train is 60 mph.
The rate of the faster train is 65 mph.

25. Strategy: Let x represent the number of ounces of pure water.

	Amount	Percent	Quantity
Pure water	x	0	$0x$
8% salt	60	0.08	$0.08(60)$
3% salt	$60 + x$	0.03	$0.03(60 + x)$

The sum of the quantities before mixing is equal to the quantity after mixing.

Solution: $0x + 0.08(60) = 0.03(60 + x)$
$$4.8 = 1.8 + 0.03x$$
$$3.0 = 0.3x$$
$$100 = x$$

100 oz of pure water must be added.

Cumulative Review Exercises

1. $-4 - (-3) - 8 + (-2) = -4 + 3 + (-8) + (-2)$
$$= -1 + (-8) + (-2)$$
$$= -9 + (-2)$$
$$= -11$$

2. $-2^2 \cdot 3^3 = -(2 \cdot 2)(3 \cdot 3 \cdot 3)$
$$= -4 \cdot 27$$
$$= -108$$

3. $4 - (2 - 5)^2 \div 3 + 2 = 4 - (-3)^2 \div 3 + 2$
$$= 4 - 9 \div 3 + 2$$
$$= 4 - 3 + 2$$
$$= 1 + 2$$
$$= 3$$

4. $4 \div \dfrac{\frac{3}{8} - 1}{5} \cdot 2 = 4 \div \dfrac{-\frac{5}{8}}{5} \cdot 2$
$$= 4 \div \left(-\dfrac{5}{8} \cdot \dfrac{1}{5}\right) \cdot 2$$
$$= 4 \div \left(-\dfrac{1}{8}\right) \cdot 2$$
$$= 4 \cdot (-8) \cdot 2$$
$$= -32 \cdot 2$$
$$= -64$$

5. $2a^2 - (b-c)^2$
$2(2)^2 - (3 - (-1))^2$
$= 2(4) - (3+1)^2$
$= 2(4) - (4)^2$
$= 2(4) - 16$
$= 8 - 16$
$= -8$

6. $\dfrac{a-b^2}{b-c}$

$\dfrac{2-(-3)^2}{-3-4} = \dfrac{2-9}{-3-4}$

$= \dfrac{-7}{-7}$

$= 1$

7. The Commutative Property of Addition

8. Let x represent the unknown number.
Three <u>times</u> the number <u>and</u> six is $3x + 6$.
$3x + (3x + 6) = 6x + 6$

9. $F = \dfrac{evB}{c}$

$Fc = evB$

$\dfrac{Fc}{ev} = B$

10. $5[y - 2(3 - 2y) + 6] = 5[y - 6 + 4y + 6]$
$ = 5[5y]$
$ = 25y$

11. $\{-4, 0\}$

12. $\{x\mid x \le 3\} \cap \{x\mid x > -1\}$

13. $Ax + By + C = 0$
$ By = -Ax - C$
$ y = \dfrac{-Ax - C}{B}$

14. $-\dfrac{5}{6}b = -\dfrac{5}{12}$

$-\dfrac{6}{5}\left(-\dfrac{5}{6}b\right) = -\dfrac{6}{5}\left(-\dfrac{5}{12}\right)$

$b = \dfrac{1}{2}$

The solution is $\dfrac{1}{2}$.

15. $2x + 5 = 5x + 2$
$ 5 = 3x + 2$
$ 3 = 3x$
$ 1 = x$
The solution is 1.

16. $\dfrac{5}{12}x - 3 = 7$

$\dfrac{5}{12}x = 10$

$\dfrac{12}{5}\left(\dfrac{5}{12}x\right) = \dfrac{12}{5}(10)$

$x = 24$
The solution is 24.

17. $2[3 - 2(3 - 2x)] = 2(3 + x)$
$2[3 - 6 + 4x] = 6 + 2x$
$2[-3 + 4x] = 6 + 2x$
$-6 + 8x = 6 + 2x$
$8x = 12 + 2x$
$6x = 12$
$x = 2$
The solution is 2.

18. $3[2x - 3(4 - x)] = 2(1 - 2x)$
$3[2x - 12 + 3x] = 2 - 4x$
$3[5x - 12] = 2 - 4x$
$15x - 36 = 2 - 4x$
$19x - 36 = 2$
$19x = 38$
$x = 2$
The solution is 2.

19. $\dfrac{1}{2}y - \dfrac{2}{3}y + \dfrac{5}{12} = \dfrac{3}{4}y - \dfrac{1}{2}$

$12\left(\dfrac{1}{2}y - \dfrac{2}{3}y + \dfrac{5}{12}\right) = 12\left(\dfrac{3}{4}y - \dfrac{1}{2}\right)$

$6y - 8y + 5 = 9y - 6$
$-2y + 5 = 9y - 6$
$-11y + 5 = -6$
$-11y = -11$
$y = 1$
The solution is 1.

20. $\dfrac{3x-1}{4} - \dfrac{4x-1}{12} = \dfrac{3+5x}{8}$

$24\left(\dfrac{3x-1}{4} - \dfrac{4x-1}{12}\right) = 24\left(\dfrac{3+5x}{8}\right)$

$6(3x-1) - 2(4x-1) = 3(3+5x)$

$18x - 6 - 8x + 2 = 9 + 15x$

$10x - 4 = 9 + 15x$

$-5x - 4 = 9$

$-5x = 13$

$x = -\dfrac{13}{5}$

The solution is $-\dfrac{13}{5}$.

21. $3 - 2(2x-1) \geq 3(2x-2) + 1$

$3 - 4x + 2 \geq 6x - 6 + 1$

$5 - 4x \geq 6x - 5$

$5 - 10x \geq -5$

$-10x \geq -10$

$x \leq 1$

$(-\infty, 1]$

22. $3x + 2 \leq 5$ and $x + 5 > 1$

$\qquad 3x \leq 3 \qquad\qquad x > -4$

$\qquad x \leq 1$

$\{x / x \leq 1\} \cap \{x / x > -4\} = \{x / -4 < x \leq 1\}$

23. $|3 - 2x| = 5$

$3 - 2x = 5 \quad$ or $\quad 3 - 2x = -5$

$\quad -2x = 2 \qquad\qquad -2x = -8$

$\quad x = -1 \qquad\qquad\quad x = 4$

The solutions are -1 and 4.

24. $3 - |2x - 3| = -8$

$-|2x - 3| = -11$

$|2x - 3| = 11$

$2x - 3 = 11 \quad$ or $\quad 2x - 3 = -11$

$\quad 2x = 14 \qquad\qquad 2x = -8$

$\quad x = 7 \qquad\qquad\quad x = -4$

The solutions are 7 and -4.

25. $|3x - 1| > 5$

$3x - 1 < -5 \quad$ or $\quad 3x - 1 > 5$

$\quad 3x < -4 \qquad\qquad 3x > 6$

$\quad x < -\dfrac{4}{3} \qquad\qquad x > 2$

$\{x \mid x < -\dfrac{4}{3}\} \qquad\qquad \{x / x > 2\}$

$\{x \mid x < -\dfrac{4}{3}\} \cup \{x / x > 2\} = \{x \mid x < -\dfrac{4}{3} \text{ or } x > 2\}$

26. $|2x - 4| < 8$

$-8 < 2x - 4 < 8$

$-8 + 4 < 2x - 4 + 4 < 8 + 4$

$-4 < 2x < 12$

$-2 < x < 6$

$\{x \mid -2 < x < 6\}$

27. Strategy: Let x represent the number of minutes used.
Cost for Plan One: $\$40 + 0.45(x - 450)$
Cost for Plan Two: $\$60$

Solution:
$40 + 0.45(x - 450) < 60$
$40 + 0.45x - 202.5 < 60$
$0.45x - 162.5 < 60$
$0.45x < 222.5$
$x < 494$

On Plan One Angelica only pays for minutes used over 450.
$494 - 450 = 44$
Plan One costs less than Plan Two. Angelica can use her phone for less than 44 minutes.

28. Strategy: Let x represent the number of ounces of pure silver.

	Amount	Cost	Value
Silver	x	15.78	$15.78x$
Alloy	100	8.26	$8.26(100)$
Mixture	$100 + x$	11.78	$11.78(100 + x)$

The sum of the values before mixing is equal to the quantity after mixing.

Solution:
$15.78x + 8.26(100) = 11.78(100 + x)$
$15.78x + 826 = 1178 + 11.78x$
$4x + 826 = 1178$
$4x = 352$
$x = 88$

88 oz of pure silver were used in the mixture.

29. Strategy: Let r represent the speed of the slower plane.

	Rate	Time	Distance
Slower plane	r	2.5	$2.5r$
Faster plane	$r + 120$	2.5	$2.5(r + 120)$

The total distance traveled by the two planes is 1400 mi.

Solution:
$$2.5r + 2.5(r + 120) = 1400$$
$$2.5r + 2.5r + 300 = 1400$$
$$5r + 300 = 1400$$
$$5r = 1100$$
$$r = 220$$
The speed of the slower plane is 220 mph.

30. Strategy: Let d represent the diameter of the bushing, T the tolerance and x the lower and upper limits of the diameter.

Solution: $|x - d| \leq T$
$$|x - 2.45| \leq 0.001$$
$$-0.001 \leq x - 2.45 \leq 0.001$$
$$-0.001 + 2.45 \leq x - 2.45 + 2.45 \leq 0.001 + 2.45$$
$$2.449 \leq x \leq 2.451$$
The lower and upper limits of the diameter of the bushing are 2.449 in. and 2.451 in.

31. Strategy: Let x represent the number of liters of 12% acid solution.

	Amount	Percent	Quantity
12% solution	x	0.12	$0.12x$
5% solution	4	0.05	$0.05(4)$
8% solution	$4 + x$	0.08	$0.08(4 + x)$

The sum of the quantities before mixing is equal to the quantity after mixing.

Solution:
$$0.12x + 0.05(4) = 0.08(4 + x)$$
$$0.12x + 0.2 = 0.32 + 0.08x$$
$$0.04x + 0.2 = 0.32$$
$$0.04x = 0.12$$
$$x = 3$$
3 L of 12% acid solution must be in the mixture.

Chapter 3: Linear Functions and Inequalities in Two Variables

Prep Test

1. $-4(x-3) = -4x + 12$

2. $\sqrt{(-6)^2 + (-8)^2} = \sqrt{36+64} = \sqrt{100} = 10$

3. $\dfrac{3-(-5)}{2-6} = \dfrac{3+5}{2-6} = \dfrac{8}{-4} = -2$

4. $-2x + 5$
   $-2(-3) + 5 = 6 + 5 = 11$

5. $\dfrac{2r}{r-1}$
   $\dfrac{2(5)}{5-1} = \dfrac{10}{4} = 2.5$

6. $2p^3 - 3p + 4$
   $2(-1)^3 - 3(-1) + 4 = 2(-1) - 3(-1) + 4$
   $\qquad\qquad\qquad = -2 + 3 + 4$
   $\qquad\qquad\qquad = 5$

7. $\dfrac{x_1 + x_2}{2}$
   $\dfrac{7 + (-5)}{2} = \dfrac{2}{2} = 1$

8. $\quad 3x - 4y = 12$
   $\quad 3x - 4(0) = 12$
   $\qquad\quad 3x = 12$
   $\qquad\qquad x = 4$

9. $2x - y = 7$
   $\quad -y = 7 - 2x$
   $\qquad y = 2x - 7$

Section 3.1

Concept Check

1. The x-coordinate of any point on the y-ais will be 0.

3. a) The ordered pair $(-2, 3)$ is located in quadrant II.
   b) The ordered pair $(4,1)$ is located in quadrant I.
   c) The ordered pair $(-3, -1)$ is located in quadrant III.
   d) The ordered pair $(5, -1)$ is located in quadrant IV.

5. If the x-coordinate of an ordered pair is positive, the ordered pair will lie in quadrant I or IV.

7. a) $y = -2x + 6$
   $\quad 2 = -2(2) + 6$
   $\quad 2 = -4 + 6$
   $\quad 2 = 2$
   Yes, the ordered pair $(2, 2)$ is a solution of the equation.
   b) $y = -2x + 6$
   $\quad 0 = -2(-3) + 6$
   $\quad 0 = 6 + 6$
   $\quad 0 \neq 12$
   No, the ordered pair $(-3, 0)$ is not a solution of the equation.
   c) $y = -2x + 6$
   $\quad 4 = -2(-1) + 6$
   $\quad 4 = 2 + 6$
   $\quad 4 \neq 8$
   No, the ordered pair $(-1, 4)$ is not a solution of the equation.
   d) $y = -2x + 6$
   $\quad 0 = -2(3) + 6$
   $\quad 0 = -6 + 6$
   $\quad 0 = 0$
   Yes, the ordered pair $(3, 0)$ is a solution of the equation.

Objective A Exercises

9. $d = \sqrt{(x_1 - x_2)^2 + (y_1 - y_2)^2}$

$d = \sqrt{(-2-6)^2 + (-9-6)^2}$

$d = \sqrt{289} = 17$

$x_m = \dfrac{-2+6}{2} = 2$

$y_m = \dfrac{-9+6}{2} = -\dfrac{3}{2}$

Length: 17; midpoint $\left(2, -\dfrac{3}{2}\right)$

11. $d = \sqrt{(x_1 - x_2)^2 + (y_1 - y_2)^2}$

$d = \sqrt{(3-5)^2 + (5-1)^2}$

$d = \sqrt{20} \approx 4.47$

$x_m = \dfrac{3+5}{2} = 4$

$y_m = \dfrac{5+1}{2} = 3$

Length: 4.47; midpoint $(4, 3)$

13. $d = \sqrt{(x_1 - x_2)^2 + (y_1 - y_2)^2}$

$d = \sqrt{(0-(-2))^2 + (3-4)^2}$

$d = \sqrt{5} \approx 2.24$

$x_m = \dfrac{0+(-2)}{2} = -1$

$y_m = \dfrac{3+4}{2} = \dfrac{7}{2}$

Length: 2.24; midpoint $\left(-1, \dfrac{7}{2}\right)$

15. $d = \sqrt{(x_1 - x_2)^2 + (y_1 - y_2)^2}$

$d = \sqrt{(-3-2)^2 + (-5-(-4))^2}$

$d = \sqrt{26} \approx 5.10$

$x_m = \dfrac{-3+2}{2} = -\dfrac{1}{2}$

$y_m = \dfrac{-5+(-4)}{2} = -\dfrac{9}{2}$

Length: 5.10; midpoint $\left(-\dfrac{1}{2}, -\dfrac{9}{2}\right)$

17. $d = \sqrt{(x_1 - x_2)^2 + (y_1 - y_2)^2}$

$d = \sqrt{(5-2)^2 + (-5-(-5))^2}$

$d = \sqrt{9} = 3$

$x_m = \dfrac{5+2}{2} = \dfrac{7}{2}$

$y_m = \dfrac{-5+(-5)}{2} = -5$

Length: 3; midpoint $\left(\dfrac{7}{2}, -5\right)$

19. $d = \sqrt{(x_1 - x_2)^2 + (y_1 - y_2)^2}$

$d = \sqrt{\left(\dfrac{3}{2} - \left(-\dfrac{1}{2}\right)\right)^2 + \left(-\dfrac{4}{3} - \dfrac{7}{3}\right)^2}$

$d = \sqrt{\dfrac{157}{9}} \approx 4.18$

$x_m = \dfrac{\dfrac{3}{2} + \left(-\dfrac{1}{2}\right)}{2} = \dfrac{1}{2}$

$y_m = \dfrac{-\dfrac{4}{3} + \dfrac{7}{3}}{2} = \dfrac{1}{2}$

Length: 4.18; midpoint $\left(\dfrac{1}{2}, \dfrac{1}{2}\right)$

21. The x-coordinates are equal.

Objective B Exercises

23. $y = 2x - 3$

x	y
−1	−5
0	−3
1	−1
2	1
3	3
4	5

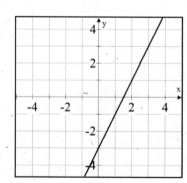

25. $y = -\dfrac{2}{3}x + 1$

x	y
−6	5
−3	3
0	1
3	−1
6	−3

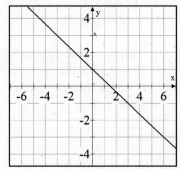

27. $y = x^2 - 4$

x	y
−3	5
−2	0
−1	−3
0	−4
1	−3
2	0
3	5

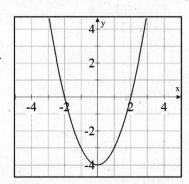

29. $y = -x^2 + 2x + 3$

x	y
−2	−5
−1	0
0	3
1	4
2	3
3	0
4	−5

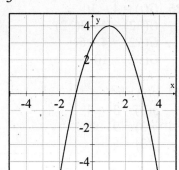

31. $y = |x + 2|$

x	y
−5	3
−4	2
−3	1
−2	0
−1	1
0	2
1	3

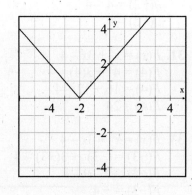

33. $y = -|x - 1| + 3$

x	y
−3	−1
−1	1
1	3
3	1
5	−1

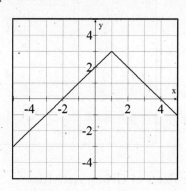

35. When $x = 1$ the equation $y = \dfrac{3}{x-1}$ has no ordered pair solution.

Critical Thinking

37. Ordered pairs: (−2, 6), (−1, 4), (0, 2), (1, 0), (2, −2), (3, −4)

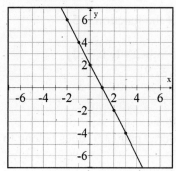

39.

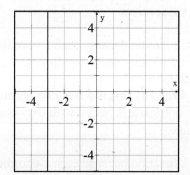

41.

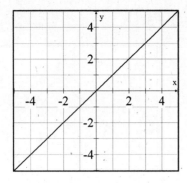

Projects of Group Activities

43. The graph of all ordered pairs (x, y) that are 5 units from the origin is a circle of radius 5 that has its center at $(0, 0)$.

Section 3.2

Concept Check

1. The value of the function is 5.

3. Domain = $\{-3, -2, -1, 1, 2, 3\}$
Range = $\{1, 4, 9\}$

5. $f(x) = x^2$
$f(3) = 3^2 = 9$
Yes, the ordered pair $(3,9)$ does belong to $f(x) = x^2$.

Objective A Exercises

7. This is a function because each x-coordinate is paired with only one y-coordinate.
D = $\{0,2,3,4,5\}$
R = $\{0,4,6,8,10\}$This is a function because each x-coordinate is paired with only one y-coordinate.

9. This is a function because each x-coordinate is paired with only one y-coordinate.
D = $\{-4,-2,0, 3\}$
R = $\{-5,-1,5\}$

11. This is a function because each x-coordinate is paired with only one y-coordinate.
D = $\{-2,-1,0,1,2\}$
R = $\{-3,3\}$

13. This is not a function because there are x-coordinates paired with two different

y-coordinates. They are $(1, 1)$, $(1, -1)$ and $(4, 2)$, $(4, -2)$.
D = $\{1,4,9\}$
R = $\{-2,-1,1,2,3\}$

15. a) Yes, this table defines a function.
  b) If $w = 3.15$, then $p = \$1.05$.
  c) If $w = 2$, then $p = \$0.65$.
  d) \$0.45

17. True

19. $f(x) = 5x - 4$
$f(3) = 5(3) - 4$
$f(3) = 15 - 4$
$f(3) = 11$

21. $f(x) = 5x - 4$
$f(0) = 5(0) - 4$
$f(0) = -4$

23. $G(t) = 4 - 3t$
$G(0) = 4 - 3(0)$
$G(0) = 4$

25. $G(t) = 4 - 3t$
$G(-2) = 4 - 3(-2)$
$G(-2) = 4 + 6$
$G(-2) = 10$

27. $q(r) = r^2 - 4$
$q(3) = 3^2 - 4$
$q(3) = 9 - 4$
$q(3) = 5$

29. $q(r) = r^2 - 4$
$q(-2) = (-2)^2 - 4$
$q(-2) = 4 - 4$
$q(-2) = 0$

31. $F(x) = x^2 + 3x - 4$
$F(4) = 4^2 + 3(4) - 4$
$F(4) = 16 + 12 - 4$
$F(4) = 24$

33. $F(x) = x^2 + 3x - 4$
$F(-3) = (-3)^2 + 3(-3) - 4$
$F(-3) = 9 - 9 - 4$
$F(-3) = -4$

35. $H(p) = \dfrac{3p}{p+2}$

$H(1) = \dfrac{3(1)}{1+2}$

$H(1) = \dfrac{3}{3}$

$H(1) = 1$

37. $H(p) = \dfrac{3p}{p+2}$

$H(t) = \dfrac{3t}{t+2}$

39. $s(t) = t^3 - 3t + 4$
$s(-1) = (-1)^3 - 3(-1) + 4$
$s(-1) = -1 + 3 + 4$
$s(-1) = 6$

41. $s(t) = t^3 - 3t + 4$
$s(a) = a^3 - 3a + 4$

43. $P(x) = 4x + 7$
$P(-2 + h) - P(-2)$
$= 4(-2 + h) + 7 - [4(-2) + 7]$
$= -8 + 4h + 7 + 8 - 7$
$= 4h$

45. 50.625 watts

47. $114.29

49. a) $4.75 per game
b) $4.00 per game

51. Values of x for which $x - 1 = 0$ are excluded from the domain of the function
$\{x | x \neq 1\}$

53. $\{x | -\infty < x < \infty\}$

55. $\{x | -\infty < x < \infty\}$

57. $\{x | -\infty < x < \infty\}$

58. $\{x | -\infty < x < \infty\}$

61. Values of x for which $x + 2 = 0$ are excluded from the domain of the function
$\{x | x \neq -2\}$

Objective B Exercises

63.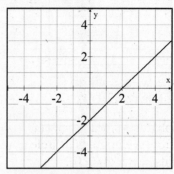

Domain $= (-\infty, \infty)$
Range $= (-\infty, \infty)$

65.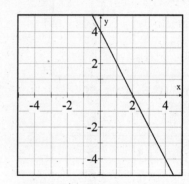

Domain $= (-\infty, \infty)$
Range $= (-\infty, \infty)$

67.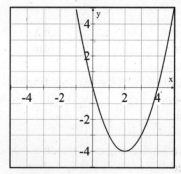

Domain $= (-\infty, \infty)$
Range $= [-4, \infty)$

69.

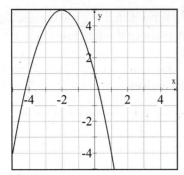

Domain = $\{x|-\infty < x < \infty\}$
Range = $\{y|y \le 5\}$

71.

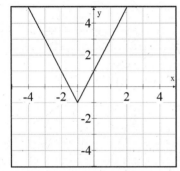

Domain = $\{x|-\infty < x < \infty\}$
Range = $\{y|y \ge -1\}$

73.

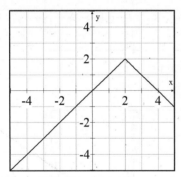

Domain = $\{x|-\infty < x < \infty\}$
Range = $\{y|y \le 2\}$

75. Domain = $(-\infty,\infty)$
Range = $[-4,4]$

77. Domain = $[-4,4]$
Range = $(-5,5)$

79. Domain = $[-5,3]$
Range = $[2,4]$

Objective B Exercises

81. Yes

83. Yes

85. Yes

Critical Thinking

87. a){(−2, −8), (−1, −1), (0, 0), (1, 1), (2, 8)}

b) Yes this set of ordered pairs defines a function because each member of the domain is assigned exactly one member of the range.

89. A relation and a function are similar in that both are sets of ordered pairs. A function is a specific type of relation. A function is a relation in which there are no two ordered pairs with the same first element.

91. a) The speed of the paratrooper 11.5 s after the beginning of the jump is 36.3 ft/s.
b) 30 ft/s

93. a) 110 beats/min
b) 75 beats/min

Projects or Group Activities

95.

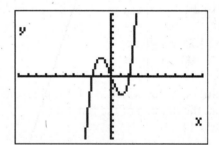

97.

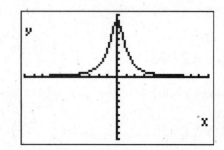

Section 3.3

Concept Check

1. If the three ordered pairs appear to lie on the same line, then it is more likely that your calculations are accurate.

3. The x coordinate of the y-intercept of a line is 0. The y coordinate of the x-intercept of a line is 0.

Objective A Exercises

5. $P = f(d) = 0.097d + 1$
$f(500) = 0.097(500) + 1 = 49.5$
The pressure is 49.5 atm.

7. $y = x - 3$

x	y
-1	-4
0	-3
3	0

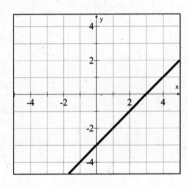

9. $y = -3x + 2$

x	y
0	2
1	-1
2	-4

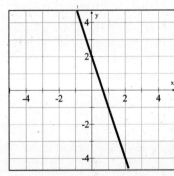

11. $f(x) = 3x - 4$

x	$f(x)$
0	-4
1	-1
2	2

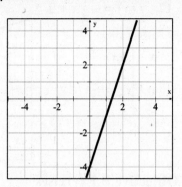

13. $f(x) = -\dfrac{2}{3}x$

x	$f(x)$
-3	2
0	0
3	-2

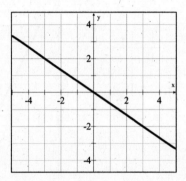

15. $y = \dfrac{2}{3}x - 4$

x	y
0	-4
3	-2
6	0

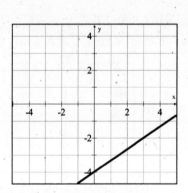

17. $f(x) = -\dfrac{1}{3}x + 2$

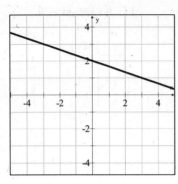

x	$f(x)$
-3	3
0	2
3	1

23. $4x + 3y = 12$

$\qquad 3y = -4x + 12$

$\qquad\quad y = -\dfrac{4}{3}x + 4$

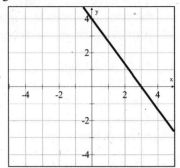

x	y
0	4
3	0
6	4

Objective B Exercises

19. $2x + y = -3$

$\qquad y = -2x - 3$

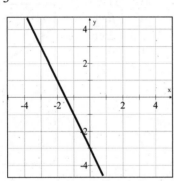

x	y
0	-3
1	-5
-1	-1

25. $x - 3y = 0$

$\qquad -3y = -x$

$\qquad\quad y = \dfrac{1}{3}x$

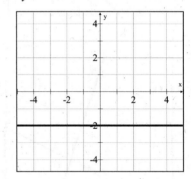

x	y
0	0
3	1
-3	-1

21. $x - 4y = 8$

$\qquad -4y = -x + 8$

$\qquad\quad y = \dfrac{1}{4}x - 2$

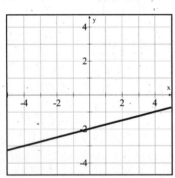

x	y
0	-2
4	-1
8	0

27. $y = -2$

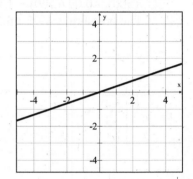

29. $x = -3$

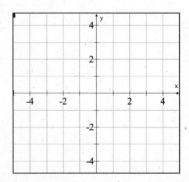

31. $3x - y = -2$
$-y = -3x - 2$
$y = 3x + 2$

x	y
0	2
1	5
-1	-1

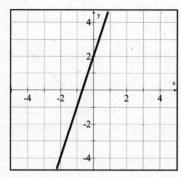

33. $3x - 2y = 8$
$-2y = -3x + 8$
$y = \dfrac{3}{2}x - 4$

x	y
0	-4
2	-1
4	2

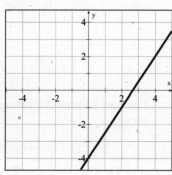

35. No. If $B = 0$, then it is not possible to solve $Ax + By = C$ for y.

Objective C Exercises

37. x-intercept:
$x - 2y = -4$
$x - 2(0) = -4$
$x = -4$
$(-4, 0)$
y-intercept:
$x - 2y = -4$
$0 - 2y = -4$
$-2y = -4$
$y = 2$
$(0, 2)$

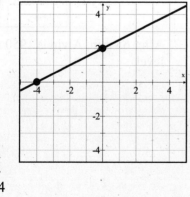

39. x-intercept:
$2x - 3y = 9$
$2x - 3(0) = 9$
$2x = 9$
$x = \dfrac{9}{2}$
$\left(\dfrac{9}{2}, 0\right)$
y-intercept:
$2x - 3y = 9$
$2(0) - 3y = 9$
$-3y = 9$
$y = -3$
$(0, -3)$

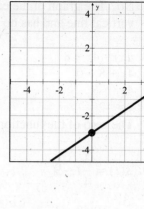

41. x-intercept:
$2x + y = 3$
$2x + 0 = 3$
$2x = 3$
$x = \dfrac{3}{2}$
$\left(\dfrac{3}{2}, 0\right)$
y-intercept:
$2x + y = 3$
$2(0) + y = 3$
$y = 3$
$(0, 3)$

43. x-intercept:
$3x + 2y = 4$
$3x + 2(0) = 4$
$3x = 4$
$x = \dfrac{4}{3}$
$\left(\dfrac{4}{3}, 0\right)$
y-intercept:
$3x + 2y = 4$
$3(0) + 2y = 4$
$2y = 4$
$y = 2$
$(0, 2)$

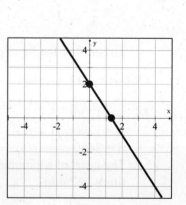

45. No. The graph of the equation $x = a$ is a vertical line and has no y-intercept.

47. $f(x) = 4x + 8$
$4x + 8 = 0$
$4x = -8$
$x = -2$
The zero of the function is -2.

49. $s(t) = \dfrac{3}{4}t - 9$

$\dfrac{3}{4}t - 9 = 0$
$3t - 36 = 0$
$3t = 36$
$t = 12$
The zero of the function is 12.

51. $f(x) = 4x$
$4x = 0$
$x = 0$
The zero of the function is 0.

53. $g(x) = \dfrac{3}{2}x - 4$

$\dfrac{3}{2}x - 4 = 0$
$3x - 8 = 0$
$x = \dfrac{8}{3}$

The zero of the function is $\dfrac{8}{3}$.

Objective D Exercises

55. $B = 1200t$
$B = 1200(7)$
$B = 8400$
The heart of a hummingbird will beat 8400 times in 7 min.

57. $W = 11t$

t	W
0	0
5	55
15	165

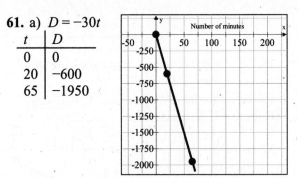

Marlys receives $165 for tutoring 15 h.

59. $C = 80n + 5000$

n	C
0	5000
50	9000
100	13000

The cost of manufacturing 50 pairs of skis is $9000.

61. a) $D = -30t$

t	D
0	0
20	-600
65	-1950

After 65 min, Alvin is 1950 m below sea level.

Critical Thinking

63. To graph the equation of a straight line by plotting points, find three ordered pair solutions of the equation. Plot these ordered pairs in a rectangular coordinate system. Draw a straight line through the points.

65. The x and y intercepts of the graph of the equation $4x + 3y = 0$ are both (0,0). A straight line is determined by two points, so we need to find another point on the line in order to graph this equation.

67. $\dfrac{x}{2} + \dfrac{y}{3} = 1$

x-intercept (2, 0)
y-intercept (0, 3)

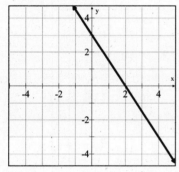

Check Your Progress: Chapter 3

1. $d = \sqrt{(x_1 - x_2)^2 + (y_1 - y_2)^2}$

$d = \sqrt{(-3 - 2)^2 + (5 - (-3))^2}$

$d = \sqrt{89} \approx 9.43$

$x_m = \dfrac{-3 + 2}{2} = -0.5$

$y_m = \dfrac{5 + (-3)}{2} = 1$

Length: 9.43; midpoint (–0.5, 1)

2. $y = x^2 + 1$

$-5 = (2)^2 + 1$

$-5 \neq 5$

No. The point (2, –5) is not on the graph of $y = x^2 + 1$.

3. $y = 2x + 2$

x	y
–4	–6
–3	–4
–2	–2
–1	0
0	2
1	4
2	6

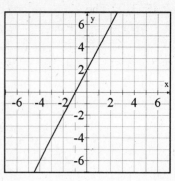

4. $y = -2|x| + 3$

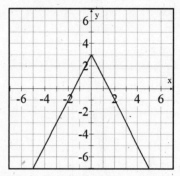

5. Yes this is a function.
Domain = {0,1,2,–5}
Range = {1,2,3,3}

6. Domain = $\{x | x \neq 0\}$

7. $s(t) = -3t^2 + 4t - 1$

$s(-3) = -3(-3)^2 + 4(-3) - 1$

$\quad\quad = -27 - 12 - 1$

$\quad\quad = -40$

8. $f(x) = 2x - 3$

$f(2 - a) = 2(2 - a) - 3$

$\quad\quad\quad = 4 - 2a - 3$

$\quad\quad\quad = 1 - 2a$

9. $s(t) = -\dfrac{3}{2}t + 3$

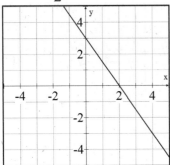

Domain: $(-\infty, \infty)$; Range: $(-\infty, \infty)$

10. $v(u) = u^2 - 6u + 5$

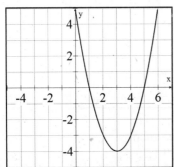

Domain: $(-\infty, \infty)$; Range: $[-4, \infty)$

11. This is not the graph of a function.
Domain: $\{x|-6 \leq x \leq 6\}$
Range: $\{y|-3 \leq y \leq 3\}$

12. This graph does represent a function.
Domain: $\{x| x \leq 2\}$
Range: $\{y| y \geq 0\}$

13.

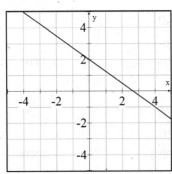

14.

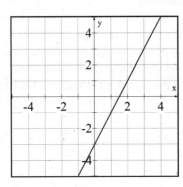

15.

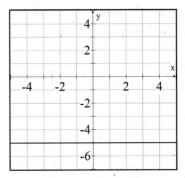

16.

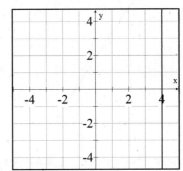

17. x-intercept:
$4x - 5y = 20$
$4x - 5(0) = 20$
$\quad\quad 4x = 20$
$\quad\quad\quad x = 5$
$(5, 0)$
y-intercept:
$4x - 5y = 20$
$3(0) - 5y = 20$
$\quad\quad -5y = 20$
$\quad\quad\quad y = -4$
$(0, -4)$

18. $g(t) = 3t + 6$

$3t + 6 = 0$

$3t = -6$

$t = -2$

The zero of the function is -2.

Section 3.4
Concept Check

1. Increases

3. slope: 3
y-intercept: $(0, 4)$

5. slope: -1
y-intercept: $(0, 1)$

Objective A Exercises

7. $(1, 3), (3, 1)$

$$m = \frac{y_2 - y_1}{x_2 - x_1} = \frac{1 - 3}{3 - 1} = \frac{-2}{2} = -1$$

The slope is -1.

9. $(-1, 4), (2, 5)$

$$m = \frac{y_2 - y_1}{x_2 - x_1} = \frac{5 - 4}{2 - (-1)} = \frac{1}{3}$$

The slope is $\frac{1}{3}$.

11. $(-1, 3), (-4, 5)$

$$m = \frac{y_2 - y_1}{x_2 - x_1} = \frac{5 - 3}{-4 - (-1)} = \frac{2}{-3} = -\frac{2}{3}$$

The slope is $-\frac{2}{3}$.

13. $(0, 3), (4, 0)$

$$m = \frac{y_2 - y_1}{x_2 - x_1} = \frac{3 - 0}{0 - 4} = \frac{3}{-4} = -\frac{3}{4}$$

The slope is $-\frac{3}{4}$.

15. $(2, 4), (2, -2)$

$$m = \frac{y_2 - y_1}{x_2 - x_1} = \frac{4 - (-2)}{2 - 2} = \frac{6}{0}$$

The slope is undefined.

17. $(2, 5), (-3, -2)$

$$m = \frac{y_2 - y_1}{x_2 - x_1} = \frac{5 - (-2)}{2 - (-3)} = \frac{7}{5}$$

The slope is $\frac{7}{5}$.

19. $(2, 3), (-1, 3)$

$$m = \frac{y_2 - y_1}{x_2 - x_1} = \frac{3 - 3}{2 - (-1)} = \frac{0}{3} = 0$$

The slope is 0.

21. $(0, 4), (-2, 5)$

$$m = \frac{y_2 - y_1}{x_2 - x_1} = \frac{5 - 4}{-2 - 0} = \frac{1}{-2} = -\frac{1}{2}$$

The slope is $-\frac{1}{2}$.

23. $(-3, -1), (-3, 4)$

$$m = \frac{y_2 - y_1}{x_2 - x_1} = \frac{4 - (-1)}{-3 - (-3)} = \frac{5}{0}$$

The slope is undefined.

25. If the slope of l is undefined, then a and c are equal.

27. $m = \dfrac{240 - 80}{6 - 2} = \dfrac{160}{4} = 40$

The average speed of the motorist is 40 mph.

29. $m = \dfrac{275 - 125}{20 - 50} = \dfrac{150}{-30} = -5$

The temperature of the oven decreases $5°$/min.

31. $m = \dfrac{13 - 6}{40 - 180} = \dfrac{7}{-140} = -0.05$

Approximately 0.05 gal of fuel is used for each mile that the car is driven.

33. $m = \dfrac{5000}{14.19} = 352.4$

The average speed of the runner was 352.4 m/min.

35. a) $\dfrac{6\ in}{5\ ft} = \dfrac{6\ in}{60\ in} = \dfrac{1}{10} > \dfrac{1}{12}$

No it does not meet the requirements for ANSI.

b) $\dfrac{12}{170} = \dfrac{6}{85} < \dfrac{1}{12}$

Yes, it does meet the requirements for ANSI.

Objective B Exercises

37. $f(x) = -x^2 + 3x$
$y_1 = f(6) = -(6)^2 + 3(6) = -18$
$P_1(6, 18)$
$y_2 = f(9) = -(9)^2 + 3(9) = -54$
$P_2 = (9, -54)$
$m = \dfrac{-54 - (-18)}{9 - 6} = -12$
The average rate of change is -12.

39. $f(x) = x^2 - 3x + 1$
$y_1 = f(-2) = (-2)^2 - 3(2) + 1 = -1$
$P_1(-2, -1)$
$y_2 = f(-1) = (-1)^2 - 3(-1) + 1 = 5$
$P_2 = (-1, 5)$
$m = \dfrac{-1 - 5}{-2 - (-1)} = -6$
The average rate of change is -6.

41. $f(x) = -2x^3 + 6x + 6$
$y_1 = f(-1) = -2(-1)^3 + 6(-1) + 6 = 2$
$P_1(-1, 2)$
$y_2 = f(1) = -2(1)^3 + 6(1) + 6 = 10$
$P_2 = (1, 10)$
$m = \dfrac{10 - 2}{1 - (-1)} = 4$
The average rate of change is 4.

43. $f(x) = 2x^3 - 5x^2 - 3x + 2$
$y_1 = f(-1) = 2(-1)^3 - 5(-1)^2 - 3(-1) + 2 = -2$
$P_1(-1, -2)$
$y_2 = f(2) = 2(2)^3 - 5(2)^2 - 3(2) + 2 = -8$
$P_2 = (2, -8)$
$m = \dfrac{-2 - (-8)}{-1 - 2} = -2$
The average rate of change is -2.

45. $P_1 (0, 34)$; $P_2(8, 58)$
$m = \dfrac{34 - 58}{0 - 8} = 3$
The average rate of change is $3°$F per hour.

47. a) The average annual rate of change in population was least during the 1860s.
b) The average annual rate of change in population was approximately 300,000 in the 1970s.

49. $P_1 (5, 1284)$; $P_2(15, 2117)$
$m = \dfrac{1284 - 2117}{5 - 15} = -83.3$
The average rate of change in the investment is $83.30 per year.

51. $P_1 (200, 16)$; $P_2(800, 3)$
$m = \dfrac{16 - 3}{200 - 800} = -0.02$
The average rate of change in the amount of power available is -0.02 watts per day.

Objective C Exercises

53. $y = \dfrac{1}{2}x + 2$

$m = \dfrac{1}{2}$

y-intercept $(0, 2)$

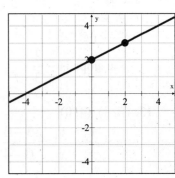

55. $y = -\dfrac{3}{2}x$

$m = -\dfrac{3}{2}$

y-intercept
$(0, 0)$

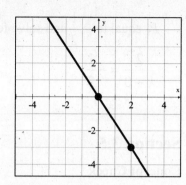

63. $x - 3y = 3$

$-3y = -x + 3$

$y = \dfrac{1}{3}x - 1$

$m = \dfrac{1}{3}$

y-intercept
$(0, -1)$

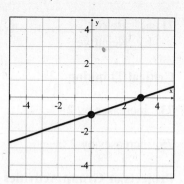

57. $y = -\dfrac{1}{2}x + 2$

$m = -\dfrac{1}{2}$

y-intercept
$(0, 2)$

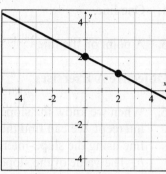

65.

59. $y = 2x - 4$

$m = 2$

y-intercept
$(0, -4)$

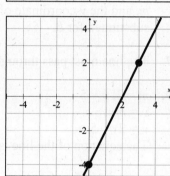

67.

61. $4x - y = 1$

$-y = -4x + 1$

$y = 4x - 1$

$m = 4$

y-intercept
$(0, -1)$

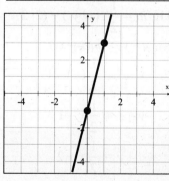

69.

71. a) Below

   b) Negative

Critical Thinking

73. increases by 2

75. increases by 2

Projects or Group Activities

77. The slope between each pair of points must be the same if all of the points line on the same line.

a) $P_1 = (2, 5)$
   $P_2 = (-1, -1)$
   $P_3 = (3, 7)$

P_1 to P_2: $m = \dfrac{5 - (-1)}{2 - (-1)} = \dfrac{6}{3} = 2$

P_2 to P_3: $m = \dfrac{7 - (-1)}{3 - (-1)} = \dfrac{8}{4} = 2$

P_3 to P_1: $m = \dfrac{7 - 5}{3 - 2} = \dfrac{2}{1} = 2$

Yes, the points all lie on the same line.

b) $P_1 = (-1, 5)$
   $P_2 = (0, 3)$
   $P_3 = (-3, 4)$

P_1 to P_2: $m = \dfrac{5 - 3}{-1 - 0} = \dfrac{2}{-1} = -2$

P_2 to P_3: $m = \dfrac{4 - 3}{-3 - 0} = \dfrac{1}{-3} = -\dfrac{1}{3}$

P_3 to P_1: $m = \dfrac{5 - 4}{-1 - (-3)} = \dfrac{1}{2}$

No, the points do not all lie on the same line.

79. $P_1 = (k, 1)$
   $P_2 = (0, -1)$
   $P_3 = (2, -2)$

P_2 to P_3: $m = -\dfrac{1}{2}$

The slope from P_1 to P_2 and from P_1 to P_3 must also be $-\dfrac{1}{2}$. Set the slope from P_1 to

P_2 equal to $-\dfrac{1}{2}$ and solve for k.

$$\dfrac{2}{k - 0} = -\dfrac{1}{2}$$
$$k = -4$$

Section 3.5

Concept Check

1. One line

3. A point on the line

Objective A Exercises

5. When we know the slope and the y-intercept, we can find the equation of the line using the slope-intercept form, $y = mx + b$. The value of the slope is substituted for m, and the y-coordinate of the y-intercept is substituted for b.

7. The line must pass through the origin, $(0, 0)$.

9. $m = 2, b = 5$
   $y = mx + b$
   $y = 2x + 5$
   The equation of the line is $y = 2x + 5$.

11. $m = \dfrac{1}{2}, (x_1, y_1) = (2, 3)$

$y - y_1 = m(x - x_1)$

$y - 3 = \dfrac{1}{2}(x - 2)$

$y - 3 = \dfrac{1}{2}x - 1$

$y = \dfrac{1}{2}x + 2$

The equation of the line is $y = \dfrac{1}{2}x + 2$.

13. $m = \dfrac{5}{4}, (x_1, y_1) = (-1, 4)$

$y - y_1 = m(x - x_1)$

$y - 4 = \dfrac{5}{4}[x - (-1)]$

$y - 4 = \dfrac{5}{4}(x + 1)$

$y - 4 = \dfrac{5}{4}x + \dfrac{5}{4}$

$y = \dfrac{5}{4}x + \dfrac{21}{4}$

The equation of the line is $y = \dfrac{5}{4}x + \dfrac{21}{4}$.

15. $m = -\dfrac{5}{3}, (x_1, y_1) = (3, 0)$

$y - y_1 = m(x - x_1)$

$y - 0 = -\dfrac{5}{3}(x - 3)$

$y = -\dfrac{5}{3}(x - 3)$

$y = -\dfrac{5}{3}x + 5$

The equation of the line is $y = -\dfrac{5}{3}x + 5$.

17. $m = -3, (x_1, y_1) = (2, 3)$

$y - y_1 = m(x - x_1)$

$y - 3 = -3(x - 2)$

$y - 3 = -3x + 6$

$y = -3x + 9$

The equation of the line is $y = -3x + 9$.

19. $m = -3, (x_1, y_1) = (-1, 7)$

$y - y_1 = m(x - x_1)$

$y - 7 = -3[x - (-1)]$

$y - 7 = -3(x + 1)$

$y - 7 = -3x - 3$

$y = -3x + 4$

The equation of the line is $y = -3x + 4$.

21. $m = \dfrac{2}{3}, (x_1, y_1) = (-1, -3)$

$y - y_1 = m(x - x_1)$

$y - (-3) = \dfrac{2}{3}[(x - (-1)]$

$y + 3 = \dfrac{2}{3}(x + 1)$

$y + 3 = \dfrac{2}{3}x + \dfrac{2}{3}$

$y = \dfrac{2}{3}x - \dfrac{7}{3}$

The equation of the line is $y = \dfrac{2}{3}x - \dfrac{7}{3}$.

23. $m = \dfrac{1}{2}, (x_1, y_1) = (0, 0)$

$y - y_1 = m(x - x_1)$

$y - 0 = \dfrac{1}{2}(x - 0)$

$y = \dfrac{1}{2}x$

The equation of the line is $y = \dfrac{1}{2}x$.

25. $m = 3, (x_1, y_1) = (2, -3)$

$y - y_1 = m(x - x_1)$

$y - (-3) = 3(x - 2)$

$y + 3 = 3x - 6$

$y = 3x - 9$

The equation of the line is $y = 3x - 9$.

27. $m = -\dfrac{2}{3}, (x_1, y_1) = (3, 5)$

$y - y_1 = m(x - x_1)$

$y - 5 = -\dfrac{2}{3}(x - 3)$

$y - 5 = -\dfrac{2}{3}x + 2$

$y = -\dfrac{2}{3}x + 7$

The equation of the line is $y = -\dfrac{2}{3}x + 7$.

29. $m = -1, b = -3$

$y = -x - 3$

The equation of the line is $y = -x - 3$.

31. $m = \dfrac{7}{5}$, $(x_1, y_1) = (1, -4)$

$y - y_1 = m(x - x_1)$

$y - (-4) = \dfrac{7}{5}(x - 1)$

$y + 4 = \dfrac{7}{5}x - \dfrac{7}{5}$

$y = \dfrac{7}{5}x - \dfrac{27}{5}$

The equation of the line is $y = \dfrac{7}{5}x - \dfrac{27}{5}$.

33. $m = -\dfrac{2}{5}$, $(x_1, y_1) = (4, -1)$

$y - y_1 = m(x - x_1)$

$y - (-1) = -\dfrac{2}{5}(x - 4)$

$y + 1 = -\dfrac{2}{5}x + \dfrac{8}{5}$

$y = -\dfrac{2}{5}x + \dfrac{3}{5}$

The equation of the line is $y = -\dfrac{2}{5}x + \dfrac{3}{5}$.

35. Slope is undefined, $(x_1, y_1) = (3, -4)$

The line is a vertical line. All points on the line have an abscissa of 3.

The equation of the line is $x = 3$.

37. $m = -\dfrac{5}{4}$, $(x_1, y_1) = (-2, -5)$

$y - y_1 = m(x - x_1)$

$y - (-5) = -\dfrac{5}{4}[x - (-2)]$

$y + 5 = -\dfrac{5}{4}(x + 2)$

$y + 5 = -\dfrac{5}{4}x - \dfrac{10}{4}$

$y = -\dfrac{5}{4}x - \dfrac{15}{2}$

The equation of the line is $y = -\dfrac{5}{4}x - \dfrac{15}{2}$.

39. $m = 0$, $(x_1, y_1) = (-2, -3)$

$y - y_1 = m(x - x_1)$

$y - (-3) = 0[x - (-2)]$

$y + 3 = 0$

$y = -3$

The equation of the line is $y = -3$.

41. $m = -2$, $(x_1, y_1) = (4, -5)$

$y - y_1 = m(x - x_1)$

$y - (-5) = -2(x - 4)$

$y + 5 = -2x + 8$

$y = -2x + 3$

The equation of the line is $y = -2x + 3$.

43. Slope is undefined, $(x_1, y_1) = (-5, -1)$

The line is a vertical line. All points on the line have an abscissa of -5.

The equation of the line is $x = -5$.

Objective B Exercises

45. Check that the coordinates of each given point are a solution of your equation.

47. $(0, 2)$, $(3, 5)$

$m = \dfrac{y_2 - y_1}{x_2 - x_1} = \dfrac{5 - 2}{3 - 0} = \dfrac{3}{3} = 1$

$y - y_1 = m(x - x_1)$

$y - 2 = 1(x - 0)$

$y - 2 = x$

$y = x + 2$

The equation of the line is $y = x + 2$.

49. $(0, -3)$, $(-4, 5)$

$m = \dfrac{y_2 - y_1}{x_2 - x_1} = \dfrac{5 - (-3)}{-4 - 0} = \dfrac{8}{-4} = -2$

$y - y_1 = m(x - x_1)$

$y - (-3) = -2(x - 0)$

$y + 3 = -2x$

$y = -2x - 3$

The equation of the line is $y = -2x - 3$.

51. $(2, 3)$, $(5, 5)$

$m = \dfrac{y_2 - y_1}{x_2 - x_1} = \dfrac{5 - 3}{5 - 2} = \dfrac{2}{3}$

$y - y_1 = m(x - x_1)$

$y - 3 = \dfrac{2}{3}(x - 2)$

$$y - 3 = \frac{2}{3}x - \frac{4}{3}$$

$$y = \frac{2}{3}x + \frac{5}{3}$$

The equation of the line is $y = \frac{2}{3}x + \frac{5}{3}$.

53. $(-1, 3), (2, 4)$

$$m = \frac{y_2 - y_1}{x_2 - x_1} = \frac{4 - 3}{2 - (-1)} = \frac{1}{3}$$

$$y - y_1 = m(x - x_1)$$

$$y - 3 = \frac{1}{3}[x - (-1)]$$

$$y - 3 = \frac{1}{3}(x + 1)$$

$$y - 3 = \frac{1}{3}x + \frac{1}{3}$$

$$y = \frac{1}{3}x + \frac{10}{3}$$

The equation of the line is $y = \frac{1}{3}x + \frac{10}{3}$.

55. $(-1, -2), (3, 4)$

$$m = \frac{y_2 - y_1}{x_2 - x_1} = \frac{4 - (-2)}{3 - (-1)} = \frac{6}{4} = \frac{3}{2}$$

$$y - y_1 = m(x - x_1)$$

$$y - 4 = \frac{3}{2}(x - 3)$$

$$y - 4 = \frac{3}{2}x - \frac{9}{2}$$

$$y = \frac{3}{2}x - \frac{1}{2}$$

The equation of the line is $y = \frac{3}{2}x - \frac{1}{2}$.

57. $(0, 3), (2, 0)$

$$m = \frac{y_2 - y_1}{x_2 - x_1} = \frac{3 - 0}{0 - 2} = -\frac{3}{2}$$

$$y - y_1 = m(x - x_1)$$

$$y = -\frac{3}{2}(x - 2)$$

$$y = -\frac{3}{2}x + 3$$

The equation of the line is $y = -\frac{3}{2}x + 3$.

59. $(-3, -1), (2, -1)$

$$m = \frac{y_2 - y_1}{x_2 - x_1} = \frac{-1 - (-1)}{2 - (-3)} = \frac{0}{5} = 0$$

$$y - y_1 = m(x - x_1)$$

$$y - (-1) = 0[(x - (-3)]$$

$$y + 1 = 0$$

$$y = -1$$

The equation of the line is $y = -1$.

61. $(-2, -3), (-1, -2)$

$$m = \frac{y_2 - y_1}{x_2 - x_1} = \frac{-3 - (-2)}{-2 - (-1)} = \frac{-1}{-1} = 1$$

$$y - y_1 = m(x - x_1)$$

$$y - (-3) = 1[x - (-2)]$$

$$y + 3 = 1(x + 2)$$

$$y = x - 1$$

The equation of the line is $y = x - 1$.

63. $(-2, 3), (2, -1)$

$$m = \frac{y_2 - y_1}{x_2 - x_1} = \frac{3 - (-1)}{-2 - 2} = \frac{4}{-4} = -1$$

$$y - y_1 = m(x - x_1)$$

$$y - (-1) = -1(x - 2)$$

$$y + 1 = -x + 2$$

$$y = -x + 1$$

The equation of the line is $y = -x + 1$.

65. $(2, 3), (5, -5)$

$$m = \frac{y_2 - y_1}{x_2 - x_1} = \frac{3 - (-5)}{2 - 5} = \frac{8}{-3} = -\frac{8}{3}$$

$$y - y_1 = m(x - x_1)$$

$$y - 3 = -\frac{8}{3}(x - 2)$$

$$y - 3 = -\frac{8}{3}x + \frac{16}{3}$$

$$y = -\frac{8}{3}x + \frac{25}{3}$$

The equation of the line is $y = -\frac{8}{3}x + \frac{25}{3}$.

67. (2, 0), (0, −1)

$$m = \frac{y_2 - y_1}{x_2 - x_1} = \frac{0 - (-1)}{2 - 0} = \frac{1}{2}$$

$$y - y_1 = m(x - x_1)$$

$$y - 0 = \frac{1}{2}(x - 2)$$

$$y = \frac{1}{2}x - 1$$

The equation of the line is $y = \frac{1}{2}x - 1$.

67. (3, −4), (−2, −4)

$$m = \frac{y_2 - y_1}{x_2 - x_1} = \frac{-4 - (-4)}{3 - (-2)} = \frac{0}{5} = 0$$

$$y - y_1 = m(x - x_1)$$

$$y - (-4) = 0(x - 3)$$

$$y + 4 = 0$$

$$y = -4$$

The equation of the line is $y = -4$.

71. (0, 0), (4, 3)

$$m = \frac{y_2 - y_1}{x_2 - x_1} = \frac{3 - 0}{4 - 0} = \frac{3}{4}$$

$$y - y_1 = m(x - x_1)$$

$$y - 0 = \frac{3}{4}(x - 0)$$

$$y = \frac{3}{4}x$$

The equation of the line is $y = \frac{3}{4}x$.

73. (2, −1), (−1, 3)

$$m = \frac{y_2 - y_1}{x_2 - x_1} = \frac{3 - (-1)}{-1 - 2} = \frac{4}{-3} = -\frac{4}{3}$$

$$y - y_1 = m(x - x_1)$$

$$y - (-1) = -\frac{4}{3}(x - 2)$$

$$y + 1 = -\frac{4}{3}x + \frac{8}{3}$$

$$y = -\frac{4}{3}x + \frac{5}{3}$$

The equation of the line is $y = -\frac{4}{3}x + \frac{5}{3}$.

75. (−2, 5), (−2, −5)

$$m = \frac{y_2 - y_1}{x_2 - x_1} = \frac{5 - (-5)}{-2 - (-2)} = \frac{10}{0}$$

The slope is undefined. The line is a vertical line. All points on the line have an abscissa of −2. The equation of the line is $x = -2$.

77. (2, 1), (−2, −3)

$$m = \frac{y_2 - y_1}{x_2 - x_1} = \frac{1 - (-3)}{2 - (-2)} = \frac{4}{4} = 1$$

$$y - y_1 = m(x - x_1)$$

$$y - 1 = 1(x - 2)$$

$$y = x - 1$$

The equation of the line is $y = x - 1$.

79. (−4, −3), (2, 5)

$$m = \frac{y_2 - y_1}{x_2 - x_1} = \frac{5 - (-3)}{2 - (-4)} = \frac{8}{6} = \frac{4}{3}$$

$$y - y_1 = m(x - x_1)$$

$$y - 5 = \frac{4}{3}(x - 2)$$

$$y - 5 = \frac{4}{3}x - \frac{8}{3}$$

$$y = \frac{4}{3}x + \frac{7}{3}$$

The equation of the line is $y = \frac{4}{3}x + \frac{7}{3}$.

81. (0, 3), (3, 0)

$$m = \frac{y_2 - y_1}{x_2 - x_1} = \frac{3 - 0}{0 - 3} = \frac{3}{-3} = -1$$

$$y - y_1 = m(x - x_1)$$

$$y - 0 = -1(x - 3)$$

$$y = -x + 3$$

The equation of the line is $y = -x + 3$.

Objective C Exercises

81. Strategy: Let x represent the number of minutes after takeoff.
Let y represent the height of the plane in feet.
Use the slope-intercept form of an equation to find the equation of the line.

Solution:
a) y-intercept $(0, 0)$; slope is 1200
$y = mx + b$
$y = 1200x + 0$
The linear function is $f(x) = 1200x$.

$0 \le x \le 26\dfrac{2}{3}$

b) Find the height of the plane 11 min after takeoff.
$y = 1200(11) = 13{,}200$
Eleven minutes after takeoff, the height of the plane will be 13,200 ft.

85. Strategy: Let x represent the year.
Let y represent the percent of trees that are hardwoods.
Use the point-slope formula to find the equation of the line.

Solution:
a) $(1964, 57)$, $(2004, 82)$
$m = \dfrac{82 - 57}{2004 - 1964} = \dfrac{25}{40} = 0.625$
$y - y_1 = m(x - x_1)$
$y - 57 = 0.625(x - 1964)$
$y - 57 = 0.625x - 1227.5$
$\quad\quad y = 0.625x - 1170.5$
The linear equation is $f(x) = 0.625x - 1170.5$

b) Predict the percent of trees that will be hardwoods in 2012.
$y = 0.625(2012) - 1170.5 = 87$
In 2012 it is predicted that 87% of the trees will be hardwoods.

87. Strategy: Let x represent the number of miles driven.
Let y represent the number of gallons of gas in the tank.
Use the slope-intercept form of an equation

to find the equation of the line.

Solution:
a) y-intercept $(0, 16)$; slope is -0.032
$y = -0.032x + 16$
Since $0 \le y \le 16$, we have
$0 \le -0.032x + 16 \le 16$
$-16 \le -0.032x \le 0$
$500 \ge x \ge 0$
The linear function is $f(x) = -0.032x + 16$, for $500 \ge x \ge 0$.
b) Find the number of gallons of gas left in the tank after driving 150 mi.
$y = -0.032(150) + 16 = 11.2$
After driving 150 mi, there are 11.2 gal of gas left in the tank.

89. Strategy: Let x represent the price of a motorcycle.
Let y represent the number of motorcycles sold.
Use the point-slope formula to find the equation of the line.

Solution:
a) $(9000, 50{,}000)$, $(8750, 55{,}000)$
$m = \dfrac{55{,}000 - 50{,}000}{8750 - 9000} = \dfrac{5000}{-250} = -20$
$y - y_1 = m(x - x_1)$
$y - 50{,}000 = -20(x - 9000)$
$y - 50{,}000 = -20x + 180{,}000)$
$\quad\quad\quad y = -20x + 230{,}000$
The linear function is $f(x) = -20x + 230{,}000$

b) Find the number of motorcycles sold when the price is $8500.
$y = -20(8500) + 230{,}000 = 60{,}000$
When the price of a motorcycle is $8500, 60,000 will be sold.

91. Strategy: Let x represent the number of ounces of lean hamburger.
Let y represent the number of calories.
Use the point-slope formula to find the equation of the line.

Solution:

a) (2, 126), (3, 189)

$$m = \frac{189 - 126}{3 - 2} = 63$$

$$y - y_1 = m(x - x_1)$$

$$y - 126 = 63(x - 2)$$

$$y - 126 = 63x - 126)$$

$$y = 63x$$

The linear function is $f(x) = 63x$

b) Find the number of calories in a 5-ounce serving.

$$y = 63(5) + 315$$

There are 315 calories in a 5-ounce serving of lean hamburger.

93. Substitute 15,000 for $f(x)$ and solve the equation for x.

95. (2, 5), (0, 3)

$$m = \frac{5 - 3}{2 - 0} = \frac{2}{2} = 1$$

$$y - y_1 = m(x - x_1)$$

$$y - 3 = 1(x - 0)$$

$$y = x + 3$$

$$f(x) = x + 3$$

97. (1, 3), (−1, 5)

$$m = \frac{5 - 3}{-1 - 1} = \frac{2}{-2} = -1$$

$$y - y_1 = m(x - x_1)$$

$$y - 3 = -1(x - 1)$$

$$y - 3 = -x + 1$$

$$y = -x + 4$$

$$f(x) = -x + 4$$

$$f(4) = -4 + 4 = 0$$

99. Given $m = \frac{4}{3}$ and a point (3, 2)

a) $y - y_1 = m(x - x_1)$

$$y - 2 = \frac{4}{3}(x - 3)$$

$$y - 2 = \frac{4}{3}x - 4$$

$$y = \frac{4}{3}x - 2$$

For $x = -6$ we have

$$y = \frac{4}{3}(-6) - 2 = -8 - 2 = -10$$

b) For $y = 6$ we have

$$6 = \frac{4}{3}x - 2$$

$$8 = \frac{4}{3}x$$

$$\frac{3}{4} \cdot 8 = \frac{3}{4} \cdot \frac{4}{3}x$$

$$6 = x$$

Critical Thinking

101. Student solutions will vary.

Find the equation of the line:

$$m = \frac{0 - 6}{6 - (-3)} = \frac{-6}{9} = -\frac{2}{3}$$

$$y - 0 = -\frac{2}{3}(x - 6)$$

$$y = -\frac{2}{3}x + 4$$

Possible answers are:

If $x = 0$, $y = -\frac{2}{3}(0) + 4 = 4$

$x = 3$, $y = -\frac{2}{3}(3) + 4 = 2$

$x = 6$, $y = -\frac{2}{3}(6) + 4 = 0$

(0,4), (3,2), (6,0)

103. Find the x- and y-coordinates for the midpoint of the line segment:

$$x_m = \frac{2 + (-4)}{2} = \frac{-2}{2} = -1$$

$$y_m = \frac{5 + 1}{2} = \frac{6}{2} = 3$$

The midpoint is (−1, 3).

Use the point-slope formula to find the equation of the line.

$$y - y_1 = m(x - x_1)$$

104.
$$y - 3 = -2[x - (-1)]$$
$$y - 3 = -2(x + 1)$$
$$y - 3 = -2x - 2$$
$$y = -2x + 1$$

Projects or Group Activities

105. $P_1 (5, 77)$; $P_2(-2, 154)$
$$m = \frac{77 - 154}{5 - (-2)} = \frac{-77}{7} = -11$$
$$y = mx + b$$
$$77 = -11(5) + b$$
$$132 = b$$
The linear function is $f(x) = -11x + 132$
$$99 = -11x + 132$$
$$-33 = -11x$$
$$x = 3$$
The steepness is $3°$ up.

Section 3.6

Concept Check

1. Two lines are parallel if they have the same slope and different y-intercepts.

3. slope

5. $m = -5$

7. $m = -\dfrac{1}{3}$

9. $m = \dfrac{2}{3}$

Objective A Exercises

13. No, the lines are not parallel because their slopes are not equal.

15. Yes, the lines are perpendicular. Their slopes are negative reciprocals of each other.

17. $2x + 3y = 2$
$$3y = -2x + 2$$

$$y = -\frac{2}{3}x + \frac{2}{3}$$
$$m_1 = -\frac{2}{3}$$
$$2x + 3y = -4$$
$$3y = -2x - 4$$
$$y = -\frac{2}{3}x - \frac{4}{3}$$
$$m_2 = -\frac{2}{3}$$
Since $m_1 = m_2 = -\dfrac{2}{3}$, the lines are parallel.

19. $x - 4y = 2$
$$-4y = -x + 2$$
$$y = \frac{1}{4}x - \frac{1}{2}$$
$$m_1 = \frac{1}{4}$$
$$4x + y = 8$$
$$y = -4x + 8$$
$$m_2 = -4$$
Since $m_1 \cdot m_2 = \dfrac{1}{4} \cdot (-4) = -1$, the lines are perpendicular.

21. $m_1 = \dfrac{6 - 2}{1 - 3} = \dfrac{4}{-2} = -2$
$$m_2 = \frac{-1 - 3}{-1 - (-1)} = \frac{-4}{0}$$
$$m_1 \neq m_2$$
The lines are not parallel.

23. $m_1 = \dfrac{-1 - 2}{4 - (-3)} = \dfrac{-3}{7} = -\dfrac{3}{7}$
$$m_2 = \frac{-4 - 3}{-2 - 1} = \frac{-7}{-3} = \frac{7}{3}$$
$$m_1 \cdot m_2 = -\frac{3}{7}\left(\frac{7}{3}\right) = -1$$
The lines are perpendicular.

25. Since the new line is parallel to $y = 2x + 1$, both lines will have the same slope. The slope of the new line is $m = 2$.

Use the point-slope formula to find the equation of the line.
$m = 2$ and $(3, -2)$
$$y - y_1 = m(x - x_1)$$
$$y - (-2) = 2(x - 3)$$
$$y + 2 = 2x - 6$$
$$y = 2x - 8$$
The equation of the line is $y = 2x - 8$.

27. Since the new line is perpendicular to $y = -\dfrac{2}{3}x - 2$, the slope of the new line must be the negative reciprocal of the slope of the given line. The slope of the new line is $m = \dfrac{3}{2}$.

Use the point-slope formula to find the equation of the line.
$m = \dfrac{3}{2}$ and $(-2, -1)$
$$y - y_1 = m(x - x_1)$$
$$y - (-1) = \frac{3}{2}[x - (-2)]$$
$$y + 1 = \frac{3}{2}(x + 2)$$
$$y + 1 = \frac{3}{2}x + 3$$
$$y = \frac{3}{2}x + 2$$
The equation of the line is $y = \dfrac{3}{2}x + 2$.

29. Since the new line is parallel to $2x - 3y = 2$, both lines will have the same slope.
$$2x - 3y = 2$$
$$-3y = -2x + 2$$
$$y = \frac{2}{3}x - \frac{2}{3}$$

The slope of the new line is $m = \dfrac{2}{3}$.

Use the point-slope formula to find the equation of the line.
$m = \dfrac{2}{3}$ and $(-2, -4)$
$$y - y_1 = m(x - x_1)$$
$$y - (-4) = \frac{2}{3}[x - (-2)]$$
$$y + 4 = \frac{2}{3}(x + 2)$$
$$y + 4 = \frac{2}{3}x + \frac{4}{3}$$
$$y = \frac{2}{3}x - \frac{8}{3}$$
The equation of the line is $y = \dfrac{2}{3}x - \dfrac{8}{3}$.

31. Since the new line is perpendicular to $y = -3x + 4$ the slope of the new line must be the negative reciprocal of the slope of the given line.
$$m_1 = -3$$
$$-3 \cdot m_2 = -1 \text{ therefore, } m_2 = \frac{1}{3}.$$

The slope of the new line is $m_2 = \dfrac{1}{3}$.

Use the point-slope formula to find the equation of the line.
$m_2 = \dfrac{1}{3}$ and $(4, 1)$
$$y - y_1 = m(x - x_1)$$
$$y - 1 = \frac{1}{3}(x - 4)$$
$$y - 1 = \frac{1}{3}x - \frac{4}{3}$$
$$y = \frac{1}{3}x - \frac{1}{3}$$
The equation of the line is $y = \dfrac{1}{3}x - \dfrac{1}{3}$.

33. Since the new line is perpendicular to $3x - 5y = 2$, the slope of the new line must be the negative reciprocal of the slope of the given line.

$3x - 5y = 2$

$-5y = -3x + 2$

$y = \dfrac{3}{5}x - \dfrac{2}{5}$

$m_1 = \dfrac{3}{5}$

$\dfrac{3}{5} \cdot m_2 = -1$

$m_2 = -\dfrac{5}{3}$

The slope of the new line is $m_2 = -\dfrac{5}{3}$.

Use the point-slope formula to find the equation of the line.

$m_2 = -\dfrac{5}{3}$ and $(-1, -3)$

$y - y_1 = m(x - x_1)$

$y - (-3) = -\dfrac{5}{3}[x - (-1)]$

$y + 3 = -\dfrac{5}{3}(x + 1)$

$y + 3 = -\dfrac{5}{3}x - \dfrac{5}{3}$

$y = -\dfrac{5}{3}x - \dfrac{14}{3}$

The equation of the line is $y = -\dfrac{5}{3}x - \dfrac{14}{3}$.

Critical Thinking

35. $P_1(3, 4)$ and $P_2(-1, 2)$.

$m = \dfrac{4 - 2}{3 - (-1)} = \dfrac{1}{2}$

To find the midpoint:

$x_m = \dfrac{3 + (-1)}{2} = 1$

$y_m = \dfrac{4 + 2}{2} = 3$

The slope of the perpendicular bisector is

-2 and the midpoint is $P_m(1,3)$.

$y - 3 = -2(x - 1)$

$y - 3 = -2x + 2$

$y = -2x + 5$

The perpendicular bisector is $y = -2x + 5$.

37. Use the points $(0, 0)$ and $(6, 3)$.

$m_1 = \dfrac{3 - 0}{6 - 0} = \dfrac{3}{6} = \dfrac{1}{2}$

$m_1 \cdot m_2 = -1$

$\dfrac{1}{2} \cdot m_2 = -1$

$m_2 = -2$

Using the point $(6, 3)$,

$y - 3 = -2(x - 6)$

$y - 3 = -2x + 12$

$y = -2x + 15$

The equation of the line is $y = -2x + 15$.

Section 3.7

Concept Check

1. A half-plane is the set of points on one side of a line in the plane.

3. $y > 2x - 7$

$0 > 2(0) - 7$

$0 > -7$

Yes, $(0, 0)$ is a solution.

5. $y < 5x + 3$

$0 < 5(0) + 3$

$0 < 3$

Yes, $(0, 0)$ is a solution.

7. $y \geq -\dfrac{3}{4}x + 9$

$0 \geq -\dfrac{3}{4}(0) + 9$

$0 \geq 9$

No, $(0, 0)$ is not a solution $y \leq \dfrac{3}{2}x - 3$

Objective A Exercises

9. $y \geq \dfrac{4}{3}x - 4$

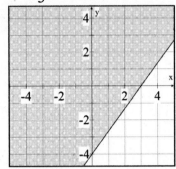

11. $y < \dfrac{3}{5}x - 3$

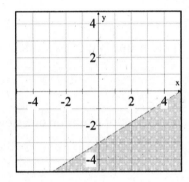

13. $4x + 3y < 9$

$3y < -4x + 9$

$y < -\dfrac{4}{3}x + 3$

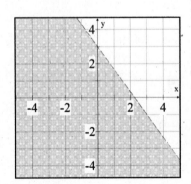

15. $2x - 5y \leq 10$

$-5y \leq -2x + 10$

$y \geq \dfrac{2}{5}x - 2$

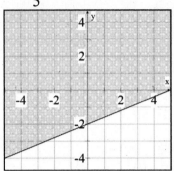

17. $3x + 2y < 4$

$2y < -3x + 4$

$y < -\dfrac{3}{2}x + 2$

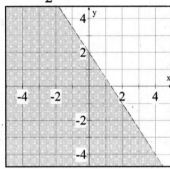

19. $-3x + 2y > 2$

$2y > 3x + 2$

$y > \dfrac{3}{2}x + 1$

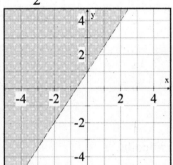

21. $x + 2 \geq 0$

$x \geq -2$

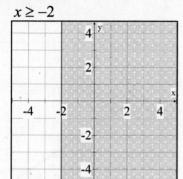

23. $3x - 5y < 10$

$-5y < -3x + 10$

$y > \dfrac{3}{5}x - 2$

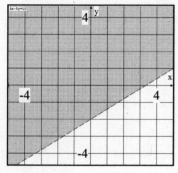

25. Quadrant I

Critical Thinking

27. The inequality $y < 3x - 1$ is not a function because given a value of x there is more than one corresponding value of y. For example, both (3, 2) and (3, −1) are ordered pairs that satisfy the inequality. This contradicts the definition of a function because there are two ordered pairs with the same first coordinate and different second coordinates.

Projects or Group Activities

29.

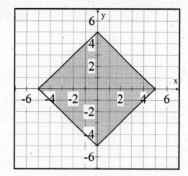

Chapter 3 Review Exercises

1. $y = \dfrac{x}{x - 2}$

$y = \dfrac{4}{4 - 2} = \dfrac{4}{2} = 2$

The ordered pair is (4, 2).

2. $P(x) = 3x + 4$

$P(-2) = 3(-2) + 4 = -6 + 4 = -2$

$P(a) = 3(a) + 4 = 3a + 4$

3. $y = x^2 - 2x - 3$

Domain: $\{x | -\infty < x < \infty\}$

Range: $\{y | y \geq -4\}$

x	y
−2	5
−1	0
0	−3
1	−4
2	−3
3	0
4	5

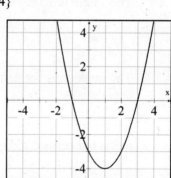

4. Domain: $[-4, \infty)$

Range: $[-5, \infty)$

5. $f(x) = -\dfrac{2}{3}x + 4$

$-\dfrac{2}{3}x + 4 = 0$

$-2x + 12 = 0$

$-2x = -12$

$x = 6$

The zero of the function is 6.

6. Domain = $\{-1, 0, 1, 5\}$

Range = $\{0, 2, 4\}$

7. $(-2, 4)$ and $(3, 5)$

$x_m = \dfrac{-2+3}{2} = \dfrac{1}{2}$

$y_m = \dfrac{4+5}{2} = \dfrac{9}{2}$

The midpoint is $\left(\dfrac{1}{2}, \dfrac{9}{2}\right)$.

Length = $\sqrt{(x_1 - x_2)^2 + (y_1 - y_2)^2}$

$= \sqrt{(3 - (-2))^2 + (5 - 4)^2}$

$= \sqrt{26} \approx 5.10$

The length is 5.10.

8. $f(x) = \dfrac{x}{x + 4}$

The function is not defined for zero in the denominator.

$x + 4 = 0$

$x = -4$

$f(x)$ is not defined for $x = -4$.

9. $y = -\dfrac{2}{3}x - 2$

x-intercept:

$0 = -\dfrac{2}{3}x - 2$

$2 = -\dfrac{2}{3}x$

$x = -3$

$(-3, 0)$

y-intercept:

$y = -\dfrac{2}{3}(0) - 2$

$y = -2$

$(0, -2)$

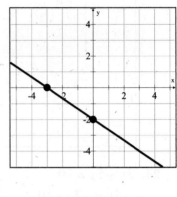

10. $3x + 2y = -6$

x-intercept:

$3x + 2(0) = -6$

$3x = -6$

$x = -2$

$(-2, 0)$

y-intercept:

$3(0) + 2y = -6$

$2y = -6$

$y = -3$

$(0, -3)$

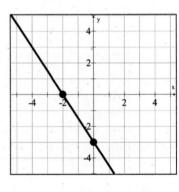

11. $y = -2x + 2$

x	y
0	2
1	0
-1	4

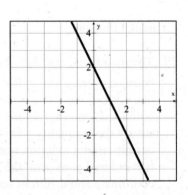

12. $4x - 3y = 12$

$\quad -3y = -4x + 12$

$\quad\quad y = \dfrac{4}{3}x - 4$

x	y
0	−4
3	0
6	4

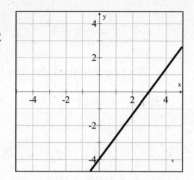

13. $(3, -2)$ and $(-1, 2)$

$m = \dfrac{y_2 - y_1}{x_2 - x_1}$

$m = \dfrac{2\ \ 2}{3 - (-1)} = \dfrac{4}{4} = -1$

14. Use the point-slope formula to find the equation of the line.

$m = \dfrac{5}{2}$ and $(-3, 4)$

$y - y_1 = m(x - x_1)$

$y - 4 = \dfrac{5}{2}[x - (-3)]$

$y - 4 = \dfrac{5}{2}(x + 3)$

$y - 4 = \dfrac{5}{2}x + \dfrac{15}{2}$

$y = \dfrac{5}{2}x + \dfrac{23}{2}$

15. $P_1 (10, 60)$; $P_2 (50, 12)$

$m = \dfrac{60 - 12}{10 - 50} = -1.2$

The average rate of change is $-1.2°F$ per minute.

16. $m = -\dfrac{1}{4}$ and $(-2, 3)$

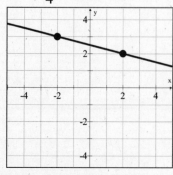

17. This graph does not represent a function.

18. $m = -\dfrac{1}{3}$ and $(-1, 4)$

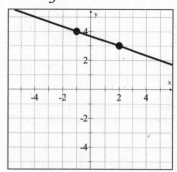

19. The slope for the parallel line is $m = -4$.

$m = -4$ and $(-2, 3)$

$y - y_1 = m(x - x_1)$

$y - 3 = -4[(x - (-2)]$

$y - 3 = -4(x + 2)$

$y - 3 = -4x - 8$

$y = -4x - 5$

The equation of the line is $y = -4x - 5$.

20. The slope for the perpendicular line is

$m = \dfrac{5}{2}$.

$m = \dfrac{5}{2}$ and $(-2, 3)$

$y - y_1 = m(x - x_1)$

$y - 3 = \dfrac{5}{2}[(x - (-2)]$

$y - 3 = \dfrac{5}{2}(x + 2)$

$y - 3 = \dfrac{5}{2}x + 5$

$y = \dfrac{5}{2}x + 8$

The equation of the line is $y = \dfrac{5}{2}x + 8$.

21.

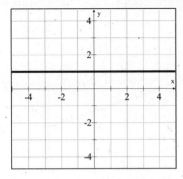

22.

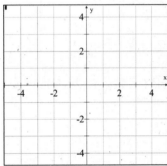

23. P(−3,3), $m = -\dfrac{2}{3}$

$$y - y_1 = m(x - x_1)$$

$$y - 3 = -\frac{2}{3}[(x - (-3)]$$

$$y - 3 = -\frac{2}{3}(x + 3)$$

$$y - 3 = -\frac{2}{3}x - 2$$

$$y = -\frac{2}{3}x + 1$$

The equation of the line is $y = -\dfrac{2}{3}x + 1$.

24. (−8, 2) and (4, 5)

$$m = \frac{5 - 2}{4 - (-8)} = \frac{3}{12} = \frac{1}{4}$$

$$y - y_1 = m(x - x_1)$$

$$y - 5 = \frac{1}{4}(x - 4)$$

$$y - 5 = \frac{1}{4}x - 1$$

$$y = \frac{1}{4}x + 4$$

The equation of the line is $y = \dfrac{1}{4}x + 4$.

25. (4, −5) and (−2, 3)

$$d = \sqrt{(x_1 - x_2)^2 + (y_1 - y_2)^2}$$

$$= \sqrt{(4 - (-2))^2 + (-5 - 3)^2}$$

$$= \sqrt{100} = 10$$

The distance is 10 .

26. (−3, 8) and (5, −2)

$$x_m = \frac{-3 + 5}{2} = \frac{2}{2} = 1$$

$$y_m = \frac{8 + (-2)}{2} = \frac{6}{2} = 3$$

The midpoint is $(1, 3)$.

27. P_1 (5,77); P_2(15,47)

$$m = \frac{77 - 47}{5 - 15} = -3$$

The average rate of change is −3°C/min.

28. $y \geq 2x - 3$

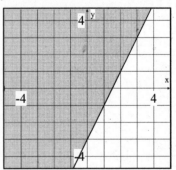

29. $3x - 2y < 6$

$-2y < -3x + 6$

$y > \dfrac{3}{2}x - 3$

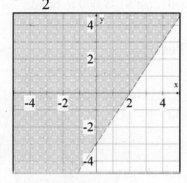

30. $(-2, 4)$ and $(4, -3)$

$m = \dfrac{4 - (-3)}{-2 - 4} = \dfrac{7}{-6} = -\dfrac{7}{6}$

$y - y_1 = m(x - x_1)$

$y - (-3) = -\dfrac{7}{6}(x - 4)$

$y + 3 = -\dfrac{7}{6}x + \dfrac{28}{6}$

$y = -\dfrac{7}{6}x + \dfrac{5}{3}$

The equation of the line is $y = -\dfrac{7}{6}x + \dfrac{5}{3}$.

31. $4x - 2y = 7$

$-2y = -4x + 7$

$y = 2x - \dfrac{7}{2}$

The slope for the parallel line is $m = 2$.

$m = 2$ and $(-2, -4)$

$y - y_1 = m(x - x_1)$

$y - (-4) = 2[(x - (-2)]$

$y + 4 = 2(x + 2)$

$y + 4 = 2x + 4$

$y = 2x$

The equation of the line is $y = 2x$.

32. The slope for the parallel line is $m = -3$.

$m = -3$ and $(3, -2)$

$y - y_1 = m(x - x_1)$

$y - (-2) = -3(x - 3)$

$y + 2 = -3x + 9$

$y = -3x + 7$

The equation of the line is $y = -3x + 7$.

33. $y = -\dfrac{2}{3}x + 6$

$m_1 = -\dfrac{2}{3}$

$m_1 \cdot m_2 = -1$

$-\dfrac{2}{3} \cdot m_2 = -1$

$m_2 = \dfrac{3}{2}$

The slope for the perpendicular line is

$m = \dfrac{3}{2}$.

$m = \dfrac{3}{2}$ and $(2, 5)$

$y - y_1 = m(x - x_1)$

$y - 5 = \dfrac{3}{2}(x - 2)$

$y - 5 = \dfrac{3}{2}x - 3$

$y = \dfrac{3}{2}x + 2$

The equation of the line is $y = \dfrac{3}{2}x + 2$.

34. Strategy: Let x represent the room rate. Let y represent the number of rooms occupied.
Use the point – slope formula to find the equation of the line.

Solution:
a) $(95, 200), (105, 190)$

$$m = \frac{200 - 190}{95 - 105} = \frac{10}{-10} = -1$$

$$y - y_1 = m(x - x_1)$$

$$y - 200 = -1(x - 95)$$

$$y - 200 = -1x + 95$$

$$y = -x + 295$$

The linear function is
$$f(x) = -x + 295, \ 0 \le x \le 295$$

b) Find the number of rooms occupied when the rate is \$120.
$$f(120) = -1(120) + 295 = 175$$
When the room rate is \$125, 175 rooms will be occupied.

35.

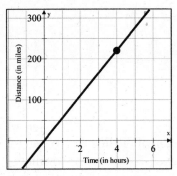

After 4 h, the car will travel 220 mi.

36. $(500, 12{,}000)$ and $(200, 6000)$

$$m = \frac{12{,}000 - 6000}{500 - 200} = \frac{6000}{300} = 20$$

The slope is 20. The manufacturing cost is \$20 per calculator.

37. a) The y-intercept is $(0, 25{,}000)$.
The slope is 80.
$$y = mx + b$$
$$y = 80x + 25{,}000$$
The linear function is $f(x) = 80x + 25{,}000$

b) Predict the cost of building a house with 2000 ft^2.
$$f(2000) = 80(2000) + 25{,}000$$
$$f(x) = 185{,}000$$
The house will cost \$185,000 to build.

Chapter 3 Test

1. $f(x) = 4 - x^2$
Domain: $(-\infty, \infty)$
Range: $(-\infty, 4]$

x	y
-2	0
-1	3
0	4
1	3
2	0
3	-5

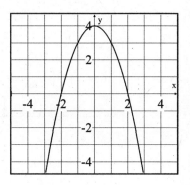

2. Yes, the graph represents a function.

3. $y = \dfrac{2}{3}x - 4$

x	y
0	-4
3	-2
6	0

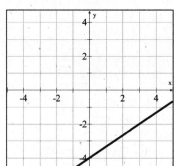

4. $2x + 3y = -3$

x	y
0	-1
3	-3
-3	1

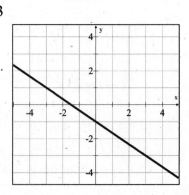

5. $f(x) = 4x - 12$
$4x - 12 = 0$
$4x = 12$
$x = 3$
The zero of the function is 3

6. $(4, 2)$ and $(-5, 8)$
Length $= \sqrt{(x_1 - x_2)^2 + (y_1 - y_2)^2}$
$= \sqrt{(4 - (-5))^2 + (2 - 8)^2}$
$= \sqrt{117} \approx 10.82$
The length is 10.82.

$x_m = \dfrac{x_1 + x_2}{2} = \dfrac{4 + (-5)}{2} = -\dfrac{1}{2}$

$y_m = \dfrac{y_1 + y_2}{2} = \dfrac{2 + 8}{2} = \dfrac{10}{2} = 5$

The midpoint is $\left(-\dfrac{1}{2}, 5\right)$.

7. $(-2, 3)$ and $(4, 2)$

$m = \dfrac{y_2 - y_1}{x_2 - x_1} = \dfrac{3 - 2}{-2 - 4} = \dfrac{1}{-6} = -\dfrac{1}{6}$

The slope of the line is $-\dfrac{1}{6}$.

8. $P(x) = 3x^2 - 2x + 1$
$P(2) = 3(2)^2 - 2(2) + 1$
$P(2) = 9$

9. $2x - 3y = 6$
x-intercept:
$2x - 3(0) = 6$
$2x = 6$
$x = 3$
$(3, 0)$
y-intercept:
$2(0) - 3y = 6$
$-3y = 6$
$y = -2$
$(0, -2)$

10. $(-2, 3)$ and $m = -\dfrac{3}{2}$

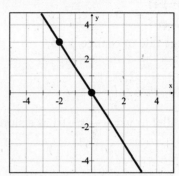

11. $m = \dfrac{2}{5}$ and $(-5, 2)$

$y - 2 = \dfrac{2}{5}[x - (-5)]$

$y - 2 = \dfrac{2}{5}(x + 5)$

$y - 2 = \dfrac{2}{5}x + 2$

$y = \dfrac{2}{5}x + 4$

The equation of the line is $y = \dfrac{2}{5}x + 4$.

12. $f(x) = \dfrac{2x + 1}{x}$
The function is not defined for zero in the denominator.
$x = 0$ is excluded from the domain of $f(x)$.

13. $(3, -4)$ and $(-2, 3)$

$m = \dfrac{3 - (-4)}{-2 - 3} = \dfrac{7}{-5} = -\dfrac{7}{5}$

$y - (-4) = -\dfrac{7}{5}(x - 3)$

$y + 4 = -\dfrac{7}{5}x + \dfrac{21}{5}$

$y = -\dfrac{7}{5}x + \dfrac{1}{5}$

The equation of the line is $y = -\dfrac{7}{5}x + \dfrac{1}{5}$.

14. $f(x) = 2 - x^2$
$f(1) = 2 - 1^2 = 1$

$f(4) = 2 - 4^2 = -14$

$P_1 (1,1); P_2 (4,-14)$

$m = \dfrac{1-(-14)}{1-4} = -5$

The average rate of change is –5.

15. Domain = $\{-4, -2, 0, 3\}$

Range = $\{0, 2, 5\}$

16. A line parallel to $y = -\dfrac{3}{2}x - 6$ has a slope

of $m = -\dfrac{3}{2}$.

$m = -\dfrac{3}{2}$ and $(1, 2)$

$y - 2 = -\dfrac{3}{2}(x - 1)$

$y - 2 = -\dfrac{3}{2}x + \dfrac{3}{2}$

$y = -\dfrac{3}{2}x + \dfrac{7}{2}$

The equation of the line is $y = -\dfrac{3}{2}x + \dfrac{7}{2}$.

17. $y = -\dfrac{1}{2}x - 3$

$m_1 = -\dfrac{1}{2}$

$m_1 \cdot m_2 = -1$

$-\dfrac{1}{2} \cdot m_2 = -1$

$m_2 = 2$

The slope of the perpendicular line is

$m = 2$.

$m = 2$ and $(-2, -3)$

$y - (-3) = 2[x - (-2)]$

$y + 3 = 2(x + 2)$

$y + 3 = 2x + 4$

$y = 2x + 1$

The equation of the line is $y = 2x + 1$.

18. $3x - 4y > 8$

$-4y > -3x + 8$

$y < \dfrac{3}{4}x - 2$

19. Strategy: Use two points on the graph to find the slope of the line.

Solution: $(3, 120,000)$ and $(12, 30,000)$

$m = \dfrac{120,000 - 30,000}{3 - 12} = \dfrac{90,000}{-9} = -10,000$

The value of the house decreases by $10,000 each year.

20. Strategy: Let x represent the tuition cost. Let y represent the number of students. Use the point – slope formula to find the equation of the line.

Solution:

a) $m = \dfrac{-6}{20} = -\dfrac{3}{10}$ and $(250, 100)$

$y - y_1 = m(x - x_1)$

$y - 100 = -\dfrac{3}{10}(x - 250)$

$y - 100 = -\dfrac{3}{10}x + 75$

$y = -\dfrac{3}{10}x + 175$

The linear function that will predict enrollment based on the cost of tuition is

$f(x) = -\dfrac{3}{10}x + 175$.

b) Find the number of students who enroll when tuition is $300.

$$f(300) = -\frac{3}{10}(300) + 175$$

$$f(300) = 85$$

When tuition is $300, 85 students will enroll.

Cumulative Review Exercises

1. Commutative Property of Multiplication

2.
$$3 - \frac{x}{2} = \frac{3}{4}$$

$$4 \cdot \left(3 - \frac{x}{2}\right) = \frac{3}{4}(4)$$

$$12 - 2x = 3$$

$$-2x = -9$$

$$x = \frac{9}{2}$$

The solution is $x = \frac{9}{2}$.

3.
$$2[y - 2(3 - y) + 4] = 4 - 3y$$

$$2[y - 6 + 2y + 4] = 4 - 3y$$

$$2(3y - 2) = 4 - 3y$$

$$6y - 4 = 4 - 3y$$

$$9y - 4 = 4$$

$$9y = 8$$

$$y = \frac{8}{9}$$

The solution is $\frac{8}{9}$.

4.
$$\frac{1 - 3x}{2} + \frac{7x - 2}{6} = \frac{4x + 2}{9}$$

$$18 \cdot \left(\frac{1 - 3x}{2} + \frac{7x - 2}{6}\right) = 18\left(\frac{4x + 2}{9}\right)$$

$$9(1 - 3x) + 3(7x - 2) = 2(4x + 2)$$

$$9 - 27x + 21x - 6 = 8x + 4$$

$$3 - 6x = 8x + 4$$

$$3 = 14x + 4$$

$$-1 = 14x$$

$$x = -\frac{1}{14}$$

The solution is $-\frac{1}{14}$.

5.
$$\begin{array}{lll} x - 3 < -4 & \text{or} & 2x + 2 > 3 \\ x < -1 & & 2x > 1 \\ & & x > \frac{1}{2} \end{array}$$

$$\{x \mid x < -1\} \qquad \{x \mid x > \frac{1}{2}\}$$

$$\{x \mid x < -1\} \quad \text{or} \quad \{x \mid x > \frac{1}{2}\}$$

$$\{x \mid x < -1\} \cup \{x \mid x > \frac{1}{2}\}$$

6.
$$8 - |2x - 1| = 4$$

$$-|2x - 1| = -4$$

$$|2x - 1| = 4$$

$$\begin{array}{lll} 2x - 1 = 4 & \text{or} & 2x - 1 = -4 \\ 2x = 5 & & 2x = -3 \\ x = \frac{5}{2} & & x = -\frac{3}{2} \end{array}$$

The solutions are $\frac{5}{2}$ and $-\frac{3}{2}$.

7.
$$|3x - 5| < 5$$

$$-5 < 3x - 5 < 5$$

$$-5 + 5 < 3x - 5 + 5 < 5 + 5$$

$$0 < 3x < 10$$

$$\frac{0}{3} < \frac{3x}{3} < \frac{10}{3}$$

$$0 < x < \frac{10}{3}$$

$$\{x \mid 0 < x < \frac{10}{3}\}$$

8.
$$4 - 2(4 - 5)^3 + 2$$

$$= 4 - 2(-1)^3 + 2$$

$$= 4 - 2(-1) + 2$$

$$= 4 + 2 + 2$$

$$= 8$$

9. $(a-b)^2 \div (ab)$

$(4-(-2))^2 \div (4(-2))$

$= (6)^2 \div (-8) = 36 \div (-8)$

$= -4.5$

10. $\{x \mid x < -2\} \cup \{x \mid x > 0\}$

$$-5 \;-4\;-3\;-2\;-1\;\;0\;\;1\;\;2\;\;3\;\;4\;\;5$$

11. $P = \dfrac{R-C}{n}$

$P \cdot n = \dfrac{R-C}{n} \cdot n$

$P \cdot n = R - C$

$P \cdot n - R = -C$

$R - P \cdot n = C$

12. $2x + 3y = 6$

$2x = -3y + 6$

$x = -\dfrac{3}{2}y + 3$

13. $3x - 1 < 4$ and $x - 2 > 2$

$3x < 5 x > 4$

$x < \dfrac{5}{3}$

$\{x \mid x < \dfrac{5}{3}\} \cap \{x \mid x > 4\} = \emptyset$

14. $P(x) = x^2 + 5$

$P(-3) = (-3)^2 + 5$

$P(-3) = 14$

15. $y = -\dfrac{5}{4}x + 3$

$y = -\dfrac{5}{4}(-8) + 3$

$y = 10 + 3$

$y = 13$

The ordered pair is $(-8, 13)$.

16. $(-1, 3)$ and $(3, -4)$

$m = \dfrac{y_2 - y_1}{x_2 - x_1} = \dfrac{3 - (-4)}{-1 - 3} = \dfrac{7}{-4} = -\dfrac{7}{4}$

17. $m = \dfrac{3}{2}$ and $(-1, 5)$

$y - y_1 = m(x - x_1)$

$y - 5 = \dfrac{3}{2}[(x - (-1)]$

$y - 5 = \dfrac{3}{2}(x + 1)$

$y - 5 = \dfrac{3}{2}x + \dfrac{3}{2}$

$y = \dfrac{3}{2}x + \dfrac{13}{2}$

The equation of the line is $y = \dfrac{3}{2}x + \dfrac{13}{2}$.

18. $(4, -2)$ and $(0, 3)$

$m = \dfrac{3 - (-2)}{0 - 4} = \dfrac{5}{-4} = -\dfrac{5}{4}$

$y - 3 = -\dfrac{5}{4}(x - 0)$

$y - 3 = -\dfrac{5}{4}x$

$y = -\dfrac{5}{4}x + 3$

The equation of the line is $y = -\dfrac{5}{4}x + 3$.

19. A line parallel to $y = -\dfrac{3}{2}x + 2$ has a slope

of $m = -\dfrac{3}{2}$.

$m = -\dfrac{3}{2}$ and $(2, 4)$

$y - 4 = -\dfrac{3}{2}(x - 2)$

$y - 4 = -\dfrac{3}{2}x + 3$

$$y = -\frac{3}{2}x + 7$$

The equation of the line is $y = -\frac{3}{2}x + 7$.

20. $3x - 2y = 5$

$-2y = -3x + 5$

$y = \frac{3}{2}x - \frac{5}{2}$

$m_1 = \frac{3}{2}$

$m_1 \cdot m_2 = -1$

$\frac{3}{2} \cdot m_2 = -1$

$m_2 = -\frac{2}{3}$

The slope of the perpendicular line is

$m = -\frac{2}{3}$.

$m = -\frac{2}{3}$ and $(4, 0)$

$y - 0 = -\frac{2}{3}(x - 4)$

$y = -\frac{2}{3}x + \frac{8}{3}$

The equation of the line is $y = -\frac{2}{3}x + \frac{8}{3}$.

21. $f(x) = -2x + 6$

$-2x + 6 = 0$

$-2x = -6$

$x = 3$

The zero of the function is 3.

22. $3x - 5y = 15$

x-intercept:

$3x - 5(0) = 15$

$3x = 15$

$x = 5$

$(5, 0)$

y-intercept:

$3(0) - 5y = 15$

$-5y = 15$

$y = -3$

$(0, -3)$

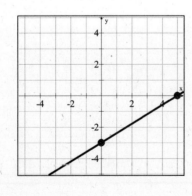

23.

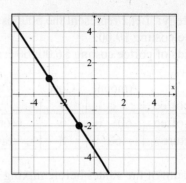

24. $3x - 2y \geq 6$

$-2y \geq -3x + 6$

$y \leq \frac{3}{2}x - 3$

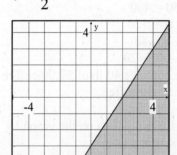

25. Strategy: Let r represent the speed of 1st plane.
The speed of the 2nd plane is $2r$.

	Rate	Time	Distance
1st plane	r	3	$3r$
2nd plane	$2r$	3	$3(2r)$

The total distance traveled by the two planes is 1800 mi.

Solution: $3r + 3(2r) = 1800$

$3r + 6r = 1800$

$9r = 1800$

$r = 200$

$2r = 2(200) = 400$

The first plane is traveling at 200 mph and the second plane is traveling at 400 mph.

26. Strategy: Let x represent the pounds of coffee costing $9.00.
Pounds of coffee costing $6.00: $60 - x$

	Amount	Cost	Value
$9 coffee	x	9	$9x$
$6 coffee	$60 - x$	6	$6(60 - x)$
Mixture	60	8	$8(60)$

The sum of the values before mixing is equal to the value after mixing.

Solution:
$$9x + 6(60 - x) = 8(60)$$
$$9x + 360 - 6x = 480$$
$$3x + 360 = 480$$
$$3x = 120$$
$$x = 40$$
$$60 - x = 60 - 40 = 20$$
The mixture contains 40 lb of $9.00 coffee and 20 lb of $6.00 coffee.

27. Strategy: Use two points on the graph to find the slope of the line.
Locate the y-intercept on the graph.
Use the slope-intercept form of an equation to write the equation of the line.

Solution: a) $(0, 30{,}000)$ and $(6, 0)$
$$m = \frac{30{,}000 - 0}{0 - 6} = \frac{30{,}000}{-6} = -5000$$
The y-intercept is $(0, 30{,}000)$.
The slope is -5000.
$$y = mx + b$$
$$y = -5000x + 30{,}000$$
The linear function is
$$f(x) = -5000x + 30{,}000$$
b) The value of the truck decreases by $5000 per year.

Chapter 4: Systems of Linear Equations and Inequalities

Prep Test

1. $10\left(\dfrac{3}{5}x + \dfrac{1}{2}y\right) = 10\left(\dfrac{3}{5}x\right) + 10\left(\dfrac{1}{2}y\right) = 6x + 5y$

2. $3x + 2y - z$
$3(-1) + 2(4) - (-2)$
$= -3 + 8 + 2$
$= 7$

3.
$$3x - 2z = 4$$
$$3x - 2(-2) = 4$$
$$3x + 4 = 4$$
$$3x = 0$$
$$x = 0$$

4. $3x + 4(-2x - 5) = -5$
$3x - 8x - 20 = -5$
$-5x - 20 = -5$
$-5x = 15$
$x = -3$
The solution is -3.

5. $0.45x + 0.06(-x + 4000) = 630$
$0.45x - 0.06x + 240 = 630$
$0.39x + 240 = 630$
$0.39x = 390$
$x = 1000$
The solution is 1000.

6. $3x - 2y = 6$
$-2y = -3x + 6$

$y = \dfrac{3}{2}x - 3$

x	y
0	-3
2	0
4	3

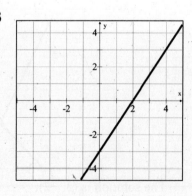

7. $y > -\dfrac{3}{5}x + 1$

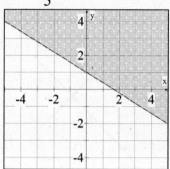

Section 4.1

Objective A Exercises

1. $x + y = 3$
$2 + 1 = 3$
$3 = 3$
$2x - 3y = 1$
$2(2) - 3(1) = 1$
$4 - 3 = 1$
$1 = 1$
Yes, $(2, 1)$ is a solution of the system of equations.

3. $3x - y = 4$
$3(1) - (-1) = 4$
$3 + 1 = 4$
$4 = 4$
$7x + 2y = -5$
$7(1) + 2(-1) \neq -5$
$7 - 2 \neq -5$
$5 \neq -5$
No, $(1, -1)$ is not a solution of the system of equations.

5. The two graphs represent the same line. The system is a dependent system of equations.

7. $(4, -1)$

Objective A Exercises

9. $x + y = 2$
$x - y = 4$

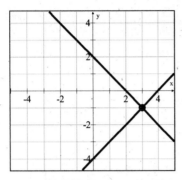

The solution is $(3, -1)$.

11. $x - y = -2$
$x + 2y = 10$

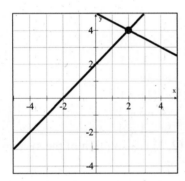

The solution is $(2, 4)$.

13. $3x - 2y = 6$
$y = 3$

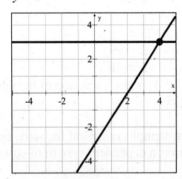

The solution is $(4, 3)$.

15. $2x + 4y = 4$
$-3x - 6y = -6$

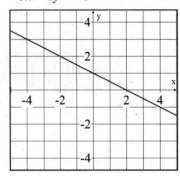

The two equations represent the same line. The system of equations is dependent. The solutions are the ordered pairs

$$\left(x, -\frac{1}{2}x + 1 \right).$$

17. $x - y = 6$
$x + y = 2$

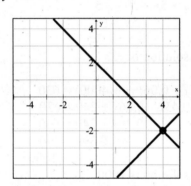

The solution is $(4, -2)$.

19. $y = x - 5$
$2x + y = 4$

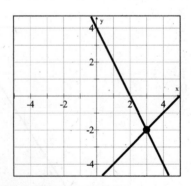

The solution is $(3, -2)$.

21. $y = \dfrac{1}{2}x - 2$

$x - 2y = 8$

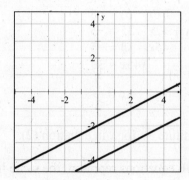

The lines are parallel and do not intersect so there is no solution.

23. $2x - 5y = 10$

$y = \dfrac{2}{5}x - 2$

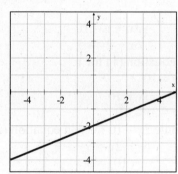

The two equations represent the same line. The system of equations is dependent.

The solutions are the ordered pairs $\left(x, \dfrac{2}{5}x - 2\right)$.

Objective B Exercises

25. (1) $x = 2y - 3$

(2) $3x + y = 5$

Substitute $2y - 3$ for x in equation (2).

$3x + y = 5$

$3(2y - 3) + y = 5$

$6y - 9 + y = 5$

$7y - 9 = 5$

$7y = 14$

$y = 2$

Substitute 2 for y in equation (1).

$x = 2y - 3$

$x = 2(2) - 3$

$x = 4 - 3$

$x = 1$

The solution is (1, 2).

27. (1) $4x - 3y = 2$

(2) $y = 2x + 1$

Substitute $2x + 1$ for y in equation (1).

$4x - 3y = 2$

$4x - 3(2x + 1) = 2$

$4x - 6x - 3 = 2$

$-2x - 3 = 2$

$-2x = 5$

$x = -\dfrac{5}{2}$

Substitute $-\dfrac{5}{2}$ for x in equation (2).

$y = 2x + 1$

$y = 2\left(-\dfrac{5}{2}\right) + 1$

$y = -5 + 1$

$y = -4$

The solution is $\left(-\dfrac{5}{2}, -4\right)$.

29. (1) $3x - 2y = -11$

(2) $x = 2y - 9$

Substitute $2y - 9$ for x in equation (1).

$3x - 2y = -11$

$3(2y - 9) - 2y = -11$

$6y - 27 - 2y = -11$

$4y - 27 = -11$

$4y = 16$

$y = 4$

Substitute 4 for y in equation (2).

$x = 2y - 9$

$x = 2(4) - 9$

$x = 8 - 9$

$x = -1$

The solution is (−1, 4).

31. (1) $3x + 2y = 4$
(2) $\quad\quad y = 1 - 2x$
Substitute $1 - 2x$ for y in equation (1).
$$3x + 2y = 4$$
$$3x + 2(1 - 2x) = 4$$
$$3x + 2 - 4x = 4$$
$$-x + 2 = 4$$
$$-x = 2$$
$$x = -2$$
Substitute -2 for x in equation (2).
$$y = 1 - 2x$$
$$y = 1 - 2(-2)$$
$$y = 1 + 4$$
$$y = 5$$
The solution is $(-2, 5)$.

33. (1) $\quad\quad 5x + 2y = 15$
(2) $\quad\quad x = 6 - y$
Substitute $6 - y$ for x in equation (1).
$$5x + 2y = 15$$
$$5(6 - y) + 2y = 15$$
$$30 - 5y + 2y = 15$$
$$30 - 3y = 15$$
$$-3y = -15$$
$$y = 5$$
Substitute 5 for y in equation (2).
$$x = 6 - y$$
$$x = 6 - 5$$
$$x = 1$$
The solution is $(1, 5)$.

45. (1) $\quad\quad 3x - 4y = 6$
(2) $\quad\quad x = 3y + 2$
Substitute $3y + 2$ for x in equation (1).
$$3x - 4y = 6$$
$$3(3y + 2) - 4y = 6$$
$$9y + 6 - 4y = 6$$
$$5y + 6 = 6$$
$$5y = 0$$
$$y = 0$$
Substitute 0 for y in equation (2).
$$x = 3y + 2$$
$$x = 2(0) + 2$$
$$x = 0 + 2$$
$$x = 2$$
The solution is $(2, 0)$.

37. (1) $\quad\quad 3x + 7y = -5$
(2) $\quad\quad y = 6x - 5$
Substitute $6x - 5$ for y in equation (1).
$$3x + 7y = -5$$
$$3x + 7(6x - 5) = -5$$
$$3x + 42x - 35 = -5$$
$$45x - 35 = -5$$
$$45x = 30$$
$$x = \frac{2}{3}$$
Substitute $\frac{2}{3}$ for x in equation (2).
$$y = 6x - 5$$
$$y = 6 \cdot \frac{2}{3} - 5$$
$$y = 4 - 5$$
$$y = -1$$
The solution is $\left(\frac{2}{3}, -1\right)$.

39. (1) $\quad\quad 3x - y = 10$
(2) $\quad\quad 6x - 2y = 5$
Solve equation (1) for y.
$$3x - y = 10$$
$$-y = -3x + 10$$
$$y = 3x - 10$$
Substitute $3x - 10$ for y in equation (2).
$$6x - 2y = 5$$
$$6x - 2(3x - 10) = 5$$
$$6x - 6x + 20 = 5$$
$$20 = 5$$
No solution. This is not a true equation. The lines are parallel and the system is inconsistent.

41. (1) $\quad\quad 3x + 4y = 14$
(2) $\quad\quad 2x + y = 1$
Solve equation (2) for y.
$$2x + y = 1$$
$$y = -2x + 1$$
Substitute $-2x + 1$ for y in equation (1).
$$3x + 4y = 14$$
$$3x + 4(-2x + 1) = 14$$
$$3x - 8x + 4 = 14$$
$$-5x + 4 = 14$$
$$-5x = 10$$
$$x = -2$$

Substitute -2 for x in equation (2).
$2x + y = 1$
$2(-2) + y = 1$
$-4 + y = 1$
$y = 5$
The solution is $(-2, 5)$.

43. (1) $3x + 5y = 0$
   (2) $x - 4y = 0$
Solve equation (2) for x.
$x - 4y = 0$
   $x = 4y$
Substitute $4y$ for x in equation (1).
   $3x + 5y = 0$
   $3(4y) + 5y = 0$
   $12y + 5y = 0$
   $17y = 0$
   $y = 0$
Substitute 0 for y in equation (2).
$x - 4y = 0$
$x - 4(0) = 0$
$x - 0 = 0$
$x = 0$
The solution is $(0, 0)$.

45. (1) $2x - 4y = 16$
   (2) $-x + 2y = -8$
Solve equation (2) for x.
$-x + 2y = -8$
   $x = 2y + 8$
Substitute $2y + 8$ for x in equation (1).
   $2x - 4y = 16$
   $2(2y + 8) - 4y = 16$
   $4y + 16 - 4y = 16$
   $16 = 16$
This is a true equation. The equations are
dependent. The solutions are the ordered
pairs $\left(x, \dfrac{1}{2}x - 4\right)$.

47. (1) $y = 3x + 2$
   (2) $y = 2x + 3$
Substitute $2x + 3$ for y in equation (1).
   $y = 3x + 2$
   $2x + 3 = 3x + 2$
   $3 = x + 2$
   $x = 1$
Substitute 1 for x in equation (2).

$y = 2x + 3$
$y = 2(1) + 3$
$y = 5$
The solution is $(1, 5)$.

49. (1) $y = 3x + 1$
   (2) $y = 6x - 1$
Substitute $6x - 1$ for y in equation (1).
   $y = 3x + 1$
   $6x - 1 = 3x + 1$
   $3x - 1 = 1$
   $3x = 2$
   $x = \dfrac{2}{3}$
Substitute $\dfrac{2}{3}$ for x in equation (2).
$y = 6x - 1$
$y = 6 \cdot \dfrac{2}{3} - 1$
$y = 3$
The solution is $\left(\dfrac{2}{3}, 3\right)$.

51. The value of $\dfrac{a}{b}$ is $\dfrac{2}{3}$.

Objective C Exercises

53. The interest rates on the two accounts are
5.5% and 7.2%.

55. Strategy: Let x represent the amount
invested at 4.2%.
$2800 is invested at 3.5%.

	Principal	Rate	Interest
Amount at 3.5%	2800	0.035	0.035(2800)
Amount at 4.2%	x	0.042	0.042x

The sum of the interest earned is $329.

Solution: $0.035(2800) + 0.042x = 329$
                $98 + 0.042x = 329$
                $0.042x = 231$
                $x = 5500$
$5500 is invested at 4.2%.

57. Strategy: Let x represent the amount invested at 6.5%.
Let y represent the total amount invested at 5%.
$6000 is invested at 4%.

	Principal	Rate	Interest
Amount at 4%	6000	0.04	0.04(6000)
Amount at 6.5%	x	0.065	0.065x
Total invested	y	0.05	0.05y

The total amount invested is y.
$y = 6000 + x$
The total interest earned is equal to 5% of the total investment.
$0.04(6000) + 0.065x = 0.05y$

Solution: (1) $y = 6000 + x$
(2) $0.04(6000) + 0.065x = 0.05y$
Substitute $6000 + x$ for y in equation (2).
$0.04(6000) + 0.065x = 0.05(6000 + x)$
$240 + 0.065x = 300 + 0.05x$
$240 + 0.015x = 300$
$0.015x = 60$
$x = 4000$
$4000 must be invested at 6.5%.

59. Strategy: Let x represent the amount invested at 3.5%.
Let y represent the amount invested at 4.5%.

	Principal	Rate	Interest
Amount at 3.5%	x	0.035	0.035x
Amount at 4.5%	y	0.045	0.045y

The total amount invested is $42,000.
$x + y = 42,000$
The interest earned from the 3.5% investment is equal to the interest earned from the 4.5% investment.
$0.035x = 0.045y$

Solution: (1) $x + y = 42,000$
(2) $0.035x = 0.045y$
Solve equation (1) for y and substitute for y in equation (2).
$y = 42,000 - x$
$0.035x = 0.045(42,000 - x)$
$0.035x = 1890 - 0.045x$
$0.080x = 1890$

$x = 23,625$
$y = 42,000 - x = 42,000 - 23,625 = 18,375$

$23,625 is invested at 3.5% and $18,375 is invested at 4.5%.

61. Strategy: Let x represent the amount invested at 4.5%.
Let y represent the amount invested at 8%.

	Principal	Rate	Interest
Amount at 4.5%	x	0.045	0.045x
Amount at 8%	y	0.08	0.08y

The total amount invested is $16,000.
$x + y = 16,000$
The total interest earned is $1070.
$0.045x + 0.08y = 1070$

Solution: (1) $x + y = 16,000$
(2) $0.045x + 0.08y = 1070$
Solve equation (1) for y and substitute for y in equation (2).
$y = 16,000 - x$
$0.045x + 0.08(16,000 - x) = 1070$
$0.045x + 1280 - 0.08x = 1070$
$-0.035x + 1280 = 1070$
$-0.035x = -210$
$x = 6000$
$6000 is invested at 4.5%.

Critical Thinking

63. The system of equations will be independent for k equal to any real number except 3.

Projects or Group Activities

65. Student answers will vary. Possible systems of equations for each case are:
a) $2x + y = -1$
   $x + y = 2$
b) $-x + y = 10$
   $-2x + 2y = 20$
c) $2x + y = 0$
   $2x + y = 10$

Section 4.2

Concept Check

1. Student answers may vary. Possible answers are 6 and −5.

Objective A Exercises

3. (1) $x - y = 5$
    (2) $x + y = 7$
    Eliminate y. Add the two equations.
    $2x = 12$
    $x = 6$
    Replace x with 6 in equation (1).
    $x - y = 5$
    $6 - y = 5$
    $-y = -1$
    $y = 1$
    The solution is (6, 1).

5. (1) $3x + y = 4$
    (2) $x + y = 2$
    Eliminate y.
    $3x + y = 4$
    $-1(x + y) = -1(2)$

    $3x + y = 4$
    $-x - y = -2$
    Add the equations.
    $2x = 2$
    $x = 1$
    Replace x with 1 in equation (2).
    $x + y = 2$
    $1 + y = 2$
    $y = 1$
    The solution is (1, 1).

7. (1) $3x + y = 7$
    (2) $x + 2y = 4$
    Eliminate y.
    $-2(3x + y) = -2(7)$
    $x + 2y = 4$

    $-6x - 2y = -14$
    $x + 2y = 4$
    Add the equations.
    $-5x = -10$
    $x = 2$
    Replace x with 2 in equation (2).
    $x + 2y = 4$

$2 + 2y = 4$
$2y = 2$
$y = 1$
The solution is (2, 1).

9. (1) $2x + 3y = -1$
    (2) $x + 5y = 3$
    Eliminate x.
    $2x + 3y = -1$
    $-2(x + 5y) = -2(3)$

    $2x + 3y = -1$
    $-2x - 10y = -6$
    Add the equations.
    $-7y = -7$
    $y = 1$
    Replace y with 1 in equation (2).
    $x + 5y = 3$
    $x + 5(1) = 3$
    $x + 5 = 3$
    $x = -2$
    The solution is (−2, 1).

11. (1) $3x - y = 4$
    (2) $6x - 2y = 8$
    Eliminate y.
    $-2(3x - y) = -2(4)$
    $-6x + 2y = -8$

    $6x - 2y = 8$
    $6x - 2y = 8$
    Add the equations.
    $0 = 0$
    This is a true equation. The equations are dependent. The solutions are the ordered pairs $(x, 3x - 4)$.

13. (1) $2x + 5y = 9$
    (2) $4x - 7y = -16$
    Eliminate x.
    $-2(2x + 5y) = -2(9)$
    $4x - 7y = -16$

    $-4x - 10y = -18$
    $4x - 7y = -16$
    Add the equations.
    $-17y = -34$
    $y = 2$

    Replace y with 2 in equation (1).
    $2x + 5y = 9$

$2x + 5(2) = 9$
$2x + 10 = 9$
$2x = -1$
$x = -\dfrac{1}{2}$

The solution is $\left(-\dfrac{1}{2}, 2\right)$.

15. (1) $4x - 6y = 5$
(2) $2x - 3y = 7$
Eliminate y.
$4x - 6y = 5$
$-2(2x - 3y) = -2(7)$

$4x - 6y = 5$
$-4x + 6y = -14$
Add the equations.
$0 = -9$
This is not a true equation. The system of equations is inconsistent and therefore has no solution.

17. (1) $3x - 5y = 7$
(2) $x - 2y = 3$
Eliminate x.
$3x - 5y = 7$
$-3(x - 2y) = -3(3)$

$3x - 5y = 7$
$-3x + 6y = -9$
Add the equations.
$y = -2$

Replace y with -2 in equation (2).
$x - 2y = 3$
$x - 2(-2) = 3$
$x + 4 = 3$
$x = -1$
The solution is $(-1, -2)$.

19. (1) $x + 3y = 7$
(2) $-2x + 3y = 22$
Eliminate x.
$2(x + 3y) = 2(7)$
$-2x + 3y = 22$

$2x + 6y = 14$
$-2x + 3y = 22$
Add the equations.
$9y = 36$

$y = 4$
Replace y with 4 in equation (1).
$x + 3y = 7$
$x + 3(4) = 7$
$x + 12 = 7$
$x = -5$
The solution is $(-5, 4)$.

21. (1) $3x + 2y = 16$
(2) $2x - 3y = -11$
Eliminate x.
$-2(3x + 2y) = -2(16)$
$3(2x - 3y) = 3(-11)$
$-6x - 4y = -32$
$6x - 9y = -33$
Add the equations,
$-13y = -65$
$y = 5$
Replace y with 5 in equation (1).
$3x + 2y = 16$
$3x + 2(5) = 16$
$3x + 10 = 16$
$3x = 6$
$x = 2$
The solution is $(2, 5)$.

23. (1) $4x + 4y = 5$
(2) $2x - 8y = -5$
Eliminate x.
$4x + 4y = 5$
$-2(2x - 8y) = -2(-5)$
$4x + 4y = 5$
$-4x + 16y = 10$
Add the equations.
$20y = 15$
$y = \dfrac{3}{4}$

Replace y with $\dfrac{3}{4}$ in equation (1).
$4x + 4y = 5$
$4x + 4\left(\dfrac{3}{4}\right) = 5$
$4x + 3 = 5$
$4x = 2$
$x = \dfrac{1}{2}$

The solution is $\left(\dfrac{1}{2}, \dfrac{3}{4}\right)$.

25. (1) $5x + 4y = 0$
(2) $3x + 7y = 0$
Eliminate x.
$-3(5x + 4y) = -3(0)$
$5(3x + 7y) = 5(0)$
$-15x - 12y = 0$
$15x + 35y = 0$
Add the equations.
$23y = 0$
$y = 0$
Replace y with 0 in equation (1).
$5x + 4y = 0$
$5x + 4(0) = 0$
$3x + 0 = 0$
$3x = 0$
$x = 0$
The solution is (0, 0).

27. (1) $5x + 2y = 1$
(2) $2x + 3y = 7$
Eliminate x.
$-2(5x + 2y) = -2(1)$
$5(2x + 3y) = 5(7)$

$-10x - 4y = -2$
$10x + 15y = 35$
Add the equations.
$11y = 33$
$y = 3$
Replace y with 3 in equation (1).
$5x + 2y = 1$
$5x + 2(3) = 1$
$5x + 6 = 1$
$5x = -5$
$x = -1$
The solution is $(-1, 3)$.

29. (1) $3x - 6y = 6$
(2) $9x - 3y = 8$
Eliminate y.
$3x - 6y = 6$
$-2(9x - 3y) = -2(8)$

$3x - 6y = 6$
$-18x + 6y = -16$
Add the equations.
$-15x = -10$
$x = \dfrac{2}{3}$

Replace x with $\dfrac{2}{3}$ in equation (1).

$3x - 6y = 6$
$3\left(\dfrac{2}{3}\right) - 6y = 6$
$2 - 6y = 6$
$-6y = 4$
$y = -\dfrac{2}{3}$

The solution is $\left(\dfrac{2}{3}, -\dfrac{2}{3}\right)$.

31. (1) $\dfrac{3}{4}x + \dfrac{1}{3}y = -\dfrac{1}{2}$
(2) $\dfrac{1}{2}x - \dfrac{5}{6}y = -\dfrac{7}{2}$
Clear the fractions.
$12\left(\dfrac{3}{4}x + \dfrac{1}{3}y\right) = 12\left(-\dfrac{1}{2}\right)$

$6\left(\dfrac{1}{2}x - \dfrac{5}{6}y\right) = 6\left(-\dfrac{7}{2}\right)$

$9x + 4y = -6$
$3x - 5y = -21$
Eliminate x.
$9x + 4y = -6$
$-3(3x - 5y) = -3(-21)$

$9x + 4y = -6$
$-9x + 15y = 63$
Add the equations.
$19y = 57$
$y = 3$
Replace y with 3 in equation (1).
$\dfrac{3}{4}x + \dfrac{1}{3}y = -\dfrac{1}{2}$
$\dfrac{3}{4}x + \dfrac{1}{3}(3) = -\dfrac{1}{2}$
$\dfrac{3}{4}x + 1 = -\dfrac{1}{2}$
$\dfrac{3}{4}x = -\dfrac{3}{2}$
$x = -2$
The solution is $(-2, 3)$.

33. (1) $\dfrac{5x}{6} + \dfrac{y}{3} = \dfrac{4}{3}$

(2) $\dfrac{2x}{3} - \dfrac{y}{2} = \dfrac{11}{6}$

Clear the fractions.

$6\left(\dfrac{5x}{6} + \dfrac{y}{3}\right) = 6\left(\dfrac{4}{3}\right)$

$6\left(\dfrac{2x}{3} - \dfrac{y}{2}\right) = 6\left(\dfrac{11}{6}\right)$

$5x + 2y = 8$
$4x - 3y = 11$
Eliminate y.
$3(5x + 2y) = 3(8)$
$2(4x - 3y) = 2(11)$

$15x + 6y = 24$
$8x - 6y = 22$
Add the equations.
$23x = 46$
$\quad x = 2$
Replace x with 2 in equation (1).
$\dfrac{5x}{6} + \dfrac{y}{3} = \dfrac{4}{3}$

$\dfrac{5(2)}{6} + \dfrac{y}{3} = \dfrac{4}{3}$

$\dfrac{5}{3} + \dfrac{y}{3} = \dfrac{4}{3}$

$\dfrac{y}{3} = -\dfrac{1}{3}$

$y = -1$
The solution is $(2, -1)$.

35. (1) $\dfrac{2x}{5} - \dfrac{y}{2} = \dfrac{13}{2}$

(2) $\dfrac{3x}{4} - \dfrac{y}{5} = \dfrac{17}{2}$

Clear the fractions.

$10\left(\dfrac{2x}{5} - \dfrac{y}{2}\right) = 10\left(\dfrac{13}{2}\right)$

$20\left(\dfrac{3x}{4} - \dfrac{y}{5}\right) = 20\left(\dfrac{17}{2}\right)$

$4x - 5y = 65$
$15x - 4y = 170$

Eliminate y.
$4(4x - 5y) = 4(65)$
$-5(15x - 4y) = -5(170)$

$16x - 20y = 260$
$-75x + 20y = -850$
Add the equations.
$-59x = -590$
$x = 10$
Replace x with 10 in equation (1).
$\dfrac{2x}{5} - \dfrac{y}{2} = \dfrac{13}{2}$

$\dfrac{2(10)}{5} - \dfrac{y}{2} = \dfrac{13}{2}$

$4 - \dfrac{y}{2} = \dfrac{13}{2}$

$-\dfrac{y}{2} = \dfrac{5}{2}$

$y = -5$
The solution is $(10, -5)$.

37. (1) $\dfrac{3x}{2} - \dfrac{y}{4} = -\dfrac{11}{12}$

(2) $\dfrac{x}{3} - y = -\dfrac{5}{6}$

Clear the fractions.

$12\left(\dfrac{3x}{2} - \dfrac{y}{4}\right) = 12\left(-\dfrac{11}{12}\right)$

$6\left(\dfrac{x}{3} - y\right) = 6\left(-\dfrac{5}{6}\right)$

$18x - 3y = -11$
$2x - 6y = -5$
Eliminate y.
$-2(18x - 3y) = -2(-11)$
$2x - 6y = -5$

$-36x + 6y = 22$
$2x - 6y = -5$
Add the equations,
$-34x = 17$

$x = -\dfrac{1}{2}$

Replace x with $-\dfrac{1}{2}$ in equation (1).

$$\frac{3x}{2} - \frac{y}{4} = -\frac{11}{12}$$

$$\frac{3}{2}\left(-\frac{1}{2}\right) - \frac{y}{4} = -\frac{11}{12}$$

$$-\frac{3}{4} - \frac{y}{4} = -\frac{11}{12}$$

$$-\frac{y}{4} = -\frac{1}{6}$$

$$y = \frac{2}{3}$$

The solution is $\left(-\frac{1}{2}, \frac{2}{3}\right)$.

39. (1) $4x - 5y = 3y + 4$
(2) $2x + 3y = 2x + 1$
Write the equations in the form $Ax + By = C$.
Solve the system of equations.
(1) $4x - 8y = 4$
(2) $3y = 1$

Solve equation (2) for y.
$3y = 1$
$$y = \frac{1}{3}$$

Replace y with $\frac{1}{3}$ in equation (1).
$4x - 5y = 3y + 4$
$$4x - 5 \cdot \frac{1}{3} = 3 \cdot \frac{1}{3} + 4$$
$$4x - \frac{5}{3} = 1 + 4$$
$$4x - \frac{5}{3} = 5$$
$$4x = \frac{20}{3}$$
$$x = \frac{5}{3}$$

The solution is $\left(\frac{5}{3}, \frac{1}{3}\right)$.

41. (1) $2x + 5y = 5x + 1$
(2) $3x - 2y = 3y + 3$
Write the equations in the form $Ax + By = C$.
Solve the system of equations.
(1) $-3x + 5y = 1$
(2) $3x - 5y = 3$

Add the equations.
$0 = 4$
This is not a true equation. The system of equations is inconsistent and therefore has no solution.

43. (1) $5x + 2y = 2x + 1$
(2) $2x - 3y = 3x + 2$
Write the equations in the form $Ax + By = C$.
Solve the system of equations.
(1) $3x + 2y = 1$
(2) $-x - 3y = 2$

Eliminate x.
$3x + 2y = 1$
$3(-x - 3y) = 3(2)$

$3x + 2y = 1$
$-3x - 9y = 6$
Add the equations.
$-7y = 7$
$y = -1$
Replace y with -1 in equation (1).
$5x + 2y = 2x + 1$
$5x + 2(-1) = 2x + 1$
$5x - 2 = 2x + 1$
$3x = 3$
$x = 1$
The solution is $(1, -1)$.

Objective B Exercises

45. (1) $x + 2y - z = 1$
(2) $2x - y + z = 6$
(3) $x + 3y - z = 2$
Eliminate z. Add equations (1) and (2).
$x + 2y - z = 1$
$2x - y + z = 6$

(4) $3x + y = 7$

Add equations (2) and (3).
$2x - y + z = 6$
$x + 3y - z = 2$

(5) $3x + 2y = 8$

Use equations (4) and (5) to solve for x and y.
$3x + y = 7$

$3x + 2y = 8$
Eliminate x.
$-1(3x + y) = -1(7)$
$3x + 2y = 8$

$-3x - y = -7$
$3x + 2y = 8$
$y = 1$
Replace y with 1 in equation (4).
$3x + y = 7$
$3x + 1 = 7$
$3x = 6$
$x = 2$
Replace x with 2 and y with 1 in equation
(1).
$x + 2y - z = 1$
$2 + 2(1) - z = 1$
$2 + 2 - z = 1$
$4 - z = 1$
$-z = -3$
$z = 3$
The solution is $(2, 1, 3)$

47. (1) $2x - y + 2z = 7$
(2) $x + y + z = 2$
(3) $3x - y + z = 6$
Eliminate y. Add equations (1) and (2).
$2x - y + 2z = 7$
$x + y + z = 2$

(4) $3x + 3z = 9$
Add equations (2) and (3).
$x + y + z = 2$
$3x - y + z = 6$

(5) $4x + 2z = 8$

Use equations (4) and (5) to solve for x and
z.
$3x + 3z = 9$
$4x + 2z = 8$
Eliminate z.
$-2(3x + 3z) = -2(9)$
$3(4x + 2z) = 3(8)$

$-6x - 6z = -18$
$12x + 6z = 24$
$6x = 6$
$x = 1$
Replace x with 1 in equation (4).
$3x + 3z = 9$

$3(1) + 3z = 9$
$3 + 3z = 9$
$3z = 6$
$z = 2$
Replace x with 1 and z with 2 in equation
(1).
$2x - y + 2z = 7$
$2(1) - y + 2(2) = 7$
$2 - y + 4 = 7$
$6 - y = 7$
$-y = 1$
$y = -1$
The solution is $(1, -1, 2)$.

49. (1) $3x + y = 5$
(2) $3y - z = 2$
(3) $x + z = 5$
Eliminate z. Add equations (2) and (3).
$3y - z = 2$
$x + z = 5$

(4) $x + 3y = 7$

Use equations (1) and (4). Solve for x and y.
$3x + y = 5$
$x + 3y = 7$

Eliminate y.
$-3(3x + y) = -3(5)$
$x + 3y = 7$

$-9x - 3y = -15$
$x + 3y = 7$
$-8x = -8$
$x = 1$
Replace x with 1 in equation (1).
$3x + y = 5$
$3(1) + y = 5$
$3 + y = 5$
$y = 2$
Replace y with 2 in equation (2).
$3y - z = 2$
$3(2) - z = 2$
$6 - z = 2$
$-z = -4$
$z = 4$
The solution is $(1, 2, 4)$.

51. (1) $x - y + z = 1$
(2) $2x + 3y - z = 3$
(3) $-x + 2y - 4z = 4$
Eliminate z. Add equations (1) and (2).
$x - y + z = 1$
$2x + 3y - z = 3$

(4) $3x + 2y = 4$

Multiply equation (1) by 4 and add to equation (3).
$4(x - y + z) = 4(1)$
$-x + 2y - 4z = 4$

$4x - 4y + 4z = 4$
$-x + 2y - 4z = 4$

(5) $3x - 2y = 8$
Use equations (4) and (5) to solve for x and y.
Add equation (4) and (5) to eliminate x.
$3x + 2y = 4$
$3x - 2y = 8$
$6x = 12$
$x = 2$
Replace x with 2 in equation (4).
$3x + 2y = 4$
$3(2) + 2y = 4$
$6 + 2y = 4$
$2y = -2$
$y = -1$

Replace x with 2 and y with -1 in equation (1).
$x - y + z = 1$
$2 - (-1) + z = 1$
$2 + 1 + z = 1$
$3 + z = 1$
$z = -2$
The solution is $(2, -1, -2)$

53. (1) $2x + 3z = 5$
(2) $3y + 2z = 3$
(3) $3x + 4y = -10$
Eliminate z. Use equations (1) and (2).
$2x + 3z = 5$
$3y + 2z = 3$

$-2(2x + 3z) = -2(5)$
$3(3y + 2z) = 3(3)$

$-4x - 6z = -10$
$9y + 6z = 9$

(4) $-4x + 9y = -1$

Use equations (3) and (4) to solve for x and y.
$3x + 4y = -10$
$-4x + 9y = -1$
Eliminate x.
$4(3x + 4y) = 4(-10)$
$3(-4x + 9y) = 3(-1)$

$12x + 16y = -40$
$-12x + 27y = -3$
$43y = -43$
$y = -1$
Replace y with -1 in equation (2).
$3y + 2z = 3$
$3(-1) + 2z = 3$
$-3 + 2z = 3$
$2z = 6$
$z = 3$
Replace z with 3 in equation (1).
$2x + 3z = 5$
$2x + 3(3) = 5$
$2x + 9 = 5$
$2x = -4$
$x = -2$
The solution is $(-2, -1, 3)$.

55. (1) $2x + 4y - 2z = 3$
(2) $x + 3y + 4z = 1$
(3) $x + 2y - z = 4$
Eliminate x. Use equations (1) and (2).
$2x + 4y - 2z = 3$
$x + 3y + 4z = 1$

$2x + 4y - 2z = 3$
$-2(x + 3y + 4z) = -2(1)$

$2x + 4y - 2z = 3$
$-2x - 6y - 8z = -2$

(4) $-2y - 10z = 1$

Use equations (2) and (3).
$x + 3y + 4z = 1$
$x + 2y - z = 4$

$x + 3y + 4z = 1$

$-1(x + 2y - z) = -1(4)$

$x + 3y + 4z = 1$
$-x - 2y + z = -4$

(5) $y + 5z = -3$

Use equations (4) and (5) to solve for y and z.
$-2y - 10z = 1$
$y + 5z = -3$
Eliminate y.
$-2y - 10z = 1$
$2(y + 5z) = 2(-3)$

$-2y - 10z = 1$
$2y + 10z = -6$
$0 = -6$
This is not a true equation. The system of equations is inconsistent and therefore has no solution.

57. (1) $2x + y - z = 5$
  (2) $x + 3y + z = 14$
  (3) $3x - y + 2z = 1$
  Eliminate z. Add equations (1) and (2).
   $2x + y - z = 5$
  $x + 3y + z = 14$

  (4) $3x + 4y = 19$

Use equations (2) and (3).
$x + 3y + z = 14$
$3x - y + 2z = 1$

$-2(x + 3y + z) = -2(14)$
$3x - y + 2z = 1$

$-2x - 6y - 2z = -28$
$3x - y + 2z = 1$

(5) $x - 7y = -27$

Use equations (4) and (5) to solve for x and y.
$3x + 4y = 19$
$x - 7y = -27$
Eliminate x.
$3x + 4y = 19$
$-3(x - 7y) = -3(-27)$

$3x + 4y = 19$
$-3x + 21y = 81$
$25y = 100$
$y = 4$
Replace y with 4 in equation (4).
$3x + 4y = 19$
$3x + 4(4) = 19$
$3x + 16 = 19$
$3x = 3$
$x = 1$
Replace x with 1 and y with 4 in equation (1).
$2x + y - z = 5$
$2(1) + 4 - z = 5$
$2 + 4 - z = 5$
$6 - z = 5$
$-z = -1$
$z = 1$
The solution is (1, 4, 1).

59. (1) $3x + y - 2z = 2$
  (2) $x + 2y + 3z = 13$
  (3) $2x - 2y + 5z = 6$
  Eliminate x. Add equations (1) and (2).
   $3x + y - 2z = 2$
  $x + 2y + 3z = 13$

  $3x + y - 2z = 2$
  $-3(x + 2y + 3z) = -3(13)$

  $3x + y - 2z = 2$
  $-3x - 6y - 9z = -39$

  (4) $-5y - 11z = -37$

Use equations (2) and (3).
$x + 2y + 3z = 13$
$2x - 2y + 5z = 6$

$-2(x + 2y + 3z) = -2(13)$
$2x - 2y + 5z = 6$

$-2x - 4y - 6z = -26$
$2x - 2y + 5z = 6$

(5) $-6y - z = -20$

Use equations (4) and (5) to solve for y and z.
$-5y - 11z = -37$
$-6y - z = -20$

Eliminate z.

$-5y - 11z = -37$

$-11(-6y - z) = -11(-20)$

$-5y - 11z = -37$

$66y + 11z = 220$

$61y = 183$

$y = 3$

Replace y with 3 in equation (4).

$-5y - 11z = -37$

$-5(3) - 11z = -37$

$-15 - 11z = -37$

$-11z = -22$

$z = 2$

Replace y with 3 and z with 2 in equation (1).

$3x + y - 2z = 2$

$3x + 3 - 2(2) = 2$

$3x + 3 - 4 = 2$

$3x - 1 = 2$

$3x = 3$

$x = 1$

The solution is $(1, 3, 2.)$

61. (1) $2x - y + z = 6$

(2) $3x + 2y + z = 4$

(3) $x - 2y + 3z = 12$

Eliminate y. Use equations (1) and (2).

$2x - y + z = 6$

$3x + 2y + z = 4$

$2(2x - y + z) = 2(6)$

$3x + 2y + z = 4$

$4x - 2y + 2z = 12$

$3x + 2y + z = 4$

(4) $7x + 3z = 16$

Add equations (2) and (3).

$3x + 2y + z = 4$

$x - 2y + 3z = 12$

(5) $4x + 4z = 16$

Use equations (4) and (5) to solve for x and z.

$7x + 3z = 16$

$4x + 4z = 16$

Eliminate z.

$4(7x + 3z) = 4(16)$

$-3(4x + 4z) = -3(16)$

$28x + 12z = 64$

$-12x - 12z = -48$

$16x = 16$

$x = 1$

Replace x with 1 in equation (4).

$7x + 3z = 16$

$7(1) + 3z = 16$

$7 + 3z = 16$

$3z = 9$

$z = 3$

Replace x with 1 and z with 3 in equation (1).

$2x - y + z = 6$

$2(1) - y + 3 = 6$

$-y + 5 = 6$

$-y = 1$

$y = -1$

The solution is $(1, -1, 3)$.

63. (1) $3x - 2y + 3z = -4$

(2) $2x + y - 3z = 2$

(3) $3x + 4y + 5z = 8$

Eliminate y. Use equations (1) and (2).

$3x - 2y + 3z = -4$

$2x + y - 3z = 2$

$3x - 2y + 3z = -4$

$2(2x + y - 3z) = 2(2)$

$3x - 2y + 3z = -4$

$4x + 2y - 6z = 4$

(4) $7x - 3z = 0$

Use equations (2) and (3).

$2x + y - 3z = 2$

$3x + 4y + 5z = 8$

$-4(2x + y - 3z) = -4(2)$

$3x + 4y + 5z = 8$

$-8x - 4y + 12z = -8$

$3x + 4y + 5z = 8$

(5) $-5x + 17z = 0$

Use equations (4) and (5) solve for x and z.

$7x - 3z = 0$

$-5x + 17z = 0$

Eliminate x.

$5(7x - 3z) = 5(0)$

$7(-5x + 17z) = 7(0)$

$35x - 15z = 0$

$-35x + 119z = 0$

$104z = 0$

$z = 0$

Replace z with 0 in equation (4).

$7x - 3z = 0$

$7x - 3(0) = 0$

$7x = 0$

$x = 0$

Replace x with 0 and z with 0 in equation (1).

$3x - 2y + 3z = -4$

$3(0) - 2y + 3(0) = -4$

$0 - 2y + 0 = -4$

$-2y = -4$

$y = 2$

The solution is (0, 2, 0).

65. (1) $3x - y + 2z = 2$

(2) $4x + 2y - 7z = 0$

(3) $2x + 3y - 5z = 7$

Eliminate y. Use equations (1) and (2).

$3x - y + 2z = 2$

$4x + 2y - 7z = 0$

$2(3x - y + 2z) = 2(2)$

$4x + 2y - 7z = 0$

$6x - 2y + 4z = 4$

$4x + 2y - 7z = 0$

(4) $10x - 3z = 4$

Use equations (1) and (3).

$3x - y + 2z = 2$

$2x + 3y - 5z = 7$

$3(3x - y + 2z) = 3(2)$

$2x + 3y - 5z = 7$

$9x - 3y + 6z = 6$

$2x + 3y - 5z = 7$

(5) $11x + z = 13$

Use equations (4) and (5) to solve for x and z.

$10x - 3z = 4$

$11x + z = 13$

Eliminate z.

$10x - 3z = 4$

$3(11x + z) = 3(13)$

$10x - 3z = 4$

$33x + 3z = 39$

$43x = 43$

$x = 1$

Replace x with 1 in equation (4).

$10x - 3z = 4$

$10(1) - 3z = 4$

$-3z = -6$

$z = 2$

Replace x with 1 and z with 2 in equation (1).

$3x - y + 2z = 2$

$3(1) - y + 2(2) = 2$

$3 - y + 4 = 2$

$7 - y = 2$

$-y = -5$

$y = 5$

The solution is (1, 5, 2).

67. (1) $2x - 3y + 7z = 0$

(2) $x + 4y - 4z = -2$

(3) $3x + 2y + 5z = 1$

Eliminate x. Use equations (1) and (2).

$2x - 3y + 7z = 0$

$x + 4y - 4z = -2$

$2x - 3y + 7z = 0$

$-2(x + 4y - 4z) = -2(-2)$

$2x - 3y + 7z = 0$

$-2x - 8y + 8z = 4$

(4) $-11y + 15z = 4$

Use equations (2) and (3).

$x + 4y - 4z = -2$

$3x + 2y + 5z = 1$

$-3(x + 4y - 4z) = -3(-2)$

$3x + 2y + 5z = 1$

$-3x - 12y + 12z = 6$

$3x + 2y + 5z = 1$

(5) $-10y + 17z = 7$

Use equations (4) and (5) to solve for y and z.

$-11y + 15z = 4$

$-10y + 17z = 7$

Eliminate y.

$10(-11y + 15z) = 10(4)$

$-11(-10y + 17z) = -11(7)$

$-110y + 150z = 40$

$110y - 187z = -77$

$-37z = -37$

$z = 1$

Replace z with 1 in equation (4).

$-11y + 15z = 4$

$-11y + 15(1) = 4$

$-11y = -11$

$y = 1$

Replace y with 1 and z with 1 in equation (1).

$2x - 3y + 7z = 0$

$2x - 3(1) + 7(1) = 0$

$2x - 3 + 7 = 0$

$2x + 4 = 0$

$2x = -4$

$x = -2$

The solution is $(-2, 1, 1)$.

69. a) (iii) no points

b) (ii) more than one point

c) (i) exactly one point

Critical Thinking

71. (1) $\dfrac{1}{x} - \dfrac{2}{y} = 3$

(2) $\dfrac{2}{x} + \dfrac{3}{y} = -1$

Clear the fractions.

$xy\left(\dfrac{1}{x} - \dfrac{2}{y}\right) = xy(3)$

$xy\left(\dfrac{2}{x} + \dfrac{3}{y}\right) = xy(-1)$

$y - 2x = 3xy$

$2y + 3x = -xy$

Eliminate y.

$-2(y - 2x) = -2(3xy)$

$2y + 3x = -xy$

$-2y + 4x = -6xy$

$2y + 3x = -xy$

$7x = -7xy$

$y = -1$

Replace y with -1 in equation (1).

$\dfrac{1}{x} - \dfrac{2}{y} = 3$

$\dfrac{1}{x} - \dfrac{2}{-1} = 3$

$\dfrac{1}{x} + 2 = 3$

$\dfrac{1}{x} = 1$

$x = 1$

The solution is $(1, -1)$.

73. (1) $\dfrac{3}{x} + \dfrac{2}{y} = 1$

(2) $\dfrac{2}{x} + \dfrac{4}{y} = -2$

Clear the fractions.

$xy\left(\dfrac{3}{x} + \dfrac{2}{y}\right) = xy(1)$

$xy\left(\dfrac{2}{x} + \dfrac{4}{y}\right) = xy(-2)$

$3y + 2x = xy$

$2y + 4x = -2xy$

Eliminate x.

$-2(3y + 2x) = -2(xy)$

$2y + 4x = -2xy$

$-6y - 4x = -2xy$

$2y + 4x = -2xy$

$-4y = -4xy$

$x = 1$

Replace x with 1 in equation (1).

$\dfrac{3}{x} + \dfrac{2}{y} = 1$

$$\frac{3}{1} + \frac{2}{y} = 1$$

$$\frac{2}{y} = -2$$

$$y = -1$$

The solution is $(1, -1)$.

75. $P(3, -2, 4)$

$$Ax + 3y + 2z = 8$$
$$A(3) + 3(-2) + 2(4) = 8$$
$$3A + 2 = 8$$
$$3A = 6$$
$$A = 2$$

$$2x + By - 3z = -12$$
$$2(3) + B(-2) - 3(4) = -12$$
$$-2B - 6 = -12$$
$$-2B = -6$$
$$B = 3$$

$$3x - 2y + Cz = 1$$
$$3(3) - 2(-2) + C(4) = 1$$
$$13 + 4C = 1$$
$$4C = -12$$
$$C = -3$$

Projects or Group Activities

77. a) The graph would be a plane which is parallel to the yz axis.
b) The graph would be a plane which is parallel to the xz axis.
c) The graph would be a plane which is parallel to the xy axis.
d) The graph would be a plane which is perpendicular to the xy plane.

Section 4.3

Concept Check

1. $(-1)\begin{bmatrix} 3 & -2 \\ -3 & 6 \end{bmatrix}$

Objective A Exercises

3. $\begin{vmatrix} 2 & -1 \\ 3 & 4 \end{vmatrix} = 2(4) - 3(-1) = 8 + 3 = 11$

5. $\begin{vmatrix} 6 & -2 \\ -3 & 4 \end{vmatrix} = 6(4) - (-3)(-2) = 24 - 6 = 18$

7. $\begin{vmatrix} 3 & 6 \\ 2 & 4 \end{vmatrix} = 3(4) - 2(6) = 12 - 12 = 0$

9. $\begin{vmatrix} 1 & -1 & 2 \\ 3 & 2 & 1 \\ 1 & 0 & 4 \end{vmatrix} = 1\begin{vmatrix} 2 & 1 \\ 0 & 4 \end{vmatrix} + 1\begin{vmatrix} 3 & 1 \\ 1 & 4 \end{vmatrix} + 2\begin{vmatrix} 3 & 2 \\ 1 & 0 \end{vmatrix}$

$$= 1(8 - 0) + 1(12 - 1) + 2(0 - 2)$$
$$= 8 + 11 - 4$$
$$= 15$$

11. $\begin{vmatrix} 3 & -1 & 2 \\ 0 & 1 & 2 \\ 3 & 2 & -2 \end{vmatrix} = 3\begin{vmatrix} 1 & 2 \\ 2 & -2 \end{vmatrix} + 1\begin{vmatrix} 0 & 2 \\ 3 & -2 \end{vmatrix} + 2\begin{vmatrix} 0 & 1 \\ 3 & 2 \end{vmatrix}$

$$= 3(-2 - 4) + 1(0 - 6) + 2(0 - 3)$$
$$= -18 - 6 - 6$$
$$= -30$$

13. $\begin{vmatrix} 4 & 2 & 6 \\ -2 & 1 & 1 \\ 2 & 1 & 3 \end{vmatrix} = 4\begin{vmatrix} 1 & 1 \\ 1 & 3 \end{vmatrix} - 2\begin{vmatrix} -2 & 1 \\ 2 & 3 \end{vmatrix} + 6\begin{vmatrix} -2 & 1 \\ 2 & 1 \end{vmatrix}$

$$= 4(3 - 1) - 2(-6 - 2) + 6(-2 - 2)$$
$$= 8 + 16 - 24$$
$$= 0$$

15. If one row of a matrix is all zeros, the value of the determinant will be 0.

Objective B Exercises

17. $2x - 5y = 26$
$5x + 3y = 3$

$$D = \begin{vmatrix} 2 & -5 \\ 5 & 3 \end{vmatrix} = 31$$

$$D_x = \begin{vmatrix} 26 & -5 \\ 3 & 3 \end{vmatrix} = 93, \; D_y = \begin{vmatrix} 2 & 26 \\ 5 & 3 \end{vmatrix} = -124$$

$$x = \frac{D_x}{D} = \frac{93}{31} = 3, \; y = \frac{D_y}{D} = \frac{-124}{31} = -4$$

The solution is $(3, -4)$.

19. $x - 4y = 8$
$3x + 7y = 5$

$$D = \begin{vmatrix} 1 & -4 \\ 3 & 7 \end{vmatrix} = 19$$

$$D_x = \begin{vmatrix} 8 & -4 \\ 5 & 7 \end{vmatrix} = 76, \; D_y = \begin{vmatrix} 1 & 8 \\ 3 & 5 \end{vmatrix} = -19$$

$$x = \frac{D_x}{D} = \frac{76}{19} = 4, \; y = \frac{D_y}{D} = \frac{-19}{19} = -1$$

The solution is $(4, -1)$.

21. $2x + 3y = 4$
$6x - 12y = -5$

$$D = \begin{vmatrix} 2 & 3 \\ 6 & -12 \end{vmatrix} = -42$$

$$D_x = \begin{vmatrix} 4 & 3 \\ -5 & -12 \end{vmatrix} = -33,$$

$$D_y = \begin{vmatrix} 2 & 4 \\ 6 & -5 \end{vmatrix} = -34$$

$$x = \frac{D_x}{D} = \frac{-33}{-42} = \frac{11}{14},$$

$$y = \frac{D_y}{D} = \frac{-34}{-42} = \frac{17}{21}$$

The solution is $\left(\dfrac{11}{14}, \dfrac{17}{21} \right)$.

23. $2x + 5y = 6$
$6x - 2y = 1$

$$D = \begin{vmatrix} 2 & 5 \\ 6 & -2 \end{vmatrix} = -34$$

$$D_x = \begin{vmatrix} 6 & 5 \\ 1 & -2 \end{vmatrix} = -17, \; D_y = \begin{vmatrix} 2 & 6 \\ 6 & 1 \end{vmatrix} = -34$$

$$x = \frac{D_x}{D} = \frac{-17}{-34} = \frac{1}{2}, \; y = \frac{D_y}{D} = \frac{-34}{-34} = 1$$

The solution is $\left(\dfrac{1}{2}, 1 \right)$.

25. $-2x + 3y = 7$
$4x - 6y = -5$

$$D = \begin{vmatrix} -2 & 3 \\ 4 & -6 \end{vmatrix} = 0$$

Since $D = 0$, $\dfrac{D_x}{D}$ is undefined. Therefore, the system of equations does not have a unique solution. The equations are not independent.

27. $2x - 5y = -2$
$3x - 7y = -3$

$$D = \begin{vmatrix} 2 & -5 \\ 3 & -7 \end{vmatrix} = 1$$

$$D_x = \begin{vmatrix} -2 & -5 \\ -3 & -7 \end{vmatrix} = -1, \; D_y = \begin{vmatrix} 2 & -2 \\ 3 & -3 \end{vmatrix} = 0$$

$$x = \frac{D_x}{D} = \frac{-1}{1} = -1, \; y = \frac{D_y}{D} = \frac{0}{1} = 0$$

The solution is $(-1, 0)$.

29. $2x - y + 3z = 9$
$x + 4y + 4z = 5$
$3x + 2y + 2z = 5$

$$D = \begin{vmatrix} 2 & -1 & 3 \\ 1 & 4 & 4 \\ 3 & 2 & 2 \end{vmatrix} = -40$$

$$D_x = \begin{vmatrix} 9 & -1 & 3 \\ 5 & 4 & 4 \\ 5 & 2 & 2 \end{vmatrix} = -40,$$

$$D_y = \begin{vmatrix} 2 & 9 & 3 \\ 1 & 5 & 4 \\ 3 & 5 & 2 \end{vmatrix} = 40$$

$$D_z = \begin{vmatrix} 2 & -1 & 9 \\ 1 & 4 & 5 \\ 3 & 2 & 5 \end{vmatrix} = -80$$

$$x = \frac{D_x}{D} = \frac{-40}{-40} = 1, \ y = \frac{D_y}{D} = \frac{40}{-40} = -1$$

$$z = \frac{D_z}{D} = \frac{-80}{-40} = 2$$

The solution is $(1, -1, 2)$.

31. $3x - y + z = 11$
   $x + 4y - 2z = -12$
   $2x + 2y - z = -3$

$$D = \begin{vmatrix} 3 & -1 & 1 \\ 1 & 4 & -2 \\ 2 & 2 & -1 \end{vmatrix} = -3$$

$$D_x = \begin{vmatrix} 11 & -1 & 1 \\ -12 & 4 & -2 \\ -3 & 2 & -1 \end{vmatrix} = -6,$$

$$D_y = \begin{vmatrix} 3 & 11 & 1 \\ 1 & -12 & -2 \\ 2 & -3 & -1 \end{vmatrix} = 6$$

$$D_z = \begin{vmatrix} 3 & -1 & 11 \\ 1 & 4 & -12 \\ 2 & 2 & -3 \end{vmatrix} = -9$$

$$x = \frac{D_x}{D} = \frac{-6}{-3} = 2, \ y = \frac{D_y}{D} = \frac{6}{-3} = -2$$

$$z = \frac{D_z}{D} = \frac{-9}{-3} = 3$$

The solution is $(2, -2, 3)$.

33. $4x - 2y + 6z = 1$
   $3x + 4y + 2z = 1$
   $2x - y + 3z = 2$

$$D = \begin{vmatrix} 4 & -2 & 6 \\ 3 & 4 & 2 \\ 2 & -1 & 3 \end{vmatrix} = 0$$

Since $D = 0$, $\dfrac{D_x}{D}$ is undefined. Therefore, the system of equations does not have a unique solution. The equations are not independent.

35. No, Cramer's Rule cannot be used to solve a dependent system of equations.

Critical Thinking

37. a) Sometimes true
   b) Always true
   c) Sometimes true

Projects or Group Activities

39. $A = \dfrac{1}{2}\left\{\begin{vmatrix} 9 & 26 \\ -3 & 6 \end{vmatrix} + \begin{vmatrix} 26 & 18 \\ 6 & 21 \end{vmatrix} + \begin{vmatrix} 18 & 16 \\ 21 & 10 \end{vmatrix} + \begin{vmatrix} 16 & 1 \\ 10 & 11 \end{vmatrix} + \begin{vmatrix} 1 & 9 \\ 11 & -3 \end{vmatrix}\right\}$

$A = \dfrac{1}{2}(132 + 438 - 156 + 166 - 102)$

$A = \dfrac{1}{2}(478)$

$= 239 \text{ ft}^2$

Check Your Progress: Chapter 4

1.

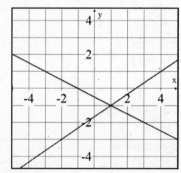

The solution is $(1, -1)$.

2.

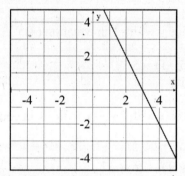

The solution is $(x, -2x + 6)$.

3. (1) $\qquad 4x - 2y = 16$
(2) $\qquad\quad 3x - y = 11$
Solve equation (2) for y.
$3x - y = 11$
$\qquad y = 3x - 11$

Substitute in equation (1).
$\qquad\quad 4x - 2y = 16$
$\quad 4x - 2(3x - 11) = 16$
$\qquad 4x - 6x + 22 = 16$

$\qquad\qquad -2x = -6$
$\qquad\qquad\quad x = 3$
Substitute 6 for x in equation (2).
$3x - y = 11$
$3(3) - y = 11$
$9 - y = 11$
$y = -2$
The solution is $(3, -2)$.

4. (1) $\qquad\quad 9x + 12y = 11$
(2) $\qquad\quad\; 6x + 8y = 9$
Solve equation (2) for x.
$6x + 8y = 9$
$6x = 9 - 8y$
$x = \dfrac{3}{2} - \dfrac{4}{3}y$

Substitute in equation (1).
$\qquad\quad 9x + 12y = 11$
$9\left(\dfrac{3}{2} - \dfrac{4}{3}y\right) + 12y = 11$

$\dfrac{27}{2} - 12y + 12y = 11$

$\dfrac{27}{2} \neq 11$

There is no solution.

5. (1) $3x + 5y = 14$
   (2) $-2x + 3y = 16$
   Eliminate x.
   $2(3x + 5y) = 2(14)$
   (3) $6x + 10y = 28$

   $3(-2x + 3y) = 3(16)$
   (4) $-6x + 9y = 48$

   Add equations (3) and (4).
   $6x + 10y = 28$
   $-6x + 9y = 48$
   $19y = 76$
   $y = 4$

   Substitute the value for y in equation (1)
   $3x + 5y = 14$
   $3x + 5(4) = 14$
   $3x = -6$
   $x = -2$
   The solution is $(-2, 4)$.

6. (1) $2x + 5y = 10$
   (2) $4x + 10y = 20$
   Eliminate x.
   $-2(2x + 5y) = -2(10)$
   $-4x - 10y = -20$

   Add the equations.
   $-4x - 10y = -20$
   $4x + 10y = 20$
   $ 0 = 0$
   This is a true equation. The equations are dependent. The solutions are the ordered

   pairs $\left(x, -\dfrac{2}{5}x + 2 \right)$.

7. (1) $x + 3y - 2z = -7$
   (2) $2x + y + z = 6$
   (3) $-3x - y + 3z = 5$
   Eliminate z. Use equations (1) and (2).
   $x + 3y - 2z = -7$
   $2x + y + z = 6$

   $2(2x + y + z) = 2(6)$
   $4x + 2y + 2z = 12$

   $x + 3y - 2z = -7$
   $4x + 2y + 2z = 12$
   (4) $5x + 5y = 5$

Use equations (2) and (3).
$2x + y + z = 6$
$-3x - y + 3z = 4$

$-3(2x + y + z) = -3(6)$
$-6x - 3y - 3z = -18$

$-6x - 3y - 3z = -18$
$-3x - y + 3z = 4$
(5) $-9x - 4y = -14$

Use equations (4) and (5) to solve for x and y.
$5x + 5y = 5$
$-9x - 4y = -14$
Eliminate x.
$5x + 5y = 5$
$9(5x + 5y) = 9(5)$
$45x + 45y = 45$

$-9x - 4y = -13$
$5(-9x - 4y) = 5(-14)$
$-45x - 20y = -70$
Add the equations
$45x + 45y = 45$
$-45x - 20y = -70$
$25y = -25$
$y = -1$
Replace y with -1 in equation (4).
$5x + 5y = 5$
$5x + 5(-1) = 5$
$5x = 10$
$x = 2$
Replace x with 2 and y with -1 in equation (1).
$x + 3y - 2z = -7$
$2 + 3(-1) - 2z = -7$
$-1 - 2z = -7$
$z = 3$
The solution is $(2, -1, 3)$.

8. (1) $4x + 5y - z = 22$
   (2) $3x - 6y + 2z = -28$
   (3) $x + 2y - 2z = 12$
   Eliminate z. Use equations (1) and (2).
   $4x + 5y - z = 22$
   $3x - 6y + 2z = -28$

   $2(4x + 5y - z) = 2(22)$
   $8x + 10y - 2z = 44$

$8x + 10y - 2z = 44$
$3x - 6y + 2z = -28$

(4) $11x + 4y = 16$

Use equations (2) and (3).
$3x - 6y + 2z = -28$
$x + 2y - 2z = 12$

(5) $4x - 4y = -16$

Use equations (4) and (5) to solve for x and y.
$11x + 4y = 16$
$4x - 4y = -16$
Eliminate y.
$15x = 0$
$x = 0$

Replace x with 0 in equation (5).
$4x - 4y = -16$
$4(0) - 4y = -16$
$-4y = -16$
$y = 4$
Replace x with 0 and y with 4 in equation (1).
$4x + 5y - z = 22$
$4(0) + 5(4) - z = 22$
$20 - z = 22$
$z = -2$
The solution is $(0, 4, -2)$.

9. $\begin{vmatrix} 5 & -3 \\ 2 & -1 \end{vmatrix} = 5(-1) - 2(-3) = -5 + 6 = 1$

10. $\begin{vmatrix} 1 & -2 & 3 \\ 4 & -2 & 1 \\ 3 & 6 & 5 \end{vmatrix} = 1\begin{vmatrix} -2 & 1 \\ 6 & 5 \end{vmatrix} + 2\begin{vmatrix} 4 & 1 \\ 3 & 5 \end{vmatrix} + 3\begin{vmatrix} 4 & -2 \\ 3 & 6 \end{vmatrix}$

$= 1(-10 - 6) + 2(20 - 3) + 3(24 + 6)$
$= -16 + 34 + 90$
$= 108$

11. $3x + 5y = 9$
$2x - 3y = 7$

$D = \begin{vmatrix} 3 & 5 \\ 2 & -3 \end{vmatrix} = -19$

$D_x = \begin{vmatrix} 9 & 5 \\ 7 & -3 \end{vmatrix} = -62, D_y = \begin{vmatrix} 3 & 9 \\ 2 & 7 \end{vmatrix} = 3$

$x = \dfrac{D_x}{D} = \dfrac{62}{19}, \ y = \dfrac{D_y}{D} = -\dfrac{3}{19}$

The solution is $\left(\dfrac{62}{19}, -\dfrac{3}{19} \right)$.

12. $x - 3y + 2z = -2$
$2x + y - z = 4$
$x - y - 5z = 17$

$D = \begin{vmatrix} 1 & -3 & 2 \\ 2 & 1 & -1 \\ 1 & -1 & -5 \end{vmatrix} = -39$

$D_x = \begin{vmatrix} -2 & -3 & 2 \\ 4 & 1 & -1 \\ 17 & -1 & -5 \end{vmatrix} = -39,$

$D_y = \begin{vmatrix} 1 & -2 & 2 \\ 2 & 4 & -1 \\ 1 & 17 & -5 \end{vmatrix} = 39$

$D_z = \begin{vmatrix} 1 & -3 & -2 \\ 2 & 1 & 4 \\ 1 & -1 & 17 \end{vmatrix} = 117$

$x = \dfrac{D_x}{D} = \dfrac{-39}{-39} = 1, \ y = \dfrac{D_y}{D} = \dfrac{39}{-39} = -1$

$z = \dfrac{D_z}{D} = \dfrac{117}{-39} = -3$

The solution is $(1, -1, -3)$.

13. Strategy: Let x represent the amount invested at 6%.
Let y represent the amount invested at 4.5%.

	Principal	Rate	Interest
Amount at 6%	x	0.06	0.06x
Amount at 4.5%	y	0.045	0.045y

The total amount invested is $20,000.
$x + y = 20{,}000$
The sum of the interest earned is $1080.
$0.06x + 0.045y = 1080$

Solution: (1) $x + y = 20{,}000$
(2) $0.06x + 0.045y = 1080$
Solve equation (1) for y and substitute for y in equation (2).
$y = 20000 - x$
$0.06x + 0.045(20000 - x) = 1080$
$0.06x + 900 - 0.045x = 1080$
$0.015x + 900 = 1080$
$0.015x = 180$
$x = 12000$
$y = 20000 - x = 20000 - 12000 = 8000$

$12,000 is invested at 6% and $8000 is invested at 4.5%.

Section 4.4

Concept Check

1. 450 mph

3. $50x + 100y$

Objective A Exercises

5. n is less than m.

7. Strategy: Let x represent the rate of the motorboat in calm water.
The rate of the current is y.

	Rate	Time	Distance
with current	$x + y$	2	2($x + y$)
against current	$x - y$	3	3($x - y$)

The distance traveled with the current is 36 mi. The distance traveled against the current

is 36 mi.
$2(x + y) = 36$
$3(x - y) = 36$

Solution: $2(x + y) = 36$
$3(x - y) = 36$
$\dfrac{1}{2} \cdot 2(x + y) = \dfrac{1}{2} \cdot 36$
$\dfrac{1}{3} \cdot 3(x - y) = \dfrac{1}{3} \cdot 36$
$x + y = 18$
$x - y = 12$
$2x = 30$
$x = 15$
$x + y = 18$
$15 + y = 18$
$y = 3$
The rate of the motorboat in calm water is 15 mph. The rate of the current is 3 mph.

9. Strategy: Let p represent the rate of the plane in calm air.
The rate of the wind is w.

	Rate	Time	Distance
with wind	$p + w$	4	4($p + w$)
against wind	$p - w$	4	4($p - w$)

The distance traveled with the wind is 2200 mi. The distance traveled against the wind is 1820 mi.
$4(p + w) = 2200$
$4(p - w) = 1820$
Solution: $4(p + w) = 2200$
$4(p - w) = 1820$
$\dfrac{1}{4} \cdot 4(p + w) = \dfrac{1}{4} \cdot 2200$
$\dfrac{1}{4} \cdot 4(p - w) = \dfrac{1}{4} \cdot 1820$
$p + w = 550$
$p - w = 455$
$2p = 1005$
$p = 502.5$
$p + w = 550$
$502.5 + w = 550$
$w = 47.5$

The rate of the plane in calm air is 502.5 mph. The rate of the wind is 47.5 mph.

11. Strategy: Let x represent the rate of the team in calm water.
The rate of the current is y.

	Rate	Time	Distance
with current	$x + y$	2	$2(x + y)$
against current	$x - y$	2	$2(x - y)$

The distance traveled with the current is 20 km. The distance traveled against the current is 12 km.
$2(x + y) = 20$
$2(x - y) = 12$

Solution: $2(x + y) = 20$
$\quad\quad\quad\quad 2(x - y) = 12$

$\dfrac{1}{2} \cdot 2(x + y) = \dfrac{1}{2} \cdot 20$

$\dfrac{1}{2} \cdot 2(x - y) = \dfrac{1}{2} \cdot 12$

$x + y = 10$
$x - y = 6$
$2x = 16$
$x = 8$

$x + y = 10$
$8 + y = 10$
$y = 2$

The rate of the team in calm water is 8 km/h.
The rate of the current is 2 km/h.

13. Strategy: Let x represent the rate of the plane in calm air.
The rate of the wind is y.

	Rate	Time	Distance
with wind	$x + y$	4	$4(x + y)$
against wind	$x - y$	5	$5(x - y)$

The distance traveled with the wind is 800 mi. The distance traveled against the wind is 800 mi.
$4(x + y) = 800$
$5(x - y) = 800$

Solution: $4(x + y) = 800$
$\quad\quad\quad\quad 5(x - y) = 800$

$\dfrac{1}{4} \cdot 4(x + y) = \dfrac{1}{4} \cdot 800$

$\dfrac{1}{5} \cdot 5(x - y) = \dfrac{1}{5} \cdot 800$

$x + y = 200$
$x - y = 160$
$2x = 360$
$x = 180$

$x + y = 200$
$180 + y = 200$
$y = 20$

The rate of the plane in calm air is 180 mph.
The rate of the wind is 20 mph.

15. Strategy: Let x represent the rate of the plane in calm air.
The rate of the wind is y.

	Rate	Time	Distance
with wind	$x + y$	5	$5(x + y)$
against wind	$x - y$	6	$6(x - y)$

The distance traveled with the wind is 600 mi. The distance traveled against the wind is 600 mi.
$5(x + y) = 800$
$6(x - y) = 800$

Solution: $5(x + y) = 600$
$\quad\quad\quad\quad 6(x - y) = 600$

$\dfrac{1}{5} \cdot 5(x + y) = \dfrac{1}{5} \cdot 600$

$\dfrac{1}{6} \cdot 6(x - y) = \dfrac{1}{6} \cdot 600$

$x + y = 120$
$x - y = 100$
$2x = 220$
$x = 110$

$x + y = 120$
$110 + y = 120$
$y = 10$

The rate of the plane in calm air is 110 mph.
The rate of the wind is 10 mph.

Objective B Exercises

17. The cost per pound of dark roast coffee is greater than the cost per pound of light roast coffee.

19. Strategy: Let x represent the cost of redwood.
The cost of pine is y.
First purchase:

	Amount	Rate	Total Value
Redwood	60	x	$60x$
Pine	80	y	$80y$

Second purchase:

	Amount	Rate	Total Value
Redwood	100	x	$100x$
Pine	60	y	$60y$

The first purchase costs $286. The second purchase costs $396.

$60x + 80y = 286$
$100x + 60y = 396$

Solution: $60x + 80y = 286$
$\qquad\qquad 100x + 60y = 396$

$3(60x + 80y) = 3(286)$
$-4(100x + 60y) = -4(396)$

$180x + 240y = 858$
$-400x - 240y = -1584$

$-220x = -726$
$x = 3.3$

$60x + 80y = 286$
$60(3.3) + 80y = 286$
$198 + 80y = 286$
$80y = 88$
$y = 1.1$
The cost of the pine is $1.10/ft. The cost of the redwood is $3.30/ft.

21. Strategy: Let x represent the cost of nylon carpet.
The cost of wool carpet is y.
First purchase:

	Amount	Rate	Total Cost
Nylon	16	x	$16x$
Wood	20	y	$20y$

Second purchase:

	Amount	Rate	Total Cost
Nylon	18	x	$18x$
Wool	25	y	$25y$

The first purchase costs $1840. The second purchase costs $2200.

$16x + 20y = 1840$
$18x + 25y = 2200$

Solution: $16x + 20y = 1840$
$\qquad\qquad 18x + 25y = 2200$

$5(16x + 20y) = 5(1840)$
$-4(18x + 25y) = -4(2200)$

$80x + 100y = 9200$
$-72x - 100y = -8800$

$8x = 400$
$x = 50$

$16x + 20y = 1840$
$16(50) + 20y = 1840$
$800 + 20y = 1840$
$20y = 1040$
$y = 52$

The cost of the wool carpet is $52/yd.

23. Strategy: Let m represent the number of mountain bikes to be manufactured. The number of trail bikes to be manufactured is t.

Cost of materials:

Type	Number	Cost	Total Cost
Mountain	m	70	$70m$
Trail	t	50	$50t$

Cost of labor:

Type	Number	Cost	Total Cost
Mountain	m	80	$80m$
Trail	t	40	$40t$

The company has budgeted $2500 for materials and $2600 for labor.

$70m + 50t = 2500$
$80m + 40t = 2600$

Solution: $70m + 50t = 2500$
$\qquad\quad 80m + 40t = 2600$

$4(70m + 50t) = 4(2500)$
$-5(80m + 40t) = -5(2600)$

$280m + 200t = 10{,}000$
$-400m - 200t = -13{,}000$

$-120m = -3000$
$m = 25$

The company plans to manufacture 25 mountain bikes during the week.

25. Strategy: Let x represent the number of miles driven in the city. The number of miles driven on the highway is $394 - x$.
Cost of hybrid driving:

	Number	Cost	Total Cost
City	x	0.09	$0.09x$
Highway	$394 \; x$	0.08	$0.08(394 - x)$

The total amount spent on gasoline was $34.74.

Solution: $0.09x + 0.08(394 - x) = 34.74$
$\qquad\quad 0.09x + 31.52 - 0.08x = 34.74$
$\qquad\qquad\qquad\qquad\quad 0.01x = 3.22$
$\qquad\qquad\qquad\qquad\qquad\; x = 322$

$394 - x = 394 - 322 = 72$

The owner drives 322 miles in the city and 72 on the highway.

27. Strategy: Let x represent the amount of the first alloy. The amount of the second alloy is y.
Gold:

	Amount	Percent	Quantity
1st alloy	x	0.10	$0.10x$
2nd alloy	y	0.30	$0.30y$

Lead:

	Amount	Percent	Quantity
1st alloy	x	0.15	$0.15x$
2nd alloy	y	0.40	$0.40y$

The resulting alloy contains 60 g of gold and 88 g of lead.
$0.10x + 0.30y = 60$
$0.15x + 0.40y = 88$

Solution: $0.10x + 0.30y = 60$
$\qquad\qquad 0.15x + 0.40y = 88$
$3(0.10x + 0.30y) = 3(60)$
$-2(0.15x + 0.40y) = -2(88)$

$0.30x + 0.90y = 180$
$-0.30x - 0.80y = -176$
$0.10y = 4$
$y = 40$

$0.10x + 0.30y = 60$
$0.10x + 0.30(40) = 60$
$0.10x + 12 = 60$
$0.10x = 48$
$x = 480$

The chemist should use 480 g of gold and 40 g of lead.

29. Strategy: Let x represent the cost of the Model II computer.
The cost of the Model IV computer is y.
The cost of the Model IX computer is z.

First shipment:

	Number	Unit Cost	Value
Model II	4	x	$4x$
Model IV	6	y	$6y$
Model IX	10	z	$10z$

Second shipment:

	Number	Unit Cost	Value
Model II	8	x	$8x$
Model IV	3	y	$3y$
Model IX	5	z	$5z$

Third shipment:

	Number	Unit Cost	Value
Model II	2	x	$2x$
Model IV	9	y	$9y$
Model IX	5	z	$5z$

The value of the first shipment was $114,000. The value of the second shipment was $72,000. The value of the third shipment was $81,000.

Solution: (1) $4x + 6y + 10z = 114{,}000$
(2) $8x + 3y + 5z = 72{,}000$
(3) $2x + 9y + 5z = 81{,}000$

Multiply equation (2) by -2 and add to equation (1).
$4x + 6y + 10z = 114{,}000$
$-16x - 6y - 10z = -144{,}000$
$-12x = -30{,}000$
$x = 2500$

Multiply equation (3) by -1 and add to equation (2).
$8x + 3y + 5z = 72{,}000$
$-2x - 9y - 5z = -81{,}000$
$6x - 6y = -9000$

$6(2500) - 6y = 9000$
$-6y = -24{,}000$
$y = 4000$

The Model IV computer costs $4000.

31. Strategy: Let x represent the amount deposited at 8%.
The amount deposited at 6% is y.
The amount deposited at 4% is z.

	Principal	Rate	Interest
8%	x	0.08	$0.08x$
6%	y	0.06	$0.06y$
4%	z	0.04	$0.04z$

The amount deposited in the 8% account is twice the amount deposited in the 6% account. The total amount invested is $25,000. The total interest earned is $1520.

Solution: (1) $x = 2y$
(2) $x + y + z = 25{,}000$
(3) $0.08x + 0.06y + 0.04z = 1520$

Substitute $2y$ for x in equation (2) and equation (3)
$2y + y + z = 25000$
(4) $3y + z = 25000$
$0.08(2y) + 0.06y + 0.04z = 1520$
(5) $0.22y + 0.04z = 1520$

Solve equation (4) for z and substitute into equation (5).
$z = 25{,}000 - 3y$
$0.22y + 0.04(25{,}000 - 3y) = 1520$
$0.22y + 1000 - 0.12y = 1520$
$0.10y = 520$
$y = 5200$

$z = 25{,}000 - 3(5200)$
$z = 9400$

$x = 2(5200)$
$x = 10{,}400$

The investor placed $10,400 in the 8% account, $5200 in the 6% account and $9400 in the 4% account.

Critical Thinking

33. Strategy: Let n represent the measure of the smaller angle.
The measure of the larger angle is m.
First relationship: $m + n = 180$
Second relationship: $m = 3n + 40$

Solution: Solve for n by substitution:
$(3n + 40) + n = 180$
$3n + 40 + n = 180$
$4n = 140$
$n = 35$
$m + n = 180$
$m + 35 = 180$
$m = 145$
The angles have measures of $35°$ and $145°$.

Projects or Group Activities

35. $w_1d_1 = w_2d_2 + w_3d_3$
$d_3 = 3d_2$

$5d_1 = d_2 + 3d_3$
$5d_1 = d_2 + 3(3d_2)$
$5d_1 = 10d_2$
$d_1 = 2d_2$

$d_1 + d_3 = 15$
$2d_2 + 3d_2 = 15$
$5d_2 = 15$
$d_2 = 3$

$d_3 = 3d_2 = 3(3) = 9$
$d_1 = 2d_2 = 2(3) = 9$

$d_1 = 6$ in, $d_2 = 3$ in, $d_3 = 9$ in

Section 4.5
Concept Check

1. $2x - y < 3$
$x - 3y \geq 6$

(i) $(5, 1)$
$2(5) - 1 < 3$
$9 < 3$ is not a true statement.

$(5, 1)$ is not a solution of the system of inequalities.

(ii) $(-3, -5)$
$2(-3) - (-5) < 3$
$-1 < 3$
$-3 - 3(-5) \geq 6$
$12 \geq 6$

$(-3, -5)$ is a solution of the system of inequalities.

Objective A Exercises

3. Solve each inequality for y.
$x - y \geq 3$
$-y \geq -x + 3$
$y \leq x - 3$

$x + y \leq 5$
$y \leq 5 - x$

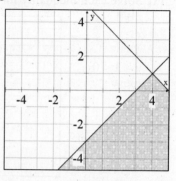

5. Solve each inequality for y.
$3x - y < 3$
$-y < -3x + 3$
$y > 3x - 3$

$2x + y \geq 2$
$y \geq 2 - 2x$

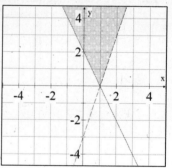

7. Solve each inequality for y.
$2x + y \geq -2$
$y \geq -2x - 2$

$6x + 3y \leq 6$
$3y \leq -6x + 6$
$y \leq -2x + 2$

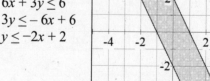

9. Solve the first inequality for y.

$3x - 2y < 6$

$-2y < -3x + 6$

$y > \dfrac{3}{2}x - 3$

$y \le 3$

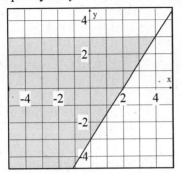

11. Solve each inequality for its variable.

$x + 1 \ge 0$

$x \ge -1$

$y - 3 \le 0$

$y \le 3$

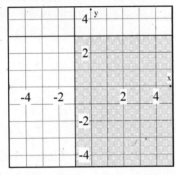

13. Solve each inequality for y.

$2x + y \ge 4$

$y \ge -2x + 4$

$3x - 2y < 6$

$-2y < -3x + 6$

$y > \dfrac{3}{2}x - 3$

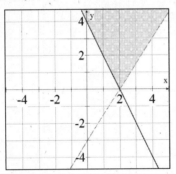

15. Solve each inequality for y.

$x - 2y \le 6$

$-2y \le -x + 6$

$y \ge \dfrac{1}{2}x - 3$

$2x + 3y \le 6$

$3y \le -2x + 6$

$y \ge -\dfrac{2}{3}x + 2$

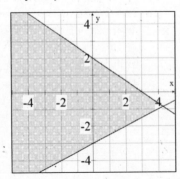

17. Solve each inequality for y.

$x - 2y \le 4$

$-2y \le -x + 4$

$y \ge \dfrac{1}{2}x - 2$

$3x + 2y \le 8$

$2y \le -3x + 8$

$y \le -\dfrac{3}{2}x + 4$

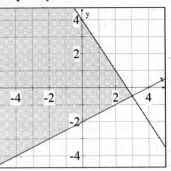

19. The solution set is the region between the parallel lines $x + y = a$ and $x + y = b$.

Critical Thinking

21. Solve each inequality for y.

$2x + 3y \le 15$

$3y \le -2x + 15$

$y \le -\dfrac{2}{3}x + 5$

$3x - y \le 6$

$-y \le -3x + 6$

$y \ge 3x - 6$

$y \ge 0$

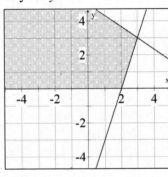

Projects or Group Activities

23. (ii) and (iii) are convex sets.

Chapter 4 Review Exercises

1. (1) $2x - 6y = 15$
 (2) $x = 4y + 8$
Substitute $4y + 8$ for x in equation (1).
$2(4y + 8) - 6y = 15$
$8y + 16 - 6y = 15$
$2y = -1$
$y = -\dfrac{1}{2}$

Substitute $-\dfrac{1}{2}$ for y in equation (2).

$x = 4\left(-\dfrac{1}{2}\right) + 8$

$x = 6$

The solution is $\left(6, -\dfrac{1}{2}\right)$.

2. (1) $3x + 2y = 2$
 (2) $x + y = 3$

Multiply equation (2) by -2 and add to equation (1).
$3x + 2y = 2$
$-2x - 2y = -6$
$x = -4$
Replace x with -4 in equation (2).
$-4 + y = 3$
$y = 7$
The solution is $(-4, 7)$.

3. $x + y = 3$
   $3x - 2y = -6$

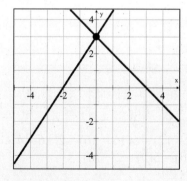

The solution is $(0, 3)$.

4. $2x - y = 4$
 $y = 2x - 4$

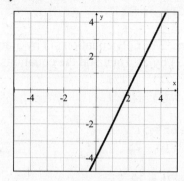

The two equations represent the same line. The system of equations is dependent. The solutions are the ordered pairs $(x, 2x - 4)$.

5. (1) $3x + 12y = 18$
 (2) $x + 4y = 6$

Solve equation (2) for x.
$x + 4y = 6$
$x = -4y + 6$
Substitute $-4y + 6$ for x in equation (1).
$3(-4y + 6) + 12y = 18$
$-12y + 18 + 12y = 18$
$18 = 18$
This is a true equation. The equations are dependent. The solutions are the ordered pairs $\left(x, -\dfrac{1}{4}x + \dfrac{3}{2}\right)$.

6. (1) $5x - 15y = 30$
 (2) $x - 3y = 6$

Multiply equation (2) by -5 and add to equation (1).
$5x - 15y = 30$
$-5x + 15y = -30$
$0 = 0$
This is a true equation. The equations are dependent. The solutions are the ordered pairs $\left(x, \dfrac{1}{3}x - 2\right)$.

7. 1) $3x - 4y - 2z = 17$
(2) $4x - 3y + 5z = 5$
(3) $5x - 5y + 3z = 14$

Eliminate z. Multiply equation (1) by 3 and equation (3) by 2. Then add the equations.
$3(3x - 4y - 2z) = 3(17)$
$2(5x - 5y + 3z) = 2(14)$

$9x - 12y - 6z = 51$
$10x - 10y + 6z = 28$

(4) $19x - 22y = 79$

Multiply equation (1) by 5 and equation (2) by 2. Then add the equations.
$5(3x - 4y - 2z) = 5(17)$
$2(4x - 3y + 5z) = 2(5)$

$15x - 20y - 10z = 85$
$8x - 6y + 10z = 10$

(5) $23x - 26y = 95$

Multiply equation (4) by 23 and equation (5) by −19. Then add the equations.

$23(19x - 22y) = 23(79)$
$-19(23x - 26y) = -19(95)$

$437x - 506y = 1817$
$-437x + 494y = -1805$
$-12y = 12$
$y = -1$

Replace y with −1 in equation (4).
$19x - 22(-1) = 79$
$19x + 22 = 79$
$19x = 57$
$x = 3$
Replace x with 3 and y with −1 in equation (1).
$3(3) - 4(-1) - 2z = 17$
$9 + 4 - 2z = 17$
$-2z = 4$
$z = -2$
The solution is $(3, -1, -2)$.

8. 1) $3x + y = 13$
(2) $2y + 3z = 5$
(3) $x + 2z = 11$

Eliminate y. Multiply equation (1) by −2 then add to equation (2).
$-2(3x + y) = -2(13)$
$2y + 3z = 5$

$-6x - 2y = -26$
$2y + 3z = 5$

(4) $-6x + 3z = -21$

Multiply equation (3) by 6 then add to equation (4).
$6(x + 2z) = 6(11)$
$-6x + 3z = -21$

$6x + 12z = 66$
$-6x + 3z = -21$
$15z = 45$
$z = 3$

Replace z with 3 in equation (3).
$x + 2(3) = 11$
$x = 5$
Replace x with 5 in equation (1).
$3(5) + y = 13$
$y = -2$
The solution is $(5, -2, 3)$.

9. $\begin{vmatrix} 6 & 1 \\ 2 & 5 \end{vmatrix} = 6(5) - 2(1) = 30 - 2 = 28$

10. $\begin{vmatrix} 1 & 5 & -2 \\ -2 & 1 & 4 \\ 4 & 3 & -8 \end{vmatrix}$

$= 1\begin{vmatrix} 1 & 4 \\ 3 & -8 \end{vmatrix} - 5\begin{vmatrix} -2 & 4 \\ 4 & -8 \end{vmatrix} - 2\begin{vmatrix} -2 & 1 \\ 4 & 3 \end{vmatrix}$
$= 1(-8 - 12) - 5(16 - 16) - 2(-6 - 4)$
$= 1(-20) - 5(0) - 2(-10)$
$= -20 + 20$
$= 0$

11. $2x - y = 7$
$3x + 2y = 7$

$$D = \begin{vmatrix} 2 & -1 \\ 3 & 2 \end{vmatrix} = 7$$

$$D_x = \begin{vmatrix} 7 & -1 \\ 7 & 2 \end{vmatrix} = 21$$

$$D_y = \begin{vmatrix} 2 & 7 \\ 3 & 7 \end{vmatrix} = -7$$

$$x = \frac{D_x}{D} = \frac{21}{7} = 3$$

$$y = \frac{D_y}{D} = \frac{-7}{7} = -1$$

The solution is $(3, -1)$.

12. $3x - 4y = 10$
$2x + 5y = 15$

$$D = \begin{vmatrix} 3 & -4 \\ 2 & 5 \end{vmatrix} = 23$$

$$D_x = \begin{vmatrix} 10 & -4 \\ 15 & 5 \end{vmatrix} = 110$$

$$D_y = \begin{vmatrix} 3 & 10 \\ 2 & 15 \end{vmatrix} = 25$$

$$x = \frac{D_x}{D} = \frac{110}{23}$$

$$y = \frac{D_y}{D} = \frac{25}{23}$$

The solution is $\left(\dfrac{110}{23}, \dfrac{25}{23} \right)$.

13. $x + y + z = 0$
$x + 2y + 3z = 5$
$2x + y + 2z = 3$

$$D = \begin{vmatrix} 1 & 1 & 1 \\ 1 & 2 & 3 \\ 2 & 1 & 2 \end{vmatrix} = 2$$

$$D_x = \begin{vmatrix} 0 & 1 & 1 \\ 5 & 2 & 3 \\ 3 & 1 & 2 \end{vmatrix} = -2$$

$$D_y = \begin{vmatrix} 1 & 0 & 1 \\ 1 & 5 & 3 \\ 2 & 3 & 2 \end{vmatrix} = -6$$

$$D_z = \begin{vmatrix} 1 & 1 & 0 \\ 1 & 2 & 5 \\ 2 & 1 & 3 \end{vmatrix} = 8$$

$$x = \frac{D_x}{D} = \frac{-2}{2} = -1$$

$$y = \frac{D_y}{D} = \frac{-6}{2} = -3$$

$$z = \frac{D_z}{D} = \frac{8}{2} = 4$$

The solution is $(-1, -3, 4)$.

14. $x + 3y + z = 6$
$2x + y - z = 12$
$x + 2y - z = 13$

$$D = \begin{vmatrix} 1 & 3 & 1 \\ 2 & 1 & -1 \\ 1 & 2 & -1 \end{vmatrix} = 7$$

$$D_x = \begin{vmatrix} 6 & 3 & 1 \\ 12 & 1 & -1 \\ 13 & 2 & -1 \end{vmatrix} = 14$$

$$D_y = \begin{vmatrix} 1 & 6 & 1 \\ 2 & 12 & -1 \\ 1 & 13 & -1 \end{vmatrix} = 21$$

$$D_z = \begin{vmatrix} 1 & 3 & 6 \\ 2 & 1 & 12 \\ 1 & 2 & 13 \end{vmatrix} = -35$$

$$x = \frac{D_x}{D} = \frac{14}{7} = 2$$

$$y = \frac{D_y}{D} = \frac{21}{7} = 3$$

$$z = \frac{D_z}{D} = \frac{-35}{7} = -5$$

The solution is $(2, 3, -5)$.

15. Solve each inequality for y.

$x + 3y \leq 6$

$3y \leq -x + 6$

$y \leq -\dfrac{1}{3}x + 2$

$2x - y \geq 4$

$-y \geq -2x + 4$

$y \leq 2x - 4$

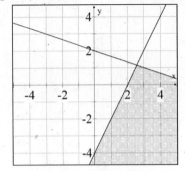

16. Solve each inequality for y.

$2x + 4y \geq 8$

$4y \geq -2x + 8$

$y \geq -\dfrac{1}{2}x + 2$

$x + y \leq 3$

$y \leq -x + 3$

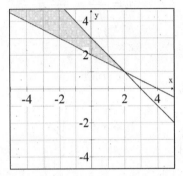

17. Strategy: Let x represent the rate of the cabin cruiser in calm water.
The rate of the current is y.

	Rate	Time	Distance
with current	$x + y$	3	$3(x + y)$
against current	$x - y$	5	$5(x - y)$

The distance traveled with the current is 60 mi. The distance traveled against the current is 60 mi.

$3(x + y) = 60$

$5(x - y) = 60$

Solution: $3(x + y) = 60$
$\qquad\qquad 5(x - y) = 60$

$\dfrac{1}{3} \cdot 3(x + y) = \dfrac{1}{3} \cdot 60$

$\dfrac{1}{5} \cdot 5(x - y) = \dfrac{1}{5} \cdot 60$

$x + y = 20$

$x - y = 12$

$2x = 32$

$x = 16$

$x + y = 20$

$16 + y = 20$

$y = 4$

The rate of the boat in calm water is 16 mph.
The rate of the current is 4 mph.

18. Strategy: Let p represent the rate of the plane in calm air.
The rate of the wind is w.

	Rate	Time	Distance
with wind	$p + w$	3	$3(p + w)$
against wind	$p - w$	4	$4(p - w)$

The distance traveled with the wind is 600 mi. The distance traveled against the wind is 600 mi.

$3(p + w) = 600$

$4(p - w) = 600$

Solution: $3(p + w) = 600$
$\qquad\qquad 4(p - w) = 600$

$\dfrac{1}{3} \cdot 3(p + w) = \dfrac{1}{3} \cdot 600$

$\dfrac{1}{4} \cdot 4(p - w) = \dfrac{1}{4} \cdot 600$

$p + w = 200$

$p - w = 150$

$2p = 350$

$p = 175$

$p + w = 200$

$175 + w = 200$

$w = 25$

The rate of the plane in calm air is 175 mph.
The rate of the wind is 25 mph.

19. Strategy: Let x represent the number of children's tickets sold.
The number of adult tickets sold is y.

Friday:

	Amount	Rate	Quantity
Children	x	5	$5x$
Adult	y	8	$8y$

Saturday:

	Amount	Rate	Quantity
Children	$3x$	5	$5(3x)$
Adult	$\frac{1}{2}y$	8	$8(\frac{1}{2})y$

The total receipts for Friday were $2500.
The total receipts for Saturday were $2500.
$5x + 8y = 2500$
$15x + 4y = 2500$

Solution: (1) $5x + 8y = 2500$
(2) $15x + 4y = 2500$
Multiply equation (2) by -2 then add to equation (1).
$5x + 8y = 2500$
$-2(15x + 4y) = -2(2500)$
$5x + 8y = 2500$
$-30x - 8y = -5000$
$-25x = -2500$
$x = 100$
On Friday, 100 children attended.

20. Strategy: Let x represent the amount invested at 3%.
The amount invested at 7% is y.

	Amount	Rate	Quantity
Amount at 3%	x	0.03	$0.03x$
Amount at 7%	y	0.07	$0.07y$

The total amount invested is $20,000.
$x + y = 20,000$
The total annual interest earned is $1200.
$0.03x + 0.07y = 1200$

Solution: (1) $x + y = 20,000$
(2) $0.03x + 0.07y = 1200$
Multiply equation (1) by -0.07 then add to equation (2).

$-0.07x - 0.07y = -1400$
$0.03x + 0.07y = 1200$

$-0.04x = -200$
$x = 5000$
Substitute 5000 for x in equation (1).
$x + y = 20,000$
$5000 + y = 20,000$
$y = 15,000$
The amount invested at 3% is $5000.
The amount invested at 7% is $15,000.

Chapter 4 Test

1. $2x - 3y = -6$
$2x - y = 2$

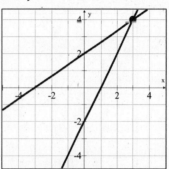

The solution is (3, 4).

2. $x - 2y = -6$
$$y = \frac{1}{2}x - 4$$

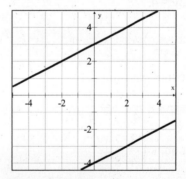

There is no solution.

3. Solve each inequality for y.

$2x - y < 3$

$-y < -2x + 3$

$y > 2x - 3$

$4x + 3y < 11$

$3y < -4x + 11$

$y < -\dfrac{4}{3}x + \dfrac{11}{3}$

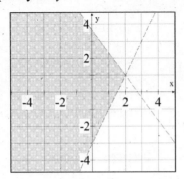

4. Solve each inequality for y.

$x + y > 2$

$y > -x + 2$

$2x - y < -1$

$-y < -2x - 1$

$y > 2x + 1$

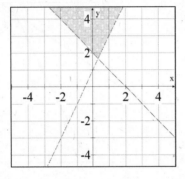

5. (1) $3x + 2y = 4$

(2) $\qquad x = 2y - 1$

Substitute $2y - 1$ for x in equation (1).

$3(2y - 1) + 2y = 4$

$6y - 3 + 2y = 4$

$8y = 7$

$y = \dfrac{7}{8}$

Substitute into equation (2).

$x = 2\left(\dfrac{7}{8}\right) - 1 = \dfrac{7}{4} - 1 = \dfrac{3}{4}$

The solution is $\left(\dfrac{3}{4}, \dfrac{7}{8}\right)$.

6. (1) $5x + 2y = -23$

(2) $2x + y = -10$

Solve equation (2) for y.

$y = -2x - 10$

Substitute $-2x - 10$ for y in equation (1).

$5x + 2(-2x - 10) = -23$

$5x - 4x - 20 = -23$

$x = -3$

Substitute in equation (2).

$2(-3) + y = -10$

$-6 + y = -10$

$y = -4$

The solution is $(-3, -4)$.

7. (1) $y = 3x - 7$

(2) $y = -2x + 3$

Substitute equation (2) into equation (1).

$-2x + 3 = 3x - 7$

$-5x + 3 = -7$

$-5x = -10$

$x = 2$

Substitute into equation (1).

$y = 3(2) - 7$

$y = -1$

The solution is $(2, -1)$.

8. (1) $3x + 4y = -2$

(2) $2x + 5y = 1$

Multiply equation (1) by 2 and equation (2) by -3 then add the new equations.

$2(3x + 4y) = 2(-2)$

$-3(2x + 5y) = -3(1)$

$6x + 8y = -4$

$-6x - 15y = -3$

$-7y = -7$

$y = 1$

Substitute into equation (1).

$3x + 4(1) = -2$

$3x = -6$

$x = -2$

The solution is $(-2, 1)$.

9. (1) $4x - 6y = 5$

(2) $6x - 9y = 4$

Multiply equation (1) by 3 and equation (2) by -2 then add the new equations.

$3(4x - 6y) = 3(5)$

$-2(6x - 9y) = -2(4)$

$12x - 18y = 15$

$-12x + 18y = -8$

$0 = 7$

This is not a true equation. The system of equations is inconsistent and therefore has no solution.

10. (1) $3x - y = 2x + y - 1$

(2) $5x + 2y = y + 6$

Write the equations in the form $Ax + By = C$.

(3) $x - 2y = -1$

(4) $5x + y = 6$

Multiply equation (4) by 2 then add to equation (3).

$x - 2y = -1$

$2(5x + y) = 2(6)$

$x - 2y = -1$

$10x - 2y = 12$

$11x = 11$

$x = 1$

Substitute in equation (4).

$5(1) + y = 6$

$y = 1$

The solution is $(1,1)$.

11. (1) $2x + 4y - z = 3$

(2) $x + 2y + z = 5$

(3) $4x + 8y - 2z = 7$

Eliminate z. Add equations (1) and (2).

$2x + 4y - z = 3$

$x + 2y + z = 5$

(4) $3x + 6y = 8$

Multiply equation (2) by 2 and add to equation (3).

$2(x + 2y + z) = 2(5)$

$4x + 8y - 2z = 7$

$2x + 4y + 2z = 10$

$4x + 8y - 2z = 7$

(5) $6x + 12y = 17$

Multiply equation (4) by -2 then add to equation (5).

$-2(3x + 6y) = -2(8)$

$6x + 12y = 17$

$-6x - 12y = -16$

$6x + 12y = 17$

$0 = 1$

This is not a true equation. The system of equations is inconsistent and therefore has no solution.

12. (1) $x - y - z = 5$

(2) $2x + z = 2$

(3) $3y - 2z = 1$

Multiply equation (1) by 3 and then add to equation (3).

$3(x - y - z) = 3(5)$

$3x - 3y - 3z = 15$

$3y - 2z = 1$

(4) $3x - 5z = 16$

Multiply equation (2) by 5 and add to equation (4).

$5(2x + z) = 5(2)$

$10x + 5z = 10$

$3x - 5z = 16$

$13x = 26$

$x = 2$

Substitute 2 in for x in equation (2).

$2(2) + z = 2$

$z = -2$

Substitute 2 in for x and -2 in for z in equation (1).

$2 - y - (-2) = 5$

$4 - y = 5$

$-y = 1$

$y = -1$

The solution is $(2, -1, -2)$.

13. $\begin{vmatrix} 3 & -1 \\ -2 & 4 \end{vmatrix} = 3(4) - (-2)(-1) = 12 - 2 = 10$

14. $\begin{vmatrix} 1 & -2 & 3 \\ 3 & 1 & 1 \\ 2 & -1 & -2 \end{vmatrix} = 1\begin{vmatrix} 1 & 1 \\ -1 & -2 \end{vmatrix} - (-2)\begin{vmatrix} 3 & 1 \\ 2 & -2 \end{vmatrix} + 3\begin{vmatrix} 3 & 1 \\ 2 & -1 \end{vmatrix}$

$= 1(-2 - (-1)) + 2(-6 - 2) + 3(-3 - 2)$

$= 1(-1) + 2(-8) + 3(-5)$

$= -1 - 16 - 15$

$= -32$

15. $x - y = 3$
$2x + y = -4$

$D = \begin{vmatrix} 1 & -1 \\ 2 & 1 \end{vmatrix} = 3$

$D_x = \begin{vmatrix} 3 & -1 \\ -4 & 1 \end{vmatrix} = -1$

$D_y = \begin{vmatrix} 1 & 3 \\ 2 & -4 \end{vmatrix} = -10$

$x = \dfrac{D_x}{D} = \dfrac{-1}{3} = -\dfrac{1}{3}$

$y = \dfrac{D_y}{D} = \dfrac{-10}{3} = -\dfrac{10}{3}$

The solution is $\left(-\dfrac{1}{3}, -\dfrac{10}{3}\right)$.

16. $5x + 2y = 9$
$3x + 5y = -7$

$D = \begin{vmatrix} 5 & 2 \\ 3 & 5 \end{vmatrix} = 19$

$D_x = \begin{vmatrix} 9 & 2 \\ -7 & 5 \end{vmatrix} = 59$

$D_y = \begin{vmatrix} 5 & 9 \\ 3 & -7 \end{vmatrix} = -62$

$x = \dfrac{D_x}{D} = \dfrac{59}{19}$

$y = \dfrac{D_y}{D} = \dfrac{-62}{19}$

The solution is $\left(\dfrac{59}{19}, -\dfrac{62}{19}\right)$.

17. $x - y + z = 2$
$2x - y - z = 1$
$x + 2y - 3z = -4$

$D = \begin{vmatrix} 1 & -1 & 1 \\ 2 & -1 & -1 \\ 1 & 2 & -3 \end{vmatrix} = 5$

$D_x = \begin{vmatrix} 2 & -1 & 1 \\ 1 & -1 & -1 \\ -4 & 2 & -3 \end{vmatrix} = 1$

$D_y = \begin{vmatrix} 1 & 2 & 1 \\ 2 & 1 & -1 \\ 1 & -4 & -3 \end{vmatrix} = -6$

$D_z = \begin{vmatrix} 1 & -1 & 2 \\ 2 & -1 & 1 \\ 1 & 2 & -4 \end{vmatrix} = 3$

$x = \dfrac{D_x}{D} = \dfrac{1}{5}$

$y = \dfrac{D_y}{D} = \dfrac{-6}{5}$

$z = \dfrac{D_z}{D} = \dfrac{3}{5}$

The solution is $\left(\dfrac{1}{5}, -\dfrac{6}{5}, \dfrac{3}{5}\right)$.

18. Strategy: Let x represent the rate of the plane in calm air.
The rate of the wind is y.

	Rate	Time	Distance
with wind	$x + y$	2	$2(x + y)$
against wind	$x - y$	2.8	$2.8(x - y)$

The distance traveled with the wind is 350 mi. The distance traveled against the wind is 350 mi.
$2(x + y) = 350$
$2.8(x - y) = 350$

Solution: $2(x + y) = 350$
$\qquad\qquad 2.8(x - y) = 350$

$\dfrac{1}{2} \cdot 2(x + y) = \dfrac{1}{2} \cdot 350$

$\dfrac{1}{2.8} \cdot 2.8(x - y) = \dfrac{1}{2.8} \cdot 350$

$x + y = 175$
$x - y = 125$
$2x = 300$
$x = 150$

$x + y = 175$
$150 + y = 175$
$y = 25$

The rate of the plane in calm air is 150 mph.
The rate of the wind is 25 mph.

19. Strategy: Let x represent the cost per yard of cotton.
The cost per yard of wool is y.
First purchase:

	Amount	Rate	Total Value
Cotton	60	x	$60x$
Wool	90	y	$90y$

Second purchase:

	Amount	Rate	Total Value
Cotton	80	x	$80x$
Wool	20	y	$20y$

The total cost of the first purchase was $1800. The total cost of the second purchase was $1000.
$60x + 90y = 1800$

$80x + 20y = 1000$

Solution: $-4(60x + 90y) = -4(1800)$
$\qquad\qquad 3(80x + 20y) = 3(1000)$

$-240x - 360y = -7200$
$240x + 60y = 3000$
$-300y = -4200$
$y = 14$

$60x + 90(14) = 1800$
$60x + 1260 = 1800$
$60x = 540$
$x = 9$

The cost of cotton is $9.00/yd.
The cost of wool is $14.00/yd.

20. Strategy: Let x represent the amount invested at 2.7%.
The amount invested at 5.1% is y.

	Amount	Rate	Quantity
Amount at 2.7%	x	0.027	$0.027x$
Amount at 5.1%	y	0.051	$0.051y$

The total amount invested is $15,000.
$x + y = 15,000$
The total annual interest earned is $549.
$0.027x + 0.051y = 549$

Solution: (1) $x + y = 15,000$
$\qquad\qquad$ (2) $0.027x + 0.051y = 549$
Multiply equation (1) by -0.051 then add to equation (2).

$-0.051x - 0.051y = -765$
$0.027x + 0.051y = 549$
$-0.024x = -216$
$x = 9000$
Substitute 9000 for x in equation (1).
$x + y = 15,000$
$9000 + y = 15,000$
$y = 6000$
The amount invested at 2.7% is $9000.
The amount invested at 5.1% is $6000.

Cumulative Review Exercises

1. $\dfrac{3}{2}x - \dfrac{3}{8} + \dfrac{1}{4}x = \dfrac{7}{12}x - \dfrac{5}{6}$

$$24\left(\dfrac{3}{2}x - \dfrac{3}{8} + \dfrac{1}{4}x\right) = 24\left(\dfrac{7}{12}x - \dfrac{5}{6}\right)$$

$$36x - 9 + 6x = 14x - 20$$
$$42x - 9 = 14x - 20$$
$$28x - 9 = -20$$
$$28x = -11$$
$$x = -\dfrac{11}{28}$$

The solution is $-\dfrac{11}{28}$.

2. $(2, -1)\,(3, 4)$

$$m = \dfrac{y_2 - y_1}{x_2 - x_1} = \dfrac{4 - (-1)}{3 - 2} = \dfrac{5}{1} = 5$$

$$y - y_1 = m(x - x_1)$$
$$y - 4 = 5(x - 3)$$
$$y - 4 = 5x - 15$$
$$y = 5x - 11$$

The equation of the line is $y = 5x - 11$.

3. $3[x - 2(5 - 2x) - 4x] + 6$
$= 3(x - 10 + 4x - 4x) + 6$
$= 3(x - 10) + 6$
$= 3x - 30 + 6$
$= 3x - 24$

4. $a + bc \div 2$
$4 + 8(-2) \div 2 = 4 - 16 \div 2 = 4 - 8 = -4$

5. $2x - 3 < 9 \quad$ or $5x - 1 < 4$
Solve each inequality.
$$2x - 3 < 9 \qquad 5x - 1 < 4$$
$$2x < 12 \qquad\quad 5x < 5$$
$$x < 6 \quad \text{or} \quad x < 1$$
$$(-\infty, 6) \cup (-\infty, 1) = (-\infty, 6)$$

6. $|x - 2| - 4 < 2$
$|x - 2| < 6$
$-6 < x - 2 < 6$
$-6 + 2 < x - 2 + 2 < 6 + 2$
$-4 < x < 8$
$\{x \mid -4 < x < 8\}$

7. $|2x - 3| > 5$
Solve each inequality.
$$2x - 3 < -5 \quad \text{or} \quad 2x - 3 > 5$$
$$2x < -2 \quad \text{or} \qquad 2x > 8$$
$$x < -1 \quad \text{or} \qquad x > 4$$
$$\{x \mid x < -1\} \cup \{x \mid x > 4\}$$

8. $f(x) = 3x^3 - 2x^2 + 1$
$f(-3) = 3(-3)^3 - 2(-3)^2 + 1$
$f(-3) = 3(-27) - 2(9) + 1$
$f(-3) = -98$

9. $f(x) = 3x^2 - 2x$
Domain $= \{x \mid -\infty < x < \infty\}$

10. $F(x) = x^2 - 3$
$F(2) = (2)^2 - 3 = 1$

11. $f(x) = 3x - 4$
$f(2 + h) = 3(2 + h) - 4 = 6 + 3h - 4 = 2 + 3h$
$f(2) = 3(2) - 4 = 6 - 4 = 2$
$f(2 + h) - f(2) = 2 + 3h - 2 = 3h$

12. $\{x \mid x \le 2\} \cap \{x \mid x > -3\}$

13. $(-2, 3),\ m = -\dfrac{2}{3}$

$$y - y_1 = m(x - x_1)$$

$$y - 3 = -\dfrac{2}{3}(x - (-2))$$

$$y - 3 = -\dfrac{2}{3}x - \dfrac{4}{3}$$

$$y = -\dfrac{2}{3}x - \dfrac{4}{3} + 3$$

$$y = -\dfrac{2}{3}x + \dfrac{5}{3}$$

14. The slope of the line $2x - 3y = 7$ is found by solving the equation for y.
$$-3y = -2x + 7$$
$$y = \dfrac{2}{3}x - \dfrac{7}{3}$$
The slope is $\dfrac{2}{3}$.

Use $(-1, 2)$ and the perpendicular slope $-\dfrac{3}{2}$

$$y - y_1 = m(x - x_1)$$

$$y - 2 = -\frac{3}{2}(x - (-1))$$

$$y - 2 = -\frac{3}{2}x - \frac{3}{2}$$

$$y = -\frac{3}{2}x - \frac{3}{2} + 2$$

$$y = -\frac{3}{2}x + \frac{1}{2}$$

15. The distance between two points is

$$d = \sqrt{(x_1 - x_2)^2 + (y_1 - y_2)^2}$$

$(-4, 2)$ and $(2, 0)$

$$d = \sqrt{(2 - (-4))^2 + (0 - 2)^2} = \sqrt{6^2 + (-2)^2}$$
$$= \sqrt{36 + 4} = \sqrt{40} \approx 6.32$$

16. $\left(\dfrac{x_1 + x_2}{2}, \dfrac{y_1 + y_2}{2}\right)$

$(-4, 3)$ and $(3, 5)$

$$\left(\frac{-4 + 3}{2}, \frac{3 + 5}{2}\right) = \left(-\frac{1}{2}, 4\right)$$

17. $2x - 5y = 10$

$-5y = -2x + 10$

$$y = \frac{2}{5}x - 2$$

slope is $\dfrac{2}{5}$; y-intercept is -2

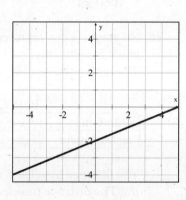

18. $3x - 4y \geq 8$

$-4y \geq -3x + 8$

$$y \leq \frac{3}{4}x - 2$$

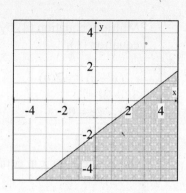

19. $5x - 2y = 10$

$3x + 2y = 6$

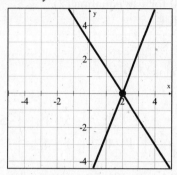

The solution is $(2, 0)$.

20. Solve each inequality for y.

$3x - 2y \geq 4$

$-2y \geq -3x + 4$

$$y \leq \frac{3}{2}x - 2$$

$x + y < 3$

$y < -x + 3$

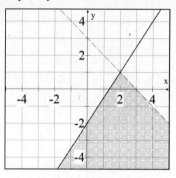

21. (1) $3x + 2z = 1$

(2) $2y - z = 1$

(3) $x + 2y = 1$

Multiply equation (2) by -1 and add to equation (3).

$-1(2y - z) = -1(1)$

$x + 2y = 1$

$-2y + z = -1$

$x + 2y = 1$

(4) $x + z = 0$

Multiply equation (4) by −2 and add to equation (1).

$-2(x + z) = -2(0)$
$3x + 2z = 1$
$-2x - 2z = 0$
$3x + 2z = 1$
$x = 1$
Substitute 1 for x in equation (3).
$1 + 2y = 1$
$2y = 0$
$y = 0$
Substitute 0 for y in equation (2).
$2(0) - z = 1$
$-z = 1$
$z = -1$
The solution is $(1, 0, -1)$.

22. $\begin{vmatrix} 2 & -5 & 1 \\ 3 & 1 & 2 \\ 6 & -1 & 4 \end{vmatrix} = 2\begin{vmatrix} 1 & 2 \\ -1 & 4 \end{vmatrix} - 3\begin{vmatrix} -5 & 1 \\ -1 & 4 \end{vmatrix} + 6\begin{vmatrix} -5 & 1 \\ 1 & 2 \end{vmatrix}$

$\qquad = 2(4 + 2) - 3(-20 + 1) + 6(-10 - 1)$
$\qquad = 2(6) - 3(-19) + 6(-11)$
$\qquad = 12 + 57 - 66$
$\qquad = 3$

23. $4x - 3y = 17$
$3x - 2y = 12$

$D = \begin{vmatrix} 4 & -3 \\ 3 & -2 \end{vmatrix} = 1$

$D_x = \begin{vmatrix} 17 & -3 \\ 12 & -2 \end{vmatrix} = 2$

$D_y = \begin{vmatrix} 4 & 17 \\ 3 & 12 \end{vmatrix} = -3$

$x = \dfrac{D_x}{D} = \dfrac{2}{1} = 2$

$y = \dfrac{D_y}{D} = \dfrac{-3}{1} = -3$

The solution is $(2, -3)$.

24. (1) $3x - 2y = 7$
(2) $\qquad y = 2x - 1$

Solve by substitution
$3x - 2(2x - 1) = 7$
$3x - 4x + 2 = 7$
$-x + 2 = 7$
$-x = 5$
$x = -5$
Substitute −5 for x in equation (2).
$y = 2(-5) - 1$
$y = -10 - 1$
$y = -11$
The solution is $(-5, -11)$.

25. Let x represent the amount of pure water.

	Amount	Percent	Quantity
Water	x	0	$0x$
4%	100	0.04	$0.04(100)$
2.5%	$100 + x$	0.025	$0.025(100 + x)$

The sum of the quantities before mixing is equal to the quantity after mixing.
$0x + 0.04(100) = 0.025(100 + x)$

Solution: $0x + 0.04(100) = 0.025(100 + x)$
$\qquad\qquad\qquad 0 + 4 = 2.5 + 0.025x$
$\qquad\qquad\qquad 1.5 = 0.025x$

$x = 60$
The amount of water that should be added is 60 ml.

26. Strategy: Let x represent the rate of the plane in calm air.
The rate of the wind is y.

	Rate	Time	Distance
with wind	$x + y$	2	$2(x + y)$
against wind	$x - y$	3	$3(x - y)$

The distance traveled with the wind is 150 mi. The distance traveled against the wind is 150 mi.

$2(x + y) = 150$
$3(x - y) = 150$

Solution: $2(x+y) = 150$
$\qquad\qquad 3(x-y) = 150$

$$\frac{1}{2} \cdot 2(x+y) = \frac{1}{2} \cdot 150$$

$$\frac{1}{3} \cdot 3(x-y) = \frac{1}{3} \cdot 150$$

$x + y = 75$
$x - y = 50$
$2x = 125$
$x = 62.5$

$x + y = 75$
$62.5 + y = 75$
$y = 12.5$

The rate of the wind is 12.5 mph.

27. Let x represent the cost per pound of hamburger.
The cost per pound of steak is y.
First purchase:

	Amount	Cost	Quantity
Hamburger	100	x	$100x$
Steak	50	y	$50y$

Second purchase:

	Amount	Cost	Quantity
Hamburger	150	x	$150x$
Steak	100	y	$100y$

The total cost of the first purchase is $541.
The total cost of the second purchase is $960.
$100x + 50y = 540$
$150x + 100y = 960$
Solution: $100x + 50y = 540$
$\qquad\qquad 150x + 100y = 960$

$-2(100x + 50y) = -2(540)$
$\quad\; 150x + 100y = 960$
$-200x - 100y = -1080$
$\;\; 150x + 100y = 960$
$-50x = -120$
$x = 2.4$

$100(2.4) + 50y = 540$
$240 + 50y = 540$
$50y = 300$
$y = 6$
The cost of steak is $6.00/lb.

28. **Strategy:** Let M represent the number of ohms, T the tolerance and r the value of the resistor. Find the tolerance and solve $|M - 12,000| \le T$ for M.

Solution: $T = 0.15 \cdot 12,000 = 1800$ ohms
$|M - 12,000| \le 1800$
$-1800 \le M - 12,000 \le 1800$
$-1800 + 12,000 \le M - 12,000 + 12,000$
$\qquad\qquad\qquad\qquad \le 1800 + 12,000$
$10,200 \le M \le 13,800$

The lower and upper limits of the resistor are 10,200 ohms and 13,800 ohms.

29. The slope of the line is
$$\frac{5000 - 1000}{100 - 0} = \frac{4000}{100} = 40.$$
The commission rate of the executive is $40 for every $1000 in sales.

Chapter 5: Polynomials

Prep Test

1. $-4(3y) = -12y$

2. $(-2)^3 = (-2)(-2)(-2) = -8$

3. $-4a - 8b + 7a = 3a - 8b$

4. $3x - 2[y - 4(x + 1) + 5]$
   $= 3x - 2(y - 4x - 4 + 5)$
   $= 3x - 2(y - 4x + 1)$
   $= 3x - 2y + 8x - 2$
   $= 11x - 2y - 2$

5. $-(x + y) = -x - y$

6. $40 = 2 \cdot 20$
   $ = 2 \cdot 4 \cdot 5$
   $ = 2 \cdot 2 \cdot 2 \cdot 5$

7. $16 = \underline{2 \cdot 2} \cdot 2 \cdot 2$
   $20 = \underline{2 \cdot 2} \cdot 5$
   $24 = \underline{2 \cdot 2} \cdot 2 \cdot 3$
   The GCF is $2 \cdot 2$ or 4.

8. $x^3 - 2x^2 + x + 5$
   $(-2)^3 - 2(-2)^2 + (-2) + 5$
   $= -8 - 2(4) - 2 + 5$
   $= -8 - 8 - 2 + 5$
   $= -13$

9. $3x + 1 = 0$
   $ 3x = -1$
   $ x = -\dfrac{1}{3}$
   The solution is $-\dfrac{1}{3}$.

Section 5.1

Concept Check

1. (i), (ii) and (iv) are monomials.

3. No, the Rule for Simplifying Powers of Products does not apply because the expression inside the parentheses is a sum, not a product.

5. $-z^0 = -1; \ z^0 = 1$

Objective A Exercises

7. $(ab^3)(a^3b) = a^4b^4$

9. $(9xy^2)(-2x^2y^2) = -18x^3y^4$

11. $(x^2y^4)^4 = x^8y^{16}$

13. $(-3x^2y^3)^4 = (-3)^4x^8y^{12} = 81x^4y^{12}$

15. $(27a^5b^3)^2 = (27)^2a^{10}b^6 = 729a^{10}b^6$

17. $[(2a^4b^3)^3]^2 = (2a^4b^3)^6 = (2)^6a^{24}b^{18} = 64a^{24}b^{18}$

19. $(x^2y^2)(xy^3)^3 = (x^2y^2)(x^3y^9) = x^5y^{11}$

21. $(-5ab)(3a^3b^2)^2 = (-5ab)(3^2a^6b^4)$
    $ = (-5ab)(9a^6b^4)$
    $ = -45a^7b^5$

23. $(3x^5y)(-4x^3)^3 = (3x^5y)((-4)^3x^9)$
    $ = (3x^5y)(-64x^9)$
    $ = -192x^{14}y$

25. $(-6a^4b^2)(-7a^2c^5) = 42a^6b^2c^5$

27. $(-2ab^2)(-3a^4b^5)^3 = (-2ab^2)((-3)^3a^{12}b^{15})$
    $ = (-2ab^2)(-27a^{12}b^{15})$
    $ = 54a^{13}b^{17}$

29. $(-3ab^3)^3(-2^2a^2b)^2 = ((-3)^3a^3b^9)(2^4a^4b^2)$
    $ = (-27a^3b^9)(16a^4b^2)$
    $ = -432a^7b^{11}$

31. $(-2x^2y^3z)(3x^2yz^4) = -6x^4y^4z^5$

33. $(2xy)(-3x^2yz)(x^2y^3z^3) = -6x^5y^5z^4$

35. $(3b^5)(2ab^2)(-2ab^2c^2) = -12a^2b^9c^2$

37. The value of n is 33.

Objective B Exercises

39. $4^{-2} = \dfrac{1}{4^2} = \dfrac{1}{16}$

41. $\dfrac{1}{2^{-7}} = 2^7 = 128$

43. $3a^{-5} = \dfrac{3}{a^5}$

45. $\dfrac{1}{3a^{-7}} = \dfrac{a^7}{3}$

47. $\dfrac{a^3}{4b^{-2}} = \dfrac{a^3b^2}{4}$

49. $xy^{-4} = \dfrac{x}{y^4}$

51. $\dfrac{1}{2x^0} = \dfrac{1}{2}$

53. $\dfrac{-3^{-2}}{(2y)^0} = \dfrac{-1}{3^2} = -\dfrac{1}{9}$

55. $(x^3y^5)^{-2} = x^{-6}y^{-10} = \dfrac{1}{x^6y^{10}}$

57. $(-3a^{-4}b^{-5})(-5a^{-2}b^4) = 15a^{-6}b^{-1} = \dfrac{15}{a^6b}$

59. $(4y^{-3}z^{-4})(-3y^3z^{-3})^{-2}$
$= (4y^{-3}z^{-4})((-3)^{-2}y^{-6}z^6)$

$= (4)(-3)^{-2}y^{-9}z^2 = \dfrac{4z^2}{(-3)^2y^9}$

$= \dfrac{4z^2}{9y^9}$

61. $(4x^{-3}y^2)^{-3}(2xy^{-3})^4$
$= (4^{-3}x^9y^{-6})(2^4x^4y^{-12})$
$= (4^{-3})(2)^4x^{13}y^{-18} = \dfrac{2^4x^{13}}{(4)^3y^{18}}$
$= \dfrac{16x^{13}}{64y^{18}} = \dfrac{x^{13}}{4y^{18}}$

63. $\dfrac{9x^5}{12x^8} = \dfrac{3}{4x^3}$

65. $\dfrac{-6x^2y}{12x^4y} = -\dfrac{1}{2x^2}$

67. $\dfrac{y^{-2}}{y^6} = y^{-2-(6)} = y^{-8} = \dfrac{1}{y^8}$

69. $\dfrac{a^6b^{-4}}{a^{-2}b^5} = a^8b^{-9} = \dfrac{a^8}{b^9}$

71. $\dfrac{-3ab^2}{(9a^2b^4)^3} = \dfrac{-3ab^2}{9^3a^6b^{12}} = \dfrac{-3ab^2}{729a^6b^{12}}$
$= -\dfrac{1}{243a^5b^{10}}$

73. $\dfrac{(3a^2b)^3}{(-6ab^3)^2} = \dfrac{3^3a^6b^3}{(-6)^2a^2b^6} = \dfrac{27a^6b^3}{36a^2b^6} = \dfrac{3a^4}{4b^3}$

75. $\dfrac{(-8x^2y^2)^4}{(16x^3y^7)^2} = \dfrac{(-8)^4x^4y^8}{(16)^2x^6y^{14}} = \dfrac{4096x^4y^8}{256x^6y^{14}}$
$= \dfrac{16x^2}{y^6}$

77. $\dfrac{(3a^4b^{-2})^{-2}}{(2a^{-3}b)^3} = \dfrac{(3)^{-2}a^{-8}b^4}{(2)^3a^{-9}b^3} = \dfrac{ab}{(2)^3(3)^2} = \dfrac{ab}{72}$

79. $\left(\dfrac{9ab^{-2}}{8a^{-2}b}\right)^{-2}\left(\dfrac{3a^{-2}b}{2a^2b^{-2}}\right)^{3}$

$= \left(\dfrac{9^{-2}a^{-2}b^{4}}{8^{-2}a^{4}b^{-2}}\right)\left(\dfrac{3^{3}a^{-6}b^{3}}{2^{3}a^{6}b^{-6}}\right)$

$= \left(\dfrac{9^{-2}b^{6}}{8^{-2}a^{6}}\right)\left(\dfrac{3^{3}b^{9}}{2^{3}a^{12}}\right)$

$= \left(\dfrac{8^{2}b^{6}}{9^{2}a^{6}}\right)\left(\dfrac{3^{3}b^{9}}{2^{3}a^{12}}\right) = \dfrac{8b^{15}}{3a^{18}}$

81. The value of $p - q$ is 0.

Objective C Exercises

83. 4.67×10^{-6}

85. 1.7×10^{-10}

87. 2×10^{11}

89. 0.000000123

91. 8,200,000,000,000,000

93. 0.039

95. $(3 \times 10^{-12})(5 \times 10^{16})$
   $= (3)(5) \times 10^{-12+16}$
   $= 15 \times 10^{4}$
   $= 150,000$

97. $(0.0000065)(3,200,000,000,000)$
   $= (6.5 \times 10^{-6})(3.2 \times 10^{12})$
   $= (6.5)(3.2) \times 10^{-6+12}$
   $= 20.8 \times 10^{6}$
   $= 20,800,000$

99. $\dfrac{9 \times 10^{-3}}{6 \times 10^{5}} = 1.5 \times 10^{-3-5}$

   $= 1.5 \times 10^{-8} = 0.000000015$

101. $\dfrac{0.0089}{500,000,000} = \dfrac{8.9 \times 10^{-3}}{5 \times 10^{8}}$

   $= 1.78 \times 10^{-3-8} = 1.78 \times 10^{-11}$

   $= 0.0000000000178$

103. $\dfrac{(3.3 \times 10^{-11})(2.7 \times 10^{15})}{8.1 \times 10^{-3}}$

   $= \dfrac{(3.3)(2.7) \times 10^{-11+15-(-3)}}{8.1}$

   $= 1.1 \times 10^{7} = 11,000,000$

105. $\dfrac{(0.00000004)(84,000)}{(0.0003)(1,400,000)}$

   $= \dfrac{4 \times 10^{-8} \times 8.4 \times 10^{4}}{3 \times 10^{-4} \times 1.4 \times 10^{6}}$

   $= \dfrac{4(8.4) \times 10^{-8+4-(-4)-6}}{3(1.4)}$

   $= 8 \times 10^{-6} = 0.000008$

107. Greater than zero

Objective D Exercises

109.   **Strategy:** To find the number of years needed to cross the galaxy, divide the width of the galaxy by the product of the rate of the space ship and the number of hours in a year.
   Solution:

   $\dfrac{5.6 \times 10^{19}}{2.5 \times 10^{4} \times 8.76 \times 10^{3}} \approx 2.6 \times 10^{11}$

   It would takes a space ship 2.6×10^{11} years to cross the galaxy.

111.   Strategy: To find the number of times larger the mass of the proton is, divide the mass of the proton by the mass of an electron.

Solution: $\dfrac{1.673 \times 10^{-27}}{9.109 \times 10^{-31}} \approx 1.837 \times 10^{3}$

The mass of a proton is approximately 1.837×10^{3} times larger than an electron.

113.   Strategy: To find how fast the radio signal travels, divide the distance from Mars to Earth by the speed of the radio signal.

Solution: $\dfrac{154{,}000{,}000}{13.8} = 11159420.29$

$\approx 1.12 \times 10^{7}$ mi/day

The radio signal travel approximately 1.12×10^{7} mi/day.

Strategy: To find the rate of the signals divide the distance by the time.

Solution:

$\dfrac{119{,}000{,}000}{11} = \dfrac{1.19 \times 10^{8}}{1.1 \times 10^{1}} = 1.08\overline{1} \times 10^{7}$

The signal travels $1.08\overline{1} \times 10^{7}$ mi/min.

115.   Strategy: Divide the mass of the sun by the mass of the Earth.

Solution: $\dfrac{2 \times 10^{30}}{5.9 \times 10^{24}} = 3.3898305 \times 10^{5}$

The sun is 3.3898305×10 times larger.

117.   Strategy: To find the number of seeds produced, divide the number of seeds by the number of pine seedlings growing.

Solution:

$\dfrac{2{,}000{,}000}{12{,}000} = \dfrac{2 \times 10^{6}}{1.2 \times 10^{4}} = 1.\overline{6} \times 10^{2}$

$1.\overline{6} \times 10^{2}$ seeds are produced.

Critical Thinking

119.   a) $3^{x^2} = 3^{2^2} = 3^4 = 81$

b) $3^{x^2} = 3^{3^2} = 3^9 = 19{,}683$

c) $3^{x^2} = 3^{0^2} = 3^0 = 1$

d) $3^{x^2} = 3^{(-2)^2} = 3^4 = 81$

121. $x^{3n} x^{4n} = x^{7n}$

123. $\dfrac{x^{n} y^{5m}}{x^{3n} y^{m}} = \dfrac{y^{4m}}{x^{2n}}$

Projects or Group Activities

125.   a)

$$1 + [1 + (1 + 2^{-1})^{-1}]^{-1} = 1 + \left(1 + \left(1 + \dfrac{1}{2}\right)^{-1}\right)^{-1}$$

$$= 1 + \left(1 + \left(\dfrac{3}{2}\right)^{-1}\right)^{-1}$$

$$= 1 + \left(1 + \dfrac{2}{3}\right)^{-1}$$

$$= 1 + \left(\dfrac{5}{3}\right)^{-1}$$

$$= 1 + \left(\dfrac{3}{5}\right)$$

$$= \dfrac{8}{5}$$

b)

$$2-[2-(2-2^{-1})^{-1}]^{-1} = 2-\left(2-\left(2-\frac{1}{2}\right)^{-1}\right)^{-1}$$

$$= 2-\left(2-\left(\frac{3}{2}\right)^{-1}\right)^{-1}$$

$$= 2-\left(2-\frac{2}{3}\right)^{-1}$$

$$= 2-\left(\frac{4}{3}\right)^{-1}$$

$$= 2-\left(\frac{3}{4}\right)$$

$$= \frac{5}{4}$$

Section 5.2
Concept Check

1. a) binomial
   b) binomial
   c) monomial
   d) trinomial
   e) none of these
   f) monomial

3. All real numbers

5. This is a polynomial.
   a) −1
   b) 8
   c) 2

7. This is not a polynomial.

9. This is a polynomial.
   a) −1
   b) 0
   c) 6

Objective A Exercises

11. $P(x) = 3x^2 - 2x - 8$
    $P(3) = 3(3)^2 - 2(3) - 8$
    $P(3) = 13$

13. $R(x) = 2x^3 - 3x^2 + 4x - 2$
    $R(2) = 2(2)^3 - 3(2)^2 + 4(2) - 2$
    $R(2) = 10$

15. $f(x) = x^4 - 2x^2 - 10$
    $f(-1) = (-1)^4 - 2(-1)^2 - 10$
    $f(-1) = -11$

17. $P(x) = x^2 - 3x - 3$

x	y
−2	7
−1	1
0	−3
1	−5
2	−5

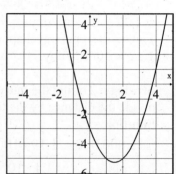

19. $P(x) = x^3 + 2$

x	y
−2	−6
−1	1
0	2
1	3
2	10

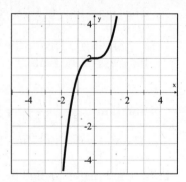

21. $f(x) = x^3 - 4x^2 - 4x + 16$

x	y
−2	0
−1	15
0	16
1	9
2	0

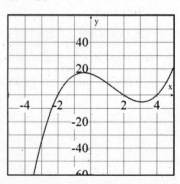

23. (a) $f(c) - g(c) > 0$
(b) $f(c) - g(c) > 0$

Objective B Exercises

25. $5x^2 + 2x - 7$
$\underline{x^2 - 8x + 12}$
$6x^2 - 6x + 5$

27. $x^2 - 3x + 8$
$\underline{-2x^2 + 3x - 7}$
$-x^2 \qquad + 1$

29. $(3y^2 - 7y) + (2y^2 - 8y + 2)$
$= (3y^2 + 2y^2) + (-7y - 8y) + 2$
$= 5y^2 - 15y + 2$

31. $(2a^2 - 3a - 7) - (-5a^2 - 2a - 9)$
$= (2a^2 + 5a^2) + (-3a + 2a) + (-7 + 9)$
$= 7a^2 - a + 2$

33. $P(x) + R(x) = (x^2 - 3xy + y^2) + (2x^2 - 3y^2)$
$= (x^2 + 2x^2) - 3xy + (y^2 - 3y^2)$
$= 3x^2 - 3xy - 2y^2$

35. $P(x) - R(x)$
$= (3x^2 + 2y^2) - (-5x^2 + 2xy - 3y^2)$
$= (3x^2 + 5x^2) - 2xy + (2y^2 + 3y^2)$
$= 8x^2 - 2xy + 5y^2$

Critical Thinking

37. $(2x^3 + 3x^2 + kx + 5) - (x^3 + x^2 - 5x - 2) = x^3 + 2x^2 + 3x + 7$
$(2x^3 - x^3) + (3x^2 - x^2) + (kx + 5x) + (5 + 2) = x^3 + 2x^2 + 3x + 7$
$x^3 + 2x^2 + (k + 5)x + 7 = x^3 + 2x^2 + 3x + 7$
$(k + 5)x = 3x$
$k + 5 = 3$
$k = -2$

39. $P(-1) = -3;\ P(x) = 4x^4 - 3x^2 + 6x + c$
$-3 = 4(-1)^4 - 3(-1)^2 + 6(-1) + c$
$-3 = 4 - 3 - 6 + c$
$-3 = -5 + c$
$2 = c$

41. If $P(x)$ is a fifth degree polynomial and $Q(x)$
is a fourth degree polynomial, then $P(x) -$
$Q(x)$ is a fifth degree polynomial.
Example: $P(x) = 3x^5 - 5x - 8$
$Q(x) = 3x^4 - 2x + 1$
$P(x) - Q(x) = 3x^5 - 3x^4 - 3x - 9$

Projects and Group Activities

43. The graph of $k(x)$ is the graph of $f(x)$ moved
2 units down.

Section 5.3

Concept Check

1. Multiply $2(2x + 1)$. By the Order of
Operation Agreement, we do multiplication
before addition.

3. FOIL is a method used to find the product of
two binomials and is based on the Distributive
Property. The letters of FOIL stand for First,
Outer, Inner, and Last.

Objective A Exercises

5. $3x^2(2x^2 - x) = 6x^4 - 3x^3$

7. $3x^2(2x^2 - x) = 6x^4 - 3x^3$

9. $3xy(2x - 3y) = 6x^2y - 9xy^2$

11. $-3xy^2(4x - 5y) = -12x^2y^2 + 15xy^3$

13. $2b + 4b(2 - b) = 2b + 8b - 4b^2 = -4b^2 + 10b$

15. $-2a^2(3a^2 - 2a + 3) = -6a^4 + 4a^3 - 6a^2$

17. $(-3y^2 - 4y + 2)(y^2) = -3y^4 - 4y^3 + 2y^2$

19. $-5x^2(4 - 3x + 3x^2 + 4x^3)$
$= -20x^2 + 15x^3 - 15x^4 - 20x^5$

21. $-2x^2y(x^2 - 3xy + 2y^2) = -2x^4y + 6x^3y^2 - 4x^2y^3$

23. $5x^3 - 4x(2x^2 + 3x - 7)$
$= 5x^3 - 8x^3 - 12x^2 + 28$
$= -3x^3 - 12x^2 + 28$

25. $2y^2 - y[3 - 2(y - 4) - y]$
$= 2y^2 - y\,(3 - 2y + 8 - y)$
$= 2y^2 - y(11 - 3y)$
$= 2y^2 - 11y + 3y^2$
$= 5y^2 - 11y$

27. $2y - 3[y - 2y(y - 3) + 4y]$
$= 2y - 3\,(y - 2y^2 + 6y + 4y)$
$= 2y - 3(-2y^2 + 11y)$
$= 2y + 6y^2 - 33y$
$= 6y^2 - 31y$

29. $P(b) = 3b$ and $Q(b) = 3b^4 - 3b^2 + 8$
$P(b)\cdot Q(b) = 3b(3b^4 - 3b^2 + 8)$
$= 9b^5 - 9b^3 + 24b$

Objective B Exercises

31. $(x - 2)(x + 7) = x^2 + 7x - 2x - 14$
$= x^2 + 5x - 14$

33. $(2y - 3)(4y + 7) = 8y^2 + 14y - 12y - 21$
$$= 8y^2 + 2y - 21$$

35. $(a + 3c)(4a - 5c) = 4a^2 - 5ac + 12ac - 15c^2$
$$= 4a^2 + 7ac - 15c^2$$

37. $(5x - 7)(5x - 7) = 25x^2 - 35x - 35x + 49$
$$= 25x^2 - 70x + 49$$

39. $2(2x - 3y)(2x + 5y)$
$$= 2(4x^2 + 10xy - 6xy - 15y^2)$$
$$= 2(4x^2 + 4xy - 15y^2)$$
$$= 8x^2 + 8xy - 30y^2$$

41. $(xy + 4)(xy - 3) = x^2y^2 - 3xy + 4xy - 12$
$$= x^2y^2 + xy - 12$$

43. $(2x^2 - 5)(x^2 - 5) = 2x^4 - 10x^2 - 5x^2 + 25$
$$= 2x^4 - 15x^2 + 25$$

51. $(2x^3 + 3x^2 - 2x + 5)(2x - 3)$
$$= (2x^3 + 3x^2 - 2x + 5)2x + (2x^3 + 3x^2 - 2x + 5)(-3)$$
$$= 4x^4 + 6x^3 - 4x^2 + 10x - 6x^3 - 9x^2 + 6x - 15$$
$$= 4x^4 - 13x^2 + 16x - 15$$

53. $(2x - 5)(2^{x4} - 3^{x3} - 2x + 9)$
$$= 2x(2^{x4} - 3^{x3} - 2x + 9) - 5(2^{x4} - 3^{x3} - 2x + 9)$$
$$= 4^{x5} - 6^{x4} - 4^{x2} + 18x - 10^{x4} + 15^{x3} + 10x - 45$$
$$= 4^{x5} - 16^{x4} + 15^{x3} - 4^{x2} + 28x - 45$$

55. $(x^2 + 2x - 3)(x^2 - 5x + 7)$
$$= x^2(x^2 - 5x + 7) + 2x(x^2 - 5x + 7) - 3(x^2 - 5x + 7)$$
$$= x^4 - 5x^3 + 7x^2 + 2x^3 - 10x^2 + 14x - 3x^2 + 15x - 21$$
$$= x^4 - 3x^3 - 6x^2 + 29x - 21$$

57. $(a - 2)(2a - 3)(a + 7)$
$$= (2a^2 - 3a - 4a + 6)(a + 7)$$
$$= (2a^2 - 7a + 6)(a + 7)$$
$$= (2a^2 - 7a + 6)a + (2a^2 - 7a + 6)7$$
$$= 2a^3 - 7a^2 + 6a + 14a^2 - 49a + 42$$
$$= 2a^3 + 7a^2 - 43a + 42$$

59. $(2x + 3)(x - 4)(3x + 5)$
$$= (2x^2 - 8x + 3x - 12)(3x + 5)$$
$$= (2x^2 - 5x - 12)(3x + 5)$$
$$= (2x^2 - 5x - 12)3x + (2x^2 - 5x - 12)5$$

45. $(5x^2 - 5y)(2x^2 - y)$
$$= 10x^4 - 5x^2y - 10x^2y + 5y^2$$
$$= 10x^4 - 15x^2y + 5y^2$$

49. $(x + 5)(x^2 - 3x + 4)$
$$= x(x^2 - 3x + 4) + 5(x^2 - 3x + 4)$$
$$= x^3 - 3x^2 + 4x + 5x^2 - 15x + 20$$
$$= x^3 + 2x^2 - 11x + 20$$

49. $(2a - 3b)(5a^2 - 6ab + 4b^2)$
$$= 2a(5a^2 - 6ab + 4b^2) - 3b(5a^2 - 6ab + 4b^2)$$
$$= 10a^3 - 12a^2b + 8ab^2 - 15a^2b + 18ab^2 - 12b^3$$
$$= 10a^3 - 27a^2b + 26ab^2 - 12b^3$$

$$= 6x^3 - 15x^2 - 36x + 10x^2 - 25x - 60$$
$$= 6x^3 - 5x^2 - 61x - 60$$

61. $P(y) = 2y^2 - 1$ and $Q(y) = y^3 - 5y^2 - 3$
$$P(y) \cdot Q(y) = (2y^2 - 1)(y^3 - 5y^2 - 3)$$
$$= 2y^2(y^3 - 5y^2 - 3) - 1(y^3 - 5y^2 - 3)$$
$$= 2y^5 - 10y^4 - 6y^2 - y^3 + 5y^2 + 3$$
$$= 2y^5 - 10y^4 - y^3 - y^2 + 3$$

63. mn

Objective C Exercises

65. $(3x - 2)(3x + 2) = 9x^2 - 4$

67. $(6 - x)(6 + x) = 36 - x^2$

69. $(2a - 3b)(2a + 3b) = 4a^2 - 9b^2$

71. $(3ab + 4)(3ab - 4) = 9a^2b^2 - 16$

73. $(x^2 + 1)(x^2 - 1) = x^4 - 1$

75. $(x - 5)^2 = x^2 - 10x + 25$

77. $(3a + 5b)^2 = 9a^2 + 30ab + 25b^2$

79. $(x^2 - 3)^2 = x^4 - 6x^2 + 9$

81. $(2x^2 - 3y^2)^2 = 4x^4 + 12x^2y^2 + 9y^4$

83. $(3mn - 5)^2 = 9m^2n^2 - 30mn + 25$

85. $y^2 - (x - y)^2 = y^2 - (x^2 - 2xy + y^2)$
$$= y^2 - x^2 + 2xy - y^2$$
$$= -x^2 + 2xy$$

87. $(x - y)^2 - (x + y)^2$
$$= x^2 - 2xy + y^2 - (x^2 + 2xy + y^2)$$
$$= x^2 - 2xy + y^2 - x^2 - 2xy - y^2$$
$$= -4xy$$

89. False

Objective D Exercises

91. ft^2

93. Strategy: To find the area, substitute the given values for L and W in the equation $A = L \cdot W$ and solve for A.

Solution: $A = L \cdot W$
$A = (3x - 2)(x + 4)$
$A = 3x^2 + 12x - 2x - 8$
$A = 3x^2 + 10x - 8$
The area is $(3x^2 + 10x - 8)$ ft^2.

95. Strategy: To find the area, add the area of the small rectangle to the area of the large rectangle.
Larger rectangle:
Length $= L_1 = x + 5$
Width $= W_1 = x - 2$
Smaller rectangle:
Length $= L_2 = 5$
Width $= W_2 = 2$

Solution: $A = L_1 \cdot W_1 + L_2 \cdot W_2$
$A = (x + 5)(x - 2) + (5)(2)$
$A = x^2 - 2x + 5x - 10 + 10$
$A = x^2 + 3x$
The area is $(x^2 + 3x)$ ft^2.

97. Strategy: To find the volume, substitute the given value for s in the equation $V = s^3$ and solve for V.

Solution: $V = s^3$
$V = (x + 3)^3$
$V = (x + 3)(x + 3)(x + 3)$
$V = (x^2 + 6x + 9)(x + 3)$
$V = x^3 + 9x^2 + 27x + 27$
The volume is $(x^3 + 9x^2 + 27x + 27)$ cm^3.

99. Strategy: To find the volume, subtract the volume of the small rectangular solid from the volume of the large rectangular solid.
Large rectangular solid:
Length $= L_1 = x + 2$
Width $= W_1 = 2x$
Height $= h_1 = x$
Small rectangular solid:
Length $= L_2 = x$
Width $= W_2 = 2x$
Height $= h_2 = 2$

Solution: $V = (L_1 \cdot W_1 \cdot h_1) - (L_2 \cdot W_2 \cdot h_2)$
$V = (x + 2)(2x)(x) - (x)(2x)(2)$
$V = (2x^2 + 4x)(x) - (2x^2)(2)$
$V = 2x^3 + 4x^2 - 4x^2$
$V = 2x^3$
The volume is $(2x^3)$ in^3.

101. **Strategy:** To find the area, substitute the given value for r into the equation $A = \pi r^2$ and solve for A.

Solution: $A = \pi r^2$
$A = 3.14(5x + 4)^2$
$A = 3.14(25x^2 + 40x + 16)$
$A = 78.5x^2 + 125.6x + 50.24$
The area is $(78.5x^2 + 125.6x + 50.24)$ in^2.

Critical Thinking

103. a) $(3x - k)(2x + k) = 6x^2 + 5x + k^2$
$6x^2 + xk - k^2 = 6x^2 + 5x + k^2$
$xk = 5x$
$k = 5$

b) $(4x + k)^2 = 16x^2 + 8x + k^2$
$16x^2 + 8xk + k^2 = 16x^2 + 8x + k^2$
$8xk = 8x$
$k = 1$

105. $(2x - 3)(x + 7) = 2x^2 + 14x - 3x - 21$
$2x^2 + 11x - 21$

Projects of Group Activities

107. Answers will vary. One example is
$(x + 3)(2x^2 - 1)$

Section 5.4

Concept Check

1. degree 4

3. If the polynomial $P(x)$ is divided by $x - a$, the remainder is $P(a)$.

Objective A Exercises

5. $\dfrac{3x^2 - 6x}{3x} = \dfrac{3x^2}{3x} - \dfrac{6x}{3x} = x - 2$

7. $\dfrac{5x^2 - 10x}{-5x} = \dfrac{5x^2}{-5x} - \dfrac{10x}{-5x} = -x + 2$

9. $\dfrac{5x^2 y^2 + 10xy}{5xy} = \dfrac{5x^2 y^2}{5xy} + \dfrac{10xy}{5xy} = xy + 2$

11. $\dfrac{x^3 + 3x^2 - 5x}{x} = \dfrac{x^3}{x} + \dfrac{3x^2}{x} - \dfrac{5x}{x} = x^2 + 3x - 5$

13. $\dfrac{9b^5 + 12b^4 + 6b^3}{3b^2} = \dfrac{9b^5}{3b^2} + \dfrac{12b^4}{3b^2} + \dfrac{6b^3}{3b^2}$
$= 3b^3 + 4b^2 + 2b$

15. $\dfrac{a^5 b - 6a^3 b + ab}{ab} = \dfrac{a^5 b}{ab} - \dfrac{6a^3 b}{ab} + \dfrac{ab}{ab}$
$= a^4 - 6a^2 + 1$

17. $P(x) = 6x^3 + 21x^2 - 15x$

Objective B Exercises

19.
$$
\begin{array}{r}
x + 8 \\
x - 5 \overline{\smash{)}\, x^2 + 3x - 40} \\
\underline{x^2 - 5x} \\
8x - 40 \\
\underline{8x - 40} \\
0
\end{array}
$$
$(x^2 + 3x - 40) \div (x - 5) = x + 8$

21.
$$
\begin{array}{r}
x^2 + 3x + 6 \\
x - 3 \overline{\smash{)}\, x^3 + 0x^2 - 3x + 2} \\
\underline{x^3 - 3x^2} \\
3x^2 - 3x \\
\underline{3x^2 - 9x} \\
6x + 2 \\
\underline{6x - 18} \\
20
\end{array}
$$
$(x^3 - 3x + 2) \div (x - 3) = x^2 + 3x + 6 + \dfrac{20}{x - 3}$

23.

$$
\begin{array}{r}
3x + 5 \\
2x+1\overline{)6x^2 + 13x + 8} \\
\underline{6x^2 + 3x} \\
10x + 8 \\
\underline{10x + 5} \\
3
\end{array}
$$

$$(6x^2 + 13x + 8) \div (2x + 1) = 3x + 5 + \dfrac{3}{2x+1}$$

25.

$$
\begin{array}{r}
5x + 7 \\
2x-1\overline{)10x^2 + 9x - 5} \\
\underline{10x^2 - 5x} \\
14x - 5 \\
\underline{14x - 7} \\
2
\end{array}
$$

$$(10x^2 + 9x - 5) \div (2x - 1) = 5x + 7 + \dfrac{2}{2x-1}$$

27.

$$
\begin{array}{r}
4x^2 + 6x + 9 \\
2x-3\overline{)8x^3 + 0x^2 + 0x - 9} \\
\underline{8x^3 - 12x^2} \\
12x^2 + 0x \\
\underline{12x^2 - 18x} \\
18x - 9 \\
\underline{18x - 27} \\
18
\end{array}
$$

$$(8x^3 - 9) \div (2x - 3) = 4x^2 + 6x + 9 + \dfrac{18}{2x-3}$$

29.

$$
\begin{array}{r}
3x^2 + 1 \\
2x^2-5\overline{)6x^4 + 0x^3 - 13x^2 + 0x - 4} \\
\underline{6x^4 \quad\quad -15x^2} \\
2x^2 + 0x - 4 \\
\underline{2x^2 \quad\quad -5} \\
1
\end{array}
$$

$$(6x^4 - 13x^2 - 4) \div (2x^2 - 5)$$

$$= 3x^2 + 1 + \dfrac{1}{2x^2 - 5}$$

31.

$$
\begin{array}{r}
x^2 - 3x - 10 \\
3x+1\overline{)3x^3 - 8x^2 - 33x - 10} \\
\underline{3x^3 + x^2} \\
-9x^2 - 33x \\
\underline{-9x^2 - 3x} \\
-30x - 10 \\
\underline{-30x - 10} \\
0
\end{array}
$$

$$\dfrac{3x^3 - 8x^2 - 33x - 10}{3x+1} = x^2 - 3x - 10$$

33.

$$
\begin{array}{r}
x^2 - 2x + 1 \\
x-3\overline{)x^3 - 5x^2 + 7x - 4} \\
\underline{x^3 - 3x^2} \\
-2x^2 + 7x \\
\underline{-2x^2 + 6x} \\
x - 4 \\
\underline{x - 3} \\
-1
\end{array}
$$

$$\dfrac{x^3 - 5x^2 + 7x - 4}{x - 3} = x^2 - 2x + 1 - \dfrac{1}{x - 3}$$

35.

$$
\begin{array}{r}
2x^3 - 3x^2 + x - 4 \\
x-5 \overline{)\, 2x^4 - 13x^3 + 16x^2 - 9x + 20} \\
\underline{2x^4 - 10x^3} \\
-3x^3 + 16x^2 \\
\underline{-3x^3 + 15x^2} \\
x^2 - 9x \\
\underline{x^2 - 5x} \\
-4x + 20 \\
\underline{-4x + 20} \\
0
\end{array}
$$

$$\frac{2x^4 - 13x^3 + 16x^2 - 9x + 20}{x-5} = 2x^3 - 3x^2 + x - 4$$

37.

$$
\begin{array}{r}
2x \\
x^2 + 2x - 1 \overline{)\, 2x^3 + 4x^2 - x + 2} \\
\underline{2x^3 + 4x^2 - 2x} \\
x + 2
\end{array}
$$

$$\frac{2x^3 + 4x^2 - x + 2}{x^2 + 2x - 1} = 2x + \frac{x+2}{x^2 + 2x - 1}$$

39.

$$
\begin{array}{r}
x^2 + 4x + 6 \\
x^2 - 2x - 1 \overline{)\, x^4 + 2x^3 - 3x^2 - 6x + 2} \\
\underline{x^4 - 2x^3 - x^2} \\
4x^3 - 2x^2 - 6x \\
\underline{4x^3 - 8x^2 - 4x} \\
6x^2 - 2x + 2 \\
\underline{6x^2 - 12x - 6} \\
10x + 8
\end{array}
$$

$$\frac{x^4 + 2x^3 - 3x^2 - 6x + 2}{x^2 - 2x - 1} = x^2 + 4x + 6 + \frac{10x + 8}{x^2 - 2x - 1}$$

41. $\dfrac{P(x)}{Q(x)} = \dfrac{2x^3 + x^2 + 8x + 7}{2x + 1}$

$$
\begin{array}{r}
x^2 + 4 \\
2x + 1 \overline{)\, 2x^3 + x^2 + 8x + 7} \\
\underline{2x^3 + x^2} \\
8x + 7 \\
\underline{8x + 4} \\
3
\end{array}
$$

$$\frac{2x^3 + x^2 + 8x + 7}{2x + 1} = x^2 + 4 + \frac{3}{2x + 1}$$

43. $6x^3 + 27x^2 + 18x - 30$

Objective C Exercise

45. 3

47.

$$\begin{array}{r|rrr} -1 & 2 & -6 & -8 \\ & & -2 & 8 \\ \hline & 2 & -8 & 0 \end{array}$$

$$(2x^2 - 6x - 8) \div (x + 1) = 2x - 8$$

49.

$$\begin{array}{r|rrr} 2 & 3 & -5 & 6 \\ & & 6 & 2 \\ \hline & 3 & 1 & 8 \end{array}$$

$$(3x^2 - 14x + 16) \div (x - 2) = 3x - 8$$

51.

$$\begin{array}{r|rrr} 1 & 3 & 0 & -4 \\ & & 3 & 3 \\ \hline & 3 & 3 & -1 \end{array}$$

$$\begin{array}{rrr} 3 & 1 & 8 \end{array}$$

$$(3x^2 - 4) \div (x - 1) = 3x + 3 - \dfrac{1}{x - 1}$$

53.

$$\begin{array}{r|rrrr} -1 & 2 & -1 & 6 & 9 \\ & & -2 & 3 & -9 \\ \hline & 2 & -3 & 9 & 0 \end{array}$$

$$(2x^3 - x^2 + 6x + 9) \div (x + 1) = 2x^2 - 3x + 9$$

55.

$$\begin{array}{r|rrrr} 2 & 4 & 0 & -1 & -18 \\ & & 8 & 16 & 30 \\ \hline & 4 & 8 & 15 & 12 \end{array}$$

$$(4x^3 - x - 18) \div (x - 2) = 4x^2 + 8x + 15 + \dfrac{12}{x - 2}$$

57.

$$\begin{array}{r|rrrr} -4 & 2 & 5 & -5 & 20 \\ & & -8 & 12 & -28 \\ \hline & 2 & -3 & 7 & -8 \end{array}$$

$$(2x^3 + 5x^2 - 5x + 20) \div (x + 4) = 2x^2 - 3x + 7 - \dfrac{8}{x + 4}$$

59.

$$\begin{array}{r|rrrrr} 2 & 3 & -4 & 8 & -5 & -5 \\ & & 6 & 4 & 24 & 38 \\ \hline & 3 & 2 & 12 & 19 & 33 \end{array}$$

$$(3x^4 - 4x^3 + 8x^2 - 5x - 5) \div (x - 2)$$

$$= 3x^3 + 2x^2 + 12x + 19 + \dfrac{33}{x - 2}$$

61.

$$\begin{array}{r|rrrrr} -1 & 3 & 3 & -1 & 3 & 2 \\ & & -3 & 0 & 1 & -4 \\ \hline & 3 & 0 & -1 & 4 & -2 \end{array}$$

$$(3x^4 + 3x^3 - x^2 + 3x + 2) \div (x + 1)$$

$$= 3x^3 - x + 4 - \dfrac{2}{x + 1}$$

63. $\dfrac{P(x)}{Q(x)} = \dfrac{3x^2 - 5x + 6}{x - 2}$

$$\begin{array}{r|rrr} 2 & 3 & -5 & 6 \\ & & 6 & 2 \\ \hline & 3 & 1 & 8 \end{array}$$

$$(3x^2 - 5x + 6) \div (x - 2) = 3x + 1 + \dfrac{8}{x - 2}$$

Objective D Exercises

65. $x - 3$

67.

$$\begin{array}{r|rrr} 3 & 2 & -3 & -1 \\ & & 6 & 9 \\ \hline & 2 & 3 & 8 \end{array}$$

$$P(3) = 8$$

69.

$$\begin{array}{r|rrrr} 4 & 1 & -2 & 3 & -1 \\ & & 4 & 8 & 44 \\ \hline & 1 & 2 & 11 & 43 \end{array}$$

$$R(4) = 43$$

71.

$$\begin{array}{r|rrrr} -2 & 2 & -4 & 3 & -1 \\ & & -4 & 16 & -38 \\ \hline & 2 & -8 & 19 & -39 \end{array}$$

$P(-2) = -39$

73.

$$\begin{array}{r|rrrr} -3 & 2 & -1 & 0 & 3 \\ & & -6 & 21 & -63 \\ \hline & 2 & -7 & 21 & -60 \end{array}$$

$Z(-3) = -60$

75.

$$\begin{array}{r|rrrrr} 2 & 1 & 3 & -2 & 4 & -9 \\ & & 2 & 10 & 16 & 40 \\ \hline & 1 & 5 & 8 & 20 & 31 \end{array}$$

$Q(2) = 31$

77.

$$\begin{array}{r|rrrrr} -3 & 2 & -1 & 0 & 2 & -5 \\ & & -6 & 21 & -63 & 183 \\ \hline & 2 & -7 & 21 & -61 & 178 \end{array}$$

$F(-3) = 178$

79.

$$\begin{array}{r|rrrr} 5 & 1 & 0 & 0 & -3 \\ & & 5 & 25 & 125 \\ \hline & 1 & 5 & 25 & 122 \end{array}$$

$P(5) = 122$

81.

$$\begin{array}{r|rrrrr} -3 & 4 & 0 & -3 & 0 & 5 \\ & & -12 & 36 & -99 & 297 \\ \hline & 4 & -12 & 33 & -99 & 302 \end{array}$$

$R(-3) = 302$

83.

$$\begin{array}{r|rrrrrr} 2 & 1 & 0 & -4 & -2 & 5 & -2 \\ & & 2 & 4 & 0 & -4 & 2 \\ \hline & 1 & 2 & 0 & -2 & 1 & 0 \end{array}$$

$Q(2) = 0$

Critical Thinking

85. Use the Remainder Theorem to find k so that
$P(3) = 0$.
$(3)^3 - 3(3)^2 - 3 + k = 0$
$27 - 27 - 3 + k = 0$
$-3 + k = 0$
$k = 3$

87. Use the Remainder Theorem to find k so that
$P(3) = 0$.
$(3)^2 + 3k - 6 = 0$
$9 + 3k - 6 = 0$
$3 + 3k = 0$
$3k = -3$
$k = -1$

89. Possible degrees of $p(x)$ and $q(x)$ are:
1, 5; 2, 4; 3, 3

Projects or Groups Activities

91. $P(-5) = (-5)^4 + (-5)^3 - 21(-5)^2 - (-5) + 20$
$\quad\quad = 0$
Yes, $x + 5$ is a factor.

Check Your Progress: Chapter 5

1. $(-12a^6b)(6a^4b^3) = -72a^{10}b^4$

2. $(-14x^6y^8)(-3x^{-1}y^{-8}) = 42x^5$

3. $(2x^3)^4(3x^2) = (16x^{12})(3x^2) = 48x^{14}$

4. $(2a^3b^{-4})^4(3a^{-3}b^2)^{-2} = \dfrac{16a^{12}b^{-16}}{9a^{-6}b^4} = \dfrac{16a^{18}}{9b^{20}}$

5. $\dfrac{x^4y}{xy^5} = \dfrac{x^3}{y^4}$

6. $\dfrac{2x^{-3}}{4x^{-5}} = \dfrac{x^2}{2}$

7. $\dfrac{3a^4 b^2 c^8}{6a^7 b^{-2} c^8} = \dfrac{b^4}{2a^3}$

8. $\dfrac{\left(3x^3 y^{-2}\right)^{-2}}{\left(2x^{-1} y^3\right)^3} = \dfrac{3^{-2} x^{-6} y^4}{8x^{-3} y^9} = \dfrac{1}{72x^3 y^5}$

9. 6.83×10^{-7}

10. $P(-2) = 4(-2)^3 - 6(-2) + 1 = -19$

11. $3x^2 - 6x + 7$
$\underline{2x^2 + x - 9}$
$5x^2 - 5x - 2$

12. $(-5x^2 + 7x - 8) - (6x^2 + 7x - 7)$
$= -5x^2 + 7x - 8 - 6x^2 - 7x + 7$
$= -11x^2 - 1$

13. $5 + 3(2x - 7) = 5 + 3(2x) - 3(7)$
$\qquad\qquad\qquad = 5 + 6x - 21$
$\qquad\qquad\qquad = 6x - 16$

14. $(3x + 4)(4x - 5)$
$= (3x)(4x) + (3x)(-5) + 4(4x) + (4)(-5)$
$= 12x^2 - 15x + 16x - 20$
$= 12x^2 + x - 20$

15. $(5x - 2)^2 = (5x - 2)(5x - 2)$
$\qquad\qquad\quad = 25x^2 - 10x - 10x + 4$
$\qquad\qquad\quad = 25x^2 - 20x + 4$

16. $(x^2 - 7x + 2)(x - 3)$
$= x(x^2 - 7x + 2) - 3(x^2 - 7x + 2)$
$= x^3 - 7x^2 + 2x - 3x^2 + 21x - 6$
$= x^3 - 10x^2 + 23x - 6$

17. $\dfrac{9x^3 + 12x^2 - 21x}{3x} = \dfrac{9x^3}{3x} + \dfrac{12x^2}{3x} - \dfrac{21x}{3x}$
$\qquad\qquad\qquad\qquad = 3x^2 + 4x - 7$

18. $\dfrac{4ab^2 + 6a^3 b^4 - 14a^2 b}{2a^2 b} = \dfrac{4ab^2}{2a^2 b} + \dfrac{6a^3 b^4}{2a^2 b} - \dfrac{14a^2 b}{2a^2 b}$

$= \dfrac{2b}{a} + 3ab^3 - 7$

19.
$$
\begin{array}{r}
2x^2 + 5x - 3 \\
x - 3 \overline{\smash{)}\, 2x^3 - x^2 - 18x + 9} \\
\underline{2x^3 - 6x^2} \\
5x^2 - 18x \\
\underline{5x^2 - 15x} \\
-3x + 9 \\
\underline{-3x + 9} \\
0
\end{array}
$$

$\dfrac{2x^3 - x^2 - 18x + 9}{x - 3} = 2x^2 + 5x - 3$

20.
$$
\begin{array}{r|rrrrr}
-2 & 3 & -2 & 0 & 1 & -5 \\
   &   & -6 & 16 & -32 & 62 \\
\hline
   & 3 & -8 & 16 & -31 & 57
\end{array}
$$

$\dfrac{3x^4 - 2x^3 + x - 5}{x + 2} = 3x^3 - 8x^2 + 16x - 31 + \dfrac{57}{x + 2}$

21. $p(-3) = 3(-3)^2 + (-3) - 4$
$\qquad\qquad = 27 - 3 - 4$
$\qquad\qquad = 20$

22.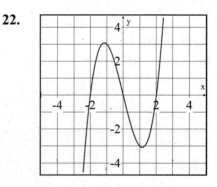

Section 5.5

Concept Check

1. The GCF of 12 and 18 is 6.

3. The GCF of $3a^2b^3$ and $6ab^4$ is $3ab^3$.

5. Yes

Objective A Exercises

7. The GCF of $6a^2$ and $15a$ is $3a$.
$6a^2 - 15a = 3a(2a - 5)$

9. The GCF of $4x^3$ and $3x^2$ is x^2.
$4x^3 - 3x^2 = x^2(4x - 3)$

11. There is no common factor.

13. The GCF of x^5, x^3 and x is x.
$x^5 - x^3 - x = x(x^4 - x^2 - 1)$

15. The GCF of $16x^2$, $12x$ and 24 is 4.
$16x^2 - 12x + 24 = 4(4x^2 - 3x + 6)$

17. The GCF of $25b^4$, $10b^3$ and $5b^2$ is $5b^2$.
$5b^2 - 10b^3 + 25b^4 = 5b^2(1 - 2b + 5b^2)$

19. The GCF of $12x^2y^2$, $18x^3y$ and $24x^2y$ is $6x^2y$.
$12x^2y^2 - 18x^3y + 24x^2y = 6x^2y(2y - 3x + 4)$

21. The GCF of $24a^3b^2$, $4a^2b^2$ and $16a^2b^4$ is $4a^2b^2$.
$24a^3b^2 - 4a^2b^2 - 16a^2b^4$
$\quad = 4a^2b^2(6a - 1 - 4b^2)$

23. $3x, 2x + 1$

Objective B Exercises

25. $x(a + 2) - 2(a + 2) = (a + 2)(x - 2)$

27. $a(x - 2) - b(2 - x) = a(x - 2) + b(x - 2)$
$\quad = (x - 2)(a + b)$

29. $x(a - 2b) + y(2b - a) = x(a - 2b) - y(a - 2b)$
$\quad = (a - 2b)(x - y)$

31. $xy + 4y - 2x - 8 = (xy + 4y) - (2x + 8)$
$\quad = y(x + 4) - 2(x + 4)$
$\quad = (x + 4)(y - 2)$

33. $ax + bx - ay - by = (ax + bx) - (ay + by)$
$\quad = x(a + b) - y(a + b)$
$\quad = (a + b)(x - y)$

35. $x^2y - 3x^2 - 2y + 6$
$\quad = (x^2y - 3x^2) - (2y - 6)$
$\quad = x^2(y - 3) - 2(y - 3)$
$\quad = (y - 3)(x^2 - 2)$

37. $6 + 2y + 3x^2 + x^2y$
$\quad = (6 + 2y) + (3x^2 + x^2y)$
$\quad = 2(3 + y) + x^2(3 + y)$
$\quad = (3 + y)(2 + x^2)$

39. $2ax^2 + bx^2 - 4ay - 2by$
$\quad = (2ax^2 + bx^2) - (4ay + 2by)$
$\quad = x^2(2a + b) - 2y(2a + b)$
$\quad = (2a + b)(x^2 - 2y)$

41. $6xb + 3ax - 4by - 2ay$
$\quad = (6xb + 3ax) - (4by + 2ay)$
$\quad = 3x(2b + a) - 2y(2b + a)$
$\quad = (2b + a)(3 - 2y)$

43. a) All three expressions are equivalent to
$x^2 - x - 6$.
b) (ii) $x^2 - 3x + 2x - 6$

Critical Thinking

45. $4x^{-2} + 6x^{-4} = 2x^{-4}(2x^2 + 3)$

47. $12x^{-1}y^{-2} - 18x^{-2}y^{-1} = 6x^{-2}y^{-2}(2x - 3y)$

Projects or Group Activities

49. Student answers will vary. One possible
polynomial is $6x^3y^2 + 15x^2y$.

51. Student answers will vary. One possible
expression is $3a(2a + b) + b(2a + b)$.

Section 5.6
Concept Check

1. (ii)

3. a) $-2, -9$
   b) $-2, 6$
   c) $-5, 4$
   d) $-3, 8$

Objective A Exercises

5. -3

7. $x^2 - 8x + 15 = (x - 3)(x - 5)$

9. $a^2 + 12a + 11 = (a + 1)(a + 11)$

11. $b^2 + 2b - 35 = (b + 7)(b - 5)$

13. $y^2 - 16y + 39 = (y - 3)(y - 13)$

15. $b^2 + 4b - 32 = (b + 8)(b - 4)$

17. $a^2 - 15a + 56 = (a - 7)(a - 8)$

19. $y^2 + 13y + 12 = (y + 1)(y + 12)$

21. $x^2 + 4x - 5 = (x + 5)(x - 1)$

23. $a^2 + 11ab + 30b^2 = (a + 5b)(a + 6b)$

25. $x^2 - 14xy + 24y^2 = (x - 2y)(x - 12y)$

27. $y^2 + 2xy - 63x^2 = (y + 9x)(y - 7x)$

29. $21 - 4x - x^2 = (7 + x)(3 - x) = -(x - 3)(x + 7)$

31. $50 + 5a - a^2 = (5 + a)(10 - a)$
    $= -(a - 10)(a + 5)$

33. Not factorable

Objective B Exercises

35. $2x^2 - 11x - 40 = (2x + 5)(x - 8)$

37. $4y^2 - 15y + 9 = (4y - 3)(y - 3)$

39. $2a^2 + 13a + 6 = (2a + 1)(a + 6)$

41. Not factorable

43. $5x^2 + 26x + 5 = (5x + 1)(x + 5)$

45. $11x^2 - 122x + 11 = (11x - 1)(x - 11)$

47. Not factorable

49. $6x^2 + 5xy - 21y^2 = (2x - 3y)(3x + 7y)$

51. $4a^2 + 43ab + 63b^2 = (4a + 7b)(a + 9b)$

53. $10x^2 - 23xy + 12y^2 = (5x - 4y)(2x - 3y)$

55. $24 + 13x - 2x^2 = (3 + 2x)(8 - x)$
    $= -(2x + 3)(x - 8)$

57. Not factorable

59. $15 - 14a - 8a^2 = (5 + 2a)(3 - 4a)$
    $= -(2a + 5)(4a - 3)$

61. $12y^3 + 22y^2 - 70y = 2y(6y^2 + 11y - 35)$
    $= 2y(3y - 5)(2y + 7)$

63. $30a^2 + 85ab + 60b^2$
    $= 5(6a^2 + 17ab + 12b^2)$
    $= 5(3a + 4b)(2a + 3b)$

65. $12x + x^2 - 6x^3 = x(12 + x - 6x^2)$
    $= x(4 + 3x)(3 - 2x)$
    $= -x(3x + 4)(2x - 3)$

67. Answers will vary. One possible answer
    is $2x^2 + 3x + 5$.

69. $2x^2 - 5x + 2 = (2x - 1)(x - 2)$
    $f(x) = 2x - 1; \ h(x) = x - 2$

71. $3a^2 + 11a - 4 = (3a - 1)(a + 4)$
    $f(a) = 3a - 1; \ g(a) = a + 4$

73. $2x^2 + 13x - 24 = (2x - 3)(x + 8)$
    $f(x) = 2x - 3; \ g(x) = x + 8$

75. $6x^2 + 7x - 5 = (2x - 1)(3x + 5)$
    $g(x) = 2x - 1; \ h(x) = 3x + 5$

77. $6t^2 - 17t - 3 = (6t + 1)(t - 3)$
    $f(t) = 6t + 1; \ g(t) = t - 3$

Critical Thinking

79. a) $x^2 + kx + 8$
  $ab = 8$
  $a + b = k$
  Factors of 8 are 2, 4; –2, –4; 1, 8; –1, –8
  Values for k: 6, –6, 9, –9
 b) $x^2 + kx - 6$
  $ab = -6$
  $a + b = k$
  Factors of –6 are 2, –3; –2, 3; 1, –6; –1, 6
  Values for k: –1, 1, –5, 5
 c) $2x^2 + kx + 3$
  $ab = 3$
  $a + 2b = k$
  Factors of 3 are 1, 3; –1, –3
  Values for k: 5, –5, 7, –7
 d) $2x^2 + kx - 5$
  $ab = -5$
  $a + 2b = k$
  Factors of –5 are 1, –5; –1, 5
  Values for k: –9, 9, –3, 3
 e) $3x^2 + kx + 5$
  $ab = 5$
  $a + 3b = k$
  Factors of 5 are 1, 5; –1, –5
  Values for k: 16, –16, 8, –8
 f) $2x^2 + kx - 3$
  $ab = -3$
  $a + 2b = k$
  Factors of –3 are 1, –3; –1, 3
  Values for k: –5, 5, –1, 1

Section 5.7

Concept Check
1. a) No
 b) Yes
 c) No
 d) Yes

3. a) Yes
 b) No
 c) Yes
 d) No

5. a) $2x^3$
 b) $3y^5$
 c) $4a^2b^6$
 d) $5c^4d$

7. a) Yes
 b) No
 c) Yes
 d) Yes

Objective A Exercises

9. $x^2 - 16 = x^2 - 4^2 = (x + 4)(x - 4)$

11. $4x^2 - 1 = (2x)^2 - 1^2 = (2x + 1)(2x - 1)$

13. $16x^2 - 121 = (4x)^2 - 11^2 = (4x + 11)(4x - 11)$

15. $1 - 9a^2 = 1^2 - (3a)^2 = (1 + 3a)(1 - 3a)$

17. $x^2y^2 - 100 = (xy)^2 - 10^2 = (xy + 10)(xy - 10)$

19. Not factorable

21. $25 - a^2b^2 = 5^2 - (ab)^2 = (5 + ab)(5 - ab)$

23. $x^2 - 12x + 36 = (x - 6)^2$

25. $b^2 - 2b + 1 = (b - 1)^2$

27. $16x^2 - 40x + 25 = (4x - 5)^2$

29. Not factorable

31. Not factorable

33. $x^2 + 6xy + 9y^2 = (x + 3y)^2$

35. $25a^2 - 40ab + 16b^2 = (5a - 4b)^2$

37. $(x - 4)^2 - 9 = [(x - 4) - 3][(x - 4) + 3]$
  $= (x - 7)(x - 1)$

39. $(x - y)^2 - (a + b)^2$
  $= [(x - y) - (a + b)][(x - y) + (a + b)]$
  $= (x - y - a - b)(x - y + a + b)$

Objective B Exercises

41. $x^3 - 27 = x^3 - 3^3$
$\quad = (x - 3)(x^2 + 3x + 9)$

43. $8x^3 - 1 = (2x)^3 - 1^3$
$\quad = (2x - 1)(4x^2 + 2x + 1)$

45. $x^3 - y^3 = (x - y)(x^2 + xy + y^2)$

47. $m^3 + n^3 = (m + n)(m^2 - mn + n^2)$

49. $64x^3 + 1 = (4x)^3 - 1^3$
$\quad = (4x + 1)(16x^2 - 4x + 1)$

51. $27x^3 - 8y^3 = (3x)^3 - (2y)^3$
$\quad = (3x - 2y)(9x^2 + 6xy + 4y^2)$

53. $x^3y^3 + 64 = (xy)^3 + (4)^3$
$\quad = (xy + 4)(x^2y^2 - 4xy + 16)$

55. Not factorable

57. Not factorable

59. $(a - b)^3 - b^3$
$\quad = [(a - b) - b][(a - b)^2 + b(a - b) + b^2]$
$\quad = (a - 2b)(a^2 - 2ab + b^2 + ab - b^2 + b^2)$
$\quad = (a - 2b)(a^2 - ab + b^2)$

Objective C Exercises

61. No. Polynomials cannot have square roots as variable terms.

63. Let $u = xy$
$\quad x^2y^2 - 8xy + 15 = u^2 - 8u + 15$
$\quad\quad = (u - 3)(u - 5)$
$\quad\quad = (xy - 3)(xy - 5)$

65. Let $u = xy$
$\quad x^2y^2 - 17xy + 60 = u^2 - 17u + 60$
$\quad\quad = (u - 12)(u - 5)$
$\quad\quad = (xy - 12)(xy - 5)$

67. Let $u = x^2$
$\quad x^4 - 9x^2 + 18 = u^2 - 9u + 18$
$\quad\quad = (u - 3)(u - 6)$
$\quad\quad = (x^2 - 3)(x^2 - 6)$

69. Let $u = b^2$
$\quad b^4 - 13b^2 - 90 = u^2 - 13u - 90$
$\quad\quad = (u + 5)(u - 18)$
$\quad\quad = (b^2 + 5)(b^2 - 18)$

71. Let $u = x^2y^2$
$\quad x^4y^4 - 8x^2y^2 + 12 = u^2 - 8u + 12$
$\quad\quad = (u - 2)(u - 6)$
$\quad\quad = (x^2y^2 - 2)(x^2y^2 - 6)$

73. Let $u = \sqrt{x}$.
$\quad x + 3\sqrt{x} + 2 = u^2 + 3u + 2$
$\quad\quad = (u + 2)(u + 1)$
$\quad\quad = (\sqrt{x} + 2)(\sqrt{x} + 1)$

75. Let $u = xy$
$\quad 3x^2y^2 - 14xy + 15 = 3u^2 - 14u + 15$
$\quad\quad = (3u - 5)(u - 3)$
$\quad\quad = (3xy - 5)(xy - 3)$

77. Let $u = ab$
$\quad 6a^2b^2 - 23ab + 21 = 6u^2 - 23u + 21$
$\quad\quad = (2u - 3)(3u - 7)$
$\quad\quad = (2ab - 3)(3ab - 7)$

79. Let $u = x^2$
$\quad 2x^4 - 13x^2 - 15 = 2u^2 - 13u - 15$
$\quad\quad = (2u - 15)(u + 1)$
$\quad\quad = (2x^2 - 15)(x^2 + 1)$

81. Let $u = x^3$.
$\quad x^6 - x^3 - 6 = u^2 - u - 6$
$\quad\quad = (u + 2)(u - 3)$
$\quad\quad = (x^3 + 2)(x^3 - 3)$

83. Let $u = xy^2$
$\quad 4x^2y^4 - 12xy^2 + 9 = 4u^2 - 12u + 9$
$\quad\quad = (2u - 3)(2u - 3)$
$\quad\quad = (2xy^2 - 3)(2xy^2 - 3)$
$\quad\quad = (2xy^2 - 3)^2$

Objective D Exercises

85. $12x^2 - 36x + 27 = 3(4x^2 - 12x + 9)$
$\quad\quad = 3(2x - 3)^2$

87. $27a^4 - a = a(27a^3 - 1)$
$= a(3a - 1)(9a^2 + 3a + 1)$

89. $20x^2 - 5 = 5(4x^2 - 1)$
$= 5(2x + 1)(2x - 1)$

91. $y^5 + 6y^4 - 55y^3 = y^3(y^2 + 6y - 55)$
$= y^3(y + 11)(y - 5)$

93. $16x^4 - 81 = (4x^2 + 9)(4x^2 - 9)$
$= (4x^2 + 9)(2x + 3)(2x - 3)$

95. $16a - 2a^4 = 2a(8 - a^3)$
$= 2a(2 - a)(4 + 2a + a^2)$

97. $a^3b^6 - b^3 = b^3(a^3b^3 - 1)$
$= b^3(ab - 1)(a^2b^2 + ab + 1)$

99. $8x^4 - 40x^3 + 50x^2 = 2x^2(4x^2 - 20x + 25)$
$= 2x^2(2x - 5)^2$

101. $x^4 - y^4 = (x^2 + y^2)(x^2 - y^2)$
$= (x^2 + y^2)(x + y)(x - y)$

103. $x^6 + y^6 = (x^2 + y^2)(x^4 - x^2y^2 + y^4)$

105. Not factorable

107. $16a^4 - 2a = 2a(8a^3 - 1)$
$= 2a(2a - 1)(4a^2 + 2a + 1)$

109. $a^4b^2 - 8a^3b^3 - 48a^2b^4$
$= a^2b^2(a^2 - 8ab - 48b^2)$
$= a^2b^2(a + 4b)(a - 12b)$

111. $x^3 - 2x^2 - 4x + 8 = x^2(x - 2) - 4(x - 2)$
$= (x - 2)(x^2 - 4)$
$= (x - 2)(x + 2)(x - 2)$
$= (x - 2)^2(x + 2)$

113. $2x^3 + x^2 - 32x - 16$
$= x^2(2x + 1) - 16(2x + 1)$
$= (2x + 1)(x^2 - 16)$
$= (2x + 1)(x + 4)(x - 4)$

115. $4x^4 - x^2 - 4x^2y^2 + y^2$
$= x^2(4x^2 - 1) - y^2(4x^2 - 1)$
$= (4x^2 - 1)(x^2 - y^2)$
$= (2x + 1)(2x - 1)(x + y)(x - y)$

117. The coefficient is 8.

Critical Thinking

119. If $9x^2 - kx + 1$ is a perfect square
trinomial then:
$(3x - 1)(3x - 1)$ or $(3x + 1)(3x + 1)$
If $(3x - 1)(3x - 1)$, then $k = 6$.
If $(3x + 1)(3x + 1)$, then $k = -6$.

121. $a^3 + (a + b)^3$
$= (a + (a + b))(a^2 - a(a + b) + (a + b)^2$
$= (2a + b)(a^2 - a^2 - ab + a^2 + 2ab + b^2)$
$= (2a + b)(a^2 + ab + b^2)$

Projects or Group Activities

123. If 3 is a zero of $P(x) = x^3 - x^2 - 3x - 9$,
then $x - 3$ is a factor of $P(x)$. Therefore
$x^3 - x^2 - 3x - 9$ is divisible by $x - 3$.

$$\frac{x^3 - x^2 - 3x - 9}{x - 3} = x^2 + 2x + 3$$
$x^3 - x^2 - 3x - 9 = (x - 3)(x^2 + 2x + 3)$

Section 5.8

Concept Check

1. If the product or two factors is zero, then at
least one of the factors is equal to zero.

3. To solve a quadratic equation using the
Principle of Zero Products we must first
write the equation in standard form, then
factor the quadratic equation. Use the
Principle of Zero Products to set each factor
equal to zero.

5. a) Yes
b) No
c) Yes
d) No
e) Yes
f) No

7. $(x - 2)(x + 6) = 0$
$x - 2 = 0 \quad x + 6 = 0$
$\quad x = 2 \qquad x = -6$
The solutions are -6 and 2.

Objective A Exercises

9. $(x + 5)(x - 3) = 0$
$x + 5 = 0 \quad x - 3 = 0$
$\quad x = -5 \qquad x = 3$
The solutions are -5 and 3.

13. $z^2 - 4z + 3 = 0$
$(z - 1)(z - 3) = 0$
$z - 1 = 0 \quad z - 3 = 0$
$\quad z = 1 \qquad z = 3$
The solutions are 1 and 3.

13. $r^2 - 10 = 3r$
$r^2 - 3r - 10 = 0$
$(r - 5)(r + 2) = 0$
$r - 5 = 0 \quad r + 2 = 0$
$\quad r = 5 \qquad r = -2$
The solutions are -2 and 5.

15. $4t^2 = 4t + 3$
$4t^2 - 4t - 3 = 0$
$(2t + 1)(2t - 3) = 0$
$2t + 1 = 0 \quad 2t - 3 = 0$
$\quad 2t = -1 \qquad 2t = 3$
$\quad t = -\dfrac{1}{2} \qquad t = \dfrac{3}{2}$
The solutions are $-\dfrac{1}{2}$ and $\dfrac{3}{2}$.

17. $4v^2 - 4v + 1 = 0$
$(2v - 1)(2v - 1) = 0$
$2v - 1 = 0 \quad 2v - 1 = 0$
$\quad 2v = 1 \qquad 2v = 1$
$\quad v = \dfrac{1}{2} \qquad v = \dfrac{1}{2}$
The solution is $\dfrac{1}{2}$.

19. $x^2 - 9 = 0$
$(x - 3)(x + 3) = 0$
$x - 3 = 0 \quad x + 3 = 0$
$\quad x = 3 \qquad x = -3$
The solutions are -3 and 3.

21. $4y^2 - 1 = 0$
$(2y - 1)(2y + 1) = 0$
$2y - 1 = 0 \quad 2y + 1 = 0$
$\quad 2y = 1 \qquad 2y = -1$
$\quad y = \dfrac{1}{2} \qquad y = -\dfrac{1}{2}$
The solutions are $-\dfrac{1}{2}$ and $\dfrac{1}{2}$.

23. $x(x - 1) = x + 15$
$x^2 - x = x + 15$
$x^2 - 2x - 15 = 0$
$(x - 5)(x + 3) = 0$
$x - 5 = 0 \quad x + 3 = 0$
$\quad x = 5 \qquad x = -3$
The solutions are -3 and 5.

25. $2x^2 - 3x - 8 = x^2 + 20$
$x^2 - 3x - 28 = 0$
$(x - 7)(x + 4) = 0$
$x - 7 = 0 \quad x + 4 = 0$
$\quad x = 7 \qquad x = -4$
The solutions are -4 and 7.

27. $(3v + 2)(v - 4) = v^2 + v + 5$
$3v^2 - 10v - 8 = v^2 + v + 5$
$2v^2 - 11x - 13 = 0$
$(2v - 13)(v + 1) = 0$
$2v - 13 = 0 \quad v + 1 = 0$
$\qquad 2v = 13 \qquad v = -1$
$\qquad v = \dfrac{13}{2}$

The solutions are -1 and $\dfrac{13}{2}$.

29. $4x^2 + x - 10 = (x - 2)(x + 1)$
$4x^2 + x - 10 = x^2 - x - 2$
$3x^2 + 2x - 8 = 0$
$(3x - 4)(x + 2) = 0$
$3x - 4 = 0 \quad x + 2 = 0$
$\quad 3x = 4 \qquad x = -2$
$\quad x = \dfrac{4}{3}$

The solutions are -2 and $\dfrac{4}{3}$.

31. $f(x) = x^2 - 6x + 5$
$x^2 - 6x + 5 = 0$
$(x - 5)(x - 1) = 0$
$x - 5 = 0 \quad x - 1 = 0$
$\quad x = 5 \qquad x = 1$
The solutions are 5 and 1.

33. $s(t) = 2t^2 - t - 1$
$2t^2 - t - 1 = 0$
$(2t + 1)(t - 1) = 0$
$2t + 1 = 0 \quad t - 1 = 0$
$\quad t = -\dfrac{1}{2} \qquad t = 1$

The solutions are $-\dfrac{1}{2}$ and 1.

35. $g(x) = 6x^2 + 13x - 5$
$6x^2 + 13x - 5 = 0$
$(2x + 5)(3x - 1) = 0$
$2x + 5 = 0 \quad 3x - 1 = 0$

$\qquad x = -\dfrac{5}{2} \qquad\qquad x = \dfrac{1}{3}$

The solutions are $-\dfrac{5}{2}$ and $\dfrac{1}{3}$.

$\quad 3x = -2 \qquad x = -2 \qquad x = 2$
$\quad x = -\dfrac{2}{3}$

The solutions are -2, $-\dfrac{2}{3}$ and 2.

Objective B Exercises

37. Strategy: The height of the box is 2 in.
The width of the box is $w - 4$.
The length of the box is
$(w + 10) - 4 = w + 6$.
The volume of a box is
length · width · height and equals 112 in³.

Solution: $2(w - 4)(w + 6) = 112$
$w^2 + 2w - 24 = 56$
$w^2 + 2w - 80 = 0$
$(w - 8)(w + 10) = 0$
$w = 8 \quad w = -10$
The width cannot be a negative number so we can eliminate -10 from consideration.
The width is 8 in.
The length is $w + 10 = 8 + 10 = 18$ in.

39. Strategy: $h = -16t^2 + 48t + 3$
$h = 35$ ft

Solution: $h = -16t^2 + 48t + 3$
$35 = -16t^2 + 48t + 3$
$16t^2 - 48t + 32 = 0$
$t^2 - 3t + 2 = 0$
$(t - 2)(t - 1) = 0$
$t = 2 \quad t = 1$
After 1 s (on the way up) and 2 s (on the way down).

41. Strategy: Let x represent the width of the Big Screen.
The length of the Big Screen is $2x + 1$.
The area of a Big Screen is
length · width $= x \cdot (2x + 1)$
The area of the Big Screen is 300 ft^2.

Solution: $x(2x + 1) = 300$
$2x^2 + x - 300 = 0$
$(2x + 25)(x - 12) = 0$
$2x + 25 = 0 \quad x - 12 = 0$
$x = -12.5 \quad x = 12$
The width cannot be a negative number so we can eliminate -12.5 from consideration.
The width is 12 ft.
The length is $2x + 1 = 2(12) + 1 = 25$ ft.

43. Larry incorrectly assumed that if $ab = 15$ then $a = 3$ and $b = 5$.

Critical Thinking

45. $a^3 + a^2 - 9a - 9 = 0$
$a^2(a + 1) - 9(a + 1) = 0$
$(a + 1)(a^2 - 9) = 0$
$(a + 1)(a + 3)(a - 3) = 0$
$a + 1 = 0 \quad a + 3 = 0 \quad a - 3 = 0$
$a = -1 \quad a = -3 \quad a = 3$
The solutions are -3, -1 and 3.

47. $3x^3 + 2x^2 - 12x - 8 = 0$
$x^2(3x + 2) - 4(3x + 2) = 0$
$(3x + 2)(x^2 - 4) = 0$
$(3x + 2)(x + 2)(x - 2) = 0$
$3x + 2 = 0 \quad x + 2 = 0 \quad x - 2 = 0$
$3x = -2 \quad x = -2 \quad x = 2$
$x = -\dfrac{2}{3}$

The solutions are -2, $-\dfrac{2}{3}$ and 2.

49. $5x^3 + 2x^2 - 20x - 8 = 0$
$x^2(5x + 2) - 4(5x + 2) = 0$
$(5x + 2)(x^2 - 4) = 0$
$(5x + 2)(x + 2)(x - 2) = 0$
$5x + 2 = 0 \quad x + 2 = 0 \quad x - 2 = 0$
$5x = -2 \quad x = -2 \quad x = 2$
$x = -\dfrac{2}{5}$

The solutions are -2, $-\dfrac{2}{5}$ and 2.

Chapter 5 Review Exercises

1. The GCF of $18a^5b^2 - 12a^3b^3 + 30a^2b$ is $6a^2b$.
$6a^2b(3a^3b - 2ab^2 + 5)$

2.
$$\begin{array}{r} 5x + 4 \\ 3x - 2 \overline{)15x^2 + 2x - 2} \\ \underline{15x^2 - 10x} \\ 12x - 2 \\ \underline{12x - 8} \\ 6 \end{array}$$

$\dfrac{15x^2 + 2x - 2}{3x - 2} = 5x + 4 + \dfrac{6}{3x - 2}$

3. $(2x^{-1}y^2z^5)^4(-3x^3yz^{-3})^2$
$= (16x^{-4}y^8z^{20})(9x^6y^2z^{-6})$
$= 144x^2y^{10}z^{14}$

4. $2ax + 4bx - 3ay - 6by$
$= 2x(a + 2b) - 3y(a + 2b)$
$= (a + 2b)(2x - 3y)$

5. $12 + x - x^2 = (4 - x)(3 + x)$
$= -(x - 4)(x + 3)$

6. $P(x) = x^3 - 2x^2 + 3x - 5$
$P(2) = (2)^3 - 2(2)^2 + 3(2) - 5$
$P(2) = 8 - 8 + 6 - 5$
$P(2) = 1$

7. $(5x^2 - 8xy + 2y^2) - (x^2 - 3y^2)$
$= (5x^2 - x^2) - 8xy + (2y^2 + 3y^2)$
$= 4x^2 - 8xy + 5y^2$

8. $24x^2 + 38x + 15 = (6x + 5)(4x + 3)$

9. $4x^2 + 12xy + 9y^2 = (2x + 3y)^2$

10. $(-2a^2b^4)(3ab^2) = -6a^3b^6$

11. $64a^3 - 27b^3 = (4a)^3 - (3b)^3$
$= (4a - 3b)(16a^2 + 12ab + 9b^2)$

12.

$$
\begin{array}{r|rrrr}
-6 & 4 & 27 & 10 & 2 \\
   &   & -24 & -18 & 48 \\
\hline
   & 4 & 3 & -8 & 50
\end{array}
$$

$$\frac{4x^3 + 27x^2 + 10x + 2}{x + 6} = 4x^2 + 3x - 8 + \frac{50}{x + 6}$$

13. $P(x) = 2x^3 - x + 7$
$P(-2) = 2(-2)^3 - (-2) + 7$
$P(-2) = -16 + 2 + 7$
$P(-2) = -7$

14. $x^2 - 3x - 40 = (x - 8)(x + 5)$

15. Let $u = xy$
$x^2y^2 - 9 = u^2 - 9 = (u + 3)(u - 3)$
$= (xy + 3)(xy - 3)$

16. $4x^2y(3x^3y^2 + 2xy - 7y^3)$
$= 12x^5y^3 + 8x^3y^2 - 28x^2y^4$

17. $x^2 - 12x + 36 = (x - 6)(x - 6) = (x - 6)^2$

18. $6x^2 + 60 = 39x$
$6x^2 - 39x + 60 = 0$
$3(2x^2 - 13x + 20) = 0$
$3(2x - 5)(x - 4) = 0$
$2x - 5 = 0 \quad x - 4 = 0$
$\quad 2x = 5 \quad\quad x = 4$

$$x = \frac{5}{2}$$

The solutions are $\dfrac{5}{2}$ and 4.

19. $5x^2 - 4x[x - 3(3x + 2) + x]$
$= 5x^2 - 4x(x - 9x - 6 + x)$
$= 5x^2 - 4x(-7x - 6)$
$= 5x^2 + 28x^2 + 24x$
$= 33x^2 + 24x$

20. $3a^6 - 15a^4 - 18a^2$
$= 3a^2(a^4 - 5a^2 - 6)$
$= 3a^2(a^2 - 6)(a^2 + 1)$

21. $(4x - 3y)^2 = 16x^2 - 24xy + 9y^2$

22.

$$
\begin{array}{r|rrrrr}
4 & 1 & 0 & 0 & 0 & -4 \\
  &   & 4 & 16 & 64 & 256 \\
\hline
  & 1 & 4 & 16 & 64 & 252
\end{array}
$$

$$\frac{x^4 - 4}{x - 4} = x^3 + 4x^2 + 16x + 64 + \frac{252}{x - 4}$$

23. $(3x^2 - 2x - 6) + (-x^2 - 3x + 4)$
$= 2x^2 - 5x - 2$

24. $(5x^2yz^4)(2xy^3z^{-1})(7x^{-2}y^{-2}z^3)$
$= (10x^3y^4z^3)(7x^{-2}y^{-2}z^3)$
$= 70xy^2z^6$

25. $\dfrac{3x^4yz^{-1}}{-12xy^3z^2} = -\dfrac{x^3}{4y^2z^3}$

26. 9.48×10^8

27. $\dfrac{3 \times 10^{-3}}{15 \times 10^2} = 0.2 \times 10^{-5} = 2 \times 10^{-6}$

28. $P(x) = -2x^3 + 2x^2 - 4$

$P(-3) = -2(-3)^3 + 2(-3)^2 - 4$

$P(-3) = 54 + 18 - 4$

$P(-3) = 68$

29. $\dfrac{16x^5 - 8x^3 + 20x}{4x} = \dfrac{16x^5}{4x} - \dfrac{8x^3}{4x} + \dfrac{20x}{4x}$

$\qquad\qquad\qquad = 4x^4 - 2x^2 + 5$

30.
$$\begin{array}{r}
2x - 3 \\
6x+1{\overline{\smash{\big)}\,12x^2 - 16x - 7}} \\
\underline{12x^2 + 2x} \\
-18x - 7 \\
\underline{-18x - 3} \\
-4
\end{array}$$

$\dfrac{12x^2 - 16x - 7}{6x+1} = 2x - 3 - \dfrac{4}{6x+1}$

31. $a^3(a^4 - 5a + 2) = a^7 - 5a^4 + 2a^3$

32. $(x + 6)(x^3 - 3x^2 - 5x + 1)$

$= x(x^3 - 3x^2 - 5x + 1) + 6(x^3 - 3x^2 - 5x + 1)$

$= x^4 - 3x^3 - 5x^2 + x + 6x^3 - 18x^2 - 30x + 6$

$= x^4 + 3x^3 - 23x^2 - 29x + 6$

33. $10a^3b^3 - 20a^2b^4 + 35ab^2$

$= 5ab^2(2a^2b - 4ab^2 + 7)$

34. $5x^5 + x^3 + 4x^2 = x^2(5x^3 + x + 4)$

35. $x(y - 3) + 4(3 - y) = x(y - 3) - 4(y - 3)$

$= (y - 3)(x - 4)$

36. $x^2 - 16x + 63 = (x - 7)(x - 9)$

37. $24x^2 + 61x - 8 = (8x - 1)(3x + 8)$

38. We are looking for two polynomials that when multiplied together give us $5x^2 + 3x - 2$.

$5x^2 + 3x - 2 = (5x - 2)(x + 1)$

$f(x) = 5x - 2$ and $g(x) = x + 1$

39. $36 - a^2 = (6 + a)(6 - a)$

40. $8 - y^3 = 2^3 - (y)^3$

$= (2 - y)(4 + 2y + y^2)$

41. Let $u = x^4$

$36x^8 - 36x^4 + 5$

$= 36u^2 - 36u + 5$

$= (6u - 5)(6u - 1)$

$= (6x^4 - 5)(6x^4 - 1)$

42. $3a^4b - 3ab^4 = 3ab(a^3 - b^3)$

$= 3ab(a - b)(a^2 + ab + b^2)$

43. Let $u = x^2$.

$x^4 - 8x^2 + 16 = u^2 - 8u + 16$

$= (u - 4)^2 = (x^2 - 4)^2$

$= (x - 2)^2(x + 2)^2$

44. $x(x - 1) = 6$

$x^2 - x - 6 = 0$

$(x - 3)(x + 2) = 0$

$x - 3 = 0 \quad x + 2 = 0$

$x = 3 \qquad\quad x = -2$

The solutions are -2 and 3.

45. $x^2 - 16x = 0$

$x(x - 16) = 0$

$x = 0 \quad x - 16 = 0$

$\qquad\qquad x = 16$

The solutions are 0 and 16.

46. $f(x) = x^2 + 7x + 6$

$x^2 + 7x + 6 = 0$

$(x + 6)(x + 1) = 0$

$x + 6 = 0 \quad x + 1 = 0$

$x = -6 \qquad\quad x = -1$

The zeros of the function are -1 and -6.

47. Let $u = x^2$.

$15x^4 + x^2 - 6 = 15u^2 + u - 6$

$= (3u + 2)(5u - 3)$

$= (3x^2 + 2)(5x^2 - 3)$

48. $\dfrac{(2a^4b^{-3}c^2)^3}{(2a^3b^2c^{-1})^4} = \dfrac{8a^{12}b^{-9}c^6}{16a^{12}b^8c^{-4}}$

$= \dfrac{c^{10}}{2b^{17}}$

49. $(x-4)(3x+2)(2x-3)$
$= (x-4)(6x^2-5x-6)$
$= x(6x^2-5x-6) - 4(6x^2-5x-6)$
$= 6x^3 - 5x^2 - 6x - 24x^2 + 20x + 24$
$= 6x^3 - 29x^2 + 14x + 24$

50. Let $u = x^2y^2$.
$21x^4y^4 + 23x^2y^2 + 6 = 21u^2 + 23u + 6$
$= (7u+3)(3u+2)$
$= (7\,x^2y^2 + 3)(3\,x^2y^2 + 2)$

51. $x^3 + 16 = x(x+16)$
$x^3 + 16 = x^2 + 16x$
$x^3 - x^2 - 16x + 16 = 0$
$x^2(x-1) - 16(x-1) = 0$
$(x-1)(x^2-16) = 0$
$(x-1)(x-4)(x+4) = 0$
$x - 1 = 0 \quad x - 4 = 0 \quad x + 4 = 0$
$\quad\quad x = 1 \quad\quad x = 4 \quad\quad x = -4$
The solutions are −4, 1 and 4.

52. $(5a+2b)(5a-2b) = 25a^2 - 4b^2$

53. 0.00254

54. $6x^2 - 31x + 18 = (3x-2)(2x-9)$

55. $y = x^2 + 1$

x	x
−2	5
−1	2
0	1
1	2
2	5

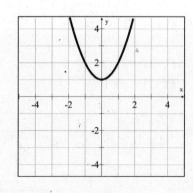

56. a) 3
 b) 8
 c) 5

57. Strategy: To find the mass of the moon, multiply the mass of the sun by 3.7×10^{-8}.

Solution:
$(2.19 \times 10^{27})(3.7 \times 10^{-8}) = 8.103 \times 10^{19}$
The mass of the moon is 8.103×10^{19} tons.

58. Strategy: Let x represent the number.
The square of the number is x^2.
The sum of the number and its square is 56.

Solution: $x + x^2 = 56$
$x^2 + x - 56 = 0$
$(x-7)(x+8) = 0$
$x = 7 \quad x = -8$
The number is 7 or −8.

59. Strategy: To find how far Earth is from the Great Galaxy of Andromeda, use the equation $d = rt$, where $r = 6.7 \times 10^8$ mph and $t = 2.2 \times 10^6$ years.
$2.2 \times 10^6 \times 24 \times 365 = 1.9272 \times 10^{10}$ hours.

Solution: $d = rt$
$d = (6.7 \times 10^8)(1.9272 \times 10^{10})$
$= 1.291224 \times 10^{19}$
The distance from Earth to the Great Galaxy of Andromeda is 1.291224×10^{19} mi.

60. Strategy: To find the area substitute the given values for L and W in the equation $A = LW$ and solve for A.

Solution: $A = LW$
$A = (5x+3)(2x-7) = 10x^2 - 29x - 21$
The area is $(10x^2 - 29x - 21)$ cm^2.

Chapter 5 Test

1. $16t^2 + 24t + 9 = (4t + 3)^2$

2. $-6rs^2(3r - 2s - 3) = -18r^2s^2 + 12rs^3 + 18rs^2$

3. $P(x) = 3x^2 - 8x + 1$
$P(2) = 3(2)^2 - 8(2) + 1$
$P(2) = 12 - 16 + 1$
$P(2) = -3$

4. $27x^3 - 8 = (3x)^3 - (2)^3$
$= (3x - 2)(9x^2 + 6x + 4)$

5. $16x^2 - 25 = (4x + 5)(4x - 5)$

6. $(3t^3 - 4t^2 + 1)(2t^2 - 5)$
$= 2t^2(3t^3 - 4t^2 + 1) - 5(3t^3 - 4t^2 + 1)$
$= 6t^5 - 8t^4 + 2t^2 - 15t^3 + 20t^2 - 5$
$= 6t^5 - 8t^4 - 15t^3 + 22t^2 - 5$

7. $-5x[3 - 2(2x - 4) - 3x]$
$= -5x(3 - 4x + 8 - 3x)$
$= -5x(-7x + 11)$
$= 35x^2 - 55x$

8. $12x^3 + 12x^2 - 45x = 3x(4x^2 + 4x - 15)$
$= 3x(2x - 3)(2x + 5)$

9. $6x^3 + x^2 - 6x - 1 = 0$
$x^2(6x + 1) - 1(6x + 1) = 0$
$(6x + 1)(x^2 - 1) = 0$
$(6x + 1)(x + 1)(x - 1) = 0$

$6x + 1 = 0 \quad x + 1 = 0 \quad x - 1 = 0$
$6x = -1 \qquad x = -1 \qquad x = 1$
$x = -\dfrac{1}{6}$

The solutions are -1, $-\dfrac{1}{6}$, 1.

10. $(6x^3 - 7x^2 + 6x - 7) - (4x^3 - 3x^2 + 7)$
$= (6x^3 - 4x^3) + (-7x^2 + 3x^2) + 6x + (-7 - 7)$
$= 2x^3 - 4x^2 + 6x - 14$

11. 5.01×10^{-7}

12.
$$\begin{array}{r} 2x + 1 \\ 7x - 3 \overline{)14x^2 + x + 1} \\ \underline{14x^2 - 6x} \\ 7x + 1 \\ \underline{7x - 3} \\ 4 \end{array}$$

$\dfrac{14x^2 + x + 1}{7x - 3} = 2x + 1 + \dfrac{4}{7x - 3}$

13. $(7 - 5x)(7 + 5x) = 49 - 25x^2$

14. Let $u = a^2$.
$6a^4 - 13a^2 - 5 = 6u^2 - 13u - 5$
$= (2u - 5)(3u + 1)$
$= (2a^2 - 5)(3a^2 + 1)$

15. $(3a + 4b)(2a - 7b) = 6a^2 - 13ab - 28b^2$

16. $3x^4 - 23x^2 - 36 = (3x^2 + 4)(x^2 - 9)$
$= (3x^2 + 4)(x + 3)(x - 3)$

17. $(-4a^2b)^3(-ab^4) = (-64a^6b^3)(-ab^4) = 64a^7b^7$

18. $6x^2 = x + 1$
$6x^2 - x - 1 = 0$
$(3x + 1)(2x - 1) = 0$
$3x + 1 = 0 \qquad 2x - 1 = 0$
$3x = -1 \qquad 2x = 1$
$x = -\dfrac{1}{3} \qquad x = \dfrac{1}{2}$

The solutions are $-\dfrac{1}{3}$ and $\dfrac{1}{2}$.

19. $P(x) = -x^3 + 4x - 8$
$P(-2) = -(-2)^3 + 4(-2) - 8$
$P(-2) = 8 - 8 - 8$
$P(-2) = -8$

20. $\dfrac{(2a^{-4}b^2)^3}{4a^{-2}b^{-1}} = \dfrac{8a^{-12}b^6}{4a^{-2}b^{-1}} = \dfrac{2b^7}{a^{10}}$

21.

$$\begin{array}{r} x^2 - 5x + 10 \\ x+3\overline{)x^3 - 2x^2 - 5x + 7} \\ \underline{x^3 + 3x^2} \\ -5x^2 - 5x \\ \underline{-5x^2 - 15x} \\ 10x + 7 \\ \underline{10x + 30} \\ -23 \end{array}$$

$$\frac{x^3 - 2x^2 - 5x + 7}{x+3} = x^2 - 5x + 10 - \frac{23}{x+3}$$

22. $12 - 17x + 6x^2 = (3x - 4)(2x - 3)$

23. $6x^2 - 4x - 3xa + 2a$
$= 2x(3x - 2) - a(3x - 2)$
$= (3x - 2)(2x - a)$

24. Strategy: To find the number of seconds in one week:
Multiply the number of seconds in a minute (60) by the number of minutes in an hour (60) by the number of hours in a day (24) by the number of days in a week (7).
Convert your answer to scientific notation.

Solution:
$(60)(60)(24)(7) = 604{,}800$
$= 6.048 \times 10^5 \text{ s}$

There are 6.048×10^5 s in one week.

25. Strategy: The distance is $h = 64$ ft.

Solution: $h = 32 + 48t - 16t^2$
$64 = 32 + 48t - 16t^2$
$0 = -32 + 48t - 16t^2$
$16t^2 - 48t + 32 = 0$
$t^2 - 3t + 2 = 0$
$(t - 2)(t - 1) = 0$
$t - 2 = 0 \quad t - 1 = 0$
$t = 2 \qquad t = 1$

The arrow will be 64 ft above the ground at 1 s and at 2 s after the arrow is released.

26. Strategy: To find the area substitute the given values for L and W in the equation $A = LW$ and solve for A.

Solution: $A = LW$
$A = (5x + 1)(2x - 1) = 10x^2 - 3x - 1$
The area is $(10x^2 - 3x - 1)$ ft^2.

Cumulative Review Exercises

1. $8 - 2[-3 - (-1)]^2 + 4$
$= 8 - 2(-3 + 1)^2 + 4$
$= 8 - 2(-2)^2 + 4$
$= 8 - 2(4) + 4$
$= 4$

2. $\dfrac{2a - b}{b - c}$

$\dfrac{2(4) - (-2)}{(-2) - 6} = \dfrac{8 + 2}{-8} = \dfrac{10}{-8} = -\dfrac{5}{4}$

3. Inverse Property of Addition

4. $2x - 4[x - 2(3 - 2x) + 4]$
$= 2x - 4(x - 6 + 4x + 4)$
$= 2x - 4(5x - 2)$
$= 2x - 20x + 8$
$= -18x + 8$

5. $\dfrac{2}{3} - y = \dfrac{5}{6}$

$\dfrac{2}{3} - y - \dfrac{2}{3} = \dfrac{5}{6} - \dfrac{2}{3}$

$-y = \dfrac{1}{6}$

$(-1)(-y) = -\dfrac{1}{6}(-1)$

$y = -\dfrac{1}{6}$

The solution is $-\dfrac{1}{6}$.

6. $8x - 3 - x = -6 + 3x - 8$

$7x - 3 = 3x - 14$

$4x - 3 = -14$

$4x = -11$

$x = -\dfrac{11}{4}$

The solution is $-\dfrac{11}{4}$.

7.

$$\begin{array}{r|rrrr} 3 & 1 & 0 & 0 & -3 \\ & & 3 & 9 & 27 \\ \hline & 1 & 3 & 9 & 24 \end{array}$$

$\dfrac{x^3 - 3}{x - 3} = x^2 + 3x + 9 + \dfrac{24}{x - 3}$

8. $3 - |2 - 3x| = -2$

$-|2 - 3x| = -5$

$|2 - 3x| = 5$

$2 - 3x = 5 \qquad 2 - 3x = -5$

$-3x = 3 \qquad\;\; -3x = -7$

$x = -1 \qquad\;\; x = \dfrac{7}{3}$

The solutions are -1 and $\dfrac{7}{3}$.

9. $P(x) = 3x^2 - 2x + 2$

$P(-2) = 3(-2)^2 - 2(-2) + 2$

$P(-2) = 3(4) + 4 + 2$

$P(-2) = 18$

10. Domain $\{x \mid x \neq -2\}$

11. $F(x) = 3x - 4$

$3x - 4 = 0$

$x = \dfrac{4}{3}$

The zero of the function is $\dfrac{4}{3}$.

12. $m = \dfrac{y_2 - y_1}{x_2 - x_1} = \dfrac{2 - 3}{4 - (-2)} = -\dfrac{1}{6}$

13. Use the point-slope formula.

$y - y_1 = m(x - x_1)$

$y - 2 = -\dfrac{3}{2}[x - (-1)]$

$y - 2 = -\dfrac{3}{2}x - \dfrac{3}{2}$

$y = -\dfrac{3}{2}x + \dfrac{1}{2}$

14. Solve the equation $3x + 2y = 4$ for y to find the slope of this line.

$3x + 2y = 4$

$2y = -3x + 4$

$y = -\dfrac{3}{2}x + 2$

$m = -\dfrac{3}{2}$

The perpendicular line will have a slope that is the negative reciprocal of $-\dfrac{3}{2}$.

$m = \dfrac{2}{3}$ and $(-2, 4)$

$y - y_1 = m(x - x_1)$

$y - 4 = \dfrac{2}{3}[x - (-2)]$

$y - 4 = \dfrac{2}{3}x + \dfrac{4}{3}$

$y = \dfrac{2}{3}x + \dfrac{16}{3}$

The equation of the perpendicular line is $y = \dfrac{2}{3}x + \dfrac{16}{3}$.

15. $2x - 3y = 2$

$x + y = -3$

$D = \begin{vmatrix} 2 & -3 \\ 1 & 1 \end{vmatrix} = 5$

$D_x = \begin{vmatrix} 2 & -3 \\ -3 & 1 \end{vmatrix} = -7$

$D_y = \begin{vmatrix} 2 & 2 \\ 1 & -3 \end{vmatrix} = -8$

$x = \dfrac{D_x}{D} = \dfrac{-7}{5} = -\dfrac{7}{5}$

$$y = \frac{D_y}{D} = \frac{-8}{5} = -\frac{8}{5}$$

The solution is $\left(-\frac{7}{5}, -\frac{8}{5}\right)$.

16. (1) $x - y + z = 0$

(2) $2x + y - 3z = -7$

(3) $-x + 2y + 2z = 5$

Add equations (1) and (3) to eliminate x.

$x - y + z = 0$

$-x + 2y + 2z = 5$

(4) $y + 3z = 5$

Add -2 times equation (1) and equation (2) to eliminate x.

$-2x + 2y - 2z = 0$

$2x + y - 3z = -7$

(5) $3y - 5z = -7$

Add -3 times equation (4) to equation (5) to eliminate y.

$-3y - 9z = -15$

$3y - 5z = -7$

$-14z = -22$

$z = \frac{11}{7}$

Substitute $\frac{11}{7}$ for z in equation (4).

$y + 3\left(\frac{11}{7}\right) = 5$

$y = \frac{2}{7}$

Substitute in values for y and z.

$x - y + z = 0$

$x - \frac{2}{7} + \frac{11}{7} = 0$

$x = -\frac{9}{7}$

The solution is $\left(-\frac{9}{7}, \frac{2}{7}, \frac{11}{7}\right)$.

17. $3x - 4y = 12$

x-intercept:

$3x - 4(0) = 12$

$3x = 12$

$x = 4$

$(4, 0)$

y-intercept:

$3(0) - 4y = 12$

$-4y = 12$

$y = -3$

$(0, -3)$

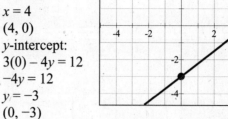

18. $-3x + 2y < 6$

$2y < 3x + 6$

$y < \frac{3}{2}x + 3$

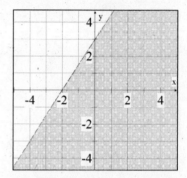

19. $x - 2y = 3$

$-2y = -x + 3$

$y = \frac{1}{2}x - \frac{3}{2}$

$-2x + y = -3$

$y = 2x - 3$

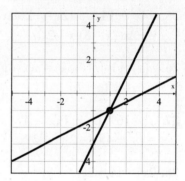

The solutions is $(1, -1)$.

20. Solve each inequality for y.

$2x + y < 3$

$y < -2x + 3$

$-2x + y \geq 1$

$y \geq 2x + 1$

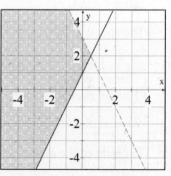

21. $(4a^{-2}b^3)(2a^3b^{-1})^{-2} = 4a^{-2}b^3(2^{-2}a^{-6}b^2)$

$= 4(2^{-2})a^{-8}b^5$

$= \dfrac{b^5}{a^8}$

22. $\dfrac{(5x^3y^{-3}z)^{-2}}{y^4z^{-2}} = \dfrac{5^{-2}x^{-6}y^6z^{-2}}{y^4z^{-2}} = \dfrac{y^2}{25x^6}$

23. $3 - (3 - 3^{-1})^{-1} = 3 - \left(3 - \dfrac{1}{3}\right)^{-1}$

$= 3 - \left(\dfrac{8}{3}\right)^{-1} = 3 - \dfrac{3}{8}$

$= \dfrac{21}{8}$

24. $(2x + 3)(2x^2 - 3x + 1)$

$= 2x(2x^2 - 3x + 1) + 3(2x^2 - 3x + 1)$

$= 4x^3 - 6x^2 + 2x + 6x^2 - 9x + 3$

$= 4x^3 - 7x + 3$

25. $-4x^3 + 14x^2 - 12x = -2x(2x^2 - 7x + 6)$

$= -2x(2x - 3)(x - 2)$

26. $a(x - y) - b(y - x) = a(x - y) + b(x - y)$

$= (x - y)(a + b)$

27. $x^4 - 16 = (x^2 + 4)(x^2 - 4)$

$= (x^2 + 4)(x - 2)(x + 2)$

28. $2x^3 - 16 = 2(x^3 - 8)$

$= 2(x - 2)(x^2 + 2x + 4)$

29. Strategy: Let x represent the speed of the faster cyclist.

The speed of the slower cyclist is $\dfrac{2}{3}x$.

	Rate	Time	Distance
Faster cyclist	x	2	$2x$
Slower cyclist	$\dfrac{2}{3}x$	2	$2\left(\dfrac{2}{3}x\right)$

The sum of the distances is 25 mi.

Solution: $2x + 2\left(\dfrac{2}{3}x\right) = 25$

$2x + \dfrac{4}{3}x = 25$

$\dfrac{10}{3}x = 25$

$x = 7.5$

$\dfrac{2}{3}x = \dfrac{2}{3}(7.5) = 5$

The faster cyclist travels at 7.5 mph and the slower cyclist travels at 5 mph.

30. Strategy: Let x represent the number of ounces of pure gold.

	Amount	Cost	Value
Pure gold	x	360	$360x$
Alloy	80	120	$80(120)$
Mixture	$x + 80$	200	$200(x + 80)$

The sum of values before mixing is equal to the value after mixing.

Solution: $360x + 80(120) = 200(x + 80)$

$360x + 9600 = 200x + 16{,}000$

$160x + 9600 = 16{,}000$

$160x = 6400$

$x = 40$

40 oz of pure gold must be mixed with the alloy.

31. $m = \dfrac{y_2 - y_1}{x_2 - x_1} = \dfrac{300 - 100}{6 - 2} = \dfrac{200}{4} = 50$

The average speed is 50 mph.

32. Strategy: To find the time, use the equation $d = rt$, where r is the speed of the space vehicle and d is the distance from Earth to the moon.

Solution: $d = rt$

$2.4 \times 10^5 = (2 \times 10^4)t$

$\dfrac{2.4 \times 10^5}{2 \times 10^4} = t$

$1.2 \times 10^1 = t$

The vehicle will reach the moon in 12 h.

Chapter 6: Rational Expressions

Prep Test

1. Multiples of 10: 10, 20, 30, 40, 50, 60, …
Multiples of 25: 25, 50, 75, …
LCM is 50.

2. $-\dfrac{3}{8} \cdot \dfrac{4}{9} = -\dfrac{3 \cdot 2 \cdot 2}{2 \cdot 2 \cdot 2 \cdot 3 \cdot 3} = -\dfrac{1}{6}$

3. $-\dfrac{4}{5} \div \dfrac{8}{15} = -\dfrac{4}{5} \cdot \dfrac{15}{8} = -\dfrac{2 \cdot 2 \cdot 3 \cdot 5}{2 \cdot 2 \cdot 2 \cdot 5} = -\dfrac{3}{2}$

4. $-\dfrac{5}{6} + \dfrac{7}{8} = -\dfrac{20}{24} + \dfrac{21}{24} = \dfrac{1}{24}$

5. $-\dfrac{3}{8} - \left(-\dfrac{7}{12}\right) = -\dfrac{3}{8} + \dfrac{7}{12} = -\dfrac{9}{24} + \dfrac{14}{24} = \dfrac{5}{24}$

6. $\dfrac{\dfrac{2}{3} - \dfrac{1}{4}}{\dfrac{1}{8} - \dfrac{2}{1}} = \dfrac{\dfrac{8}{12} - \dfrac{3}{12}}{\dfrac{1}{8} - \dfrac{16}{8}} = \dfrac{\dfrac{5}{12}}{-\dfrac{15}{8}} = \dfrac{5}{12} \div \dfrac{-15}{8}$

$= \dfrac{5}{12} \cdot \left(-\dfrac{8}{15}\right) = -\dfrac{2 \cdot 2 \cdot 2 \cdot 5}{2 \cdot 2 \cdot 3 \cdot 3 \cdot 5} = -\dfrac{2}{9}$

7. $\dfrac{2x - 3}{x^2 - x + 1}$

$\dfrac{2(2) - 3}{2^2 - 2 + 1} = \dfrac{4 - 3}{4 - 2 + 1} = \dfrac{1}{3}$

8. $4(2x + 1) = 3(x - 2)$
$8x + 4 = 3x - 6$
$5x = -10$
$x = -2$
The solution is -2.

9. $10\left(\dfrac{t}{2} + \dfrac{t}{5}\right) = 10(1)$

$\dfrac{10t}{2} + \dfrac{10t}{5} = 10$

$5t + 2t = 10$

$7t = 10$

$t = \dfrac{10}{7}$

The solution is $\dfrac{10}{7}$.

10. **Strategy:** Let x represent the rate of the second plane.
The rate of the first plane is $x - 20$.

	Rate	Time	Distance
First plane	$x - 20$	2	$2(x - 20)$
Second plane	x	2	$2x$

The planes are 480 mi apart. The total distance traveled by the two planes is $2(x - 20) + 2x = 480$.
Solution:
$2(x - 20) + 2x = 480$
$2x - 40 + 2x = 480$
$4x - 40 = 480$
$4x = 520$
$x = 130$
$x - 20 = 130 - 20 = 110$
The rate of the first plane is 110 mph and the rate of the second plane is 130 mph.

Section 6.1

Concept Check

1. (i) and (iii) are rational functions.

3. A rational expression is in simplest form when the numerator and denominator have no common factors.

Objective A Exercises

5. $f(x) = \dfrac{2}{x-3}$

$f(4) = \dfrac{2}{4-3} = \dfrac{2}{1} = 2$

7. $f(x) = \dfrac{x-2}{x+4}$

$f(-2) = \dfrac{-2-2}{-2+4} = \dfrac{-4}{2} = -2$

9. $f(x) = \dfrac{1}{x^2 - 2x + 1}$

$f(-2) = \dfrac{1}{(-2)^2 - 2(-2) + 1} = \dfrac{1}{9}$

11. $f(x) = \dfrac{x-2}{2x^2 + 3x + 8}$

$f(3) = \dfrac{3-2}{2(3)^2 + 3(3) + 8} = \dfrac{1}{35}$

13. $f(x) = \dfrac{x^2 - 2x}{x^3 - x + 4}$

$f(-1) = \dfrac{(-1)^2 - 2(-1)}{(-1)^3 - (-1) + 4} = \dfrac{3}{4}$

15. $f(x) = \dfrac{4}{x-3}$

$x - 3 = 0$

$x = 3$

The domain is $\{x | x \ne 3\}$.

17. $H(x) = \dfrac{x}{x+4}$

$x + 4 = 0$

$x = -4$

The domain is $\{x | x \ne -4\}$.

19. $h(x) = \dfrac{5x}{3x+9}$

$3x + 9 = 0$

$3x = -9$

$x = -3$

The domain is $\{x | x \ne -3\}$.

21. $q(x) = \dfrac{4-x}{(x-4)(3x-2)}$

$(x-4)(3x-2) = 0$

$x - 4 = 0 \quad 3x - 2 = 0$

$x = 4 \qquad 3x = 2$

$x = \dfrac{2}{3}$

The domain is $\left\{ x \,|\, x \ne \dfrac{2}{3}, 4 \right\}$.

23. $f(x) = \dfrac{2x-1}{x^2 + x - 6}$

$f(x) = \dfrac{2x-1}{(x+3)(x-2)}$

$(x+3)(x-2) = 0$

$x + 3 = 0 \quad x - 2 = 0$

$x = -3 \qquad x = 2$

The domain is $\{x | x \ne -3, 2\}$.

25. $f(x) = \dfrac{x+1}{x^2 + 1}$

The domain is $\{x | x \in \text{real numbers}\}$..

27. Yes, the domain of each function is all real numbers.

Objective B Exercises

29. The expression $\dfrac{x+3}{x+3} = 1$ is true for all

values of x. Any number over itself is equal to 1.

31. $\dfrac{4-8x}{4} = \dfrac{4(1-2x)}{4} = 1 - 2x$

33. $\dfrac{6x^2 - 2x}{2x} = \dfrac{2x(3x-1)}{2x} = 3x - 1$

35. $\dfrac{3x^3 y^3 - 12x^2 y^2 + 15xy}{3xy} = \dfrac{3xy(x^2 y^2 - 4xy + 5)}{3xy}$

$= x^2 y^2 - 4xy + 5$

37. $\dfrac{8x^2(x-3)}{4x(x-3)} = \dfrac{8x^2}{4x} = 2x$

39. $\dfrac{-36a^2 - 48a}{18a^3 + 24a^2} = \dfrac{-12a(3a+4)}{6a^2(3a+4)}$

$= \dfrac{-12a}{6a^2} = -\dfrac{2}{a}$

41. $\dfrac{3x-6}{x^2 + 2x}$

This expression is in simplest form.

43. $\dfrac{x^2 - 7x + 12}{x^2 - 9x + 20} = \dfrac{(x-3)(x-4)}{(x-5)(x-4)} = \dfrac{x-3}{x-5}$

45. $\dfrac{2x^2 - 5x - 3}{2x^2 - 3x - 9} = \dfrac{(2x+1)(x-3)}{(2x+3)(x-3)} = \dfrac{2x+1}{2x+3}$

47. $\dfrac{3x^2 + 10x - 8}{8 - 14x + 3x^2} = \dfrac{(3x-2)(x+4)}{(2-3x)(4-x)}$

$= \dfrac{(3x-2)(x+4)}{-(3x-2)(4-x)} = -\dfrac{x+4}{4-x} = \dfrac{x+4}{x-4}$

49. $\dfrac{a^2 - b^2}{a^2 + b^3} = \dfrac{(a+b)(a-b)}{(a+b)(a^2 - ab + b^2)}$

$= \dfrac{a-b}{a^2 - ab + b^2}$

51. $\dfrac{8x^3 - y^3}{4x^2 - y^2} = \dfrac{(2x-y)(4x^2 + 2xy + y^2)}{(2x-y)(2x+y)}$

$= \dfrac{4x^2 + 2xy + y^2}{2x+y}$

53. $\dfrac{x^2(a-2) - a + 2}{ax^2 - ax} = \dfrac{x^2(a-2) - (a-2)}{ax(x-1)}$

$\dfrac{(a-2)(x^2 - 1)}{ax(x-1)} = \dfrac{(a-2)(x-1)(x+1)}{ax(x-1)}$

$= \dfrac{(a-2)(x+1)}{ax}$

55. $\dfrac{x^4 - 2x^2 - 3}{x^4 + 2x^2 + 1} = \dfrac{(x^2+1)(x^2-3)}{(x^2+1)(x^2+1)} = \dfrac{x^2-3}{x^2+1}$

57. $\dfrac{6x^2 y^2 + 11xy + 4}{9x^2 y^2 + 9xy - 4} = \dfrac{(3xy+4)(2xy+1)}{(3xy+4)(3xy-1)}$

$= \dfrac{2xy+1}{3xy-1}$

57. $\dfrac{a^{2n} + a^n - 12}{a^{2n} - 2a^n - 3} = \dfrac{(a^n-3)(a^n+4)}{(a^n-3)(a^n+1)} = \dfrac{a^n+4}{a^n+1}$

59. $k = 4$

Critical Thinking

61. $\dfrac{x^2 + 7x + 12}{x^3 + 4x^2 - 9x - 36} = \dfrac{(x+3)(x+4)}{(x+4)(x^2-9)}$

$= \dfrac{(x+3)(x+4)}{(x+4)(x-3)(x+3)} = \dfrac{1}{x-3}$

63. $\dfrac{2x^2 + ax - 10x - 5a}{2x^2 + ax - 6x - 3a} = \dfrac{2x^2 - 10x + ax - 5a}{2x^2 - 6x + ax - 3a}$

$= \dfrac{2x(x-5) + a(x-5)}{2x(x-3) + a(x-3)} = \dfrac{(2x+a)(x-5)}{(2x+a)(x-3)}$

$= \dfrac{x-5}{x-3}$

Projects or Group Activities

65. $h(x) = \dfrac{x+2}{x+3}$

$h(3.1) = 0.83607$

$h(3.01) = 0.83361$

$h(3.001) = 0.83336$

$h(3.0001) = 0.8333$

The values of $h(x)$ decrease.

Section 6.2

Concept Check

1. a) $\dfrac{x}{x+1}$

b) $\dfrac{x^2-4}{3x-4}$

c) $x+5$

3. $\dfrac{a}{b} \cdot \dfrac{d}{c}$

Objective A Exercises

5. $\dfrac{27a^2b^5}{16xy^2} \cdot \dfrac{20x^2y^3}{9a^2b} = \dfrac{27 \cdot 20a^2b^5x^2y^3}{16 \cdot 9xy^2a^2b}$

$= \dfrac{15b^4xy}{4}$

7. $\dfrac{3x-15}{4x^2-2x} \cdot \dfrac{20x^2-10x}{15x-75}$

$= \dfrac{3(x-5)}{2x(2x-1)} \cdot \dfrac{10x(2x-1)}{15(x-5)}$

$= \dfrac{3(x-5) \cdot 10x(2x-1)}{2x(2x-1) \cdot 15(x-5)}$

$= \dfrac{3 \cdot 10x}{2x \cdot 15} = 1$

9. $\dfrac{x^2y^3}{x^2-4x-5} \cdot \dfrac{2x^2-13x+15}{x^4y^3}$

$= \dfrac{x^2y^3}{(x+1)(x-5)} \cdot \dfrac{(2x-3)(x-5)}{x^4y^3}$

$= \dfrac{x^2y^3 \cdot (2x-3)(x-5)}{(x+1)(x-5) \cdot x^4y^3}$

$= \dfrac{2x-3}{x^2(x+1)}$

11. $\dfrac{x^2-3x+2}{x^2-8x+15} \cdot \dfrac{x^2+x-12}{8-2x-x^2}$

$= \dfrac{(x-2)(x-1)}{(x-3)(x-5)} \cdot \dfrac{(x+4)(x-3)}{(4+x)(2-x)}$

$= \dfrac{(x-2)(x-1) \cdot (x+4)(x-3)}{(x-3)(x-5) \cdot (4+x)(2-x)}$

$= -\dfrac{x-1}{x-5}$

13. $\dfrac{x^3-y^3}{2x^2+xy-3y^2} \cdot \dfrac{2x^2+5xy+3y^2}{x^2+xy+y^2}$

$= \dfrac{(x-y)(x^2+xy+y^2)}{(2x+3y)(x-y)} \cdot \dfrac{(2x+3y)(x+y)}{x^2+xy+y^2}$

$= \dfrac{(x-y)(x^2+xy+y^2) \cdot (2x+3y)(x+y)}{(2x+3y)(x-y) \cdot (x^2+xy+y^2)}$

$= x+y$

15. $n = 8$

Objective B Exercises

17. $\dfrac{6x^2y^4}{35a^2b^5} \div \dfrac{12x^3y^3}{7a^4b^5} = \dfrac{6x^2y^4}{35a^2b^5} \cdot \dfrac{7a^4b^5}{12x^3y^3}$

$= \dfrac{6 \cdot 7x^2y^4a^4b^5}{35 \cdot 12x^3y^3a^2b^5} = \dfrac{a^2y}{10x}$

19. $\dfrac{2x-6}{6x^2-15x} \div \dfrac{4x^2-12x}{18x^3-45x^2}$

$= \dfrac{2x-6}{6x^2-15x} \cdot \dfrac{18x^3-45x^2}{4x^2-12x}$

$= \dfrac{2(x-3)}{3x(2x-5)} \cdot \dfrac{9x^2(2x-5)}{4x(x-3)}$

$= \dfrac{2(x-3) \cdot 9x^2(2x-5)}{3x(2x-5) \cdot 4x(x-3)}$

$= \dfrac{2 \cdot 9x^2}{3x \cdot 4x} = \dfrac{3}{2}$

21. $\dfrac{2x^2 - 2y^2}{14x^2 y^4} \div \dfrac{x^2 + 2xy + y^2}{35xy^3}$

$= \dfrac{2x^2 - 2y^2}{14x^2 y^4} \cdot \dfrac{35xy^3}{x^2 + 2xy + y^2}$

$= \dfrac{2(x-y)(x+y)}{14x^2 y^4} \cdot \dfrac{35xy^3}{(x+y)(x+y)}$

$= \dfrac{2(x-y)(x+y) \cdot 35xy^3}{14x^2 y^4 \cdot (x+y)(x+y)}$

$= \dfrac{70xy^3 (x-y)}{14x^2 y^4 (x+y)}$

$= \dfrac{5(x-y)}{xy(x+y)}$

23. $\dfrac{x^2 - 8x + 15}{x^2 + 2x - 35} \div \dfrac{15 - 2x - x^2}{x^2 + 9x + 14}$

$= \dfrac{x^2 - 8x + 15}{x^2 + 2x - 35} \cdot \dfrac{x^2 + 9x + 14}{15 - 2x - x^2}$

$= \dfrac{(x-3)(x-5)}{(x+7)(x-5)} \cdot \dfrac{(x+2)(x+7)}{(3-x)(5+x)}$

$= \dfrac{(x-3)(x-5) \cdot (x+2)(x+7)}{(x+7)(x-5) \cdot (3-x)(5+x)}$

$= -\dfrac{x+2}{x+5}$

25. $\dfrac{14 + 17x - 6x^2}{3x^2 + 14x + 8} \div \dfrac{4x^2 - 49}{2x^2 + 15x + 28}$

$= \dfrac{14 + 17x - 6x^2}{3x^2 + 14x + 8} \cdot \dfrac{2x^2 + 15x + 28}{4x^2 - 49}$

$= \dfrac{(7-2x)(2+3x)}{(3x+2)(x+4)} \cdot \dfrac{(2x+7)(x+4)}{(2x+7)(2x-7)}$

$= \dfrac{(7-2x)(2+3x) \cdot (2x+7)(x+4)}{(3x+2)(x+4) \cdot (2x+7)(2x-7)}$

$= -1$

27. $\dfrac{6x^2 + 6x}{3x + 6x^2 + 3x^3} \div \dfrac{x^2 - 1}{1 - x^3}$

$= \dfrac{6x^2 + 6x}{3x + 6x^2 + 3x^3} \cdot \dfrac{1 - x^3}{x^2 - 1}$

$= \dfrac{6x(x+1)}{3x(1+x)(1+x)} \cdot \dfrac{(1-x)(1+x+x^2)}{(x+1)(x-1)}$

$= \dfrac{6x(x+1) \cdot (1-x)(1+x+x^2)}{3x(1+x)(1+x) \cdot (x+1)(x-1)}$

$= -\dfrac{2(1+x+x^2)}{(x+1)^2}$

29. $\dfrac{a}{b}$

Critical Thinking

31.

$\dfrac{x^2 - x - 12}{x^2 - 6x + 5} \div \dfrac{x^2 + 3x - 28}{x^2 + 5x - 6} \cdot \dfrac{x^2 + 12x + 35}{x^2 + 11x + 30}$

$= \dfrac{x^2 - x - 12}{x^2 - 6x + 5} \cdot \dfrac{x^2 + 5x - 6}{x^2 + 3x - 28} \cdot \dfrac{x^2 + 12x + 35}{x^2 + 11x + 30}$

$= \dfrac{(x-4)(x+3)}{(x-5)(x-1)} \cdot \dfrac{(x+6)(x-1)}{(x+7)(x-4)} \cdot \dfrac{(x+7)(x+5)}{(x+5)(x+6)}$

$= \dfrac{x+3}{x-5}$

33. $\dfrac{x^2 + x - 30}{x^2 + 2x - 3} \div \dfrac{x^2 + 8x + 12}{x^2 + x - 2} \div \dfrac{x^2 - x - 20}{x^2 + x - 2}$

$= \dfrac{x^2 + x - 30}{x^2 + 2x - 3} \cdot \dfrac{x^2 + x - 2}{x^2 + 8x + 12} \cdot \dfrac{x^2 + x - 2}{x^2 - x - 20}$

$= \dfrac{(x+6)(x-5)}{(x+3)(x-1)} \cdot \dfrac{(x+2)(x-1)}{(x+6)(x+2)} \cdot \dfrac{(x+2)(x-1)}{(x-5)(x+4)}$

$= \dfrac{(x-1)(x+2)}{(x+3)(x+4)}$

Projects or Group Activities

35. $\dfrac{(x+4)}{(x-2)} \cdot \dfrac{(x-3)}{(x-1)} = \dfrac{x^2+x-12}{x^2-3x+2}$

$P(x) = \dfrac{x^2+x-12}{x^2-3x+2}$

37. $\dfrac{x^2-x-6}{x^2-1} \div \dfrac{x^2-4}{x^2+5x+4}$

$= \dfrac{x^2-x-6}{x^2-1} \cdot \dfrac{x^2+5x+4}{x^2-4}$

$= \dfrac{(x-3)(x+2)}{(x+1)(x-1)} \cdot \dfrac{(x+4)(x+1)}{(x+2)(x-2)}$

$= \dfrac{(x-3)(x+4)}{(x-1)(x-2)} = \dfrac{x^2+x-12}{x^2-3x+2}$

$P(x) = \dfrac{x^2+x-12}{x^2-3x+2}$

Section 6.3

Concept Check

1. a) $14(x+3)$
   b) $(x+4)(x-6)$
   c) $(x-2)(x+2)(x+5)$

Objective A Exercises

3. a) There are six factors of a.
   b) There are four factors of b.

5. The LCM is $12x^2y^4$.

$\dfrac{3}{4x^2y} = \dfrac{3}{4x^2y} \cdot \dfrac{3y^3}{3y^3} = \dfrac{9y^3}{12x^2y^4}$

$\dfrac{17}{12xy^4} = \dfrac{17}{12xy^4} \cdot \dfrac{x}{x} = \dfrac{17x}{12x^2y^4}$

7. The LCM is $6x^2(x-2)$.

$\dfrac{x-2}{3x(x-2)} = \dfrac{x-2}{3x(x-2)} \cdot \dfrac{2x}{2x} = \dfrac{2x^2-4x}{6x^2(x-2)}$

$\dfrac{3}{6x^2} = \dfrac{3}{6x^2} \cdot \dfrac{x-2}{x-2} = \dfrac{3x-6}{6x^2(x-2)}$

9. The LCM is $2x(x-5)$.

$\dfrac{3x-1}{2x(x-5)}$

$-3x = \dfrac{-3x}{1} = \dfrac{-3x}{1} \cdot \dfrac{2x(x-5)}{2x(x-5)} = -\dfrac{6x^3-30x^2}{2x(x-5)}$

11. The LCM is $(2x-3)(2x+3)$.

$\dfrac{3x}{2x-3} = \dfrac{3x}{2x-3} \cdot \dfrac{2x+3}{2x+3} = \dfrac{6x^2+9x}{(2x-3)(2x+3)}$

$\dfrac{5x}{2x+3} = \dfrac{5x}{2x+3} \cdot \dfrac{2x-3}{2x-3} = \dfrac{10x^2-15x}{(2x-3)(2x+3)}$

13. The LCM is $(x-3)(x+3)$.

$\dfrac{2x}{x^2-9} = \dfrac{2x}{(x-3)(x+3)}$

$\dfrac{x+1}{x-3} = \dfrac{x+1}{x-3} \cdot \dfrac{x+3}{x+3} = \dfrac{x^2+4x+3}{(x-3)(x+3)}$

15. $3x^2-12y^2 = 3(x+2y)(x-2y)$
$6x-12y = 6(x-2y)$
The LCM is $6(x+2y)(x-2y)$.

$\dfrac{3}{3x^2-12y^2} = \dfrac{3}{3(x+2y)(x-2y)} \cdot \dfrac{2}{2}$

$= \dfrac{6}{6(x+2y)(x-2y)}$

$\dfrac{5}{6x-12y} = \dfrac{5}{6(x-2y)} \cdot \dfrac{x+2y}{x+2y}$

$= \dfrac{5x+10y}{6(x-2y)(x+2y)}$

17. $x^2 - 1 = (x+1)(x-1)$.

$x^2 - 2x + 1 = (x-1)(x-1)$

The LCM is

$(x+1)(x-1)(x-1) = (x+1)(x-1)^2$

$\dfrac{3x}{x^2-1} = \dfrac{3x}{(x+1)(x-1)} \cdot \dfrac{x-1}{x-1}$

$= \dfrac{3x^2 - 3x}{(x+1)(x-1)^2}$

$\dfrac{5x}{x^2 - 2x + 1} = \dfrac{5x}{(x-1)(x-1)} \cdot \dfrac{x+1}{x+1}$

$= \dfrac{5x^2 + 5x}{(x+1)(x-1)^2}$

19. $8 - x^3 = -(x^3 - 8) = -(x-2)(x^2 + 2x + 4)$

The LCM is $-(x-2)(x^2 + 2x + 4)$.

$\dfrac{x-3}{8-x^3} = -\dfrac{x-3}{(x-2)(x^2+2x+4)}$

$\dfrac{2}{4+2x+x^2} = \dfrac{2}{x^2+2x+4} \cdot \dfrac{x-2}{x-2}$

$= \dfrac{2x-4}{(x-2)(x^2+2x+4)}$

21. $x^2 + 2x - 3 = (x+3)(x-1)$

$x^2 + 6x + 9 = (x+3)(x+3)$

The LCM is

$(x-1)(x+3)(x+3) = (x-1)(x+3)^2$.

$\dfrac{2x}{x^2+2x-3} = \dfrac{2x}{(x+3)(x-1)} \cdot \dfrac{x+3}{x+3}$

$= \dfrac{2x^2 + 6x}{(x-1)(x+3)^2}$

$\dfrac{-x}{x^2+6x+9} = \dfrac{-x}{(x+3)(x+3)} \cdot \dfrac{x-1}{x-1}$

$= -\dfrac{x^2 - x}{(x-1)(x+3)^3}$

23. $4x^2 - 16x + 15 = (2x-3)(2x-5)$

$6x^2 - 19x + 10 = (2x-5)(3x-2)$

The LCM is $(2x-3)(2x-5)(3x-2)$.

$\dfrac{-4x}{4x^2-16x+15} = \dfrac{-4x}{(2x-3)(2x-5)} \cdot \dfrac{3x-2}{3x-2}$

$= -\dfrac{12x^2 - 8x}{(2x-3)(2x-5)(3x-2)}$

$\dfrac{3x}{6x^2-19x+10} = \dfrac{3x}{(2x-5)(3x-2)} \cdot \dfrac{2x-3}{2x-3}$

$= \dfrac{6x^2 - 9x}{(2x-3)(2x-5)(3x-2)}$

25. $6x^2 - 17x + 12 = (3x-4)(2x-3)$

$4 - 3x = -(3x-4)$

The LCM is $(3x-4)(2x-3)$.

$\dfrac{5}{6x^2-17x+12} = \dfrac{5}{(3x-4)(2x-3)}$

$\dfrac{2x}{4-3x} = -\dfrac{2x}{3x-4} \cdot \dfrac{2x-3}{2x-3}$

$= -\dfrac{4x^2 - 6x}{(3x-4)(2x-3)}$

$\dfrac{x+1}{2x-3} = \dfrac{x+1}{2x-3} \cdot \dfrac{3x-4}{3x-4} = \dfrac{3x^2 - x - 4}{(3x-4)(2x-3)}$

27. $15 - 2x - x^2 = -(x^2 + 2x - 15)$

$= -(x+5)(x-3)$

The LCM is $(x+5)(x-3)$.

$\dfrac{2x}{x-3} = \dfrac{2x}{x-3} \cdot \dfrac{x+5}{x+5} = \dfrac{2x^2 + 10x}{(x-3)(x+5)}$

$\dfrac{-2}{x+5} = \dfrac{-2}{x+5} \cdot \dfrac{x-3}{x-3} = -\dfrac{2x-6}{(x-3)(x+5)}$

$\dfrac{x-1}{20-x-x^2} = -\dfrac{x-1}{(x-3)(x+5)}$

Objective B Exercises

29. True

31. The LCM is $4x^2$.

$$-\frac{3}{4x^2}+\frac{8}{4x^2}-\frac{3}{4x^2}=\frac{-3+8-3}{4x^2}$$

$$=\frac{2}{4x^2}=\frac{1}{2x^2}$$

33. The LCM is $3x^2+x-10$.

$$\frac{3x}{3x^2+x-10}-\frac{5}{3x^2+x-10}=\frac{3x-5}{3x^2+x-10}$$

$$=\frac{3x-5}{(3x-5)(x+2)}=\frac{1}{x+2}$$

35. The LCM is $30a^2b^2$.

$$\frac{2}{5ab}-\frac{3}{10a^2b}+\frac{4}{15ab^2}$$

$$=\frac{2}{5ab}\cdot\frac{6ab}{6ab}-\frac{3}{10a^2b}\cdot\frac{3b}{3b}+\frac{4}{15ab^2}\cdot\frac{2a}{2a}$$

$$=\frac{12ab-9b+8a}{30a^2b^2}$$

37. The LCM is $40ab$.

$$\frac{3}{4ab}-\frac{2}{5a}+\frac{3}{10b}-\frac{5}{8ab}$$

$$=\frac{3}{4ab}\cdot\frac{10}{10}-\frac{2}{5a}\cdot\frac{8b}{8b}+\frac{3}{10b}\cdot\frac{4a}{4a}-\frac{5}{8ab}\cdot\frac{5}{5}$$

$$=\frac{30-16b+12a-25}{30ab}$$

$$=\frac{5-16b+12a}{40ab}$$

39. The LCM is $12x$.

$$\frac{3x-4}{6x}-\frac{2x-5}{4x}=\frac{3x-4}{6x}\cdot\frac{2}{2}-\frac{2x-5}{4x}\cdot\frac{3}{3}$$

$$=\frac{2(3x-4)-3(2x-5)}{12x}=\frac{6x-8-6x+15}{12x}$$

$$=\frac{7}{12x}$$

41. The LCM is $10x^2y^2$.

$$\frac{2y-4}{5xy^2}+\frac{3-2x}{10x^2y}=\frac{2y-4}{5xy^2}\cdot\frac{2x}{2x}+\frac{3-2x}{10x^2y}\cdot\frac{y}{y}$$

$$=\frac{2x(2y-4)+y(3-2x)}{10x^2y^2}=\frac{4xy-8x+3y-2xy}{10x^2y^2}$$

$$=\frac{2xy-8x+3y}{10x^2y^2}$$

43. The LCM is $(a-2)(a+1)$.

$$\frac{3a}{a-2}-\frac{5a}{a+1}=\frac{3a}{a-2}\cdot\frac{a+1}{a+1}-\frac{5a}{a+1}\cdot\frac{a-2}{a-2}$$

$$=\frac{3a(a+1)-5a(a-2)}{(a-2)(a+1)}=\frac{3a^2+3a-5a^2+10a}{(a-2)(a+1)}$$

$$=\frac{-2a^2+13a}{(a-2)(a+1)}=-\frac{a(2a+13)}{(a-2)(a+1)}$$

45. The LCM is $(2x-5)(5x-2)$.

$$\frac{x}{2x-5}-\frac{2}{5x-2}=\frac{x}{2x-5}\cdot\frac{5x-2}{5x-2}-\frac{2}{5x-2}\cdot\frac{2x-5}{2x-5}$$

$$=\frac{x(5x-2)-2(2x-5)}{(2x-5)(5x-2)}=\frac{5x^2-2x-4x+10}{(2x-5)(5x-2)}$$

$$=\frac{5x^2-6x+10}{(2x-5)(5x-2)}$$

47. The LCM is $b(a-b)$.

$$\frac{1}{a-b}+\frac{1}{b}=\frac{1}{a-b}\cdot\frac{b}{b}+\frac{1}{b}\cdot\frac{a-b}{a-b}$$

$$=\frac{b+a-b}{b(a-b)}=\frac{a}{b(a-b)}$$

49. The LCM is $a(a-3)$.

$$\frac{6a}{a-3}-5+\frac{3}{a}=\frac{6a}{a-3}\cdot\frac{a}{a}-\frac{5}{1}\cdot\frac{a(a-3)}{a(a-3)}+\frac{3}{a}\cdot\frac{a-3}{a-3}$$

$$=\frac{6a^2-5a^2+15a+3a-9}{a(a-3)}$$

$$=\frac{a^2+18a-9}{a(a-3)}$$

51. The LCM is $x(6x-5)$.

$$\frac{5}{x}-\frac{5x}{5-6x}+2$$

$$=\frac{5}{x}\cdot\frac{6x-5}{6x-5}-\frac{-5x}{6x-5}\cdot\frac{x}{x}+\frac{2}{1}\cdot\frac{x(6x-5)}{x(6x-5)}$$

$$=\frac{5(6x-5)+5x(x)+2(x)(6x-5)}{x(6x-5)}$$

$$=\frac{30x-25+5x^2+12x^2-10x}{x(6x-5)}$$

$$=\frac{17x^2+20x-25}{x(6x-5)}$$

53. $x^2-6x+9=(x-3)(x-3)$

$x^2-9=(x+3)(x-3)$

The LCM is

$(x+3)(x-3)(x-3)=(x+3)(x-3)^2$.

$$\frac{1}{x^2-6x+9}-\frac{1}{x^2-9}$$

$$=\frac{1}{(x-3)(x-3)}\cdot\frac{x+3}{x+3}-\frac{1}{(x+3)(x-3)}\cdot\frac{x-3}{x-3}$$

$$=\frac{x+3-x+3}{(x+3)(x-3)^2}$$

$$=\frac{6}{(x+3)(x-3)^2}$$

55. $x^2+4x+4=(x+2)(x+2)$

The LCM is $(x+2)(x+2)=(x+2)^2$.

$$\frac{1}{x+2}-\frac{3x}{x^3+4x+4}$$

$$=\frac{1}{(x+2)}\cdot\frac{x+2}{x+2}-\frac{3x}{(x+2)(x+2)}$$

$$=\frac{x+2-3x}{(x+2)(x+2)}=\frac{-2x+2}{(x+2)^2}$$

$$=-\frac{2(x-1)}{(x+2)^2}$$

57. $x^2+2x-8=(x+4)(x-2)$

The LCM is $(x+4)(x-2)$.

$$\frac{-3x^2+8x+2}{x^2+2x-8}-\frac{2x-5}{x+4}$$

$$=\frac{-3x^2+8x+2}{(x+4)(x-2)}-\frac{2x-5}{x+4}\cdot\frac{x-2}{x-2}$$

$$=\frac{-3x^2+8x+2-2x^2+9x-10}{(x+4)(x-2)}$$

$$=\frac{-5x^2+17x-8}{(x+4)(x-2)}$$

$$=-\frac{5x^2-17x+8}{(x+4)(x-2)}$$

59. $4x^2-36=4(x^2-9)=4(x-3)(x+3)$

The LCM is $4(x-3)(x+3)$.

$$\frac{x^2+4}{4x^2-36}-\frac{13}{x+3}$$

$$=\frac{x^2+4}{4(x-3)(x+3)}-\frac{13}{x+3}\cdot\frac{4(x-3)}{4(x-3)}$$

$$=\frac{x^2+4-13(4)(x-3)}{4(x-3)(x+3)}=\frac{x^2+4-52x+156}{4(x-3)(x+3)}$$

$$=\frac{x^2-52x+160}{4(x-3)(x+3)}$$

61. $4x^2+9x+2=(4x+1)(x+2)$

The LCM is $(4x+1)(x+2)$.

$$\frac{3x-4}{4x+1}+\frac{3x+6}{4x^2+9x+2}$$

$$=\frac{3x-4}{4x+1}\cdot\frac{x+2}{x+2}+\frac{3x+6}{(4x+1)(x+2)}$$

$$= \frac{3x^2 + 2x - 8 + 3x + 6}{(4x+1)(x+2)} = \frac{3x^2 + 5x - 2}{(4x+1)(x+2)}$$

$$= \frac{(3x-1)(x+2)}{(4x+1)(x+2)}$$

$$= \frac{3x-1}{4x+1}$$

63. $x^2 + x - 12 = (x+4)(x-3)$

$x^2 + 7x + 12 = (x+4)(x+3)$

The LCM is $(x+4)(x-3)(x+3)$.

$$\frac{x+1}{x^2+x-12} - \frac{x-3}{x^2+7x+12}$$

$$= \frac{x+1}{(x+4)(x-3)} \cdot \frac{x+3}{x+3} - \frac{x-3}{(x+4)(x+3)} \cdot \frac{x-3}{x-3}$$

$$= \frac{x^2 + 4x + 3 - x^2 + 6x - 9}{(x+4)(x-3)(x+3)}$$

$$= \frac{10x - 6}{(x+4)(x-3)(x+3)}$$

65. $x^2 - 2x - 15 = (x-5)(x+3)$

$5 - x = -(x-5)$

The LCM is $(x-5)(x+3)$.

$$\frac{2x^2 - 2x}{x^2 - 2x - 15} - \frac{2}{x+3} + \frac{x}{5-x}$$

$$= \frac{2x^2 - 2x}{(x-5)(x+3)} - \frac{2}{x+3} \cdot \frac{x-5}{x-5} + \frac{-(x)}{x-5} \cdot \frac{x+3}{x+3}$$

$$= \frac{2x^2 - 2x - 2x + 10 - x^2 - 3x}{(x-5)(x+3)}$$

$$= \frac{x^2 - 7x + 10}{(x-5)(x+3)} = \frac{(x-5)(x-2)}{(x-5)(x+3)}$$

$$= \frac{x-2}{x+3}$$

67. $3x^2 - 11x - 20 = (3x+4)(x-5)$

The LCM is $(3x+4)(x-5)$

$$\frac{x}{3x+4} + \frac{3x+2}{x-5} - \frac{7x^2 + 24x + 28}{3x^2 - 11x - 20}$$

$$= \frac{x}{3x+4} \cdot \frac{x-5}{x-5} + \frac{3x+2}{x-5} \cdot \frac{3x+4}{3x+4} - \frac{7x^2 + 24x + 28}{(3x+4)(x-5)}$$

$$= \frac{x^2 - 5x + 9x^2 + 18x + 8 - 7x^2 - 24x - 28}{(3x+4)(x-5)}$$

$$= \frac{3x^2 - 11x - 20}{(2x-5)(x-2)} = \frac{(3x+4)(x-5)}{(3x+4)(x-5)}$$

$$= 1$$

69. $8x^2 - 10x + 3 = (4x-3)(2x-1)$

$1 - 2x = -(2x-1)$

The LCM is $(4x-3)(2x-1)$.

$$\frac{x+1}{1-2x} - \frac{x+3}{4x-3} + \frac{10x^2 + 7x - 9}{8x^2 - 10x + 3}$$

$$= \frac{-(x+1)}{2x-1} \cdot \frac{4x-3}{4x-3} - \frac{x+3}{4x-3} \cdot \frac{2x-1}{2x-1} + \frac{10x^2 + 7x - 9}{(4x-3)(2x-1)}$$

$$= \frac{-4x^2 - x + 3 - 2x^2 - 5x + 3 + 10x^2 + 7x - 9}{(4x-3)(2x-1)}$$

$$= \frac{4x^2 + x - 3}{(4x-3)(2x-1)} = \frac{(4x-3)(x+1)}{(4x-3)(2x-1)}$$

$$= \frac{x+1}{2x-1}$$

71. $8x^3 - 1 = (2x-1)(4x^2 + 2x + 1)$

The LCM is $(2x-1)(4x^2 + 2x + 1)$.

$$\frac{2x}{4x^2 + 2x + 1} + \frac{4x+1}{8x^3 - 1}$$

$$= \frac{2x}{4x^2 + 2x + 1} \cdot \frac{2x-1}{2x-1} + \frac{4x+1}{(2x-1)(4x^2 + 2x + 1)}$$

$$= \frac{4x^2 - 2x + 4x + 1}{(2x-1)(4x^2 + 2x + 1)} = \frac{4x^2 + 2x + 1}{(2x-1)(4x^2 + 2x + 1)}$$

$$= \frac{1}{2x-1}$$

73. $x^4 - 16 = (x^2 + 4)(x^2 - 4)$

The LCM is $(x^2 + 4)(x^2 - 4)$.

$\dfrac{x^2 - 12}{x^4 - 16} + \dfrac{1}{x^2 - 4} - \dfrac{1}{x^2 + 4}$

$= \dfrac{x^2 - 12}{(x^2 + 4)(x^2 - 4)} + \dfrac{1}{x^2 - 4} \cdot \dfrac{x^2 + 4}{x^2 + 4} - \dfrac{1}{x^2 + 4} \cdot \dfrac{x^2 - 4}{x^2 - 4}$

$= \dfrac{x^2 - 12 + x^2 + 4 - x^2 + 4}{(x^2 + 4)(x^2 - 4)}$

$= \dfrac{x^2 - 4}{(x^2 + 4)(x^2 - 4)}$

$= \dfrac{1}{x^2 + 4}$

Critical Thinking

75. $\dfrac{x^2 - 4x + 4}{2x + 1} \cdot \dfrac{2x^2 + x}{x^3 - 4x} - \dfrac{3x - 2}{x + 1}$

$= \dfrac{(x - 2)(x - 2)}{2x + 1} \cdot \dfrac{x(2x + 1)}{x(x + 2)(x - 2)} - \dfrac{3x - 2}{x + 1}$

$= \dfrac{x - 2}{x + 2} - \dfrac{3x - 2}{x + 1}$

The LCM is $(x + 2)(x + 1)$.

$\dfrac{x - 2}{x + 2} - \dfrac{3x - 2}{x + 1}$

$= \dfrac{x - 2}{x + 2} \cdot \dfrac{x + 1}{x + 1} - \dfrac{3x - 2}{x + 1} \cdot \dfrac{x + 2}{x + 2}$

$= \dfrac{x^2 - x - 2 - 3x^2 - 4x + 4}{(x + 2)(x + 1)} = \dfrac{-2x^2 - 5x + 2}{(x + 2)(x + 1)}$

$= -\dfrac{2x^2 + 5x - 2}{(x + 2)(x + 1)}$

77. The LCM is ab.

$\left(\dfrac{a - 2b}{b} + \dfrac{b}{a} \right) \div \left(\dfrac{b + a}{a} - \dfrac{2a}{b} \right)$

$= \left(\dfrac{a - 2b}{b} \cdot \dfrac{a}{a} + \dfrac{b}{a} \cdot \dfrac{b}{b} \right) \div \left(\dfrac{b + a}{a} \cdot \dfrac{b}{b} - \dfrac{2a}{b} \cdot \dfrac{a}{a} \right)$

$= \left(\dfrac{a^2 - 2ab + b^2}{ab} \right) \div \left(\dfrac{b^2 + ab - 2a^2}{ab} \right)$

$= \dfrac{(a - b)(a - b)}{ab} \cdot \dfrac{ab}{(b + 2a)(b - a)}$

$= \dfrac{a - b}{-(b + 2a)}$

$= -\dfrac{a - b}{b + 2a} = \dfrac{b - a}{b + 2a}$

79. $\dfrac{2x}{x^2 - x - 6} - \dfrac{6x - 6}{2x^2 - 9x + 9} \div \dfrac{x^2 + x - 2}{2x - 3}$

$= \dfrac{2x}{x^2 - x - 6} - \dfrac{6(x - 1)}{(2x - 3)(x - 3)} \div \dfrac{(x + 2)(x - 1)}{2x - 3}$

$= \dfrac{2x}{(x - 3)(x + 2)} - \dfrac{6(x - 1)}{(2x - 3)(x - 3)} \cdot \dfrac{2x - 3}{(x + 2)(x - 1)}$

$= \dfrac{2x}{(x - 3)(x + 2)} - \dfrac{6}{(x - 3)(x + 2)}$

$= \dfrac{2x - 6}{(x - 3)(x + 2)} = \dfrac{2(x - 3)}{(x - 3)(x + 2)}$

$= \dfrac{2}{x + 2}$

Check Your Progress: Chapter 6

1. $\dfrac{x^2 - 2x - 8}{x^2 - 8x + 16} = \dfrac{(x - 4)(x + 2)}{(x - 4)(x - 4)} = \dfrac{x + 2}{x - 4}$

2. $\dfrac{2x^2 - 11x - 40}{6x^2 - x - 40} = \dfrac{(2x + 5)(x - 8)}{(2x + 5)(3x - 8)} = \dfrac{x - 8}{3x - 8}$

3. $\dfrac{x^2-3x-18}{x^2-5x-24}\cdot\dfrac{x^2-2x-15}{x^2+12x+27}$

$=\dfrac{(x-6)(x+3)}{(x-8)(x+3)}\cdot\dfrac{(x-5)(x+3)}{(x+3)(x+9)}$

$=\dfrac{(x-6)(x-5)}{(x-8)(x+9)}$

4. $\dfrac{x^2+x-72}{x^2+14x+45}\cdot\dfrac{2x^2+15x+25}{3x^2-15x+72}$

$=\dfrac{(x-8)(x+9)}{(x+5)(x+9)}\cdot\dfrac{(2x+5)(x+5)}{(3x+9)(x-8)}$

$=\dfrac{2x+5}{3x+9}$

5. $\dfrac{2x^2-3x-27}{x^2+4x-12}\div\dfrac{6x^2-23x-18}{x^2+15x+54}$

$=\dfrac{2x^2-3x-27}{x^2+4x-12}\cdot\dfrac{x^2+15x+54}{6x^2-23x-18}$

$=\dfrac{(2x-9)(x+3)}{(x+6)(x-2)}\cdot\dfrac{(x+9)(x+6)}{(3x+2)(2x-9)}$

$=\dfrac{(x+3)(x+9)}{(x-2)(3x+2)}$

6. $\dfrac{3x^2+17x-28}{x^2+2x-15}\div\dfrac{12x^2-13x-4}{x^2-6x+9}$

$=\dfrac{3x^2+17x-28}{x^2+2x-15}\cdot\dfrac{x^2-6x+9}{12x^2-13x-4}$

$=\dfrac{(3x-4)(x+7)}{(x+5)(x-3)}\cdot\dfrac{(x-3)(x-3)}{(4x+1)(3x-4)}$

$=\dfrac{(x+7)(x-3)}{(x+5)(4x+1)}$

7. $x^2+4x=x(x+4)$
$x^2+9x+20=(x+4)(x+5)$
The LCM is $x(x+4)(x+5)$

8. $x^2-4=(x+2)(x-2)$
$x^2+2x-8=(x+4)(x-2)$
The LCM is $(x+2)(x-2)(x+4)$

9. The LCM is $(x-1)(x-4)$

$\dfrac{x+6}{x-1}+\dfrac{4x+5}{x-4}$

$=\dfrac{x+6}{x-1}\cdot\dfrac{x-4}{x-4}+\dfrac{4x+5}{x-4}\cdot\dfrac{x-1}{x-1}$

$=\dfrac{(x+6)(x-4)+(4x+5)(x-1)}{(x-1)(x-4)}$

$=\dfrac{x^2+2x-24+4x^2+x-5}{(x-1)(x-4)}$

$=\dfrac{5x^2+3x-29}{(x-1)(x-4)}$

10. $x+3$
$x^2+12x+27=(x+3)(x+9)$
The LCM is $(x+3)(x+9)$

$\dfrac{x+9}{x+3}+\dfrac{3x+4}{x^2+12x+27}$

$=\dfrac{x+9}{x+3}\cdot\dfrac{x+9}{x+9}+\dfrac{3x+4}{(x+3)(x+9)}$

$=\dfrac{x^2+18x+81+3x+4}{(x+3)(x+9)}$

$=\dfrac{x^2+21x+85}{(x+3)(x+9)}$

11. The LCM is $(3x+4)(x+1)$

$\dfrac{x+9}{3x+4}-\dfrac{x+3}{x+1}$

$=\dfrac{x+9}{3x+4}\cdot\dfrac{x+1}{x+1}-\dfrac{x+3}{x+1}\cdot\dfrac{3x+4}{3x+4}$

$=\dfrac{(x+9)(x+1)-(x+3)(3x+4)}{(3x+4)(x+1)}$

$=\dfrac{x^2+10x+9-3x^2-13x-12}{(3x+4)(x+1)}$

$=\dfrac{-2x^2-3x-3}{(3x+4)(x+1)}=-\dfrac{2x^2+3x+3}{(3x+4)(x+1)}$

12. $3x^2 + 20x - 63 = (3x - 7)(x + 9)$

The LCM is $(3x - 7)(x + 9)$.

$$\frac{x-8}{3x^2+20x-63} - \frac{x+2}{3x-7}$$

$$= \frac{x-8}{(3x-7)(x+9)} - \frac{x+2}{3x-7}$$

$$= \frac{x-8}{(3x-7)(x+9)} - \frac{x+2}{3x-7} \cdot \frac{x+9}{x+9}$$

$$= \frac{x-8-(x+2)(x+9)}{(3x-7)(x+9)}$$

$$= \frac{x-8-x^2-11x-18}{(3x-7)(x-9)}$$

$$= \frac{-x^2-10x-26}{(3x-7)(x+9)} = -\frac{x^2+10x+26}{(3x-7)(x+9)}$$

Section 6.4

Concept Check

1. A complex fraction is a fraction whose numerator or denominator contains one or more fractions.

3. The LCM is 3.

$$\frac{2-\dfrac{1}{3}}{4+\dfrac{11}{3}} = \frac{2-\dfrac{1}{3}}{4+\dfrac{11}{3}} \cdot \frac{3}{3} = \frac{2\cdot3-\dfrac{1}{3}\cdot3}{4\cdot3+\dfrac{11}{3}\cdot3}$$

$$= \frac{6-1}{12+11} = \frac{5}{23}$$

5. The LCM is 6.

$$\frac{3-\dfrac{2}{3}}{5+\dfrac{5}{6}} = \frac{3-\dfrac{2}{3}}{5+\dfrac{5}{6}} \cdot \frac{6}{6} = \frac{3\cdot6-\dfrac{2}{3}\cdot6}{5\cdot6+\dfrac{5}{6}\cdot6}$$

$$= \frac{18-4}{30+5} = \frac{14}{35} = \frac{2}{5}$$

Objective A Exercises

7. The LCM is x^2.

$$\frac{1+\dfrac{1}{x}}{1-\dfrac{1}{x^2}} = \frac{1+\dfrac{1}{x}}{1-\dfrac{1}{x^2}} \cdot \frac{x^2}{x^2} = \frac{1\cdot x^2 + \dfrac{1}{x}\cdot x^2}{1\cdot x^2 - \dfrac{1}{x^2}\cdot x^2}$$

$$= \frac{x^2+x}{x^2-1} = \frac{x(x+1)}{(x-1)(x+1)}$$

$$= \frac{x}{x-1}$$

9. The LCM is a.

$$\frac{a-2}{\dfrac{4}{a}-a} = \frac{a-2}{\dfrac{4}{a}-a} \cdot \frac{a}{a} = \frac{a\cdot a - 2\cdot a}{\dfrac{4}{a}\cdot a - a\cdot a}$$

$$= \frac{a^2-2a}{4-a^2} = \frac{a(a-2)}{(2-a)(2+a)}$$

$$= -\frac{a}{a+2}$$

11. The LCM is a^2.

$$\frac{2+\dfrac{1}{a}}{4-\dfrac{1}{a^2}} = \frac{2+\dfrac{1}{a}}{4-\dfrac{1}{a^2}} \cdot \frac{a^2}{a^2} = \frac{2\cdot x^2 + \dfrac{1}{a}\cdot a^2}{4\cdot a^2 - \dfrac{1}{a^2}\cdot a^2}$$

$$= \frac{2a^2+a}{4a^2-1} = \frac{a(2a+1)}{(2a-1)(2a+1)}$$

$$= \frac{a}{2a-1}$$

13. The LCM is x.

$$\frac{x-\dfrac{1}{x}}{x+\dfrac{1}{x}} = \frac{x-\dfrac{1}{x}}{x+\dfrac{1}{x}} \cdot \frac{x}{x}$$

$$= \frac{x^2-1}{x^2+1}$$

15. The LCM is a^2.

$$\dfrac{\dfrac{1}{a^2} - \dfrac{1}{a}}{\dfrac{1}{a^2} + \dfrac{1}{a}} = \dfrac{\dfrac{1}{a^2} - \dfrac{1}{a}}{\dfrac{1}{a^2} + \dfrac{1}{a}} \cdot \dfrac{a^2}{a^2}$$

$$= \dfrac{1-a}{1+a}$$

17. The LCM is $x + 2$.

$$\dfrac{2 - \dfrac{4}{x+2}}{5 - \dfrac{10}{x+2}} = \dfrac{2 - \dfrac{4}{x+2}}{5 - \dfrac{10}{x+2}} \cdot \dfrac{x+2}{x+2}$$

$$= \dfrac{2(x+2) - 4}{5(x+2) - 10} = \dfrac{2x+4-4}{5x+10-10}$$

$$= \dfrac{2x}{5x} = \dfrac{2}{5}$$

19. The LCM is $2a - 3$.

$$\dfrac{\dfrac{3}{2a-3} + 2}{\dfrac{-6}{2a-3} - 4} = \dfrac{\dfrac{3}{2a-3} + 2}{\dfrac{-6}{2a-3} - 4} \cdot \dfrac{2a-3}{2a-3}$$

$$= \dfrac{3 + 2(2a-3)}{-6 - 4(2a-3)} = \dfrac{3 + 4a - 6}{-6 - 8a + 12}$$

$$= \dfrac{4a-3}{-8a+6} = \dfrac{4a-3}{-2(4a-3)}$$

$$= -\dfrac{1}{2}$$

21. The LCM is $(x - 4)(x + 1)$.

$$\dfrac{1 - \dfrac{1}{x-4}}{1 - \dfrac{6}{x+1}} = \dfrac{1 - \dfrac{1}{x-4}}{1 - \dfrac{6}{x+1}} \cdot \dfrac{(x-4)(x+1)}{(x-4)(x+1)}$$

$$= \dfrac{(x-4)(x+1) - (x+1)}{(x-4)(x+1) - 6(x-4)}$$

$$= \dfrac{x^2 - 3x - 4 - x - 1}{x^2 - 3x - 4 - 6x + 24}$$

$$= \dfrac{x^2 - 4x - 5}{x^2 - 9x + 20} = \dfrac{(x-5)(x+1)}{(x-5)(x-4)}$$

$$= \dfrac{x+1}{x-4}$$

23. The LCM is $(x - 3)(2 - x)$.

$$\dfrac{1 - \dfrac{2}{x-3}}{1 + \dfrac{3}{2-x}} = \dfrac{1 - \dfrac{2}{x-3}}{1 + \dfrac{3}{2-x}} \cdot \dfrac{(x-3)(2-x)}{(x-3)(2-x)}$$

$$= \dfrac{(x-3)(2-x) - 2(2-x)}{(x-3)(2-x) + 3(x-3)}$$

$$= \dfrac{2x - x^2 - 6 + 3x - 4 + 2x}{2x - x^2 - 6 + 3x + 3x - 9}$$

$$= \dfrac{-x^2 + 7x - 10}{-x^2 + 8x - 15} = \dfrac{-(x^2 - 7x + 10)}{-(x^2 - 8x + 15)}$$

$$= \dfrac{(x-5)(x-2)}{(x-5)(x-3)}$$

$$= \dfrac{x-2}{x-3}$$

25. The LCM is $(2x + 3)$.

$$\frac{x-4+\dfrac{9}{2x+3}}{x+3-\dfrac{5}{2x+3}} = \frac{x-4+\dfrac{9}{2x+3}}{x+3-\dfrac{5}{2x+3}} \cdot \frac{2x+3}{2x+3}$$

$$= \frac{x(2x+3)-4(2x+3)+9}{x(2x+3)+3(2x+3)-5}$$

$$= \frac{2x^2+3x-8x-12+9}{2x^2+3x+6x+9-5}$$

$$= \frac{2x^2-5x-3}{2x^2+9x+4} = \frac{(2x+1)(x-3)}{(2x+1)(x+4)}$$

$$= \frac{x-3}{x+4}$$

27. The LCM is $(x+4)(x-3)$.

$$\frac{x-3+\dfrac{10}{x+4}}{x+7+\dfrac{16}{x-3}} = \frac{x-3+\dfrac{10}{x+4}}{x+7+\dfrac{16}{x-3}} \cdot \frac{(x+4)(x-3)}{(x+4)(x-3)}$$

$$= \frac{x(x+4)(x-3)-3(x+4)(x-3)+10(x-3)}{x(x+4)(x-3)+7(x+4)(x-3)+16(x+4)}$$

$$= \frac{(x-3)(x^2+4x-3x-12+10)}{(x+4)(x^2-3x+7x-21+16)}$$

$$= \frac{(x-3)(x^2+x-2)}{(x+4)(x^2+4x-5)} = \frac{(x-3)(x+2)(x-1)}{(x+4)(x+5)(x-1)}$$

$$= \frac{(x-3)(x+2)}{(x+4)(x+5)}$$

29. The LCM is x^2.

$$\frac{1-\dfrac{1}{x}-\dfrac{6}{x^2}}{1-\dfrac{4}{x}+\dfrac{3}{x^2}} = \frac{1-\dfrac{1}{x}-\dfrac{6}{x^2}}{1-\dfrac{4}{x}+\dfrac{3}{x^2}} \cdot \frac{x^2}{x^2}$$

$$= \frac{x^2-x-6}{x^2-4x+3} = \frac{(x-3)(x+2)}{(x-3)(x-1)}$$

$$= \frac{x+2}{x-1}$$

31. The LCM is x^2.

$$\frac{1+\dfrac{1}{x}-\dfrac{12}{x^2}}{\dfrac{9}{x^2}+\dfrac{3}{x}-2} = \frac{1+\dfrac{1}{x}-\dfrac{12}{x^2}}{\dfrac{9}{x^2}+\dfrac{3}{x}-2} \cdot \frac{x^2}{x^2}$$

$$= \frac{x^2+x-12}{9+3x-2x^2} = \frac{(x-3)(x+4)}{(3-x)(3+2x)}$$

$$= -\frac{x+4}{2x+3}$$

33. The LCM is x^2y^2.

$$\frac{\dfrac{1}{y^2}-\dfrac{1}{xy}-\dfrac{2}{x^2}}{\dfrac{1}{y^2}-\dfrac{3}{xy}+\dfrac{2}{x^2}} = \frac{\dfrac{1}{y^2}-\dfrac{1}{xy}-\dfrac{2}{x^2}}{\dfrac{1}{y^2}-\dfrac{3}{xy}+\dfrac{2}{x^2}} \cdot \frac{x^2y^2}{x^2y^2}$$

$$= \frac{x^2-xy-2y^2}{x^2-3xy+2y^2} = \frac{(x+y)(x-2y)}{(x-y)(x-2y)}$$

$$= \frac{x+y}{x-y}$$

35. The LCM is $x(x+1)$.

$$\frac{\dfrac{x}{x+1}-\dfrac{1}{x}}{\dfrac{x}{x+1}+\dfrac{1}{x}} = \frac{\dfrac{x}{x+1}-\dfrac{1}{x}}{\dfrac{x}{x+1}+\dfrac{1}{x}} \cdot \frac{x(x+1)}{x(x+1)}$$

$$= \frac{x^2-(x+1)}{x^2+(x+1)} = \frac{x^2-x-1}{x^2+x+1}$$

37. The LCM is $a(a-2)$.

$$\frac{\dfrac{1}{a}-\dfrac{3}{a-2}}{\dfrac{2}{a}+\dfrac{5}{a-2}} = \frac{\dfrac{1}{a}-\dfrac{3}{a-2}}{\dfrac{2}{a}+\dfrac{5}{a-2}} \cdot \frac{a(a-2)}{a(a-2)}$$

$$= \frac{a-2-3a}{2a-4+5a} = \frac{-2a-2}{7a-4}$$

$$= -\frac{2a+2}{7a-4} = -\frac{2(a+1)}{7a-4}$$

39. The LCM is $(x-1)(x+1)$.

$$\dfrac{\dfrac{x-1}{x+1}-\dfrac{x+1}{x-1}}{\dfrac{x-1}{x+1}+\dfrac{x+1}{x-1}}=\dfrac{\dfrac{x-1}{x+1}-\dfrac{x+1}{x-1}}{\dfrac{x-1}{x+1}+\dfrac{x+1}{x-1}}\cdot\dfrac{(x-1)(x+1)}{(x-1)(x+1)}$$

$$=\dfrac{(x-1)(x-1)-(x+1)(x+1)}{(x-1)(x-1)+(x+1)(x+1)}$$

$$=\dfrac{x^2-2x+1-x^2-2x-1}{x^2-2x+1+x^2+2x+1}=\dfrac{-4x}{2x^2+2}$$

$$=\dfrac{-4x}{2(x^2+1)}=-\dfrac{2x}{x^2+1}$$

41. The LCM is a.

$$a+\dfrac{a}{a+\dfrac{1}{a}}=a+\dfrac{a}{a+\dfrac{1}{a}}\cdot\dfrac{a}{a}$$

$$=a+\dfrac{a^2}{a^2+1}$$

The LCM is a^2+1.

$$a+\dfrac{a^2}{a^2+1}=a\cdot\dfrac{a^2+1}{a^2+1}+\dfrac{a^2}{a^2+1}$$

$$=\dfrac{a^3+a+a^2}{a^2+1}$$

$$=\dfrac{a^3+a^2+a}{a^2+1}$$

43. $\dfrac{1}{1-\dfrac{1}{a}}=\dfrac{1}{1-\dfrac{1}{a}}\cdot\dfrac{a}{a}=\dfrac{a}{a-1}$

The reciprocal is $\dfrac{a-1}{a}$.

Critical Thinking

45. $\dfrac{x^{-1}}{y^{-1}}+\dfrac{y}{x}=\dfrac{\dfrac{1}{x}}{\dfrac{1}{y}}+\dfrac{y}{x}$

$$=\dfrac{1}{x}\cdot\dfrac{y}{1}+\dfrac{y}{x}=\dfrac{y}{x}+\dfrac{y}{x}$$

$$=\dfrac{2y}{x}$$

47. $\dfrac{\dfrac{1}{x+h}-\dfrac{1}{x}}{h}=\dfrac{\dfrac{x-x-h}{x(x+h)}}{h}=\dfrac{\dfrac{-h}{x(x+h)}}{h}$

$$=\dfrac{-h}{x(x+h)}\cdot\dfrac{1}{h}=\dfrac{-1}{x(x+h)}$$

Projects or Group Activities

49. a) $P(x)=\dfrac{Cx}{\left[1-\dfrac{1}{(x+1)^{60}}\right]}$

$$=\dfrac{Cx}{\dfrac{(x+1)^{60}}{(x+1)^{60}}-\dfrac{1}{(x+1)^{60}}}=\dfrac{Cx}{\dfrac{(x+1)^{60}-1}{(x+1)^{60}}}$$

$$=\dfrac{Cx(x+1)^{60}}{(x+1)^{60}-1}$$

b) monthly interest rate: 0.08/12

$$P\left(\dfrac{0.08}{12}\right)=\dfrac{20{,}000\left(\dfrac{0.08}{12}\right)\left(\dfrac{0.08}{12}+1\right)^{60}}{\left(\dfrac{0.08}{12}+1\right)^{60}-1}$$

$$=\$405.53$$

Section 6.5

Concept Check

1. A ratio is the quotient of two quantities that have the same unit. A rate is the quotient of two quantities that have different units.

Objective A Exercises

3. $\dfrac{x}{30} = \dfrac{3}{10}$.

$\dfrac{x}{30} \cdot 30 = \dfrac{3}{10} \cdot 30$

$x = 9$

The solution is 9.

5. $\dfrac{2}{x} = \dfrac{8}{30}$

$\dfrac{2}{x} \cdot 30x = \dfrac{8}{30} \cdot 30x$

$60 = 8x$

$\dfrac{15}{2} = x$

The solution is $\dfrac{15}{2}$.

7. $\dfrac{x+1}{10} = \dfrac{2}{5}$

$\dfrac{x+1}{10} \cdot 10 = \dfrac{2}{5} \cdot 10$

$x + 1 = 4$

$x = 3$

The solution is 3.

9. $\dfrac{4}{x+2} = \dfrac{3}{4}$

$\dfrac{4}{x+2} \cdot 4(x+2) = \dfrac{3}{4} \cdot 4(x+2)$

$16 = 3(x+2)$

$16 = 3x + 6$

$10 = 3x$

$x = \dfrac{10}{3}$

The solution is $\dfrac{10}{3}$.

11. $\dfrac{x}{4} = \dfrac{x-2}{8}$

$\dfrac{x}{4} \cdot 8 = \dfrac{x-2}{8} \cdot 8$

$2x = x - 2$

$x = -2$

The solution is -2.

13. $\dfrac{16}{2-x} = \dfrac{4}{x}$

$\dfrac{16}{2-x} \cdot x(2-x) = \dfrac{4}{x} \cdot x(2-x)$

$16x = 4(2-x)$

$16x = 8 - 4x$

$20x = 8$

$x = \dfrac{8}{20} = \dfrac{2}{5}$

The solution is $\dfrac{2}{5}$.

15. $\dfrac{8}{x-2} = \dfrac{4}{x+1}$

$\dfrac{8}{x-2} \cdot (x+1)(x+2) = \dfrac{4}{x+1} \cdot (x+1)(x-2)$

$8(x+1) = 4(x-2)$

$8x + 8 = 4x - 8$

$4x = -16$

$x = -4$

The solution is -4.

17. $\dfrac{x}{3} = \dfrac{x+1}{7}$

$\dfrac{x}{3} \cdot 21 = \dfrac{x+1}{7} \cdot 21$

$7x = 3(x+1)$

$7x = 3x + 3$

$4x = 3$

$x = \dfrac{3}{4}$

The solution is $\dfrac{3}{4}$.

19. $\dfrac{8}{3x-2} = \dfrac{2}{2x+1}$

$\dfrac{8}{3x-2} \cdot (3x-2)(2x+1) = \dfrac{2}{2x+1} \cdot (3x-2)(2x+1)$

$8(2x+1) = 2(3x-2)$

$16x + 8 = 6x - 4$

$10x = -12$

$x = \dfrac{-12}{10} = -\dfrac{6}{5}$

The solution is $-\dfrac{6}{5}$.

21. $\dfrac{3x+1}{3x-4} = \dfrac{x}{x-2}$

$\dfrac{3x+1}{3x-4} \cdot (3x-4)(x-2) = \dfrac{x}{x-2} \cdot (3x-4)(x-2)$

$(3x+1)(x-2) = x(3x-4)$

$3x^2 - 5x - 2 = 3x^2 - 4x$

$-2 = x$

The solution is -2.

23. True

Objective B Exercises

25. Strategy: Let x represent the number of grams of protein in a 454–gram box of pasta. Write and solve a proportion.

Solution: $\dfrac{56}{7} = \dfrac{454}{x}$

$\dfrac{56}{7} \cdot 7x = \dfrac{454}{x} \cdot 7x$

$56x = 3178$

$x = 56.75$

There are 56.75 g of protein in a 454-gram box of pasta.

27. Strategy: Let x represent the number of computers with defective USB ports. Write and solve a proportion.

Solution: $\dfrac{300}{15} = \dfrac{70,000}{x}$

$\dfrac{300}{15} \cdot 15x = \dfrac{70,000}{x} \cdot 15x$

$300x = 1,050,000$

$x = 3500$

There are 3500 computers with defective USB ports.

29. Strategy: Let x represent the cost in dollars. Write and solve a proportion.

Solution: $\dfrac{1}{0.346} = \dfrac{11}{x}$

$\dfrac{1}{0.346} \cdot 0.346x = \dfrac{11}{x} \cdot 0.346x$

$x = 3.81$

A gallon of milk would cost $3.81.

31. Strategy: Let w represent the width of the room. Let l represent the length of the room. Write and solve two proportions to determine the dimensions of the room.

Solution: $\dfrac{\frac{1}{4}}{1} = \dfrac{\frac{9}{2}}{w}, \quad \dfrac{\frac{1}{4}}{1} = \dfrac{6}{l}$

$\dfrac{\frac{1}{4}}{1} \cdot w = \dfrac{\frac{9}{2}}{w} \cdot w \qquad \dfrac{\frac{1}{4}}{1} \cdot l = \dfrac{6}{l} \cdot l$

$\dfrac{1}{4}w = \dfrac{9}{2} \qquad\qquad \dfrac{1}{4}l = 6$

$w = 18 \qquad\qquad\qquad l = 24$

The room dimensions are 18 ft by 24 ft.

33. Strategy: Let x represent the number of miles needed to walk to lose one pound. Write and solve a proportion.

Solution:

$\dfrac{4}{650} = \dfrac{x}{3500}$

$\dfrac{4}{650} \cdot (650)(3500) = \dfrac{x}{3500} \cdot (650)(3500)$

$14000 = 650x$

$21.54 = x$

A person must walk 21.54 mi to lose one pound.

35. Strategy: Let x represent the distance the nerve message traveled in the shrew. Write and solve a proportion.

Solution: $\dfrac{9}{100} = \dfrac{x}{1}$

$\dfrac{9}{100} \cdot (100) = \dfrac{x}{1} \cdot (100)$

$9 = 100x$

$0.09 = x$

It would travel 0.09 ft.

$0.09(12) = 1$ in.

Critical Thinking

37. Strategy: Let x represent the value of 1 U.S. dollar in Euros. Write and solve a proportion.

Solution: $\dfrac{1}{0.59} = \dfrac{1.21}{x}$

$\dfrac{1}{0.59} \cdot (0.59)(x) = \dfrac{1.21}{x} \cdot (0.59)(x)$

$x = 0.59(1.21) = 0.71$

The value of one U.S. dollar in Euros is $.71.

Section 6.6

Concept Check

1. -4 and 1

3. Henry can mow $\dfrac{1}{4}$ of the lawn in 10 min.

Objective A Exercises

5. $\dfrac{x}{2} + \dfrac{5}{6} = \dfrac{x}{3}$

$6\left(\dfrac{x}{2} + \dfrac{5}{6}\right) = 6\left(\dfrac{x}{3}\right)$

$3x + 5 = 2x$

$5 = -x$

$-5 = x$

The solution is -5.

7. $\dfrac{8}{2x-1} = 2$

$(2x-1)\left(\dfrac{8}{2x-1}\right) = (2x-1)(2)$

$8 = 4x - 2$

$10 = 4x$

$x = \dfrac{10}{4} = \dfrac{5}{2}$

The solution is $\dfrac{5}{2}$.

9. $1 - \dfrac{3}{y} = 4$

$y\left(1 - \dfrac{3}{y}\right) = 4(y)$

$y - 3 = 4y$

$-3 = 3y$

$-1 = y$

The solution is -1.

11. $\dfrac{3}{x-2} = \dfrac{4}{x}$

$x(x-2)\left(\dfrac{3}{x-2}\right) = x(x-2)\left(\dfrac{4}{x}\right)$

$3x = 4(x-2)$

$3x = 4x - 8$

$8 = x$

The solution is 8.

13. $\dfrac{6}{2y+3} = \dfrac{6}{y}$

$y(2y+3)\left(\dfrac{6}{2y+3}\right) = y(2y+3)\left(\dfrac{6}{y}\right)$

$6y = 6(2y+3)$

$6y = 12y + 18$

$-6y = 18$

$y = -3$

The solution is -3.

15. $\dfrac{5}{y+3} - 2 = \dfrac{7}{y+3}$

$(y+3)\left(\dfrac{5}{y+3} - 2\right) = (y+3)\left(\dfrac{7}{y+3}\right)$

$5 - 2(y+3) = 7$

$5 - 2y - 6 = 7$

$-2y - 1 = 7$

$-2y = 8$

$y = -4$

The solution is -4.

17. $\dfrac{-4}{a-4} = 3 - \dfrac{a}{a-4}$

$(a-4)\left(\dfrac{-4}{a-4}\right) = (a-4)\left(3 - \dfrac{a}{a-4}\right)$

$-4 = 3a - 12 - a$

$-4 = 2a - 12$

$8 = 2a$

$4 = a$

$a = 4$ is not in the domain of the rational expressions. The equation has no solution.

19. $\dfrac{4}{x-5} = 9 - \dfrac{2x}{x-5}$

$(x-5)\left(\dfrac{4}{x-5}\right) = (x-5)\left(9 - \dfrac{2x}{x-5}\right)$

$4 = 9(x-5) - 2x$

$4 = 9x - 45 - 2x$

$4 = 7x - 45$

$49 = 7x$

$7 = x$

The solution is 7.

21. $\dfrac{4}{x-2} - 4 = \dfrac{-12}{x+2}$

$(x+2)(x-2)\left(\dfrac{4}{x-2} - 4\right) = (x+2)(x-2)\left(\dfrac{-12}{x+2}\right)$

$4(x+2) - 4(x+2)(x-2) = -12(x-2)$

$4x + 8 - 4x^2 + 16 = -12x + 24$

$-4x^2 + 4x + 24 = -12x + 24$

$0 = 4x^2 - 16x$

$0 = 4x(x-4)$

$4x = 0 \qquad x - 4 = 0$

$x = 0 \qquad\quad x = 4$

The solutions are 0 and 4.

23. $\dfrac{4}{x-4}=x+\dfrac{x}{x-4}$

$(x-4)\left(\dfrac{4}{x-4}\right)=(x-4)\left(x+\dfrac{x}{x-4}\right)$

$4=x(x-4)+x$

$4=x^2-4x+x$

$0=x^2-3x-4$

$0=(x+1)(x-4)$

$x+1=0 \qquad x-4=0$

$x=-1 \qquad x=4$

$x=4$ is not in the domain of the rational expressions. The solution is -1.

25. $\dfrac{2x}{x+2}+3x=\dfrac{-5}{x+2}$

$(x+2)\left(\dfrac{2x}{x+2}+3x\right)=(x+2)\left(\dfrac{-5}{x+2}\right)$

$2x+3x(x+2)=-5$

$2x+3x^2+6x=-5$

$3x^2+8x+5=0$

$(3x+5)(x+1)=0$

$3x+5=0 \qquad x+1=0$

$3x=-5 \qquad\quad x=-1$

$x=-\dfrac{5}{3}$

The solutions are $-\dfrac{5}{3}$ and -1.

27. $\dfrac{x}{2x-9}-3x=\dfrac{10}{9-2x}$

$(2x-9)\left(\dfrac{x}{2x-9}-3x\right)=(2x-9)\left(\dfrac{10}{9-2x}\right)$

$x-3x(2x-9)=-10$

$x-6x^2+27x=-10$

$-6x^2+28x+10=0$

$-2(3x^2-14x-5)=0$

$-2(3x+1)(x-5)=0$

$3x+1=0 \qquad x-5=0$

$3x=-1 \qquad\quad x=5$

$x=-\dfrac{1}{3}$

The solutions are $-\dfrac{1}{3}$ and 5.

29. $\dfrac{5}{x-2} - \dfrac{2}{x+2} = \dfrac{3}{x^2-4}$

$\dfrac{5}{x-2} - \dfrac{2}{x+2} = \dfrac{3}{(x-2)(x+2)}$

$(x-2)(x+2)\left(\dfrac{5}{x-2} - \dfrac{2}{x+2}\right) = (x-2)(x+2)\left(\dfrac{3}{(x-2)(x+2)}\right)$

$5(x+2) - 2(x-2) = 3$

$5x + 10 - 2x + 4 = 3$

$3x + 14 = 3$

$3x = -11$

$x = -\dfrac{11}{3}$

The solution is $-\dfrac{11}{3}$.

31. $\dfrac{9}{x^2+7x+10} = \dfrac{5}{x+2} - \dfrac{3}{x+5}$

$\dfrac{9}{(x+2)(x+5)} = \dfrac{5}{x+2} - \dfrac{3}{x+5}$

$(x+2)(x+5)\left(\dfrac{9}{(x+2)(x+5)}\right) = (x+2)(x+5)\left(\dfrac{5}{x+2} - \dfrac{3}{x+5}\right)$

$9 = 5(x+5) - 3(x+2)$

$9 = 5x + 25 - 3x - 6$

$9 = 2x + 19$

$-10 = 2x$

$-5 = x$

$x = -5$ is not in the domain of the rational expressions. The equation has no solution.

Objective B Exercises

33. Strategy: Let t represent the time required for an experienced bricklayer to do the job. time required for an inexperienced bricklayer is $2t$.

	Rate	Time	Part
Experienced bricklayer	$\dfrac{1}{t}$	6	$\dfrac{6}{t}$
Inexperienced bricklayer	$\dfrac{1}{2t}$	16	$\dfrac{16}{2t}$

The sum of the parts of the task completed by each bricklayer must equal 1.

Solution:

$$\frac{6}{t}+\frac{16}{2t}=1$$

$$2t\left(\frac{6}{t}+\frac{16}{2t}\right)=1(2t)$$

$$12+16=2t$$

$$28=2t$$

$$14=t$$

Working alone the experienced bricklayer can do the job in 14 h.

35. Strategy: Let t represent the time required by the second machine working alone.

	Rate	Time	Part
1st machine	$\dfrac{1}{40}$	30	$\dfrac{30}{40}$
2nd machine	$\dfrac{1}{t}$	15	$\dfrac{15}{t}$

The sum of the part of the task completed by the 1st machine and the part completed by the 2nd machine must equal 1.

Solution: $\dfrac{30}{40}+\dfrac{15}{t}=1$

$$40t\left(\frac{30}{40}+\frac{15}{t}\right)=1(40t)$$

$$30t+600=40t$$

$$600=10t$$

$$60=t$$

It would take the second machine 60 min.

37. Strategy: Let t represent the time required to fill the bottles working together.

	Rate	Time	Part
1st machine	$\dfrac{1}{10}$	t	$\dfrac{t}{10}$
2nd machine	$\dfrac{1}{12}$	t	$\dfrac{t}{12}$
3rd machine	$\dfrac{1}{15}$	t	$\dfrac{t}{15}$

The sum of the parts of the task completed by each machine must equal 1.

Solution: $\dfrac{t}{10}+\dfrac{t}{12}+\dfrac{t}{15}=1$

$$60\left(\frac{t}{10}+\frac{t}{12}+\frac{t}{15}\right)=1(60)$$

$$6t+5t+4t=60$$

$$15t=60$$

$$t=4$$

When all three machines are working together, it takes 4 h to fill the bottles.

37. Strategy: Let t represent the time required to empty the tank working together.

	Rate	Time	Part
Inlet pipe	$\dfrac{1}{45}$	t	$\dfrac{t}{45}$
Outlet pipe	$\dfrac{1}{30}$	t	$\dfrac{t}{30}$

The part completed by the outlet pipe minus the part complete by the inlet pipe equals 1.

Solution: $\dfrac{t}{30} - \dfrac{t}{45} = 1$

$90\left(\dfrac{t}{30} - \dfrac{t}{45}\right) = 1(90)$

$3t - 2t = 90$

$t = 90$

It would take 90 min to empty the tank.

41. Strategy: Let t represent the time required for the clowns to blow up 76 balloons.

	Rate	Time	Part
1st clown	$\dfrac{1}{2}$	t	$\dfrac{t}{2}$
2nd clown	$\dfrac{1}{3}$	t	$\dfrac{t}{3}$
Balloon popping	$\dfrac{1}{5}$	t	$\dfrac{t}{5}$

The sum of the tasks completed by the two clowns minus the balloons popping is equal to 76. This gives us the number of balloons filled after t minutes.

Solution: $\dfrac{t}{2} + \dfrac{t}{3} - \dfrac{t}{5} = 76$

$30\left(\dfrac{t}{2} + \dfrac{t}{3} - \dfrac{t}{5}\right) = 76(30)$

$15t + 10t - 6t = 2280$

$19t = 2280$

$t = 120$

It will take 120 min to have 76 balloons.

43. Strategy: Let t represent the time required to address 140 envelopes.

	Rate	Time	Part
1st clerk	$\dfrac{1}{70}$	t	$\dfrac{t}{70}$
2nd clerk	$\dfrac{3}{280}$	t	$\dfrac{3t}{280}$

The part of the task completed by the two clerks must equal 1.

Solution: $\dfrac{t}{70} + \dfrac{3t}{280} = 1$

$280\left(\dfrac{t}{70} + \dfrac{3t}{280}\right) = 1(280)$

$4t + 3t = 280$

$7t = 280$

$t = 40$

With both clerks working together, it will take 40 min (2400 s) to complete the task.

45. t is less than n.

Objective C Exercises

47. Strategy: Let r represent the rate of the runner.
The rate of the bicyclist is $r + 7$.

	Distance	Rate	Time
Runner	16	r	$\dfrac{16}{r}$
Bicyclist	30	$r + 7$	$\dfrac{30}{r + 7}$

The time traveled by the runner equals the time traveled by the bicyclist.

Solution: $\dfrac{16}{r} = \dfrac{30}{r + 7}$

$r(r + 7)\left(\dfrac{16}{r}\right) = r(r + 7)\left(\dfrac{30}{r + 7}\right)$

$16(r + 7) = 30r$

$16r + 112 = 30r$

$112 = 14r$

$8 = r$

The rate of the runner is 8 mph.

49. Strategy: Let r represent the rate of the tortoise.

The rate of the hare is $180r$.

	Distance	Rate	Time
Tortoise	360	r	$\dfrac{360}{r}$
Hare	360	$180r$	$\dfrac{360}{180r}$

The time for the tortoise is 14 min 55 s (or 895 s) more than the hare.

Solution: $\dfrac{360}{180r} + 895 = \dfrac{360}{r}$

$\dfrac{2}{r} + 895 = \dfrac{360}{r}$

$r\left(\dfrac{2}{r} + 895\right) = r\left(\dfrac{360}{r}\right)$

$2 + 895r = 360$

$895r = 358$

$r = 0.4$

$180r = 180(0.4) = 72$

The rate of the tortoise is 0.4 ft/s.
The rate of the hare is 72 ft/s.

51. Strategy: Let r represent the rate of the jogger.

The rate of the cyclist is $2r$.

	Distance	Rate	Time
Runner	30	r	$\dfrac{30}{r}$
Cyclist	30	$2r$	$\dfrac{30}{2r}$

The time for the jogger is 3 h more than the time for the cyclist.

Solution: $\dfrac{30}{r} = \dfrac{30}{2r} + 3$

$2r\left(\dfrac{30}{r}\right) = 2r\left(\dfrac{30}{2r} + 3\right)$

$60 = 30 + 6r$

$30 = 6r$

$5 = r$

$2r = 2(5) = 10$

The rate of the cyclist is 10 mph.

53. Strategy: Let r represent the rate of the helicopter.

The rate of the commercial jet is $4r$.

	Distance	Rate	Time
Helicopter	105	r	$\dfrac{105}{r}$
Commercial jet	735	$4r$	$\dfrac{735}{4r}$

The total time was 2.2 h.

Solution: $\dfrac{105}{r} + \dfrac{735}{4r} = 2.2$

$4r\left(\dfrac{105}{r} + \dfrac{735}{4r}\right) = 4r(2.2)$

$420 + 735 = 8.8r$

$1155 = 8.8r$

$r = 131.25$

$4r = 4(131.25) = 525$

The rate of the jet was 525 mph.

55. Strategy: Let r represent the rate of the current.

	Distance	Rate	Time
With current	20	$7 + r$	$\dfrac{20}{7+r}$
Against current	8	$7 - r$	$\dfrac{8}{7-r}$

The time traveling with the current equals the time traveling against the current.

Solution: $\dfrac{20}{7+r} = \dfrac{8}{7-r}$

$(7+r)(7-r)\left(\dfrac{20}{7+r}\right) = (7+r)(7-r)\left(\dfrac{8}{7-r}\right)$

$20(7-r) = 8(7+r)$

$140 - 20r = 56 + 8r$

$84 = 28r$

$3 = r$

The rate of the current is 3 mph.

57. Strategy: Let r represent the rate of the wind.

	Distance	Rate	Time
With wind	3059	$550 + r$	$\dfrac{3059}{550+r}$
Against wind	2450	$550 - r$	$\dfrac{2450}{550-r}$

The time traveling with the wind equals the time traveling against the wind.

Solution: $\dfrac{3059}{550+r} = \dfrac{2450}{550-r}$

$(550+r)(550-r)\left(\dfrac{3059}{550+r}\right) = (550+r)(550-r)\left(\dfrac{2450}{550-r}\right)$

$3059(550-r) = 2450(550+r)$

$1{,}682{,}450 - 3059r = 1{,}347{,}500 + 2450r$

$1{,}682{,}450 = 1{,}347{,}500 + 5509r$

$334{,}950 = 5509r$

$r = 60.80$

The rate of the wind is 60.80 mph.

59. Strategy: Let r represent the rate of the current.

	Distance	Rate	Time
With current	16	$6+r$	$\dfrac{16}{6+r}$
Against current	16	$6-r$	$\dfrac{16}{6-r}$

The total time is 6 h.

Solution: $\dfrac{16}{6+r}+\dfrac{16}{6-r}=6$

$(6+r)(6-r)\left(\dfrac{16}{6+r}+\dfrac{16}{6-r}\right)=(6+r)(6-r)(6)$

$16(6-r)+16(6+r)=6(36-r^2)$

$96-16r+96+16r=216-6r^2$

$192=216-6r^2$

$0=24-6r^2$

$0=6(4-r^2)$

$0=6(2-r)(2+r)$

$2-r=0 \quad 2+r=0$

$r=2 \qquad r=-2$

The rate of the current cannot be a negative number. The rate of the current is 2 mph.

Critical Thinking

61. Strategy: Let r represent the rate of the bus.

	Distance	Rate	Time
Usual conditions	165	r	$\dfrac{165}{r}$
Bad weather	165	$r-5$	$\dfrac{165}{r-5}$

The bus was 15 min later in bad weather
The solution is 3.

Solution: $\dfrac{165}{r-5}-\dfrac{165}{r}=\dfrac{1}{4}$

$4r(r-5)\left(\dfrac{165}{r-5}-\dfrac{165}{r}\right)=4r(r-5)\left(\dfrac{1}{4}\right)$

$660r-660(r-5)=r^2-5r$

$660r-660r+3300=r^2-5r$

$r^2-5r-3300=0$

$(r-60)(r+55)=0$

$r-60=0 \qquad r+55=0$

$r=60 \qquad r=-55$

The rate of the bus cannot be a negative number. The usual rate of the bus is 60 mph.

Projects or Group Activities

63. a) Fractions for which the numerator is 1.

b) $\dfrac{1}{4}+\dfrac{1}{8}$

c) Student answers will vary. One example is $\dfrac{1}{2}+\dfrac{1}{10}$

d) $\dfrac{1}{2}+\dfrac{1}{12}$ and $\dfrac{1}{4}+\dfrac{1}{3}$

Section 6.7

Concept Check

1. $y=kx$

3. $z=kxy$

Objective A Exercises

5. Strategy: Write the basic direct variation equation replacing the variables with the given values. Solve for k.
Write the direct variation equation replacing k with its value. Substitute 5000 for s and solve for P.

Solution:

$P = ks$

$4000 = 250k$

$16 = k$

$P = 16s = 16(5000) = 80,000$

When the company sells 5000 products, its profit will be \$80,000.

7. **Strategy:** Write the basic direct variation equation replacing the variables with the given values. Solve for k.
Write the direct variation equation replacing k with its value. Substitute 15 for d and solve for p.
Solution:

$p = kd$

$4.5 = 10k$

$0.45 = k$

$p = 0.45d = 0.45(15) = 6.75$

The pressure is 6.75 lb/in^2.

9. **Strategy:** Write the basic direct variation equation replacing the variables with the given values. Solve for k.
Write the direct variation equation replacing k with its value. Substitute 10 for t and solve for d.

Solution:

$d = kt^2$

$144 = k(3)^2$

$144 = 9k$

$16 = k$

$d = 16t^2 = 16(10^2) = 16(100) = 1600$

In 10 s the object will fall 1600 ft.

11. **Strategy:** Write the basic inverse variation equation replacing the variables with the given values. Solve for k.
Write the inverse variation equation replacing k with its value. Substitute 5 for n and solve for T.

Solution:

$T = \dfrac{k}{n}$

$500 = \dfrac{k}{1}$

$500 = k$

$T = \dfrac{500}{n} = \dfrac{500}{5} = 100$

It will take 5 computers 100 s to solve the same problem.

13. **Strategy:** Write the basic direct variation equation replacing the variables with the given values. Solve for k.
Write the direct variation equation replacing k with its value. Substitute 15 for v and solve for L.

Solution:

$L = kv^2$

$640 = k(20)^2$

$640 = 400k$

$1.6 = k$

$L = 1.6v^2 = 1.6(15)^2 = 360$

The load on the sail will be 360 lbs.

15. **Strategy:** Write the basic direct variation equation replacing the variables with the given values. Solve for k.
Write the direct variation equation replacing k with its value. Substitute 230,000 for p and solve for c.

Solution:

$c = kp$

$15600 = 260,000k$

$0.06 = k$

$c = 0.06p = 0.06(230,000) = 13,800$

The commission on a \$230,000 home is \$13,800.

17. Strategy: Write the basic combined variation equation replacing the variables with the given values. Solve for k. Write the combined variation equation replacing k with its value. Substitute 24 for r, 180 for v and solve for I.

Solution:

$$I = \frac{kv}{r}$$

$$10 = \frac{110k}{11}$$

$$110 = 110k$$

$$1 = k$$

$$I = \frac{1v}{r} = \frac{180}{24} = 7.5$$

The current is 7.5 amps.

19. Strategy: Write the basic inverse variation equation replacing the variables with the given values. Solve for k. Write the inverse variation equation replacing k with its value. Substitute 5 for d and solve for I.

Solution:

$$I = \frac{k}{d^2}$$

$$12 = \frac{k}{10^2}$$

$$12 = \frac{k}{100}$$

$$1200 = k$$

$$I = \frac{1200}{d^2} = \frac{1200}{5^2} = \frac{1200}{25} = 48$$

The intensity is 48 foot-candles when the distance is 5 ft.

Critical Thinking

21. If x is doubled then y is doubled.

Projects or Group Activities

23. If x is doubled then y is doubled.

25. If x is doubled, y is divided by 4.

Chapter 6 Review Exercises

1. $$\frac{a^6b^5 + a^5b^6}{a^5b^4 - a^4b^4} \cdot \frac{a-b}{a^2 - b^2}$$

    $$= \frac{a^5b^5(a+b)}{a^4b^4(a-1)} \cdot \frac{a-b}{(a+b)(a-b)}$$

    $$= \frac{a^5b^5(a+b)(a-b)}{a^4b^4(a-1)(a+b)(a-b)}$$

    $$= \frac{ab}{u-1}$$

2. The LCM is $(x-3)(x+2)$

    $$\frac{x}{x-3} - 4 - \frac{2x-5}{x+2}$$

    $$= \frac{x}{x-3} \cdot \frac{x+2}{x+2} - 4 \cdot \frac{(x-3)(x+2)}{(x-3)(x+2)} - \frac{2x-5}{x+2} \cdot \frac{x-3}{x-3}$$

    $$= \frac{x(x+2) - 4(x-3)(x+2) - (2x-5)(x-3)}{(x-3)(x+2)}$$

    $$= \frac{x^2 + 2x - 4x^2 + 4x + 24 - 2x^2 + 11x - 15}{(x-3)(x+2)}$$

    $$= \frac{x^2 + 2x - 4x^2 + 4x + 24 - 2x^2 + 11x - 15}{(x-3)(x+2)}$$

    $$= \frac{-5x^2 + 17x + 9}{(x-3)(x+2)}$$

    $$= -\frac{5x^2 - 17x - 9}{(x-3)(x+2)}$$

3. $$P(x) = \frac{x}{x-3}$$

    $$P(4) = \frac{4}{4-3} = \frac{4}{1} = 4$$

4. $\dfrac{3x-2}{x+6} = \dfrac{3x+1}{x+9}$

$(x+6)(x+9)\left(\dfrac{3x-2}{x+6}\right) = (x+6)(x+9)\left(\dfrac{3x+1}{x+9}\right)$

$(3x-2)(x+9) = (3x+1)(x+6)$

$3x^2 + 25x - 18 = 3x^2 + 19x + 6$

$25x - 18 = 19x + 6$

$6x = 24$

$x = 4$

The solution is 4.

5. $\dfrac{\dfrac{3x+4}{3x-4} + \dfrac{3x-4}{3x+4}}{\dfrac{3x-4}{3x+4} - \dfrac{3x+4}{3x-4}}$

The LCM is $(3x-4)(3x+4)$.

$\dfrac{\dfrac{3x+4}{3x-4} + \dfrac{3x-4}{3x+4}}{\dfrac{3x-4}{3x+4} - \dfrac{3x+4}{3x-4}} \cdot \dfrac{(3x-4)(3x+4)}{(3x-4)(3x+4)}$

$= \dfrac{(3x+4)^2 + (3x-4)^2}{(3x-4)^2 - (3x+4)^2}$

$= \dfrac{9x^2 + 24x + 16 + 9x^2 - 24x + 16}{9x^2 - 24x + 16 - 9x^2 - 24x - 16}$

$= \dfrac{18x^2 + 32}{-48x} = \dfrac{2(9x^2 + 16)}{-48x}$

$= -\dfrac{9x^2 + 16}{24x}$

6. The LCM is $(4x-1)(4x+1)$.

$\dfrac{4x}{4x-1} \cdot \dfrac{4x+1}{4x+1} = \dfrac{16x^2 + 4x}{(4x-1)(4x+1)}$

$\dfrac{3x-1}{4x+1} \cdot \dfrac{4x-1}{4x-1} = \dfrac{12x^2 - 7x + 1}{(4x-1)(4x+1)}$

7. $r = \dfrac{2}{3-r}$

$r(3-r) = \dfrac{2}{3-r} \cdot (3-r)$

$3r - r^2 = 2$

$0 = r^2 - 3r + 2$

$0 = (r-2)(r-1)$

$0 = r-2 \qquad 0 = r-1$

$2 = r \qquad\quad 1 = r$

The solutions are 1 and 2.

8. $P(x) = \dfrac{x^2 - 2}{3x^2 - 2x + 5}$

$P(-2) = \dfrac{(-2)^2 - 2}{3(-2)^2 - 2(-2) + 5}$

$= \dfrac{4 - 2}{12 + 4 + 5}$

$= \dfrac{2}{21}$

9. $\dfrac{10}{5x+3} = \dfrac{2}{10x-3}$

$(5x+3)(10x-3)\left(\dfrac{10}{5x+3}\right) = (5x+3)(10x-3)\left(\dfrac{2}{10x-3}\right)$

$10(10x-3) = 2(5x+3)$

$100x - 30 = 10x + 6$

$90x = 36$

$x = \dfrac{36}{90} = \dfrac{2}{5}$

10. $f(x) = \dfrac{2x-7}{3x^2 + 3x - 18}$

$3x^2 + 3x - 18 = 0$

$3(x^2 + x - 6) = 0$

$3(x+3)(x-2) = 0$

$x + 3 = 0 \quad x - 2 = 0$

$x = -3 \qquad x = 2$

The domain is $\{x|\, x \neq -3, 2\}$.

11. $\dfrac{3x^4 + 11x^2 - 4}{3x^4 + 13x^2 + 4} = \dfrac{(3x^2 - 1)(x^2 + 4)}{(3x^2 + 1)(x^2 + 4)}$

$\qquad = \dfrac{3x^2 - 1}{3x^2 + 1}$

12. $g(x) = \dfrac{2x}{x - 3}$

$x - 3 = 0$

$x = 3$

The domain is $\{x \mid x \neq 3\}$.

13. $\dfrac{x^3 - 8}{x^3 + 2x^2 + 4x} \cdot \dfrac{x^3 + 2x^2}{x^2 - 4}$

$= \dfrac{(x - 2)(x^2 + 2x + 4)}{x(x^2 + 2x + 4)} \cdot \dfrac{x^2(x + 2)}{(x - 2)(x + 2)}$

$= \dfrac{(x - 2)(x^2 + 2x + 4)x^2(x + 2)}{x(x^2 + 2x + 4)(x - 2)(x + 2)}$

$= x$

14. The LCM is $x^2 - 4$.

$\dfrac{3x^2 + 2}{x^2 - 4} - \dfrac{9 - x^2}{x^2 - 4} = \dfrac{3x^2 + 2 - 9x + x^2}{x^2 - 4}$

$= \dfrac{4x^2 - 9x + 2}{x^2 - 4} = \dfrac{(4x - 1)(x - 2)}{(x + 2)(x - 2)}$

$= \dfrac{4x - 1}{x + 2}$

15. $\dfrac{4}{x - 3} = \dfrac{5}{x}$

$\dfrac{4}{x - 3} \cdot x(x - 3) = \dfrac{5}{x} \cdot x(x - 3)$

$4x = 5(x - 3)$

$4x = 5x - 15$

$-x = -15$

$x = 15$

The solution is 15.

16. $\dfrac{30}{x^2 + 5x + 4} + \dfrac{10}{x + 4} = \dfrac{4}{x + 1}$

$\dfrac{30}{(x + 4)(x + 1)} + \dfrac{10}{x + 4} = \dfrac{4}{x + 1}$

$(x + 4)(x + 1)\left(\dfrac{30}{(x + 4)(x + 1)} + \dfrac{10}{x + 4} \right) = (x + 4)(x + 1)\left(\dfrac{4}{x + 1} \right)$

$30 + 10(x + 1) = 4(x + 4)$

$30 + 10x + 10 = 4x + 16$

$10x + 40 = 4x + 16$

$6x = -24$

$x = -4$

$x = -4$ is not in the domain of the rational expressions. The equation has no solution.

17. $\dfrac{x+2+\dfrac{3}{x-4}}{x-3-\dfrac{2}{x-4}}$

$\dfrac{x+2+\dfrac{3}{x-4}}{x-3-\dfrac{2}{x-4}} \cdot \dfrac{x-4}{x-4} = \dfrac{(x+2)(x-4)+3}{(x-3)(x-4)-2}$

$= \dfrac{x^2-2x-8+3}{x^2-7x+12-2}$

$= \dfrac{x^2-2x-5}{x^2-7x+10}$

18. $x^2-9x+20=(x-4)(x-5)$

$4-x=-(x-4)$

The LCM is $(x-5)(x-4)$.

$\dfrac{x-3}{x-5} \cdot \dfrac{x-4}{x-4} = \dfrac{x^2-7x+12}{(x-5)(x-4)}$

$\dfrac{x}{x^2-9x+20} = \dfrac{x}{(x-5)(x-4)}$

$\dfrac{1}{4-x} = \dfrac{-1}{x-4} \cdot \dfrac{x-5}{x-5} = \dfrac{-x+5}{(x-5)(x-4)}$

$= -\dfrac{x-5}{(x-5)(x-4)}$

19. $\dfrac{6}{2x-3} = \dfrac{5}{x+5} + \dfrac{5}{2x^2+7x-15}$

$\dfrac{6}{2x-3} = \dfrac{5}{x+5} + \dfrac{5}{(2x-3)(x+5)}$

$(2x-3)(x+5)\left(\dfrac{6}{2x-3}\right) = (2x-3)(x+5)\left(\dfrac{5}{x+5} + \dfrac{5}{(2x-3)(x+5)}\right)$

$6(x+5) = 5(2x-3)+5$

$6x+30 = 10x-15+5$

$6x+30 = 10x-10$

$40 = 4x$

$10 = x$

The solution is 10.

20. $\dfrac{5}{x-4} - \dfrac{2x+1}{x^2-3x-4}$

$= \dfrac{5}{x-4} - \dfrac{2x+1}{(x-4)(x+1)}$

$= \dfrac{5}{x-4} \cdot \dfrac{x+1}{x+1} - \dfrac{2x+1}{(x-4)(x+1)}$

$= \dfrac{5x+5-2x-1}{(x-4)(x+1)} = \dfrac{3x+4}{(x-4)(x+1)}$

21. $\dfrac{27x^3-8}{9x^3+6x^2+4x} \div \dfrac{9x^2-12x+4}{9x^2-4}$

$= \dfrac{27x^3-8}{9x^3+6x^2+4x} \cdot \dfrac{9x^2-4}{9x^2-12x+4}$

$= \dfrac{(3x-2)(9x^2+6x+4)}{x(9x^2+6x+4)} \cdot \dfrac{(3x+2)(3x-2)}{(3x-2)(3x-2)}$

$= \dfrac{(3x-2)(9x^2+6x+4)(3x+2)(3x-2)}{x(9x^2+6x+4)(3x-2)(3x-2)}$

$= \dfrac{3x+2}{x}$

22. $3x^2-7x+2 = (3x-1)(x-2)$

The LCM is $(3x-1)(x-2)$.

$\dfrac{6x}{3x^2-7x+2} - \dfrac{2}{3x-1} + \dfrac{3x}{x-2}$

$= \dfrac{6x}{(3x-1)(x-2)} - \dfrac{2}{3x-1} \cdot \dfrac{x-2}{x-2} + \dfrac{3x}{x-2} \cdot \dfrac{3x-1}{3x-1}$

$= \dfrac{6x-2(x-2)+3x(3x-1)}{(3x-1)(x-2)}$

$= \dfrac{6x-2x+4+9x^2-3x}{(3x-1)(x-2)}$

$= \dfrac{9x^2+x+4}{(3x-1)(x-2)}$

23. $\dfrac{x^3-27}{x^2-9} = \dfrac{(x-3)(x^2+3x+9)}{(x-3)(x+3)} = \dfrac{x^2+3x+9}{x+3}$

24. The LCM is $(x-3)(x+2)$.

$\dfrac{2x}{x-3} - \dfrac{5}{x+2}$

$= \dfrac{2x}{x-3} \cdot \dfrac{x+2}{x+2} - \dfrac{5}{x+2} \cdot \dfrac{x-3}{x-3}$

$= \dfrac{2x^2+4x-5x+15}{(x-3)(x+2)}$

$= \dfrac{2x^2-x+15}{(x-3)(x+2)}$

25. Since $3x^2+4 > 0$ for all x, the domain is $\{x \mid x \in \text{real numbers}\}$.

26. The LCM is $(x+1)(x-2)$.

$\dfrac{x-4}{x+1} + \dfrac{x-3}{x-2}$

$= \dfrac{x-4}{x+1} \cdot \dfrac{x-2}{x-2} + \dfrac{x-3}{x-2} \cdot \dfrac{x+1}{x+1}$

$= \dfrac{x^2-6x+8+x^2-2x-3}{(x-2)(x+1)}$

$= \dfrac{2x^2-8x+5}{(x-2)(x+1)}$

27. $\dfrac{16-x^2}{x^3-2x^2-8x} = \dfrac{(4-x)(4+x)}{x(x^2-2x-8)}$

$= \dfrac{(4-x)(4+x)}{x(x-4)(x+2)}$

$= -\dfrac{x+4}{x(x+2)}$

28. $\dfrac{8x^3-27}{4x^2-9} = \dfrac{(2x-3)(4x^2+6x+9)}{(2x-3)(2x+3)}$

$= \dfrac{4x^2+6x+9}{2x+3}$

29. $\dfrac{16-x^2}{6x+12} \cdot \dfrac{x^2+5x+6}{x^2-8x+16}$

$= \dfrac{(4-x)(4+x)}{6(x+2)} \cdot \dfrac{(x+3)(x+2)}{(x-4)(x-4)}$

$= \dfrac{(4-x)(4+x)(x+3)(x+2)}{6(x+2)(x-4)(x-4)}$

$= -\dfrac{(x+4)(x+3)}{6(x-4)}$

30. $\dfrac{x^2-5x+4}{x^2-2x-8} \div \dfrac{x^2-4x+3}{x^2+8x+12}$

$= \dfrac{x^2-5x+4}{x^2-2x-8} \cdot \dfrac{x^2+8x+12}{x^2-4x+3}$

$= \dfrac{(x-4)(x-1)}{(x-4)(x+2)} \cdot \dfrac{(x+2)(x+6)}{(x-3)(x-1)}$

$= \dfrac{(x-4)(x-1)(x+2)(x+6)}{(x-4)(x+2)(x-3)(x-1)}$

$= \dfrac{x+6}{x-3}$

31. $\dfrac{8x^3-64}{4x^3+4x^2+x} \div \dfrac{x^2+2x+4}{4x^2-1}$

$= \dfrac{8x^3-64}{4x^3+4x^2+x} \cdot \dfrac{4x^2-1}{x^2+2x+4}$

$= \dfrac{8(x-2)(x^2+2x+4)}{x(2x+1)(2x+1)} \cdot \dfrac{(2x+1)(2x-1)}{x^2+2x+4}$

$= \dfrac{8(x-2)(x^2+2x+4)(2x+1)(2x-1)}{x(2x+1)(2x+1)(x^2+2x+4)}$

$= \dfrac{8(x-2)(2x-1)}{x(2x+1)}$

32. $\dfrac{3-x}{x^2+3x+9} \div \dfrac{x^2-9}{x^3-27}$

$= \dfrac{3-x}{x^2+3x+9} \cdot \dfrac{x^3-27}{x^2-9}$

$= \dfrac{3-x}{x^2+3x+9} \cdot \dfrac{(x-3)(x^2+3x+9)}{(x-3)(x+3)}$

$= \dfrac{(3-x)(x-3)(x^2+3x+9)}{(x^2+3x+9)(x-3)(x+3)}$

$= -\dfrac{x-3}{x+3}$

33. The LCM is $6a^3b^2$.

$\dfrac{3}{2a^3b^2} + \dfrac{5}{6a^2b}$

$\dfrac{3}{2a^3b^2} \cdot \dfrac{3}{3} + \dfrac{5}{6a^2b} \cdot \dfrac{ab}{ab} = \dfrac{9}{6a^3b^2} + \dfrac{5ab}{6a^3b^2}$

$= \dfrac{9+5ab}{6a^3b^2}$

34. $9x^2-4 = (3x-2)(3x+2)$

The LCM is $(3x-2)(3x+2)$.

$\dfrac{8}{9x^2-4} + \dfrac{5}{3x-2} - \dfrac{4}{3x+2}$

$= \dfrac{8}{(3x-3)(3x+2)} + \dfrac{5}{3x-2} \cdot \dfrac{3x+2}{3x+2} - \dfrac{4}{3x+2} \cdot \dfrac{3x-2}{3x-2}$

$= \dfrac{8+5(3x+2)-4(3x-2)}{(3x-2)(3x+2)}$

$= \dfrac{8+15x+10-12x+8}{(3x-2)(3x+2)}$

$= \dfrac{3x+26}{(3x-2)(3x+2)}$

35. $\dfrac{x-6+\dfrac{6}{x-1}}{x+3-\dfrac{12}{x-1}}$

$=\dfrac{x-6+\dfrac{6}{x-1}}{x+3-\dfrac{12}{x-1}}\cdot\dfrac{x-1}{x-1}=\dfrac{(x-6)(x-1)+6}{(x+3)(x-1)-12}$

$=\dfrac{x^2-7x+6+6}{x^2+2x-3-12}$

$=\dfrac{x^2-7x+12}{x^2+2x-15}=\dfrac{(x-3)(x-4)}{(x-3)(x+5)}$

$=\dfrac{x-4}{x+5}$

36. $\dfrac{x+\dfrac{3}{x-4}}{3+\dfrac{x}{x-4}}$

$=\dfrac{x+\dfrac{3}{x-4}}{3+\dfrac{x}{x-4}}\cdot\dfrac{x-4}{x-4}=\dfrac{x(x-4)+3}{3(x-4)+x}$

$=\dfrac{x^2-4x+3}{4x-12}=\dfrac{(x-3)(x-1)}{4(x-3)}$

$=\dfrac{x-1}{4}$

37. The LCM is $(x-3)(x+1)$.

$\dfrac{x+2}{x-3}=\dfrac{2x-5}{x+1}$

$(x-3)(x+1)\left(\dfrac{x+2}{x-3}\right)=(x-3)(x+1)\left(\dfrac{2x-5}{x+1}\right)$

$(x+1)(x+2)=(x-3)(2x-5)$

$x^2+3x+2=2x^2-11x+15$

$0=(x-13)(x-1)$

$x-13=0 \quad x-1=0$

$x=13 \qquad x=1$

The solutions are 1 and 13.

38. $\dfrac{5x}{2x-3}+4=\dfrac{3}{2x-3}$

$(2x-3)\left(\dfrac{5x}{2x-3}+4\right)=(2x-3)\left(\dfrac{3}{2x-3}\right)$

$5x+4(2x-3)=3$

$5x+8x-12=3$

$13x=15$

$x=\dfrac{15}{13}$

The solution is $\dfrac{15}{13}$.

39. $I=\dfrac{V}{R}$

$I\cdot R=\dfrac{V}{R}\cdot R$

$IR=V$

$R=\dfrac{V}{I}$

40. $\dfrac{6}{x-3} - \dfrac{1}{x+3} = \dfrac{51}{x^2-9}$

$(x-3)(x+3)\left(\dfrac{6}{x-3} - \dfrac{1}{x+3}\right) = (x^2-9)\left(\dfrac{51}{x^2-9}\right)$

$6(x+3) - (x-3) = 51$

$6x + 18 - x + 3 = 51$

$5x + 21 = 51$

$5x = 30$

$x = 6$

The solution is 6.

41. Strategy: Let t represent the time required to empty the tub.

	Rate	Time	Part
Inlet pipe	$\dfrac{1}{24}$	t	$\dfrac{t}{24}$
Drain pipe	$\dfrac{1}{15}$	t	$\dfrac{t}{15}$

The tub is empty when the difference between the drain pipe part and the inlet pipe part equals 1.

Solution: $\dfrac{t}{15} - \dfrac{t}{24} = 1$

$120\left(\dfrac{t}{15} - \dfrac{t}{24}\right) = 1(120)$

$8t - 5t = 120$

$3t = 120$

$t = 40$

It would take 40 min to empty the tub with both pipes open.

42. Strategy: Let r represent the rate of the cyclist.
The rate of the bus is $3r$.

	Distance	Rate	Time
Cyclist	90	r	$\dfrac{90}{r}$
Bus	90	$3r$	$\dfrac{90}{3r}$

The difference in the time is 4 h.

Solution: $\dfrac{90}{r} - \dfrac{90}{3r} = 4$

$3r\left(\dfrac{90}{r}\right) - 3r\left(\dfrac{90}{3r}\right) = 4(3r)$

$270 - 90 = 12r$

$180 = 12r$

$r = 15$

$3r = 3(15) = 45$

The rate of the bus is 45 mph.

43. Strategy: Let r represent the rate of the helicopter.
The rate of the airplane is $r + 20$.

	Distance	Rate	Time
Helicopter	9	r	$\dfrac{9}{r}$
Airplane	10	$r + 20$	$\dfrac{10}{r+20}$

The time traveled by the helicopter equals the time traveled by the airplane.

Solution: $\dfrac{9}{r} = \dfrac{10}{r+20}$

$r(r+20)\left(\dfrac{9}{r}\right) = r(r+20)\left(\dfrac{10}{r+20}\right)$

$9(r+20) = 10r$

$9r + 180 = 10r$

$r = 180$

The rate of the helicopter is 180 mph.

44. Strategy: Let x represent the time. Write and solve a proportional equation.

Solution: $\dfrac{2}{5} = \dfrac{150}{x}$

$\dfrac{2}{5} \cdot 5x = \dfrac{150}{x} \cdot 5x$

$2x = 750$

$x = 375$

It will take 375 min to read 150 pages.

45. Strategy: Write the basic inverse variation equation replacing the variable with the given values. Solve for k.
Write the inverse variation equation replacing k with its value. Substitute 100 for R and solve for I.

Solution:

$I = \dfrac{k}{R}$

$4 = \dfrac{k}{50}$

$200 = k$

$I = \dfrac{200}{R} = \dfrac{200}{100} = 2$

The current is 2 amps.

46. Strategy: Let x represent the number of miles. Write and solve a proportion.

Solution: $\dfrac{2.5}{10} = \dfrac{12}{x}$

$\dfrac{2.5}{10} \cdot 10x = \dfrac{12}{x} \cdot 10x$

$2.5x = 120$

$x = 48$

48 mi would be represented.

47. Strategy: Write the basic direct variation equation replacing the variable with the given values. Solve for k.
Write the direct variation equation replacing k with its value. Substitute 65 for v and solve for 2.

Solution:

$s = kv^2$

$170 = k(50)^2$

$170 = 2500k$

$0.068 = k$

$s = 0.068v^2 = 0.068(65^2) = 0.068(4225)$

$s = 287.3$

The stopping distance for a car traveling at 65 mph is 287.3 ft.

48. Strategy: Let t represent the time required for an apprentice working alone to install a fan.

	Rate	Time	Part
Electrician	$\dfrac{1}{65}$	40	$\dfrac{40}{65}$
Apprentice	$\dfrac{1}{t}$	40	$\dfrac{40}{t}$

The sum of the part of the task completed by the electrician and the part completed by the apprentice equals 1.

Solution: $\dfrac{40}{65} + \dfrac{40}{t} = 1$

$65t\left(\dfrac{40}{65} + \dfrac{40}{t}\right) = 1(65t)$

$40t + 2600 = 65t$

$2600 = 25t$

$t = 104$

It would take the apprentice 104 min to compete the job alone.

Chapter 6 Test

1.
$$\frac{3}{x+1} = \frac{2}{x}$$

$$x(x+1)\left(\frac{3}{x+1}\right) = x(x+1)\left(\frac{2}{x}\right)$$

$$3x = 2x + 2$$

$$x = 2$$

The solution is 2.

2.
$$\frac{x^2 + x - 6}{x^2 + 7x + 12} \div \frac{x^2 - 3x + 2}{x^2 + 6x + 8}$$

$$= \frac{x^2 + x - 6}{x^2 + 7x + 12} \cdot \frac{x^2 + 6x + 8}{x^2 - 3x + 2}$$

$$= \frac{(x+3)(x-2)}{(x+3)(x+4)} \cdot \frac{(x+2)(x+4)}{(x-2)(x-1)}$$

$$= \frac{(x+3)(x-2)(x+2)(x+4)}{(x+3)(x+4)(x-2)(x-1)}$$

$$= \frac{x+2}{x-1}$$

3. The LCM is $(x+2)(x-3)$.

$$\frac{2x-1}{x+2} - \frac{x}{x-3}$$

$$= \frac{2x-1}{x+2} \cdot \frac{x-3}{x-3} - \frac{x}{x-3} \cdot \frac{x+2}{x+2}$$

$$= \frac{2x^2 - 7x + 3 - x^2 - 2x}{(x+2)(x-3)}$$

$$= \frac{x^2 - 9x + 3}{(x+2)(x-3)}$$

4. $x^2 + x - 6 = (x+3)(x-2)$

$x^2 - 9 = (x+3)(x-3)$

The LCM is $(x-2)(x-3)(x+3)$.

$$\frac{x+1}{x^2+x-6} = \frac{x+1}{(x-2)(x+3)} \cdot \frac{x-3}{x-3}$$

$$= \frac{x^2-2x-3}{(x-2)(x-3)(x+3)}$$

$$\frac{2x}{x^2-9} = \frac{2x}{(x-3)(x+3)} \cdot \frac{x-2}{x-2}$$

$$= \frac{2x^2-4x}{(x-2)(x-3)(x+3)}$$

5. $\dfrac{4x}{2x-1} = 2 - \dfrac{1}{2x-1}$

$$(2x-1)\left(\frac{4x}{2x-1}\right) = (2x-1)\left(2 - \frac{1}{2x-1}\right)$$

$$4x = 4x - 2 - 1$$

$$4x = 4x - 3$$

$$0 = -3$$

There is no solution.

6. $x^2 + 3x - 10 = (x+5)(x-2)$

The LCM is $(x+5)(x-2)$.

$$\frac{3}{x+5} + \frac{2x}{x^2+3x-10}$$

$$= \frac{3}{x+5} \cdot \frac{x-2}{x-2} + \frac{2x}{(x+5)(x-2)}$$

$$= \frac{3x-6+2x}{(x+5)(x-2)}$$

$$= \frac{5x-6}{(x+5)(x-2)}$$

7. $\dfrac{3x^2-12}{5x-15} \cdot \dfrac{2x^2-18}{x^2+5x+6}$

$$= \frac{3(x+2)(x-2)}{5(x-3)} \cdot \frac{2(x+3)(x-3)}{(x+3)(x+2)}$$

$$= \frac{3(x+2)(x-2)2(x+3)(x-3)}{5(x-3)(x+3)(x+2)}$$

$$= \frac{6(x-2)}{5}$$

8. $f(x) = \dfrac{3x^2-x+1}{x^2-9}$

$$x^2 - 9 = 0$$

$$(x+3)(x-3) = 0$$

$$x + 3 = 0 \quad x - 3 = 0$$

$$x = -3 \qquad x = 3$$

The domain is $\{x \mid x \neq -3, 3\}$.

9. $\dfrac{1 - \dfrac{1}{x} - \dfrac{12}{x^2}}{1 + \dfrac{6}{x} + \dfrac{9}{x^2}}$

$$= \frac{1 - \dfrac{1}{x} - \dfrac{12}{x^2}}{1 + \dfrac{6}{x} + \dfrac{9}{x^2}} \cdot \frac{x^2}{x^2} = \frac{x^2-x-12}{x^2+6x+9}$$

$$= \frac{(x-4)(x+3)}{(x+3)(x+3)}$$

$$= \frac{x-4}{x+3}$$

10. $\dfrac{1-\dfrac{1}{x+2}}{1-\dfrac{3}{x+4}}$

$=\dfrac{1-\dfrac{1}{x+2}}{1-\dfrac{3}{x+4}}\cdot\dfrac{(x+2)(x+4)}{(x+2)(x+4)}$

$=\dfrac{(x+2)(x+4)-(x+4)}{(x+2)(x+4)-3(x+2)}$

$=\dfrac{(x+4)(x+2-1)}{(x+2)(x+4-3)}$

$=\dfrac{(x+4)(x+1)}{(x+2)(x+1)}$

$=\dfrac{x+4}{x+2}$

11. $\dfrac{2x^2-x-3}{2x^2-5x+3}\div\dfrac{3x^2-x-4}{x^2-1}$

$=\dfrac{2x^2-x-3}{2x^2-5x+3}\cdot\dfrac{x^2-1}{3x^2-x-4}$

$=\dfrac{(2x-3)(x+1)}{(2x-3)(x-1)}\cdot\dfrac{(x+1)(x-1)}{(3x-4)(x+1)}$

$=\dfrac{(2x-3)(x+1)(x+1)(x-1)}{(2x-3)(x-1)(3x-4)(x+1)}$

$=\dfrac{x+1}{3x-4}$

12. The LCM is $(x+1)$.

$\dfrac{4x}{x+1}-x=\dfrac{2}{x+1}$

$(x+1)\left(\dfrac{4x}{x+1}-x\right)=(x+1)\left(\dfrac{2}{x+1}\right)$

$4x-x(x+1)=2$

$4x-x^2-x=2$

$x^2-3x+2=0$

$(x-2)(x-1)=0$

$x-2=0 \quad x-1=0$

$x=2 \qquad x=1$

The solutions are 1 and 2.

13. $\dfrac{2a^2-8a+8}{4+4a-3a^2}$

$=\dfrac{2(a^2-4a+4)}{(2-a)(2+3a)}=\dfrac{2(a-2)(a-2)}{(2-a)(2+3a)}$

$=-\dfrac{2(a-2)}{3a+2}$

14. $x-\dfrac{12}{x-3}=\dfrac{x}{x-3}$

$\left(x-\dfrac{12}{x-3}\right)\cdot(x-3)=\dfrac{x}{x-3}\cdot(x-3)$

$x(x-3)-12=x$

$x^2-3x-12=x$

$x^2-4x-12=0$

$(x-6)(x+2)=0$

$x-6=0 \qquad x+2=0$

$x=6 \qquad\quad x=-2$

The solutions are -2 and 6.

15. $f(x) = \dfrac{3 - x^2}{x^3 - 2x^2 + 4}$

$f(-1) = \dfrac{3 - (-1)^2}{(-1)^3 - 2(-1)^2 + 4}$

$f(-1) = \dfrac{3 - 1}{-1 - 2 + 4}$

$f(-1) = 2$

16. $x^2 + 3x - 4 = (x + 4)(x - 1)$

$x^2 - 1 = (x + 1)(x - 1)$

The LCM is $(x + 4)(x + 1)(x - 1)$.

$\dfrac{x + 2}{x^2 + 3x - 4} - \dfrac{2x}{x^2 - 1}$

$= \dfrac{x + 2}{(x + 4)(x - 1)} \cdot \dfrac{x + 1}{x + 1} - \dfrac{2x}{(x + 1)(x - 1)} \cdot \dfrac{x + 4}{x + 4}$

$= \dfrac{x^2 + 3x + 2 - 2x^2 - 8x}{(x + 4)(x + 1)(x - 1)}$

$= \dfrac{-x^2 - 5x + 2}{(x + 4)(x + 1)(x - 1)}$

17. $\dfrac{x + 1}{2x + 5} = \dfrac{x - 3}{x}$

$x(2x + 5)\left(\dfrac{x + 1}{2x + 5}\right) = x(2x + 5)\left(\dfrac{x - 3}{x}\right)$

$x(x + 1) = (2x + 5)(x - 3)$

$x^2 + x = 2x^2 - x - 15$

$0 = x^2 - 2x - 15$

$0 = (x - 5)(x + 3)$

$x - 5 = 0 \quad x + 3 = 0$

$x = 5 \qquad x = -3$

The solutions are −3 and 5.

18. Strategy: Let r represent the rate of the hiker.
The rate of the cyclist is $r + 7$.

	Distance	Rate	Time
Hiker	6	r	$\dfrac{6}{r}$
Cyclist	20	$r + 7$	$\dfrac{20}{r + 7}$

The time traveled by the hiker equals the time traveled by the cyclist.

Solution: $\dfrac{6}{r} = \dfrac{20}{r + 7}$

$r(r + 7)\left(\dfrac{6}{r}\right) = r(r + 7)\left(\dfrac{20}{r + 7}\right)$

$6(r + 7) = 20r$

$6r + 42 = 20r$

$42 = 14r$

$r = 3$

$r + 7 = 3 + 7 = 10$

The rate of the cyclist is 10 mph.

19. Strategy: Write the basic combined variation equation replacing the variables with the given values. Solve for k. Write the combined variation equation replacing k with its value. Substitute 8000 for l, $\frac{1}{2}$ for d and solve for r.

Solution:

$$r = \frac{kl}{d^2}$$

$$3.2 = \frac{16,000k}{\left(\frac{1}{4}\right)^2}$$

$$0.0000125 = k$$

$$r = \frac{0.0000125l}{d^2} = \frac{0.0000125(8000)}{\left(\frac{1}{2}\right)^2} = 0.4$$

The resistance is 0.4 ohms.

20. Strategy: Let x represent the number of rolls. Write and solve a proportion.

Solution: $\dfrac{2}{45} = \dfrac{x}{315}$

$$\frac{2}{45} \cdot 315 = \frac{x}{315} \cdot 315$$

$$x = 14$$

The office requires 14 rolls of wallpaper.

21. Strategy: Let t represent the time required for both landscapers working together.

	Rate	Time	Part
1st landscaper	$\dfrac{1}{30}$	t	$\dfrac{t}{30}$
2nd landscaper	$\dfrac{1}{15}$	t	$\dfrac{t}{15}$

The sum of the parts of the task completed by each landscaper equals 1.

Solution: $\dfrac{t}{30} + \dfrac{t}{15} = 1$

$$30\left(\frac{t}{30} + \frac{t}{15}\right) = 1(30)$$

$$t + 2t = 30$$

$$3t = 30$$

$$t = 10$$

It would take 10 min to complete the task when both landscapers work together.

22. Strategy: Write the basic inverse variation equation replacing the variables with the given values. Solve for k. Write the inverse variation equation replacing k with its value. Substitute 5 for d and solve for I.

Solution:

$$I = \frac{k}{d^2}$$

$$50 = \frac{k}{8^2}$$

$$50 = \frac{k}{64}$$

$$3200 = k$$

$$I = \frac{3200}{d^2} = \frac{3200}{5^2} = \frac{3200}{25} = 128$$

The intensity is 128 decibels when the distance is 5 m.

Cumulative Review Exercises

1. $8 - 4[-3 - (-2)]^2 \div 5$

$= 8 - 4[-3 + 2]^2 \div 5 = 8 - 4[-1]^2 \div 5$

$= 8 - 4(1) \div 5 = 8 - \dfrac{4}{5}$

$= \dfrac{36}{5}$

2. $\dfrac{2x-3}{6}-\dfrac{x}{9}=\dfrac{x-4}{3}$

$18\left(\dfrac{2x-3}{6}-\dfrac{x}{9}\right)=18\left(\dfrac{x-4}{3}\right)$

$3(2x-3)-2x=6(x-4)$

$6x-9-2x=6x-24$

$4x-9=6x-24$

$2x=15$

$x=\dfrac{15}{2}$

The solution is $\dfrac{15}{2}$.

3. $5-|x-4|=2$

$-|x-4|=-3$

$|x-4|=3$

$x-4=3 \qquad x-4=-3$

$x=7 \qquad\ \ x=1$

The solutions are 1 and 7.

4. $f(x)=\dfrac{x}{x-3}$

$x-3=0$

$x=3$

The domain is $\{x|\,x\neq 3\}$.

5. $P(x)=\dfrac{x-1}{2x-3}$

$P(-2)=\dfrac{-2-1}{2(-2)-3}=\dfrac{-3}{-7}$

$P(-2)=\dfrac{3}{7}$

6. 3.5×10^{-8}

7. $\dfrac{(2a^{-2}b^{3})^{-2}}{(4a)^{-1}}=\dfrac{2^{-2}a^{4}b^{-6}}{4^{-1}a^{-1}}=2^{-2}4a^{4-(-1)}b^{-6}$

$=\dfrac{1}{4}4a^{5}b^{-6}=\dfrac{a^{5}}{b^{6}}$

8. $x-3(1-2x)\geq 1-4(2-2x)$

$x-3+6x\geq 1-8+8x$

$7x-3\geq -7+8x$

$-x-3\geq -7$

$-x\geq -4$

$x\leq 4$

$(-\infty,4]$

9. $(2a^{2}-3a+1)(-2a^{2})=-4a^{4}+6a^{3}-2a^{2}$

10. $2x^{2}+3x-2=(2x-1)(x+2)$

11. $x^{3}y^{3}-27=(xy)^{3}-3^{3}$

$=(xy-3)(x^{2}y^{2}+3xy+9)$

12. $\dfrac{x^{4}+x^{3}y-6x^{2}y^{2}}{x^{3}-2x^{2}y}=\dfrac{x^{2}(x^{2}+xy-6y^{2})}{x^{2}(x-2y)}$

$=\dfrac{x^{2}(x-2y)(x+3y)}{x^{2}(x-2y)}$

$=x+3y$

13. $3x-2y=6$

$-2y=-3x+6$

$y=\dfrac{3}{2}x-3$

$m=\dfrac{3}{2}$ and $(-2,-1)$

$y-y_{1}=m(x-x_{1})$

$y-(-1)=\dfrac{3}{2}[x-(-2)]$

$y+1=\dfrac{3}{2}(x+2)$

$y+1=\dfrac{3}{2}x+3$

$y=\dfrac{3}{2}x+2$

14. $8x^2 - 6x - 9 = 0$

$(4x + 3)(2x - 3) = 0$

$4x + 3 = 0 \quad 2x - 3 = 0$

$4x = -3 \qquad 2x = 3$

$x = -\dfrac{3}{4} \qquad x = \dfrac{3}{2}$

The solutions are $-\dfrac{3}{4}$ and $\dfrac{3}{2}$.

15. $\dfrac{4x^3 + 2x^2 - 10x + 1}{x - 2}$

$$
\begin{array}{r}
4x^2 + 10x + 10 \\
x - 2 \overline{\smash{)}\, 4x^3 + 2x^2 - 10x + 1} \\
\underline{4x^3 - 8x^2} \\
10x^2 - 10x \\
\underline{10x^2 - 20x} \\
10x + 1 \\
\underline{10x - 20} \\
21
\end{array}
$$

$\dfrac{4x^3 + 2x^2 - 10x + 1}{x - 2} = 4x^2 + 10x + 10 + \dfrac{21}{x - 2}$

16. $\dfrac{16x^2 - 9y^2}{16x^2 y - 12xy^2} \div \dfrac{4x^2 - xy - 3y^2}{12x^2 y^2}$

$\dfrac{16x^2 - 9y^2}{16x^2 y - 12xy^2} \cdot \dfrac{12x^2 y^2}{4x^2 - xy - 3y^2}$

$= \dfrac{(4x - 3y)(4x + 3y)}{4xy(4x - 3y)} \cdot \dfrac{12x^2 y^2}{(4x + 3y)(x - y)}$

$= \dfrac{(4x - 3y)(4x + 3y)12x^2 y^2}{4xy(4x - 3y)(4x + 3y)(x - y)}$

$= \dfrac{3xy}{x - y}$

17. $2x^2 + 2x = 2x(x + 1)$

$2x^4 - 2x^3 - 4x^2 = 2x^2(x^2 - x - 2)$

$\qquad = 2x^2(x - 2)(x + 1)$

The LCM is $2x^2(x - 2)(x + 1)$.

$\dfrac{xy}{2x^2 + 2x} = \dfrac{2xy}{2x(x + 1)} \cdot \dfrac{x(x - 2)}{x(x - 2)}$

$= \dfrac{x^2 y(x - 2)}{2x^2(x - 2)(x + 1)} = \dfrac{x^3 y - 2x^2 y}{2x^2(x - 2)(x + 1)}$

$\dfrac{2}{2x^4 - 2x^3 - 4x^2} = \dfrac{2}{2x^2(x - 2)(x + 1)}$

18. $3x^2 - x - 2 = (3x + 2)(x - 1)$

$x^2 - 1 = (x + 1)(x - 1)$

The LCM is $(3x + 2)(x + 1)(x - 1)$.

$\dfrac{5x}{3x^2 - x - 2} - \dfrac{2x}{x^2 - 1}$

$= \dfrac{5x}{(3x + 2)(x - 1)} \cdot \dfrac{x + 1}{x + 1} - \dfrac{2x}{(x + 1)(x - 1)} \cdot \dfrac{3x + 2}{3x + 2}$

$\dfrac{5x^2 + 5x - 6x^2 - 4x}{(3x + 2)(x + 1)(x - 1)}$

$= \dfrac{-x^2 + x}{(3x + 2)(x + 1)(x - 1)}$

$= \dfrac{-x(x - 1)}{(3x + 2)(x + 1)(x - 1)}$

$= -\dfrac{x}{(3x + 2)(x + 1)(x - 1)}$

19. $-3x + 5y = -15$

x-intercept:

$(5, 0)$

y-intercept:

$(0, -3)$

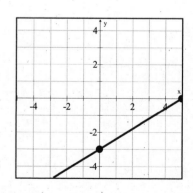

20. $x + y \leq 3$ $-2x + y > 4$
     $y \leq 3 - x$ $y > 2x + 4$

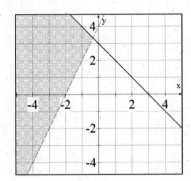

21. $\begin{vmatrix} 6 & 5 \\ 2 & -3 \end{vmatrix} = 6(-3) - 5(2) = -18 - 10 = -28$

22. $\dfrac{x - 4 + \dfrac{5}{x+2}}{x + 2 - \dfrac{1}{x+2}}$

$= \dfrac{x - 4 + \dfrac{5}{x+2}}{x + 2 - \dfrac{1}{x+2}} \cdot \dfrac{x+2}{x+2} = \dfrac{(x-4)(x+2) + 5}{(x+2)(x+2) - 1}$

$= \dfrac{x^2 - 2x - 8 + 5}{x^2 + 4x + 4 - 1}$

$= \dfrac{x^2 - 2x - 3}{x^2 + 4x + 3} = \dfrac{(x-3)(x+1)}{(x+3)(x+1)}$

$= \dfrac{x-3}{x+3}$

23. (1) $x + y + z = 3$
     (2) $-2x + y + 3z = 2$
     (3) $2x - 4y + z = -1$
Eliminate x. Multiply equation (1) by 2 and add to equation (2).
    $2x + 2y + 2z = 6$
   $-2x + y + 3z = 2$

    (4) $3y + 5z = 8$
Add equations (2) and (3).
   $-2x + y + 3z = 2$
    $2x - 4y + z = -1$

    (5) $-3y + 4z = 1$
Use equations (4) and (5) to solve for y and z.
   $3y + 5z = 8$
  $-3y + 4z = 1$
Eliminate y.
   $3y + 5z = 8$
  $-3y + 4z = 1$
       $9z = 9$
        $z = 1$
Replace z with 1 in equation (4).
   $3y + 5z = 8$
  $3y + 5(1) = 8$
    $3y - 3$
     $y = 1$
Replace y with 1 and z with 1 in equation (1).
   $x + y + z = 3$
   $x + 1 + 1 = 3$
    $x + 2 = 3$
      $x = 1$
The solution is (1, 1, 1).

24. $|3x - 2| > 4$
   $3x - 2 < -4$ $3x - 2 > 4$
   $3x < -2$ $3x > 6$
   $x < -\dfrac{2}{3}$ $x > 2$

$\left\{ x \mid x < -\dfrac{2}{3} \right\} \cup \left\{ x \mid x > 2 \right\}$

25. $\dfrac{2}{x - 3} = \dfrac{5}{2x - 3}$

$(2x - 3)(x - 3)\left(\dfrac{2}{x - 3} \right) = (2x - 3)(x - 3)\left(\dfrac{5}{2x - 3} \right)$

$2(2x - 3) = 5(x - 3)$

$4x - 6 = 5x - 15$

$-6 = x - 15$

$x = 9$

The solution is 9.

26. $x^2 - 36 = (x+6)(x-6)$
The LCM is $(x+6)(x-6)$.

$$\frac{3}{x^2 - 36} = \frac{2}{x-6} - \frac{5}{x+6}$$

$$\frac{3}{(x+6)(x-6)} = \frac{2}{x-6} - \frac{5}{x+6}$$

$$(x+6)(x-6)\left(\frac{3}{(x+6)(x-6)}\right) = (x+6)(x-6)\left(\frac{2}{x-6} - \frac{5}{x+6}\right)$$

$$3 = 2(x+6) - 5(x-6)$$

$$3 = 2x + 12 - 5x + 30$$

$$3 = -3x + 42$$

$$-39 = -3x$$

$$x = 13$$

The solution is 13.

27. $I = \dfrac{E}{R+r}$

$$I \cdot (R+r) = \frac{E}{R+r} \cdot (R+r)$$

$$IR + Ir = E$$

$$Ir = E - IR$$

$$r = \frac{E - IR}{I}$$

28. $(1 - x^{-1})^{-1} = \left(1 - \dfrac{1}{x}\right)^{-1} = \left(\dfrac{x-1}{x}\right)^{-1}$

$$= \frac{x}{x-1}$$

29. Strategy: Let x represent the number of pounds of almonds.

	Amount	Cost	Value
Almonds	x	5.40	$5.40x$
Peanuts	50	2.60	$2.60(50)$
Mixture	$50+x$	4.00	$4(50+x)$

The sum of the values before mixing equals the value after mixing.

Solution: $5.40x + 2.60(50) = 4(50 + x)$
$$5.4x + 130 = 4x + 200$$
$$1.4x + 130 = 200$$
$$1.4x = 70$$
$$x = 50$$

50 lb of almonds must be mixed with peanuts.

30. Strategy: Let x represent the number of people expected to vote. Write and solve a proportion.

Solution: $\dfrac{3}{5} = \dfrac{x}{125,000}$

$$125,000 \cdot \frac{3}{5} = \frac{x}{125,000} \cdot 125,000$$

$$x = 75,000$$

75,000 people are expected to vote.

31. Strategy: Let t represent the time required by the new computer.
The time required by the old computer is $6t$.

	Rate	Time	Part
New computer	$\dfrac{1}{t}$	12	$\dfrac{12}{t}$
Older computer	$\dfrac{1}{6t}$	12	$\dfrac{12}{6t}$

The sum of the parts of the task completed by each computer equals 1.

Solution: $\dfrac{12}{t} + \dfrac{12}{6t} = 1$

$$6t\left(\dfrac{12}{t} + \dfrac{12}{6t}\right) = 1(6t)$$

$$72 + 12 = 6t$$

$$84 = 6t$$

$$t = 14$$

It would take the new computer 14 min to complete the task when working alone.

32. Strategy: Let r represent the rate of the wind.

	Distance	Rate	Time
With the wind	900	$300 + r$	$\dfrac{900}{300+r}$
Against the wind	600	$300 - r$	$\dfrac{600}{300-r}$

The time traveled with the wind equals the time against the wind.

Solution: $\dfrac{900}{300+r} = \dfrac{600}{300-r}$

$$(300+r)(300-r)\left(\dfrac{900}{300+r}\right) = (300+r)(300-r)\left(\dfrac{600}{300-r}\right)$$

$$900(300-r) = 600(300+r)$$

$$270{,}000 - 900r = 180{,}000 + 600r$$

$$90{,}000 = 1500r$$

$$60 = r$$

The rate of the wind is 60 mph.

33. Strategy: To find the distance apart, calculate the number of times around the track each person has gone and find the difference in the distance.

Solution: The walker travels 3 mi in 1 h.

$$\dfrac{3\ mi}{1\ h} \cdot \dfrac{1\ lap}{0.25\ mi} = \dfrac{3}{0.25} = 12\ laps$$

The jogger travels 5 miles in 1 hour.

$$\dfrac{5\ mi}{1\ h} \cdot \dfrac{1\ lap}{0.25\ mi} = \dfrac{5}{0.25} = 20\ laps$$

Since the walker and jogger have completed integer multiples of laps, they are both at the starting point after one hour. Therefore they have no distance between them after 1 h.

Chapter 7: Exponents and Radicals

Prep Test

1. $48 = ? \cdot 3$
What number multiplied by 3 equals 48?
$? = 16$

2. $2^5 = 2 \cdot 2 \cdot 2 \cdot 2 \cdot 2 = 32$

3. $6\left(\dfrac{3}{2}\right) = \dfrac{6}{1}\left(\dfrac{3}{2}\right) = \dfrac{3 \cdot 2}{1}\left(\dfrac{3}{2}\right) = \dfrac{3 \cdot 3}{1 \cdot 1} = 9$

4. $\dfrac{1}{2} - \dfrac{2}{3} + \dfrac{1}{4} = \dfrac{6}{12} - \dfrac{8}{12} + \dfrac{3}{12}$
$= \dfrac{6 - 8 + 3}{12}$
$= \dfrac{1}{12}$

5. $(3 - 7x) - (4 - 2x)$
$= 3 - 7x - 4 + 2x$
$= -5x - 1$

6. $\dfrac{3x^5 y^6}{12x^4 y} = \dfrac{xy^5}{4}$

7. $(3x - 2)^2 = (3x - 2)(3x - 2)$
$= 9x^2 - 6x - 6x + 4$
$= 9x^2 - 12x + 4$

8. $(2 + 4x)(5 - 3x)$
$= 10 - 6x + 20x - 12x^2$
$= -12x^2 + 14x + 10$

9. $(6x - 1)(6x + 1)$
$= 36x^2 + 6x - 6x - 1$
$= 36x^2 - 1$

10. $x^2 - 14x - 5 = 10$
$x^2 - 14x - 15 = 0$
$(x - 15)(x + 1) = 0$
$x - 15 = 0 \quad x + 1 = 0$
$\qquad x = 15 \qquad x = -1$
The solutions are -1 and 15.

Section 7.1

Concept Check

1. 125

3. -32

5. 9

7. 3

Objective A Exercises

9. (i) and (iii) are not real numbers.

11. $8^{1/3} = (2^3)^{1/3} = 2$

13. $9^{3/2} = (3^2)^{3/2} = 3^3 = 27$

15. $27^{-2/3} = (3^3)^{-2/3} = 3^{-2} = \dfrac{1}{3^2} = \dfrac{1}{9}$

17. $32^{2/5} = (2^5)^{2/5} = 2^2 = 4$

19. $(-25)^{5/2}$
This is not a real number.
The base of the exponent expression is a negative number and the denominator of the exponent is a positive even number.

21. $\left(\dfrac{25}{49}\right)^{-3/2}$
$= \left(\dfrac{5^2}{7^2}\right)^{-3/2} = \left(\left(\dfrac{5}{7}\right)^2\right)^{-3/2}$
$= \left(\dfrac{5}{7}\right)^{-3} = \dfrac{5^{-3}}{7^{-3}} = \dfrac{7^3}{5^3}$
$= \dfrac{343}{125}$

23. $x^{1/2} x^{1/2} = x$

25. $y^{-1/4} y^{3/4} = y^{1/2}$

27. $x^{-2/3} \cdot x^{3/4} = x^{1/12}$

29. $a^{1/3} \cdot a^{3/4} \cdot a^{-1/2} = a^{7/12}$

31. $\dfrac{a^{1/2}}{a^{3/2}} = a^{-1} = \dfrac{1}{a}$

33. $\dfrac{y^{-3/4}}{y^{1/4}} = y^{-1} = \dfrac{1}{y}$

35. $\dfrac{y^{2/3}}{y^{-5/6}} = y^{9/6} = y^{3/2}$

37. $(x^2)^{-1/2} = x^{-1} = \dfrac{1}{x}$

39. $(x^{-2/3})^6 = x^{-4} = \dfrac{1}{x^4}$

41. $(a^{-1/2})^{-2} = a$

43. $(x^{-3/8})^{-4/5} = x^{3/10}$

45. $(a^{1/2} \cdot a)^2 = (a^{3/2})^2 = a^3$

47. $(x^{-1/2} \cdot x^{3/4})^{-2} = (x^{1/4})^{-2} = x^{-1/2} = \dfrac{1}{x^{1/2}}$

49. $(y^{-1/2} \cdot y^{2/3})^{2/3} = (y^{1/6})^{2/3} = y^{1/9}$

51. $(x^{-3}y^6)^{-1/3} = xy^{-2} = \dfrac{x}{y^2}$

53. $(x^{-2}y^{1/3})^{-3/4} = x^{3/2}y^{-1/4} = \dfrac{x^{3/2}}{y^{1/4}}$

55. $\left(\dfrac{x^{1/2}}{y^2}\right)^4 = \dfrac{x^2}{y^8}$

57. $\dfrac{x^{1/4} \cdot x^{-1/2}}{x^{2/3}} = \dfrac{x^{-1/4}}{x^{2/3}} = x^{-11/12} = \dfrac{1}{x^{11/12}}$

59. $\left(\dfrac{y^{2/3} \cdot y^{-5/6}}{y^{1/9}}\right)^9 = \left(\dfrac{y^{-1/6}}{y^{1/9}}\right)^9 = \left(y^{-5/18}\right)^9$

$= y^{-5/2} = \dfrac{1}{y^{5/2}}$

61. $\left(\dfrac{b^2 \cdot b^{-3/4}}{b^{-1/2}}\right)^{-1/2} = \left(\dfrac{b^{5/4}}{b^{-1/2}}\right)^{-1/2} = \left(b^{7/4}\right)^{-1/2}$

$= b^{-7/8} = \dfrac{1}{b^{7/8}}$

63. $(a^{2/3}b^2)^6(a^3b^3)^{1/3} = (a^4b^{12})(ab) = a^5b^{13}$

65. $(16x^{-2}y^4)^{-1/2}(xy^{1/2}) = (16)^{-1/2}(xy^{-2})(xy^{1/2})$

$= (16)^{-1/2}x^2y^{-3/2}$

$= \dfrac{x^2}{16^{1/2}y^{3/2}}$

$= \dfrac{x^2}{4y^{3/2}}$

67. $(x^{-2/3}y^{-3})^3(27x^{-3}y^6)^{-1/3} = (x^{-2}y^{-9})(27)^{-1/3}(xy^{-2})$

$= (27)^{-1/3}x^{-1}y^{-11}$

$= \dfrac{1}{27^{1/3}xy^{11}}$

$= \dfrac{1}{3xy^{11}}$

69. $\dfrac{\left(4a^{4/3}b^{-2}\right)^{-1/2}}{(a^{1/6}b^{-3/2})^2} = \dfrac{\left((4)^{-1/2}a^{-2/3}b\right)}{a^{1/3}b^{-3}} = \dfrac{b^4}{2a}$

71. $\left(\dfrac{x^{1/2}y^{-3/4}}{y^{2/3}}\right)^{-6} = \left(x^{1/2}y^{-17/12}\right)^{-6}$

$= x^{-3}y^{17/2} = \dfrac{y^{17/2}}{x^3}$

73. $\left(\dfrac{b^{-3}}{64a^{-1/2}}\right)^{-2/3} = \dfrac{b^2}{64^{-2/3}a^{1/3}}$

$= \dfrac{64^{2/3}b^2}{a^{1/3}} = \dfrac{16b^2}{a^{1/3}}$

75. $y^{3/2}(y^{1/2} - y^{1/2}) = y^{4/2} - y^{2/2} = y^2 - y$

77. $a^{-1/4}(a^{5/4} - a^{9/4}) = a^{4/4} - a^{8/4} = a - a^2$

Objective B Exercises

79. False

81. $3^{1/4} = \sqrt[4]{3}$

83. $a^{3/2} = (a^3)^{1/2} = \sqrt{a^3}$

85. $(2t)^{5/2} = \sqrt{(2t)^5} = \sqrt{32t^5}$

87. $-2x^{2/3} = -2(x)^{2/3} = -2\sqrt[3]{x^2}$

89. $(a^2b)^{2/3} = \sqrt[3]{(a^2b)^2} = \sqrt[3]{a^4b^2}$

91. $(a^2b^4)^{3/5} = \sqrt[5]{(a^2b^4)^3} = \sqrt[5]{a^6b^{12}}$

93. $(4x-3)^{3/4} = \sqrt[4]{(4x-3)^3}$

95. $x^{-2/3} = \dfrac{1}{x^{2/3}} = \dfrac{1}{\sqrt[3]{x^2}}$

97. $\sqrt{14} = 14^{1/2}$

99. $\sqrt[3]{x} = x^{1/3}$

101. $\sqrt[3]{x^4} = x^{4/3}$

103. $\sqrt[5]{b^3} = b^{3/5}$

105. $\sqrt[3]{2x^2} = (2x^2)^{1/3}$

107. $-\sqrt{3x^5} = -(3x^5)^{1/2}$

109. $3x\sqrt[3]{y^2} = 3xy^{2/3}$

111. $\sqrt{a^2-2} = (a^2-2)^{1/2}$

Objective C Exercises

113. Positive

115. Not a real number

117. $\sqrt{x^{16}} = x^8$

119. $-\sqrt{x^8} = -x^4$

121. $\sqrt[3]{x^3y^9} = xy^3$

123. $-\sqrt[3]{x^{15}y^3} = -x^5y$

125. $\sqrt{16a^4b^{12}} = 4a^2b^6$

127. The square root of a negative number is not a real number.

129. $\sqrt[3]{27x^9} = 3x^3$

131. $\sqrt[3]{-64x^9y^{12}} = -4x^3y^4$

133. $-\sqrt[4]{x^8y^{12}} = -x^2y^3$

135. $\sqrt[5]{x^{20}y^{10}} = x^4y^2$

137. $\sqrt[4]{81x^4y^{20}} = 3xy^5$

139. $\sqrt[5]{32a^5b^{10}} = 2ab^2$

Critical Thinking

141. No. If $x \geq 0$, the statement is true. However, if $x < 0$ then $\sqrt{x^2} = |x|$.

Projects or Group Activities

143.
$$\sqrt{2} \approx 1 + \cfrac{1}{2 + \cfrac{1}{2 + \cfrac{1}{2 + \cfrac{1}{2}}}}$$

$$= 1 + \cfrac{1}{2 + \cfrac{1}{2 + \cfrac{1}{2.5}}} = 1 + \cfrac{1}{2 + \cfrac{1}{2.4}}$$

$$\approx 1 + \frac{1}{2.417} \approx 1.414$$

Section 7.2

Concept Check

1. Neither

3. Perfect square

5. Perfect cube

7. Perfect cube

9. No

11. No

Objective A Exercises

11. $\sqrt{x^4 y^3 z^5} = \sqrt{x^4 y^2 z^4 (yz)}$

$$= \sqrt{x^4 y^2 z^4} \sqrt{yz}$$

$$= x^2 yz^2 \sqrt{yz}$$

15. $\sqrt{8a^3 b^8} = \sqrt{4a^2 b^8 (2a)}$

$$= \sqrt{4a^2 b^8} \sqrt{2a}$$

$$= 2ab^4 \sqrt{2a}$$

17. $\sqrt{45x^2 y^3 z^5} = \sqrt{9x^2 y^2 z^4 (5yz)}$

$$= \sqrt{9x^2 y^2 z^4} \sqrt{5yz}$$

$$= 3xyz^2 \sqrt{5yz}$$

19. $\sqrt[4]{48x^4 y^5 z^6} = \sqrt[4]{16x^4 y^4 z^4 (3yz^2)}$

$$= \sqrt[4]{16x^4 y^4 z^4} \sqrt[4]{3yz^2}$$

$$= 2xyz \sqrt[4]{3yz^2}$$

21. $\sqrt[3]{a^{16} b^8} = \sqrt[3]{a^{15} b^6 (ab^2)}$

$$= \sqrt[3]{a^{15} b^6} \sqrt[3]{ab^2}$$

$$= a^5 b^2 \sqrt[3]{ab^2}$$

23. $\sqrt[3]{-125x^2 y^4} = \sqrt[3]{-125y^3 (x^2 y)}$

$$= \sqrt[3]{-125y^3} \sqrt[3]{x^2 y}$$

$$= -5y \sqrt[3]{x^2 y}$$

25. $\sqrt[3]{a^4 b^5 c^6} = \sqrt[3]{a^3 b^3 c^6 (ab^2)}$

$$= \sqrt[3]{a^3 b^3 c^6} \sqrt[3]{ab^2}$$

$$= abc^2 \sqrt[3]{ab^2}$$

27. $\sqrt[4]{16x^9 y^5} = \sqrt[4]{16x^8 y^4 (xy)}$

$$= \sqrt[4]{16x^8 y^4} \sqrt[4]{xy}$$

$$= 2x^2 y \sqrt[4]{xy}$$

Objective B Exercises

29. True

31. $2\sqrt{x} - 8\sqrt{x} = -6\sqrt{x}$

33. $\sqrt{8x} - \sqrt{32x} = \sqrt{4 \cdot 2x} - \sqrt{16 \cdot 2x}$

$$= \sqrt{4} \sqrt{2x} - \sqrt{16} \sqrt{2x}$$

$$= 2\sqrt{2x} - 4\sqrt{2x}$$

$$= -2\sqrt{2x}$$

35. $\sqrt{18b} + \sqrt{75b} = \sqrt{9 \cdot 2b} + \sqrt{25 \cdot 3b}$

$\qquad = \sqrt{9}\sqrt{2b} + \sqrt{25}\sqrt{3b}$

$\qquad = 3\sqrt{2b} + 5\sqrt{3b}$

37. $3\sqrt{8x^2y^3} - 2x\sqrt{32y^3} = 3\sqrt{4 \cdot 2x^2y^3} - 2x\sqrt{16 \cdot 2y^3}$

$\qquad = 3\sqrt{4x^2y^2}\sqrt{2y} - 2x\sqrt{16y^2}\sqrt{2y}$

$\qquad = 3 \cdot 2xy\sqrt{2y} - 2x \cdot 4y\sqrt{2y}$

$\qquad = 6xy\sqrt{2y} - 8xy\sqrt{2y}$

$\qquad = -2xy\sqrt{2y}$

39. $2a\sqrt{27ab^5} + 3b\sqrt{3a^3b} = 2a\sqrt{3^3ab^5} + 3b\sqrt{3a^3b}$

$\qquad = 2a\sqrt{3^2b^4}\sqrt{3ab} + 3b\sqrt{a^2}\sqrt{3ab}$

$\qquad = 2a \cdot 3b^2\sqrt{3ab} + 3ab\sqrt{3ab}$

$\qquad = 6ab^2\sqrt{3ab} + 3ab\sqrt{3ab}$

41. $\sqrt[3]{16} - \sqrt[3]{54} = \sqrt[3]{8 \cdot 2} - \sqrt[3]{27 \cdot 2}$

$\qquad = \sqrt[3]{8}\sqrt[3]{2} - \sqrt[3]{27}\sqrt[3]{2}$

$\qquad = 2\sqrt[3]{2} - 3\sqrt[3]{2}$

$\qquad = -\sqrt[3]{2}$

43. $2b\sqrt[3]{16b^2} + \sqrt[3]{128b^5} = 2b\sqrt[3]{8 \cdot 2b^2} + \sqrt[3]{64b^3 \cdot 2b^2}$

$\qquad = 2b\sqrt[3]{8}\sqrt[3]{2b^2} + \sqrt[3]{64b^3}\sqrt[3]{2b^2}$

$\qquad = 4b\sqrt[3]{2b^2} + 4b\sqrt[3]{2b^2}$

$\qquad = 8b\sqrt[3]{2b^2}$

45. $3\sqrt[4]{32a^5} - a\sqrt[4]{162a} = 3\sqrt[4]{16a^4 \cdot 2a} - a\sqrt[4]{81 \cdot 2a}$

$\qquad = 3\sqrt[4]{16a^4}\sqrt[4]{2a} - a\sqrt[4]{81}\sqrt[4]{2a}$

$\qquad = 3 \cdot 2a\sqrt[4]{2a} - a \cdot 3\sqrt[4]{2a}$

$\qquad = 6a\sqrt[4]{2a} - 3a\sqrt[4]{2a}$

$\qquad = 3a\sqrt[4]{2a}$

47. $2\sqrt{50} - 3\sqrt{125} + \sqrt{98}$

$\qquad = 2\sqrt{25 \cdot 2} - 3\sqrt{25 \cdot 5} + \sqrt{49 \cdot 2}$

$\qquad = 2\sqrt{25}\sqrt{2} - 3\sqrt{25}\sqrt{5} + \sqrt{49}\sqrt{2}$

$\qquad = 10\sqrt{2} - 15\sqrt{5} + 7\sqrt{2}$

$\qquad = 17\sqrt{2} - 15\sqrt{5}$

49. $\sqrt{9b^3} - \sqrt{25b^3} + \sqrt{49b^3}$

$\qquad = \sqrt{9b^2 \cdot b} - \sqrt{25b^2 \cdot b} + \sqrt{49b^2 \cdot b}$

$\qquad = \sqrt{9b^2}\sqrt{b} - \sqrt{25b^2}\sqrt{b} + \sqrt{49b^2}\sqrt{b}$

$\qquad = 3b\sqrt{b} - 5b\sqrt{b} + 7b\sqrt{b}$

$\qquad = 5b\sqrt{b}$

51. $2x\sqrt{8xy^2} - 3y\sqrt{32x^3} + \sqrt{4x^3y^3}$

$\qquad = 2x\sqrt{4y^2 \cdot 2x} - 3y\sqrt{16x^2 \cdot 2x} + \sqrt{4x^2y^2 \cdot xy}$

$\qquad = 2x\sqrt{4y^2}\sqrt{2x} - 3y\sqrt{16x^2}\sqrt{2x} + \sqrt{4x^2y^2}\sqrt{xy}$

$\qquad = 4xy\sqrt{2x} - 12xy\sqrt{2x} + 2xy\sqrt{xy}$

$\qquad = -8xy\sqrt{2x} + 2xy\sqrt{xy}$

Critical Thinking

53. $\sqrt[3]{54xy^3} - 5\sqrt[3]{2xy^3} + y\sqrt[3]{128x}$

$\qquad = \sqrt[3]{27y^3 \cdot 2x} - 5\sqrt[3]{y^3 \cdot 2x} + y\sqrt[3]{64 \cdot 2x}$

$\qquad = \sqrt[3]{27y^3}\sqrt[3]{2x} - 5\sqrt[3]{y^3}\sqrt[3]{2x} + y\sqrt[3]{64}\sqrt[3]{2x}$

$\qquad = 3y\sqrt[3]{2x} - 5y\sqrt[3]{2x} + 4y\sqrt[3]{2x}$

$\qquad = 2y\sqrt[3]{2x}$

55. $2a\sqrt[4]{32b^5} - 3b\sqrt[4]{162a^4b} + \sqrt[4]{2a^4b^5}$

$\qquad = 2a\sqrt[4]{16b^4 \cdot 2b} - 3b\sqrt[4]{81a^4 \cdot 2b} + \sqrt[4]{a^4b^4 \cdot 2b}$

$\qquad = 2a\sqrt[4]{16b^4}\sqrt[4]{2b} - 3b\sqrt[4]{81a^4}\sqrt[4]{2b} + \sqrt[4]{a^4b^4}\sqrt[4]{2b}$

$\qquad = 4ab\sqrt[4]{2b} - 9ab\sqrt[4]{2b} + ab\sqrt[4]{2b}$

$\qquad = -4ab\sqrt[4]{2b}$

Projects or Group Activities

57. Domain: $(-\infty, 3]$

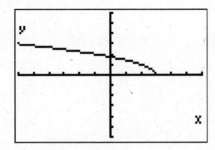

59. Domain: $(-\infty, \infty)$

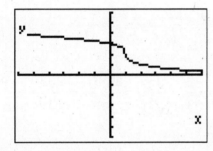

Section 7.3

Concept Check

1. If $\sqrt[n]{a}$ and $\sqrt[n]{b}$ are real numbers, then
$\sqrt[n]{a} \cdot \sqrt[n]{b} = \sqrt[n]{ab}$.

3. $3 - \sqrt{5}$

5. $4 + 3\sqrt{11}$

7. (i) and (iii) can be simplified using the Product Property of Radicals.

Objective A Exercises

9. $\sqrt{8}\sqrt{32} = \sqrt{256} = 16$

11. $\sqrt[3]{4}\sqrt[3]{8} = 2\sqrt[3]{4}$

13. $\sqrt{x^2 y^5}\sqrt{xy} = \sqrt{x^3 y^6}$
$= \sqrt{x^2 y^6 \cdot x} = xy^3\sqrt{x}$

15. $\sqrt{2x^2 y}\sqrt{32xy} = \sqrt{64x^3 y^2}$
$= \sqrt{64x^2 y^2 \cdot x} = 8xy\sqrt{x}$

17. $\sqrt[3]{x^2 y}\sqrt[3]{16x^4 y^2} = \sqrt[3]{16x^6 y^3}$
$= \sqrt[3]{8x^6 y^3 \cdot 2} = 2x^2 y\sqrt[3]{2}$

19. $\sqrt[4]{12ab^3}\sqrt[4]{4a^5 b^2} = \sqrt[4]{48a^6 b^5}$
$= \sqrt[4]{16a^4 b^4 \cdot 3a^2 b} = 2ab\sqrt[4]{3a^2 b}$

21. $\sqrt{3}(\sqrt{27} - \sqrt{3}) = \sqrt{81} - \sqrt{9} = 9 - 3 = 6$

23. $\sqrt{x}(\sqrt{x} - \sqrt{2}) = \sqrt{x^2} - \sqrt{2x} = x - \sqrt{2x}$

25. $\sqrt{2x}(\sqrt{8x} - \sqrt{32}) = \sqrt{16x^2} - \sqrt{64x}$
$= \sqrt{16x^2} - \sqrt{64 \cdot x} = 4x - 8\sqrt{x}$

27. $(3 - 2\sqrt{5})(2 + \sqrt{5}) = 6 + 3\sqrt{5} - 4\sqrt{5} - 2(\sqrt{5})^2$
$= 6 + 3\sqrt{5} - 4\sqrt{5} - 10 = -4 - \sqrt{5}$

29. $(-2 + \sqrt{7})(3 + 5\sqrt{7}) = -6 - 10\sqrt{7} + 3\sqrt{7} + 5(\sqrt{7})^2$
$= -6 - 10\sqrt{7} + 3\sqrt{7} + 35 = 29 - 7\sqrt{7}$

31. $(6+3\sqrt{2})(4-2\sqrt{2})=24-12\sqrt{2}+12\sqrt{2}-6(\sqrt{2})^2$
$=24-12\sqrt{2}+12\sqrt{2}-12=12$

33. $(5-2\sqrt{7})(5+2\sqrt{7})=25+10\sqrt{7}-10\sqrt{7}-4(\sqrt{7})^2$
$=25+10\sqrt{7}-10\sqrt{7}-28=-3$

35. $(3-\sqrt{2x})(1+5\sqrt{2x})$
$=3+15\sqrt{2x}-\sqrt{2x}-5(\sqrt{2x})^2$
$=3+15\sqrt{2x}-\sqrt{2x}-10x$
$=-10x+14\sqrt{2x}+3$

37. $(2+\sqrt{x})^2=(2+\sqrt{x})(2+\sqrt{x})$
$=4+2\sqrt{x}+2\sqrt{x}+(\sqrt{x})^2$
$=x+4\sqrt{x}+4$

39. $(\sqrt{3x}-5)^2=(\sqrt{3x}-5)(\sqrt{3x}-5)$
$=(\sqrt{3x})^2-5\sqrt{3x}-5\sqrt{3x}+25$
$=3x-10\sqrt{3x}+25$

41. $(4-\sqrt{2x+1})^2=(4-\sqrt{2x+1})(4-\sqrt{2x+1})$
$=16-4\sqrt{2x+1}-4\sqrt{2x+1}+(\sqrt{2x+1})^2$
$=2x+1-8\sqrt{2x+1}+16$
$=2x-8\sqrt{2x+1}+17$

43. True

Objective B Exercises

45. To rationalize the denominator of a radical expression means to rewrite the expression with no radicals in the denominator. To do this, multiply both the numerator and the denominator by the same expression, one that removes the radicals(s) from the denominator of the original expression.

47. $\sqrt[3]{4x}$

49. $\sqrt{3}+x$

51. $\dfrac{\sqrt{60y^4}}{\sqrt{12y}}=\sqrt{\dfrac{60y^4}{12y}}$
$=\sqrt{5y^3}=\sqrt{y^2\cdot 5y}$
$=y\sqrt{5y}$

53. $\dfrac{\sqrt{65ab^4}}{\sqrt{5ab}}=\sqrt{\dfrac{65ab^4}{5ab}}$
$=\sqrt{13b^3}=\sqrt{b^2\cdot 13b}$
$=b\sqrt{13b}$

55. $\dfrac{1}{\sqrt{2}}=\dfrac{1}{\sqrt{2}}\cdot\dfrac{\sqrt{2}}{\sqrt{2}}=\dfrac{\sqrt{2}}{\sqrt{2^2}}=\dfrac{\sqrt{2}}{2}$

57. $\dfrac{2}{\sqrt{3y}}=\dfrac{2}{\sqrt{3y}}\cdot\dfrac{\sqrt{3y}}{\sqrt{3y}}=\dfrac{2\sqrt{3y}}{\sqrt{9y^2}}=\dfrac{2\sqrt{3y}}{3y}$

59. $\dfrac{9}{\sqrt{3a}}=\dfrac{9}{\sqrt{3a}}\cdot\dfrac{\sqrt{3a}}{\sqrt{3a}}=\dfrac{9\sqrt{3a}}{\sqrt{9a^2}}$
$=\dfrac{9\sqrt{3a}}{3a}=\dfrac{3\sqrt{3a}}{a}$

61. $\sqrt{\dfrac{y}{2}}=\dfrac{\sqrt{y}}{\sqrt{2}}=\dfrac{\sqrt{y}}{\sqrt{2}}\cdot\dfrac{\sqrt{2}}{\sqrt{2}}=\dfrac{\sqrt{2y}}{\sqrt{4}}=\dfrac{\sqrt{2y}}{2}$

63. $\dfrac{5}{\sqrt[3]{9}}=\dfrac{5}{\sqrt[3]{9}}\cdot\dfrac{\sqrt[3]{3}}{\sqrt[3]{3}}=\dfrac{5\sqrt[3]{3}}{\sqrt[3]{27}}=\dfrac{5\sqrt[3]{3}}{3}$

65. $\dfrac{5}{\sqrt[3]{3y}}=\dfrac{5}{\sqrt[3]{3y}}\cdot\dfrac{\sqrt[3]{9y^2}}{\sqrt[3]{9y^2}}=\dfrac{5\sqrt[3]{9y^2}}{\sqrt[3]{27y^3}}=\dfrac{5\sqrt[3]{9y^2}}{3y}$

67. $\dfrac{6x}{\sqrt[4]{9x}}=\dfrac{6x}{\sqrt[4]{9x}}\cdot\dfrac{\sqrt[4]{9x^3}}{\sqrt[4]{9x^3}}=\dfrac{6x\sqrt[4]{9x^3}}{\sqrt[4]{81x^4}}=\dfrac{6x\sqrt[4]{9x^3}}{3x}=2\sqrt[4]{9x^3}$

69. $\dfrac{9x^2}{\sqrt[5]{27x}} = \dfrac{9x^2}{\sqrt[5]{27x}} \cdot \dfrac{\sqrt[5]{9x^4}}{\sqrt[5]{9x^4}} = \dfrac{9x^2\sqrt[5]{9x^4}}{\sqrt[5]{243x^5}}$

$\quad = \dfrac{9x^2\sqrt[5]{9x^4}}{3x} = 3x\sqrt[5]{9x^4}$

71. $\dfrac{\sqrt{15a^2b^5}}{\sqrt{30a^5b^3}} = \sqrt{\dfrac{15a^2b^5}{30a^5b^3}} = \sqrt{\dfrac{b^2}{2a^3}} = \dfrac{\sqrt{b^2}}{\sqrt{a^2 \cdot 2a}}$

$\quad = \dfrac{b}{a\sqrt{2a}} = \dfrac{b}{a\sqrt{2a}} \cdot \dfrac{\sqrt{2a}}{\sqrt{2a}} = \dfrac{b\sqrt{2a}}{a\sqrt{4a^2}}$

$\quad = \dfrac{b\sqrt{2a}}{2a^2}$

73. $\dfrac{\sqrt{12x^3y}}{\sqrt{20x^4y}} = \sqrt{\dfrac{12x^3y}{20x^4y}} = \sqrt{\dfrac{3}{5x}}$

$\quad = \dfrac{\sqrt{3}}{\sqrt{5x}} = \dfrac{\sqrt{3}}{\sqrt{5x}} \cdot \dfrac{\sqrt{5x}}{\sqrt{5x}} = \dfrac{\sqrt{15x}}{\sqrt{25x^2}}$

$\quad = \dfrac{\sqrt{15x}}{5x}$

75. $\dfrac{-2}{1-\sqrt{2}} = \dfrac{-2}{1-\sqrt{2}} \cdot \dfrac{1+\sqrt{2}}{1+\sqrt{2}}$

$\quad = \dfrac{-2(1+\sqrt{2})}{1^2-(\sqrt{2})^2} = \dfrac{-2-2\sqrt{2}}{1-2} = \dfrac{-2-2\sqrt{2}}{-1}$

$\quad = 2+2\sqrt{2}$

77. $\dfrac{-4}{3-\sqrt{2}} = \dfrac{-4}{3-\sqrt{2}} \cdot \dfrac{3+\sqrt{2}}{3+\sqrt{2}}$

$\quad = \dfrac{-4(3+\sqrt{2})}{3^2-(\sqrt{2})^2} = \dfrac{-12-4\sqrt{2}}{9-2} = \dfrac{-12-4\sqrt{2}}{7}$

79. $\dfrac{5}{2-\sqrt{7}} = \dfrac{5}{2-\sqrt{7}} \cdot \dfrac{2+\sqrt{7}}{2+\sqrt{7}}$

$\quad = \dfrac{5(2+\sqrt{7})}{2^2-(\sqrt{7})^2} = \dfrac{10+5\sqrt{7}}{4-7} = \dfrac{10+5\sqrt{7}}{-3}$

$\quad = -\dfrac{10+5\sqrt{7}}{3}$

81. $\dfrac{-7}{\sqrt{x}-3} = \dfrac{-7}{\sqrt{x}-3} \cdot \dfrac{\sqrt{x}+3}{\sqrt{x}+3}$

$\quad = \dfrac{-7(\sqrt{x}+3)}{(\sqrt{x})^2-3^2} = \dfrac{-(7\sqrt{x}+21)}{x-9}$

$\quad = -\dfrac{7\sqrt{x}+21}{x-9}$

83. $\dfrac{\sqrt{3}+\sqrt{4}}{\sqrt{2}+\sqrt{3}} = \dfrac{\sqrt{3}+\sqrt{2^2}}{\sqrt{2}+\sqrt{3}} = \dfrac{\sqrt{3}+2}{\sqrt{2}+\sqrt{3}} \cdot \dfrac{\sqrt{2}-\sqrt{3}}{\sqrt{2}-\sqrt{3}}$

$\quad = \dfrac{\sqrt{6}-(\sqrt{3})^2+2\sqrt{2}-2\sqrt{3}}{(\sqrt{2})^2-(\sqrt{3})^2}$

$\quad = \dfrac{\sqrt{6}-3+2\sqrt{2}-2\sqrt{3}}{2-3}$

$\quad = \dfrac{\sqrt{6}-3+2\sqrt{2}-2\sqrt{3}}{-1}$

$\quad = -\sqrt{6}+3-2\sqrt{2}+2\sqrt{3}$

85. $\dfrac{2+3\sqrt{5}}{1-\sqrt{5}} = \dfrac{2+3\sqrt{5}}{1-\sqrt{5}} \cdot \dfrac{1+\sqrt{5}}{1+\sqrt{5}}$

$\quad = \dfrac{2+2\sqrt{5}+3\sqrt{5}+3(\sqrt{5})^2}{1^2-(\sqrt{5})^2}$

$\quad = \dfrac{2+5\sqrt{5}+15}{1-4}$

$\quad = \dfrac{17+5\sqrt{4}}{-3}$

$\quad = -\dfrac{17+5\sqrt{4}}{3}$

87. $\dfrac{2\sqrt{a}-\sqrt{b}}{4\sqrt{a}+3\sqrt{b}} = \dfrac{2\sqrt{a}-\sqrt{b}}{4\sqrt{a}+3\sqrt{b}} \cdot \dfrac{4\sqrt{a}-3\sqrt{b}}{4\sqrt{a}-3\sqrt{b}}$

$\quad = \dfrac{8\sqrt{a^2}-6\sqrt{ab}-4\sqrt{ab}+3\sqrt{b^2}}{(4\sqrt{a})^2-(3\sqrt{b})^2}$

$\quad = \dfrac{8a-10\sqrt{ab}+3b}{16a-9b}$

89. $\dfrac{3\sqrt{y}-y}{\sqrt{y}+2y} = \dfrac{3\sqrt{y}-y}{\sqrt{y}+2y} \cdot \dfrac{\sqrt{y}-2y}{\sqrt{y}-2y}$

$= \dfrac{3(\sqrt{y})^2 - 6y\sqrt{y} - y\sqrt{y} + 2y^2}{(\sqrt{y})^2 - (2y)^2}$

$= \dfrac{3y - 7y\sqrt{y} + 2y^2}{y - 4y^2},$

$= \dfrac{3 - 7\sqrt{y} + 2y}{1 - 4y}$

Critical Thinking

91. $(\sqrt{8} - \sqrt{2})^3 = (2\sqrt{2} - \sqrt{2})^3 = (\sqrt{2}(2-1))^3$

$= (\sqrt{2})^3 = \sqrt{8} = 2\sqrt{2}$

93. $(\sqrt{2} - \sqrt{3})^2 = (\sqrt{2} - \sqrt{3})(\sqrt{2} - \sqrt{3})$

$= 2 - \sqrt{6} - \sqrt{6} + 3 = 5 - 2\sqrt{6}$

95. $\dfrac{\sqrt{9+h} - 3}{h} \cdot \dfrac{\sqrt{9+h} + 3}{\sqrt{9+h} + 3} = \dfrac{9+h-9}{h(\sqrt{9+h} + 3)}$

$= \dfrac{1}{\sqrt{9+h} + 3}$

Projects or Group Activities

97. $\dfrac{1}{(3+\sqrt[3]{2})} \cdot \dfrac{9 - 3\sqrt[3]{2} + \sqrt[3]{4}}{9 - 3\sqrt[3]{2} + \sqrt[3]{4}} = \dfrac{9 - 3\sqrt[3]{2} + \sqrt[3]{4}}{29}$

Check Your Progress: Chapter 7

1. $(32)^{4/5} = (32^{1/5})^4 = 2^4 = 16$

2. $(16)^{-3/4} = (16^{1/4})^{-3} = 2^{-3} = \dfrac{1}{2^3} = \dfrac{1}{8}$

3. $\left(\dfrac{27}{8}\right)^{-1/3} = \left(\dfrac{27^{1/3}}{8^{1/3}}\right)^{-1} = \left(\dfrac{3}{2}\right)^{-1} = \dfrac{2}{3}$

4. $\left(\dfrac{64}{81}\right)^{-3/2} = \left(\dfrac{64^{1/2}}{81^{1/2}}\right)^{-3} = \left(\dfrac{8}{9}\right)^{-3} = \left(\dfrac{9}{8}\right)^3$

$= \dfrac{729}{512}$

5. $x^{3/4} \cdot x^{-1/2} = x^{3/4 + (-1/2)} = x^{1/4}$

6. $(8x^6)^{2/3} = 8^{2/3} \cdot (x^6)^{2/3} = 4x^4$

7. $\dfrac{z^{5/6}}{z^{3/4}} = z^{5/6 - 3/4} = z^{1/12}$

8. $\left(\dfrac{a^{-1/3}b^{3/2}}{c^{2/3}}\right)^{-6} = \dfrac{(a^{-1/3})^{-6}(b^{3/2})^{-6}}{(c^{2/3})^{-6}} = \dfrac{a^2 b^{-9}}{c^{-4}}$

$= \dfrac{a^2 c^4}{b^9}$

9. $\sqrt{9x^{12}} = (9x^{12})^{1/2} = 3x^6$

10. $\sqrt[5]{-32a^5 b^{15}} = (-32a^5 b^{15})^{1/5} = -2ab^3$

11. $\sqrt{72a^3 b^{10}} = (72a^3 b^{10})^{1/2} = (36 \cdot 2 \cdot a^2 \cdot a \cdot b^{10})^{1/2}$

$= 3ab^5 (2a)^{1/2} = 6ab^5 \sqrt{2a}$

12. $\sqrt[3]{16x^7 y^3 z^{11}} = (16x^7 y^3 z^{11})^{1/3}$

$= (8 \cdot 2 \cdot x^6 \cdot x \cdot y^3 \cdot z^9 \cdot z^2)^{1/3} = 2x^2 yz^3 (2xz^2)^{1/3}$

$= 2x^2 yz^3 \sqrt[3]{2xz^2}$

13. $6\sqrt{8a^2 b^3} - 4a\sqrt{32b^3} = 6\sqrt{4 \cdot 2a^2 b^3} - 4a\sqrt{16 \cdot 2b^3}$

$= 6\sqrt{4a^2 b^2} \sqrt{2b} - 4a\sqrt{16b^2} \sqrt{2b}$

$= 6 \cdot 2ab\sqrt{2b} - 4a \cdot 4b\sqrt{2b}$

$= 12ab\sqrt{2b} - 16ab\sqrt{2b}$

$= -4ab\sqrt{2b}$

14. $3\sqrt{50} - 9\sqrt{72} + 6\sqrt{98}$

$= 3\sqrt{25 \cdot 2} - 9\sqrt{36 \cdot 2} + 6\sqrt{49 \cdot 2}$

$= 3 \cdot 5\sqrt{2} - 9 \cdot 6\sqrt{2} + 6 \cdot 7\sqrt{2}$

$= 15\sqrt{2} - 54\sqrt{2} + 42\sqrt{2}$

$= 3\sqrt{2}$

15. $x\sqrt{3x^3} + 2x^2\sqrt{27x} - \sqrt{75x^5}$

$= x\sqrt{3x^2 \cdot x} + 2x^2\sqrt{9 \cdot 3x} - \sqrt{25 \cdot 3x^4 \cdot x}$

$= x \cdot x\sqrt{3x} + 2 \cdot 3x^2\sqrt{3x} - 5x^2\sqrt{3x}$

$= x^2\sqrt{3x} + 6x^2\sqrt{3x} - 5x^2\sqrt{3x}$

$= 2x^2\sqrt{3x}$

16. $\left(\sqrt[3]{4x^2 y}\right)\left(\sqrt[3]{6xy^4}\right) = \sqrt[3]{24x^3 y^5}$

$= \sqrt[3]{8x^3 y^3 \cdot 3y^2} = 2xy\sqrt[3]{3y^2}$

17. $\sqrt{3x}\left(2\sqrt{6x^3} - \sqrt{12x}\right)$

$= 2\sqrt{18x^4} - \sqrt{36x^2} = 2\sqrt{9 \cdot 2 \cdot x^4} - \sqrt{36x^2}$

$= 6x^2\sqrt{2} - 6x$

18. $(2\sqrt{5} + 7)(3\sqrt{5} - 1)$

$= 6\sqrt{25} - 2\sqrt{5} + 21\sqrt{5} - 7$

$= 30 + 19\sqrt{5} - 7$

$= 23 + 19\sqrt{5}$

19. $(2\sqrt{x} - 3)^2 = (2\sqrt{x} - 3)(2\sqrt{x} - 3)$

$= 4\sqrt{x^2} - 6\sqrt{x} - 6\sqrt{x} + 9$

$= 4x - 12\sqrt{x} + 9$

20. $\dfrac{6}{\sqrt{8}} \cdot \dfrac{\sqrt{8}}{\sqrt{8}} = \dfrac{6\sqrt{8}}{\sqrt{64}} = \dfrac{6\sqrt{4 \cdot 2}}{8}$

$= \dfrac{12\sqrt{2}}{8} = \dfrac{3\sqrt{2}}{2}$

21. $\sqrt[3]{\dfrac{3}{4}} = \dfrac{\sqrt[3]{3}}{\sqrt[3]{4}} \cdot \dfrac{\sqrt[3]{2}}{\sqrt[3]{2}} = \dfrac{\sqrt[3]{6}}{\sqrt[3]{8}} = \dfrac{\sqrt[3]{6}}{2}$

22. $\dfrac{7}{2\sqrt{3} + 3} \cdot \dfrac{2\sqrt{3} - 3}{2\sqrt{3} - 3} = \dfrac{7(2\sqrt{3} - 3)}{4\sqrt{9} - 9}$

$\dfrac{14\sqrt{3} - 21}{12 - 9} = \dfrac{14\sqrt{3} - 21}{3}$

23. $\dfrac{2\sqrt{x}}{\sqrt{x} - 2} \cdot \dfrac{\sqrt{x} + 2}{\sqrt{x} + 2} = \dfrac{2\sqrt{x}(\sqrt{x} + 2)}{(\sqrt{x} - 2)(\sqrt{x} + 2)}$

$= \dfrac{2\sqrt{x^2} + 4\sqrt{x}}{\sqrt{x^2} - 4} = \dfrac{2x + 4\sqrt{x}}{x - 4}$

24. $\dfrac{3\sqrt{2} + 5}{2\sqrt{2} - 1} \cdot \dfrac{2\sqrt{2} + 1}{2\sqrt{2} + 1} = \dfrac{(3\sqrt{2} + 5)(2\sqrt{2} + 1)}{(2\sqrt{2} - 1)(2\sqrt{2} + 1)}$

$= \dfrac{6\sqrt{4} + 3\sqrt{2} + 10\sqrt{2} + 5}{4\sqrt{4} - 1} = \dfrac{12 + 13\sqrt{2} + 5}{8 - 1}$

$= \dfrac{17 + 13\sqrt{2}}{7}$

Section 7.4

Concept Check

1. Sometimes true

3. No

Objective A Exercises

5. $\sqrt[3]{4x} = -2$

$\left(\sqrt[3]{4x}\right)^3 = (-2)^3$

$4x = -8$

$x = -2$

The solution is -2.

7.
$$\sqrt{3x-2} = 5$$
$$(\sqrt{3x-2})^2 = (5)^2$$
$$3x - 2 = 25$$
$$3x = 27$$
$$x = 9$$
The solution is 9.

9.
$$\sqrt{4x-3} + 9 = 4$$
$$\sqrt{4x-3} = -5$$
No solution

11.
$$\sqrt[3]{2x-6} = 4$$
$$(\sqrt[3]{2x-6})^3 = (4)^3$$
$$2x - 6 = 64$$
$$2x = 70$$
$$x = 35$$
The solution is 35.

13.
$$\sqrt[4]{3x} + 2 = 5$$
$$\sqrt[4]{3x} = 3$$
$$(\sqrt[4]{3x})^4 = (3)^4$$
$$3x = 81$$
$$x = 27$$
The solution is 27.

15.
$$\sqrt[3]{2x-3} + 5 = 2$$
$$\sqrt[3]{2x-3} = -3$$
$$(\sqrt[3]{2x-3})^3 = (-3)^3$$
$$2x - 3 = -27$$
$$2x = -24$$
$$x = -12$$
The solution is −12.

17.
$$4\sqrt{x-2} + 2 = x + 3$$
$$4\sqrt{x-2} = x + 1$$
$$\left(4\sqrt{x-2}\right)^2 = (x+1)^2$$
$$16(x-2) = x^2 + 2x + 1$$
$$16x - 32 = x^2 + 2x + 1$$
$$x^2 - 14x + 33 = 0$$
$$(x-3)(x-11) = 0$$
$$x - 3 = 0 \quad x - 11 = 0$$
$$x = 3 \qquad x = 11$$

Check both solutions in the original equation
$$4\sqrt{x-2} + 2 = x + 3$$
$$4\sqrt{3-2} + 2 = 3 + 3$$
$$4\sqrt{1} + 2 = 6$$
$$6 = 6$$
$$4\sqrt{x-2} + 2 = x + 3$$
$$4\sqrt{11-2} + 2 = 11 + 3$$
$$4\sqrt{9} + 2 = 14$$
$$14 = 14$$
The solution is 3 and 11.

19.
$$\sqrt{x} + \sqrt{x-5} = 5$$
$$\sqrt{x} = 5 - \sqrt{x-5}$$
$$(\sqrt{x})^2 = (5 - \sqrt{x-5})^2$$
$$x = 25 - 10\sqrt{x-5} + x - 5$$
$$0 = 20 - 10\sqrt{x-5}$$
$$-20 = -10\sqrt{x-5}$$
$$2 = \sqrt{x-5}$$
$$(2)^2 = (\sqrt{x-5})^2$$
$$4 = x - 5$$
$$x = 9$$
The solution is 9.

21.
$$\sqrt{2x+5} - \sqrt{2x} = 1$$
$$\sqrt{2x+5} = 1 + \sqrt{2x}$$
$$(\sqrt{2x+5})^2 = (1 + \sqrt{2x})^2$$
$$2x + 5 = 1 + 2\sqrt{2x} + 2x$$
$$5 = 1 + 2\sqrt{2x}$$
$$4 = 2\sqrt{2x}$$
$$2 = \sqrt{2x}$$
$$(2)^2 = (\sqrt{2x})^2$$
$$4 = 2x$$
$$x = 2$$
The solution is 2.

23.
$$\sqrt{2x} - \sqrt{x-1} = 1$$
$$\sqrt{2x} = 1 + \sqrt{x-1}$$
$$(\sqrt{2x})^2 = (1 + \sqrt{x-1})^2$$
$$2x = 1 + 2\sqrt{x-1} + x - 1$$
$$x = 2\sqrt{x-1}$$
$$(x)^2 = (2\sqrt{x-1})^2$$
$$x^2 = 4(x-1)$$
$$x^2 = 4x - 4$$
$$x^2 - 4x + 4 = 0$$
$$(x-2)(x-2) = 0$$
$$x - 2 = 0 \quad x - 2 = 0$$
$$x = 2 \qquad x = 2$$
The solution is 2.

25.
$$\sqrt{2x+2} + \sqrt{x} = 3$$
$$\sqrt{2x+2} = 3 - \sqrt{x}$$
$$(\sqrt{2x+2})^2 = (3 - \sqrt{x})^2$$
$$2x + 2 = 9 - 6\sqrt{x} + x$$
$$x + 2 = 9 - 6\sqrt{x}$$
$$x - 7 = -6\sqrt{x}$$
$$(x-7)^2 = (-6\sqrt{x})^2$$
$$x^2 - 14x + 49 = 36x$$
$$x^2 - 50x + 49 = 0$$
$$(x-49)(x-1) = 0$$
$$x - 49 = 0 \quad x - 1 = 0$$
$$x = 49 \qquad x = 1$$
Check both solutions in the original equation

$$\sqrt{2x+2} + \sqrt{x} \neq 3$$
$$\sqrt{2(49)+2} + \sqrt{49} \neq 3$$
$$\sqrt{100} + \sqrt{49} \neq 3$$
$$10 + 7 \neq 3$$
$$17 \neq 3$$
$$\sqrt{2(1)+2} + \sqrt{1} = 3$$
$$\sqrt{4} + \sqrt{1} = 3$$
$$2 + 1 = 3$$
$$3 = 3$$
The solution is 1.

27. $\sqrt{x} < \sqrt{x+5}$.
Therefore $\sqrt{x} - \sqrt{x+5} < 0$ and cannot equal a positive number.

Objective B Exercises

29. Strategy: To find the distance the object will fall, substitute the given values for t and g in the equation and solve for d.

Solution: $t = \sqrt{\dfrac{2d}{g}}$

$$3 = \sqrt{\dfrac{2d}{5.5}}$$
$$(3)^2 = \left(\sqrt{\dfrac{2d}{5.5}}\right)^2$$
$$9 = \dfrac{2d}{5.5}$$
$$49.5 = 2d$$
$$24.75 = d$$
On the moon, the object will fall 24.75 ft in 3 s.

29. a) **Strategy:** To find the height of the water, evaluate the function for $t = 10$.

Solution: $h(t) = (88.18 - 3.18t)^{2/5}$

$h(10) = (88.18 - 3.18(10))^{2/5}$

$h(10) = (88.18 - 31.8)^{2/5}$

$h(10) = (56.38)^{2/5}$

$h(10) = 5.0$

The height of the water is 5.0 ft.

b) **Strategy:** To find how long it will take to empty the tank substitute the given value for h in the equation and solve for t.

Solution: $h(t) = (88.18 - 3.18t)^{2/5}$

$0 = (88.18 - 3.18t)^{2/5}$

$0 = ((88.18 - 3.18t)^{2/5})^{5/2}$

$0 = 88.18 - 3.18t$

$3.18t = 88.16$

$t = 27.7$

The tank will empty in 27.7 s.

33. Strategy: To find the length of the pendulum, substitute the given value for T in the equation and solve for L.

Solution: $T = 2\pi\sqrt{\dfrac{L}{32}}$

$3 = 2\pi\sqrt{\dfrac{L}{32}}$

$\left(\dfrac{3}{2\pi}\right)^2 = \left(\sqrt{\dfrac{L}{32}}\right)^2$

$\left(\dfrac{3}{2\pi}\right)^2 = \dfrac{L}{32}$

$32\left(\dfrac{3}{2\pi}\right)^2 = L$

$7.3 = L$

The length of the pendulum is 7.30 ft.

35. Strategy: Find the difference in the widths. Use the Pythagorean Theorem to find the width of the screen of a regular TV and then repeat the process to find the width of the screen for HDTV. Subtract the width of the regular TV from the width of the HDTV.

Solution: $c^2 = a^2 + b^2$

For the regular TV:

$27^2 = 16.2^2 + b^2$

$729 = 262.44 + b^2$

$466.56 = b^2$

$(466.56)^{1/2} = (b^2)^{1/2}$

$21.6 = b$

For the HDTV

$33^2 = 16.2^2 + b^2$

$1089 = 262.44 + b^2$

$826.56 = b^2$

$(826.56)^{1/2} = (b^2)^{1/2}$

$28.75 = b$

$28.75 - 21.6 = 7.15$

The HDTV is approximately 7.15 in. wider.

Critical Thinking

37. Strategy: Use the Pythagorean Theorem to determine the longest pole that can be placed in the box.

Solution:

Find the length diagonal of the bottom of the box.

$c^2 = a^2 + b^2$

$c^2 = 2^2 + 3^2$

$c^2 = 4 + 9$

$c^2 = 13$

This represents the value one of the legs of the right triangle needed to find the length of the pole.

$c^2 = a^2 + b^2$

$c^2 = 13 + 4^2$

$c^2 = 13 + 16$

$c^2 = 19$

$c = 5.4$

The longest pole can be 5.4 ft.

Projects or Group Activities

39. No. One way to see this is to calculate the distance of the bottom of the ladder from the wall at one second intervals. A second way is to note that it takes 4 s for the top of the ladder to reach the ground. In that 4 s, the bottom of the ladder has moved 4 ft so the average speed is 1 ft/s.

Section 7.5

Concept Check

1. An imaginary number is a number whose square is a negative number.
A complex number is a number of the form $a + bi$, where a and b are real numbers and $i = \sqrt{-1}$.

3. $3; 7$

5. $7; 0$

Objective A Exercises

7. $\sqrt{-25} = i\sqrt{25} = 5i$

9. $\sqrt{-98} = i\sqrt{98} = i\sqrt{49 \cdot 2} = 7i\sqrt{2}$

11. $\dfrac{6 + \sqrt{-4}}{2} = \dfrac{6 + i\sqrt{4}}{2} = \dfrac{6 + 2i}{2} = 3 + i$

13. $\dfrac{6 - 5\sqrt{-8}}{4} = \dfrac{6 - 5i\sqrt{4 \cdot 2}}{4} = \dfrac{6 - 10i\sqrt{2}}{4}$

$= \dfrac{3}{2} - \dfrac{5i\sqrt{2}}{2}$

15. $-b + \sqrt{b^2 - 4ac}$

$-4 + \sqrt{(4)^2 - 4(1)(5)} = -4 + \sqrt{16 - 20}$

$= -4 + \sqrt{-4} = -4 + i\sqrt{4}$

$= -4 + 2i$

17. $-b + \sqrt{b^2 - 4ac}$

$-(-4) + \sqrt{(-4)^2 - 4(2)(10)} = 4 + \sqrt{16 - 80}$

$= 4 + \sqrt{-64} = 4 + i\sqrt{64}$

$= 4 + 8i$

19. $-b + \sqrt{b^2 - 4ac}$

$-(-8) + \sqrt{(-8)^2 - 4(3)(6)} = 8 + \sqrt{64 - 72}$

$= 8 + \sqrt{-8} = 8 + i\sqrt{4 \cdot 2}$

$= 8 + 2i\sqrt{2}$

21. $-b + \sqrt{b^2 - 4ac}$

$-2 + \sqrt{(2)^2 - 4(4)(7)} = -2 + \sqrt{4 - 112}$

$= -2 + \sqrt{-108} = -2 + i\sqrt{36 \cdot 3}$

$= -2 + 6i\sqrt{3}$

23. $-b + \sqrt{b^2 - 4ac}$

$-5 + \sqrt{(5)^2 - 4(-2)(-6)} = -5 + \sqrt{25 - 48}$

$= -5 + \sqrt{-23}$

$= -5 + i\sqrt{23}$

25. $-b + \sqrt{b^2 - 4ac}$

$-4 + \sqrt{(4)^2 - 4(-3)(-6)} = -4 + \sqrt{16 - 72}$

$= -4 + \sqrt{-56} = -4 + i\sqrt{4 \cdot 14}$

$= -4 + 2i\sqrt{14}$

Objective B Exercises

27. $(2 + 4i) + (6 - 5i) = 8 - i$

29. $(-2 - 4i) - (6 - 8i) = -8 + 4i$

31. $(8 - 2i) - (2 + 4i) = 6 - 6i$

33. $5 + (6 - 4i) = 11 - 4i$

35. $3i - (6 + 5i) = -6 - 2i$

37. The real parts of the complex numbers are additive inverses.

Objective C Exercises

39. $(7i)(-9i) = -63i^2 = -63(-1) = 63$

41. $\sqrt{-2}\sqrt{-8} = i\sqrt{2} \cdot i\sqrt{8} = i^2\sqrt{16} = -\sqrt{16} = -4$

43. $(5+2i)(5-2i) = 25 - 10i + 10i - 4i^2$
$= 25 - 4i^2 = 25 - 4(-1) = 29$

45. $2i(6+2i) = 12i + 4i^2 = 12i + 4(-1)$
$= -4 + 12i$

47. $-i(4-3i) = -4i + 3i^2 = -4i + 3(-1)$
$= -3 - 4i$

49. $(5-2i)(3+i) = 15 + 5i - 6i - 2i^2$
$= 15 - i - 2i^2$
$= 15 - i - 2(-1)$
$= 17 - i$

51. $(6+5i)(3+2i) = 18 + 12i + 15i + 10i^2$
$= 18 + 27i + 10i^2$
$= 18 + 27i + 10(-1)$
$= 8 + 27i$

53. $(2+5i)^2 = 4 + 20i + 25i^2$
$= 4 + 20i + 25(-1)$
$= -21 + 20i$

55. $\left(\dfrac{6}{5} + \dfrac{3}{5}i\right)\left(\dfrac{2}{3} - \dfrac{1}{3}i\right) = \dfrac{4}{5} - \dfrac{2}{5}i + \dfrac{2}{5}i - \dfrac{1}{5}i^2$

$= \dfrac{4}{5} - \dfrac{1}{5}i^2$

$= \dfrac{4}{5} - \dfrac{1}{5}(-1)$

$= \dfrac{4}{5} + \dfrac{1}{5} = 1$

57. True

Objective D Exercises

59. $\dfrac{3}{i} = \dfrac{3}{i} \cdot \dfrac{i}{i} = \dfrac{3i}{i^2} = \dfrac{3i}{-1} = -3i$

61. $\dfrac{2-3i}{-4i} = \dfrac{2-3i}{-4i} \cdot \dfrac{i}{i} = \dfrac{2i - 3i^2}{-4i^2}$

$= \dfrac{2i - 3(-1)}{-4(-1)} = \dfrac{2i + 3}{4}$

$= \dfrac{3}{4} + \dfrac{1}{2}i$

63. $\dfrac{4}{5+i} = \dfrac{4}{5+i} \cdot \dfrac{5-i}{5-i} = \dfrac{20-4i}{25-i^2}$

$= \dfrac{20-4i}{25-(-1)} = \dfrac{20-4i}{26}$

$= \dfrac{10}{13} - \dfrac{2}{13}i$

65. $\dfrac{2}{2-i} = \dfrac{2}{2-i} \cdot \dfrac{2+i}{2+i} = \dfrac{4+2i}{4-i^2}$

$= \dfrac{4+2i}{4-(-1)} = \dfrac{4+2i}{5}$

$= \dfrac{4}{5} + \dfrac{2}{5}i$

67. $\dfrac{1-3i}{3+i} = \dfrac{1-3i}{3+i} \cdot \dfrac{3-i}{3-i} = \dfrac{3 - i - 9i + 3i^2}{9 - i^2}$

$= \dfrac{3 - 10i + 3(-1)}{9 - (-1)} = \dfrac{-10i}{10}$

$= -i$

69. $\dfrac{3i}{1+4i} = \dfrac{3i}{1+4i} \cdot \dfrac{1-4i}{1-4i} = \dfrac{3i - 12i^2}{1 - 16i^2}$

$= \dfrac{3i - 12(-1)}{1 - 16(-1)} = \dfrac{3i + 12}{17}$

$= \dfrac{12}{17} + \dfrac{3}{17}i$

71. $\dfrac{2-3i}{3+i} = \dfrac{2-3i}{3+i} \cdot \dfrac{3-i}{3-i} = \dfrac{6-2i-9i+3i^2}{9-i^2}$

$= \dfrac{6-11i+3(-1)}{9-(-1)} = \dfrac{3-11i}{10}$

$= \dfrac{3}{10} - \dfrac{11}{10}i$

73. $\dfrac{5+3i}{3-i} = \dfrac{5+3i}{3-i} \cdot \dfrac{3+i}{3+i} = \dfrac{15+5i+9i+3i^2}{9-i^2}$

$= \dfrac{15+14i+3(-1)}{9-(-1)} = \dfrac{12+14i}{10}$

$= \dfrac{6+7i}{5}$

$= \dfrac{6}{5} + \dfrac{7}{5}i$

75. True

Critical Thinking

77. $x^2 - 10x + 29 = 0$

$(5-3i)^2 - 10(5-3i) + 29 = 0$

$25 - 30i - 9 - 50 + 30i + 29 = 0$

$\qquad\qquad\qquad\qquad\qquad -5 = 0$

No, $5-3i$ is not a solution.

79. $x^2 - 2x + 4 = 0$

$(1-i\sqrt{3})^2 - 2(1-i\sqrt{3}) + 4 = 0$

$1 - 2i\sqrt{3} - 3 - 2 + 2i\sqrt{3} + 4 = 0$

$\qquad\qquad\qquad\qquad\qquad 0 = 0$

Yes, $1-i\sqrt{3}$ is a solution.

Projects or Group Activities

81. 1

83. I

Chapter 7 Review Exercises

1. $(16x^{-4}y^{12})^{1/4}(100x^6y^{-2})^{1/2}$

$= (16)^{1/4}x^{-1}y^3 \cdot 100^{1/2}\,x^3y^{-1}$

$= 20x^2y^2$

2. $\sqrt[4]{3x-5} = 2$

$(\sqrt[4]{3x-5})^4 = 2^4$

$3x - 5 = 16$

$3x = 21$

$x = 7$

The solution is 7.

3. $(6-5i)(4+3i) = 24 + 18i - 20i - 15i^2$

$= 24 - 2i - 15(-1)$

$= 39 - 2i$

4. $7y\sqrt[3]{x^2} = 7x^{2/3}y$

5. $(\sqrt{3}+8)(\sqrt{3}-2) = \sqrt{3}^2 - 2\sqrt{3} + 8\sqrt{3} - 16$

$= 3 + 6\sqrt{3} - 16$

$= 6\sqrt{3} - 13$

6. $\sqrt{4x+9} + 10 = 11$

$\sqrt{4x+9} = 1$

$(\sqrt{4x+9})^2 = 1^2$

$4x + 9 = 1$

$4x = -8$

$x = -2$

The solution is -2.

7. $\dfrac{x^{-3/2}}{x^{7/2}} = x^{-10/2} = x^{-5} = \dfrac{1}{x^5}$

8. $\dfrac{8}{\sqrt{3y}} = \dfrac{8}{\sqrt{3y}} \cdot \dfrac{\sqrt{3y}}{\sqrt{3y}} = \dfrac{8\sqrt{3y}}{\sqrt{3^2 y^2}} = \dfrac{8\sqrt{3y}}{3y}$

9. $\sqrt[3]{-8a^6b^{12}} = -2a^2b^4$

10. $\sqrt{50a^4b^3} - ab\sqrt{18a^2b}$

$= \sqrt{25a^4b^2 \cdot 2b} - ab\sqrt{9a^2 \cdot 2b}$

$= 5a^2b\sqrt{2b} - 3a^2b\sqrt{2b}$

$= 2a^2b\sqrt{2b}$

11. $\dfrac{14}{4-\sqrt{2}} = \dfrac{14}{4-\sqrt{2}} \cdot \dfrac{4+\sqrt{2}}{4+\sqrt{2}} = \dfrac{56+14\sqrt{2}}{16-\sqrt{2}^2}$

$= \dfrac{56+14\sqrt{2}}{16-2} = \dfrac{56+14\sqrt{2}}{14}$

$= 4+\sqrt{2}$

12. $\dfrac{5+2i}{3i} = \dfrac{5+2i}{3i} \cdot \dfrac{-3i}{-3i} = \dfrac{-15i-6i^2}{-9i^2}$

$= \dfrac{-15i-6(-1)}{-9(-1)} = \dfrac{-15i+6}{9}$

$= \dfrac{2}{3} - \dfrac{5}{3}i$

13. $\sqrt{18a^3b^6} = \sqrt{9a^2b^6 \cdot 2a} = 3ab^3\sqrt{2a}$

14. $(17+8i) - (15-4i) = 2+12i$

15. $3x\sqrt[3]{54x^8y^{10}} - 2x^2y\sqrt[3]{16x^5y^7}$

$= 3x\sqrt[3]{27x^6y^9 \cdot 2x^2y} - 2x^2y\sqrt[3]{8x^3y^6 \cdot 2x^2y}$

$= 9x^3y^3\sqrt[3]{2x^2y} - 4x^3y^3\sqrt[3]{2x^2y}$

$= 5x^3y^3\sqrt[3]{2x^2y}$

16. $\sqrt[3]{16x^4y}\sqrt[3]{4xy^5} = \sqrt[3]{64x^5y^6}$

$= \sqrt[3]{64x^3y^6 \cdot x^2} = 4xy^2\sqrt[3]{x^2}$

17. $i(3-7i) = 3i - 7i^2 = 3i - 7(-1) = 7+3i$

18. $\dfrac{(4a^{-2/3}b^4)^{-1/2}}{(a^{-1/6}b^{3/2})^2} = \dfrac{(4)^{-1/2}a^{1/3}b^{-2}}{a^{-1/3}b^3}$

$= \dfrac{a^{2/3}}{2b^5}$

19. $\sqrt[5]{-64a^8b^{12}} = \sqrt[5]{-32a^5b^{10} \cdot 2a^3b^2}$

$= -2ab^2\sqrt[5]{2a^3b^2}$

20. $\dfrac{5+9i}{1-i} = \dfrac{5+9i}{1-i} \cdot \dfrac{1+i}{1+i} = \dfrac{5+5i+9i+9i^2}{1-i^2}$

$= \dfrac{5+14i+9(-1)}{1-(-1)} = \dfrac{-4+14i}{2}$

$= -2+7i$

21. $\sqrt{-12}\sqrt{-6} = i\sqrt{12} \cdot i\sqrt{6} = i^2\sqrt{72}$

$= -1\sqrt{36 \cdot 2} = -6\sqrt{2}$

22. $\sqrt{x-5} + \sqrt{x+6} = 11$

$\sqrt{x-5} = 11 - \sqrt{x+6}$

$(\sqrt{x-5})^2 = (11-\sqrt{x+6})^2$

$x-5 = 121 - 22\sqrt{x+6} + x+6$

$-5 = 127 - 22\sqrt{x+6}$

$-132 = -22\sqrt{x+6}$

$6 = \sqrt{x+6}$

$(6)^2 = (\sqrt{x+6})^2$

$36 = x+6$

$30 = x$

The solution is 30.

23. $\sqrt[4]{81a^8b^{12}} = 3a^2b^3$

24. $\dfrac{9}{\sqrt[3]{3x}} = \dfrac{9}{\sqrt[3]{3x}} \cdot \dfrac{\sqrt[3]{9x^2}}{\sqrt[3]{9x^2}} = \dfrac{9\sqrt[3]{9x^2}}{\sqrt[3]{27x^3}}$

$= \dfrac{9\sqrt[3]{9x^2}}{3x} = \dfrac{3\sqrt[3]{9x^2}}{x}$

25. $(-8+3i) - (4-7i) = -12+10i$

26. $(2+\sqrt{2x-1})^2 = 4 + 4\sqrt{2x-1} + 2x-1$

$= 2x + 4\sqrt{2x-1} + 3$

27. $4x\sqrt{12x^2y} + \sqrt{3x^4y} - x^2\sqrt{27y}$

$= 4x\sqrt{4x^2 \cdot 3y} + \sqrt{x^4 \cdot 3y} - x^2\sqrt{9 \cdot 3y}$

$= 8x^2\sqrt{3y} + x^2\sqrt{3y} - 3x^2\sqrt{3y}$

$= 6x^2\sqrt{3y}$

28. $81^{-1/4} = (3^4)^{-1/4} = 3^{-1} = \dfrac{1}{3}$

29. $(a^{16})^{-5/8} = a^{-10} = \dfrac{1}{a^{10}}$

30. $-\sqrt{49x^6y^{16}} - 7x^3y^8$

31. $4a^{2/3} = 4\sqrt[3]{a^2}$

32. $(9x^2y^4)^{-1/2}(x^6y^6)^{1/3} = 9^{-1/2}x^{-1}y^{-2}x^2y^2$

$= 9^{-1/2}x^1y^0 = 9^{-1/2}x$

$= \dfrac{x}{9^{1/2}} = \dfrac{x}{3}$

33. $\sqrt[4]{x^6y^8z^{10}} = \sqrt[4]{x^4y^8z^8 \cdot x^2z^2}$

$= xy^2z^2\sqrt[4]{x^2z^2}$

34. $\sqrt{54} + \sqrt{24} = \sqrt{9 \cdot 6} + \sqrt{4 \cdot 6}$

$= 3\sqrt{6} + 2\sqrt{6} = 5\sqrt{6}$

35. $\sqrt{48x^5y} - x\sqrt{80x^2y} = \sqrt{16x^4 \cdot 3xy} - x\sqrt{16x^2 \cdot y}$

$= 4x^2\sqrt{3xy} - 4x^2\sqrt{y}$

36. $\sqrt{32}\sqrt{50} = \sqrt{1600} = 40$

37. $\sqrt{3x}(3 + \sqrt{3x}) = 3\sqrt{3x} + \sqrt{(3x)^2}$

$= 3\sqrt{3x} + 3x$

38. $\dfrac{\sqrt{125x^6}}{\sqrt{5x^3}} = \sqrt{\dfrac{125x^6}{5x^3}} = \sqrt{25x^3} = 5x\sqrt{x}$

39. $\dfrac{2 - 3\sqrt{7}}{6 - \sqrt{7}} = \dfrac{2 - 3\sqrt{7}}{6 - \sqrt{7}} \cdot \dfrac{6 + \sqrt{7}}{6 + \sqrt{7}}$

$= \dfrac{12 + 2\sqrt{7} - 18\sqrt{7} - 3(\sqrt{7})^2}{6^2 - (\sqrt{7})^2}$

$= \dfrac{12 - 16\sqrt{7} - 21}{36 - 7}$

$= \dfrac{-9 - 16\sqrt{7}}{29}$

40. $\sqrt{-36} = i\sqrt{36} = 6i$

41. $-b + \sqrt{b^2 - 4ac}$

$-(-8) + \sqrt{(-8)^2 - 4(1)(25)} = 8 + \sqrt{64 - 100}$

$= 8 + \sqrt{-36} = 8 + i\sqrt{36}$

$= 8 + 6i$

42. $-b + \sqrt{b^2 - 4ac}$

$-2 + \sqrt{2^2 - 4(1)(9)} = -2 + \sqrt{4 - 36}$

$= -2 + \sqrt{-32} = -2 + i\sqrt{16 \cdot 2}$

$= -2 + 4i\sqrt{2}$

43. $(5 + 2i) + (4 - 3i) = 9 - i$

44. $(3 + 2\sqrt{5})(3 - 2\sqrt{5})$

$= 9 - 6\sqrt{5} + 6\sqrt{5} - 4(\sqrt{5})^2$

$= 9 - 20 = -11$

45. $(3 - 9i) - 7 = -4 - 9i$

46. $(4 - i)^2 = (4 - i)(4 - i)$

$= 16 - 8i + i^2 = 16 - 8i + (-1)$

$= 15 - 8i$

47. $\dfrac{-6}{i} = \dfrac{-6}{i} \cdot \dfrac{i}{i} = \dfrac{-6i}{i^2} = \dfrac{-6i}{-1} = 6i$

48. $\dfrac{7}{2-i} = \dfrac{7}{2-i} \cdot \dfrac{2+i}{2+i} = \dfrac{14+7i}{4-i^2}$

$= \dfrac{14+7i}{4-(-1)} = \dfrac{14+7i}{5}$

$= \dfrac{14}{5} + \dfrac{7}{5}i$

49. $\sqrt{2x-7} + 2 = 5$

$\sqrt{2x-7} = 3$

$\left(\sqrt{2x-7}\right)^2 = 3^2$

$2x - 7 = 9$

$2x = 16$

$x = 8$

The solution is 8.

50. $\sqrt[3]{9x} = -6$

$\left(\sqrt[3]{9x}\right)^3 = (-6)^3$

$9x = -216$

$x = -24$

The solution is −24.

51. Strategy: Use the Pythagorean Theorem to find the width of the rectangle.

Solution: $c^2 = a^2 + b^2$

$13^2 = 12^2 + b^2$

$169 = 144 + b^2$

$25 = b^2$

$(25)^{1/2} = (b^2)^{1/2}$

$5 = b$

The width of the rectangle is 5 in.

52. Strategy: To find the amount of power, substitute the given value for v in the equation and solve for P.

Solution: $v = 4.05\sqrt[3]{P}$

$20 = 4.05\sqrt[3]{P}$

$4.94 = \sqrt[3]{P}$

$(4.94)^3 = (\sqrt[3]{P})^3$

$120 \approx P$

The amount of power is 120 watts.

53. Strategy: To find the distance required, substitute the given values for v and a in the equation and solve for s.

Solution: $v = \sqrt{2as}$

$88 = \sqrt{2(16s)}$

$(88)^2 = (\sqrt{32s})^2$

$7744 = 32s$

$242 = s$

The distance required is 242 ft.

54. Strategy: To find the distance, use the Pythagorean Theorem.

Solution: $c^2 = a^2 + b^2$

$12^2 = 10^2 + b^2$

$144 = 100 + b^2$

$44 = b^2$

$(44)^{1/2} = (b^2)^{1/2}$

$6.63 = b$

The distance is 6.63 ft.

Chapter 7 Test

1. $\dfrac{1}{2}\sqrt[4]{x^3} = \dfrac{1}{2}x^{3/4}$

2. $\sqrt[3]{54x^7y^3} - x\sqrt[3]{128x^4y^3} - x^2\sqrt[3]{2xy^3}$

$= \sqrt[3]{27x^6y^3 \cdot 2x} - x\sqrt[3]{64x^3y^3 \cdot 2x} - x^2\sqrt[3]{y^3 \cdot 2x}$

$= 3x^2y\sqrt[3]{2x} - 4x^2y\sqrt[3]{2x} - x^2y\sqrt[3]{2x}$

$= -2x^2y\sqrt[3]{2x}$

3. $3y^{2/5} = 3\sqrt[5]{y^2}$

4. $(2+5i)(4-2i) = 8 - 4i + 20i - 10i^2$

$= 8 + 16i - 10(-1)$

$= 18 + 16i$

5. $(3-2\sqrt{x})^2 = (3-2\sqrt{x})(3-2\sqrt{x})$

$\qquad = 9-12\sqrt{x}+4(\sqrt{x})^2$

$\qquad = 4x-12\sqrt{x}+9$

6. $\dfrac{r^{2/3}r^{-1}}{r^{-1/2}} = \dfrac{r^{-1/3}}{r^{-1/2}} = r^{1/6}$

7. $\sqrt{x+12}-\sqrt{x}=2$

$\sqrt{x+12}=2+\sqrt{x}$

$(\sqrt{x+12})^2 = (2+\sqrt{x})^2$

$x+12 = 4+4\sqrt{x}+x$

$12 = 4+4\sqrt{x}$

$8 = 4\sqrt{x}$

$2 = \sqrt{x}$

$(2)^2 = (\sqrt{x})^2$

$4 = x$

The solution is 4.

8. $\sqrt[4]{4a^5b^3}\sqrt[4]{8a^3b^7} = \sqrt[4]{32a^8b^{10}}$

$\qquad = \sqrt[4]{16a^8b^8 \cdot 2b^2} = 2a^2b^2\sqrt[4]{2b^2}$

9. $\sqrt{3x}(\sqrt{x}-\sqrt{25x}) = \sqrt{3x^2}-\sqrt{75x^2}$

$\qquad = \sqrt{x^2 \cdot 3}-\sqrt{25x^2 \cdot 3} = x\sqrt{3}-5x\sqrt{3}$

$\qquad = -4x\sqrt{3}$

10. $(5-2i)-(8-4i) = -3+2i$

11. $\sqrt{32x^4y^7} = \sqrt{16x^4y^6 \cdot 2y} = 4x^2y^3\sqrt{2y}$

12. $(2\sqrt{3}+4)(3\sqrt{3}-1)$

$\qquad = 6\sqrt{3^2}-2\sqrt{3}+12\sqrt{3}-4$

$\qquad = 18+10\sqrt{3}-4$

$\qquad = 14+10\sqrt{3}$

13. $\sqrt{-5}\cdot\sqrt{-20} = i\sqrt{5}\cdot i\sqrt{20} = i^2\sqrt{100}$

$\qquad = -1\sqrt{100} = -10$

14. $\dfrac{4-2\sqrt{5}}{2-\sqrt{5}} = \dfrac{4-2\sqrt{5}}{2-\sqrt{5}}\cdot\dfrac{2+\sqrt{5}}{2+\sqrt{5}}$

$\qquad = \dfrac{8+4\sqrt{5}-4\sqrt{5}-2(\sqrt{5})^2}{2^2-(\sqrt{5})^2}$

$\qquad = \dfrac{8-10}{4-5} = \dfrac{-2}{-1} = 2$

15. $\sqrt{18a^3}+a\sqrt{50a} = \sqrt{9a^2 \cdot 2a}+a\sqrt{25 \cdot 2a}$

$\qquad = 3a\sqrt{2a}+5a\sqrt{2a}$

$\qquad = 8a\sqrt{2a}$

16. $(\sqrt{a}-3\sqrt{b})(2\sqrt{a}+5\sqrt{b})$

$\qquad = 2\sqrt{a^2}+5\sqrt{ab}-6\sqrt{ab}-15\sqrt{b^2}$

$\qquad = 2a-\sqrt{ab}-15b$

17. $\dfrac{(2x^{1/3}y^{-2/3})^6}{(x^{-4}y^8)^{1/4}} = \dfrac{2^6x^2y^{-4}}{x^{-1}y^2} = \dfrac{64x^3}{y^6}$

18. $\dfrac{10x}{\sqrt[3]{5x^2}} = \dfrac{10x}{\sqrt[3]{5x^2}}\cdot\dfrac{\sqrt[3]{25x}}{\sqrt[3]{25x}} = \dfrac{10x\sqrt[3]{25x}}{\sqrt[3]{125x^3}}$

$\qquad = \dfrac{10x\sqrt[3]{25x}}{5x} = 2\sqrt[3]{25x}$

19. $\dfrac{2+3i}{1-2i} = \dfrac{2+3i}{1-2i}\cdot\dfrac{1+2i}{1+2i} = \dfrac{2+4i+3i+6i^2}{1-4i^2}$

$\qquad = \dfrac{2+7i+6(-1)}{1-4(-1)} = \dfrac{-4+7i}{5}$

$\qquad = -\dfrac{4}{5}+\dfrac{7}{5}i$

20. $\sqrt[3]{2x-2}+4=2$

$\sqrt[3]{2x-2}=-2$

$(\sqrt[3]{2x-2})^3 = (-2)^3$

$2x-2 = -8$

$2x = -6$

$x = -3$

The solution is −3.

21. $\left(\dfrac{4a^4}{b^2}\right)^{-3/2} = \dfrac{(4a^4)^{-3/2}}{(b^2)^{-3/2}} = \dfrac{4^{-3/2}\,a^{-6}}{b^{-3}} = \dfrac{b^3}{8a^6}$

22. $\sqrt[3]{27a^4b^3c^7} = \sqrt[3]{27a^3b^3c^6 \cdot ac}$

$= 3abc^2\sqrt[3]{ac}$

23. $\dfrac{\sqrt{32x^5y}}{\sqrt{2xy^3}} = \sqrt{\dfrac{32x^5y}{2xy^3}} = \sqrt{\dfrac{16x^4}{y^2}} = \dfrac{4x^2}{y}$

24. $\dfrac{5x}{\sqrt{5x}} = \dfrac{5x}{\sqrt{5x}} \cdot \dfrac{\sqrt{5x}}{\sqrt{5x}} = \dfrac{5x\sqrt{5x}}{\left(\sqrt{5x}\right)^2}$

$= \dfrac{5x\sqrt{5x}}{5x} = \sqrt{5x}$

25. Strategy: To find the distance the object has fallen substitute the value for v in the equation and solve for d.

Solution: $v = \sqrt{64d}$

$192 = \sqrt{64d}$

$(192)^2 = \left(\sqrt{64d}\right)^2$

$36{,}864 = 64d$

$576 = d$

The object has fallen 576 ft.

Cumulative Review Exercises

1. The Distributive Property

2. $f(x) = 3x^2 - 2x + 1$
$f(-3) = 3(-3)^2 - 2(-3) + 1$
$f(-3) = 27 + 6 + 1$
$f(-3) = 34$

3. $5 - \dfrac{2}{3}x = 4$

$5 - \dfrac{2}{3}x - 5 = 4 - 5$

$-\dfrac{2}{3}x = -1$

$\left(-\dfrac{3}{2}\right)\left(-\dfrac{2}{3}x\right) = -1\left(-\dfrac{3}{2}\right)$

$x = \dfrac{3}{2}$

The solution is $\dfrac{3}{2}$.

4. $2[4 - 2(3 - 2x)] = 4(1 - x)$
$2[4 - 6 + 4x] = 4 - 4x$
$2[-2 + 4x] = 4 - 4x$
$-4 + 8x = 4 - 4x$
$-4 + 12x = 4$
$12x = 8$
$x = \dfrac{2}{3}$

The solution is $\dfrac{2}{3}$.

5. $2 + |4 - 3x| = 5$
$|4 - 3x| = 3$

$4 - 3x = -3 \qquad 4 - 3x = 3$
$-3x = -7 \qquad\;\; -3x = -1$
$x = \dfrac{7}{3} \qquad\quad\;\; x = \dfrac{1}{3}$

The solutions are $\dfrac{1}{3}$ and $\dfrac{7}{3}$.

6. $6x - 3(2x + 2) > 3 - 3(x + 2)$
$6x - 6x - 6 > 3 - 3x - 6$
$-6 > -3x - 3$
$-3 > -3x$
$1 < x$
$\{x \mid x > 1\}$

7. $|2x + 3| \le 9$
$-9 \le 2x + 3 \le 9$
$-9 - 3 \le 2x + 3 - 3 \le 9 - 3$
$-12 \le 2x \le 6$
$-6 \le x \le 3$
$\{x \mid -6 \le x \le 3\}$

8. $81x^2 - y^2 = (9x + y)(9x - y)$

9. $x^5 + 2x^3 - 3x = x(x^4 + 2x^2 - 3)$
$$= x(x^2 + 3)(x^2 - 1)$$
$$= x(x^2 + 3)(x + 1)(x - 1)$$

10. Find the slope of the line.
$$m = \frac{y_2 - y_1}{x_2 - x_1} = \frac{2 - 3}{-1 - 2} = \frac{-1}{-3} = \frac{1}{3}$$

Use the point-slope formula to find the equation of the line.
$$y - y_1 = m(x - x_1)$$
$$y - 3 = \frac{1}{3}(x - 2)$$
$$y - 3 = \frac{1}{3}x - \frac{2}{3}$$
$$y = \frac{1}{3}x + \frac{7}{3}$$

The equation of the line is $y = \frac{1}{3}x + \frac{7}{3}$.

11. $\begin{vmatrix} 1 & 2 & -3 \\ 0 & -1 & 2 \\ 3 & 1 & -2 \end{vmatrix} = 1\begin{vmatrix} -1 & 2 \\ 1 & -2 \end{vmatrix} - 2\begin{vmatrix} 0 & 2 \\ 3 & -2 \end{vmatrix} - 3\begin{vmatrix} 0 & -1 \\ 3 & 1 \end{vmatrix}$

$$= 1(0) - 2(-6) - 3(3)$$
$$= 3$$

12. $P = \dfrac{R - C}{n}$
$$P \cdot n = \frac{R - C}{n} \cdot n$$
$$nP = R - C$$
$$nP + C = R$$
$$C = R - nP$$

13. $(2^{-1}x^2 y^{-6})(2^{-1}y^{-4})^{-2}$
$$= (2^{-1}x^2 y^{-6})(2^2 y^8)$$
$$= 2x^2 y^2$$

14. $\dfrac{x^2 y^3}{x^2 + 2x - 8} \cdot \dfrac{2x^2 - 7x + 6}{xy^4}$

$$= \frac{x^2 y^3}{(x + 4)(x - 2)} \cdot \frac{(2x - 3)(x - 2)}{xy^4}$$

$$= \frac{x^2 y^3 (2x - 3)(x - 2)}{xy^4 (x + 4)(x - 2)}$$

$$= \frac{x(2x - 3)}{y(x + 4)}$$

15. $\sqrt{40x^3} - x\sqrt{90x} = \sqrt{4x^2 \cdot 10x} - x\sqrt{9 \cdot 10x}$
$$= 2x\sqrt{10x} - 3x\sqrt{10x}$$
$$= -x\sqrt{10x}$$

16. $\dfrac{x}{x - 2} - 2x = \dfrac{-3}{x - 2}$

$$(x - 2)\left(\frac{x}{x - 2} - 2x\right) = (x - 2)\left(\frac{-3}{x - 2}\right)$$
$$x - 2x(x - 2) = -3$$
$$x - 2x^2 + 4x = -3$$
$$-2x^2 + 5x + 3 = 0$$
$$2x^2 - 5x - 3 = 0$$
$$(2x + 1)(x - 3) = 0$$
$$2x + 1 = 0 \qquad x - 3 = 0$$
$$2x = -1 \qquad\qquad x = 3$$
$$x = -\frac{1}{2}$$

The solutions are $-\dfrac{1}{2}$ and 3.

17. Solve $3x - 2y = -6$ for y.
$$3x - 2y = -6$$
$$-2y = -3x - 6$$
$$y = \frac{3}{2}x + 3$$

The y-intercept is $(0, 3)$.

The slope is $\dfrac{3}{2}$.

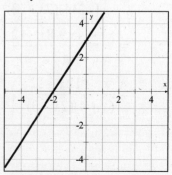

18. $3x + 2y \le 4$

$2y \le -3x + 4$

$y \le -\dfrac{3}{2}x + 2$

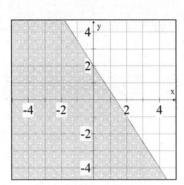

19. $\dfrac{2i}{3-i} = \dfrac{2i}{3-i} \cdot \dfrac{3+i}{3+i} = \dfrac{6i + 2i^2}{9 - i^2}$

$= \dfrac{6i + 2(-1)}{9 - (-1)} = \dfrac{6i - 2}{10}$

$= \dfrac{-1 + 3i}{5}$

$= -\dfrac{1}{5} + \dfrac{3}{5}i$

20. $\sqrt[3]{3x - 4} + 5 = 1$

$\sqrt[3]{3x - 4} = -4$

$(\sqrt[3]{3x - 4})^3 = (-4)^3$

$3x - 4 = -64$

$3x = -60$

$x = -20$

The solution is -20.

21. $\dfrac{x}{2x - 3} + \dfrac{4}{x + 4}$

$= \dfrac{x}{2x - 3} \cdot \dfrac{x + 4}{x + 4} + \dfrac{4}{x + 4} \cdot \dfrac{2x - 3}{2x - 3}$

$= \dfrac{x(x + 4) + 4(2x - 3)}{(x + 4)(2x - 3)}$

$= \dfrac{x^2 + 4x + 8x - 12}{(x + 4)(2x - 3)}$

$= \dfrac{x^2 + 12x - 12}{(x + 4)(2x - 3)}$

22. $2x - y = 4$

$-2x + 3y = 5$

$D = \begin{vmatrix} 2 & -1 \\ -2 & 3 \end{vmatrix} = 4$

$D_x = \begin{vmatrix} 4 & -1 \\ 5 & 3 \end{vmatrix} = 17$

$D_y = \begin{vmatrix} 2 & 4 \\ -2 & 5 \end{vmatrix} = 18$

$x = \dfrac{D_x}{D} = \dfrac{17}{4}$

$y = \dfrac{D_y}{D} = \dfrac{18}{4} = \dfrac{9}{2}$

The solution is $\left(\dfrac{17}{4}, \dfrac{9}{2}\right)$.

23. Strategy: Let x represent the unknown rate of the car.

The unknown rate of the plane is $5x$.

	Distance	Rate	Time
car	25	x	$\dfrac{25}{x}$
Plane	625	$5x$	$\dfrac{625}{5x}$

The total time of the trip was 3 h.

$\dfrac{25}{x} + \dfrac{625}{5x} = 3$

Solution: $\dfrac{25}{x} + \dfrac{625}{5x} = 3$

$5x\left(\dfrac{25}{x} + \dfrac{625}{5x}\right) = 5x(3)$

$125 + 625 = 15x$

$750 = 15x$

$50 = x$

$250 = 5x$

The rate of the plane is 250 mph.

24. Strategy: To find the time it takes light to travel from Earth to the moon, use the formula $D = RT$. Substitute in the given values for R and D and solve for T.

Solution: $D = RT$

$1.86 \times 10^5 \cdot T = 232,500$

$T = 1.25 \times 10^0$

$T = 1.25$

The time is 1.25 s.

25. Strategy: To find the height of the periscope, substitute in the given value for d and solve for h.

Solution: $d = \sqrt{1.5h}$

$7 = \sqrt{1.5h}$

$(7)^2 = (\sqrt{1.5h})^2$

$49 = 1.5h$

$32.7 \approx h$

The height of the periscope is 32.7 ft.

26. $m = \dfrac{y_2 - y_1}{x_2 - x_1} = \dfrac{400 - 0}{5000 - 0} = \dfrac{400}{5000} = 0.08$

The annual income is 8% of the investment.

Chapter 8: Quadratic Equations

Prep Test

1. $\sqrt{18} = \sqrt{9 \cdot 2} = 3\sqrt{2}$

2. $3i$

3. $\dfrac{3x-2}{x-1} - 1 = \dfrac{3x-2}{x-1} - \dfrac{x-1}{x-1}$

$= \dfrac{3x-2-x+1}{x-1}$

$= \dfrac{2x-1}{x-1}$

4. $b^2 - 4ac$

$(-4)^2 - 4(2)(1) = 16 - 8$

$= 8$

5. $4x^2 + 28x + 49 = (2x+7)(2x+7)$

Yes, it is a perfect square trinomial.

6. $4x^2 - 4x + 1 = (2x-1)(2x-1) = (2x-1)^2$

7. $9x^2 - 4 = (3x+2)(3x-2)$

8. $\{x|\ x < -1\} \cap \{x|\ x < 4\}$

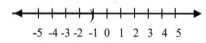

$$-5\ -4\ -3\ -2\ -1\ 0\ 1\ 2\ 3\ 4\ 5$$

9.

$x(x-1) = x + 15$

$x^2 - x = x + 15$

$x^2 - 2x - 15 = 0$

$(x-5)(x+3) = 0$

$x - 5 = 0 \quad x + 3 = 0$

$x = 5 \qquad x = -3$

The solutions are -3 and 5.

10.

$$\dfrac{4}{x-3} = \dfrac{16}{x}$$

$$x(x-3)\left(\dfrac{4}{x-3}\right) = x(x-3)\left(\dfrac{16}{x}\right)$$

$$4x = 16(x-3)$$

$$4x = 16x - 48$$

$$-12x = -48$$

$$x = 4$$

The solution is 4.

Section 8.1

Concept Check

1. (i) and (iii) are not quadratic equations.

3. The Principle of Zero Products states that if the product of two factors is zero, then at least one of the factors must be zero. This principle is used to solve quadratic equations when they are written as a product of factors.

5. $5(x+4) = 0$

$x + 4 = 0$

$x = -4$

The solution is -4.

7. $2x(x-4) = 0$

$2x = 0 \quad x - 4 = 0$

$x = 0 \qquad x = 4$

The solutions are 0 and 4.

Objective A Exercises

9. $x^2 - 4x = 0$

$x(x-4) = 0$

$x = 0 \quad x - 4 = 0$

$\qquad\qquad x = 4$

The solutions are 0 and 4.

11. $t^2 - 25 = 0$

$(t+5)(t-5) = 0$

$t+5 = 0 \quad t-5 = 0$

$t = -5 \qquad t = 5$

The solutions are −5 and 5.

13. $s^2 - s - 6 = 0$

$(s-3)(s+2) = 0$

$s-3 = 0 \quad s+2 = 0$

$s = 3 \qquad s = -2$

The solutions are −2 and 3.

15. $y^2 - 6y + 9 = 0$

$(y-3)(y-3) = 0$

$y-3 = 0 \quad y-3 = 0$

$y = 3 \qquad y = 3$

The solution is 3.

17. $9z^2 - 18z = 0$

$9z(z-2) = 0$

$9z = 0 \quad z-2 = 0$

$z = 0 \qquad z = 2$

The solutions are 0 and 2.

19. $r^2 - 3r = 10$

$r^2 - 3r - 10 = 0$

$(r-5)(r+2) = 0$

$r-5 = 0 \quad r+2 = 0$

$r = 5 \qquad r = -2$

The solutions are −2 and 5.

21. $v^2 + 10 = 7v$

$v^2 - 7v + 10 = 0$

$(v-5)(v-2) = 0$

$v-5 = 0 \quad v-2 = 0$

$v = 5 \qquad v = 2$

The solutions are 2 and 5.

23. $2x^2 - 9x - 18 = 0$

$(2x+3)(x-6) = 0$

$2x+3 = 0 \quad x-6 = 0$

$2x = -3 \qquad x = 6$

$x = -\dfrac{3}{2}$

The solutions are $-\dfrac{3}{2}$ and 6.

25. $4z^2 - 9z + 2 = 0$

$(4z-1)(z-2) = 0$

$4z-1 = 0 \quad z-2 = 0$

$4z = 1 \qquad z = 2$

$z = \dfrac{1}{4}$

The solutions are $\dfrac{1}{4}$ and 2.

27. $3w^2 + 11w = 4$

$3w^2 + 11w - 4 = 0$

$(3w-1)(w+4) = 0$

$3w-1 = 0 \quad w+4 = 0$

$3w = 1 \qquad w = -4$

$w = \dfrac{1}{3}$

The solutions are −4 and $\dfrac{1}{3}$.

29. $6x^2 = 23x + 18$

$6x^2 - 23x - 18 = 0$

$(2x-9)(3x+2) = 0$

$2x-9 = 0 \quad 3x+2 = 0$

$2x = 9 \qquad 3x = -2$

$x = \dfrac{9}{2} \qquad x = -\dfrac{2}{3}$

The solutions are $-\dfrac{2}{3}$ and $\dfrac{9}{2}$.

31. $4 - 15u - 4u^2 = 0$

$(1 - 4u)(4 + u) = 0$

$1 - 4u = 0 \quad 4 + u = 0$

$-4u = -1 \qquad u = -4$

$u = \dfrac{1}{4}$

The solutions are $\dfrac{1}{4}$ and -4.

33. $x + 18 = x(x - 6)$

$x + 18 = x^2 - 6x$

$0 = x^2 - 7x - 18$

$0 = (x - 9)(x + 2)$

$x - 9 = 0 \quad x + 2 = 0$

$x = 9 \qquad x = -2$

The solutions are -2 and 9.

35. $4s(s + 3) = s - 6$

$4s^2 + 12s = s - 6$

$4s^2 + 11s + 6 = 0$

$(4s + 3)(s + 2) = 0$

$4s + 3 = 0 \quad s + 2 = 0$

$4s = -3 \qquad s = -2$

$s = -\dfrac{3}{4}$

The solutions are -2 and $-\dfrac{3}{4}$.

37. $u^2 - 2u + 4 = (2u - 3)(u + 2)$

$u^2 - 2u + 4 = 2u^2 + u - 6$

$0 = u^2 + 3u - 10$

$0 = (u + 5)(u - 2)$

$u + 5 = 0 \quad u - 2 = 0$

$u = -5 \quad u = 2$

The solutions are -5 and 2.

39. $(3x - 4)(x + 4) = x^2 - 3x - 28$

$3x^2 + 8x - 16 = x^2 - 3x - 28$

$2x^2 + 11x + 12 = 0$

$(2x + 3)(x + 4) = 0$

$2x + 3 = 0 \quad x + 4 = 0$

$2x = -3 \qquad x = -4$

$x = -\dfrac{3}{2}$

The solutions are -4 and $-\dfrac{3}{2}$.

41. $(x - r_1)(x - r_2) = 0$

$(x - 2)(x - 5) = 0$

$x^2 - 7x + 10 = 0$

43. $(x - r_1)(x - r_2) = 0$

$[x - (-2)][x - (-4)] = 0$

$(x + 2)(x + 4) = 0$

$x^2 + 6x + 8 = 0$

45. $(x - r_1)(x - r_2) = 0$

$(x - 6)[x - (-1)] = 0$

$(x - 6)(x + 1) = 0$

$x^2 - 5x - 6 = 0$

47. $(x - r_1)(x - r_2) = 0$

$(x - 3)[x - (-3)] = 0$

$(x - 3)(x + 3) = 0$

$x^2 - 9 = 0$

49. $(x - r_1)(x - r_2) = 0$

$(x - 4)(x - 4) = 0$

$x^2 - 8x + 16 = 0$

51. $(x - r_1)(x - r_2) = 0$

$(x - 0)(x - 5) = 0$

$x^2 - 5x = 0$

53. $(x - r_1)(x - r_2) = 0$

$(x - 3)\left(x - \dfrac{1}{2}\right) = 0$

$x^2 - \dfrac{7}{2}x + \dfrac{3}{2} = 0$

$2\left(x^2 - \dfrac{7}{2}x + \dfrac{3}{2}\right) = 0 \cdot 2$

$2x^2 - 7x + 3 = 0$

55. $(x - r_1)(x - r_2) = 0$

$\left(x - \left(-\dfrac{3}{4}\right)\right)(x - 2) = 0$

$\left(x + \dfrac{3}{4}\right)(x - 2) = 0$

$x^2 - \dfrac{5}{4}x - \dfrac{3}{2} = 0$

$4\left(x^2 - \dfrac{5}{4}x - \dfrac{3}{2}\right) = 0 \cdot 4$

$4x^2 - 5x - 6 = 0$

57. $(x - r_1)(x - r_2) = 0$

$\left(x - \left(-\dfrac{5}{3}\right)\right)[x - (-2)] = 0$

$\left(x + \dfrac{5}{3}\right)(x + 2) = 0$

$x^2 + \dfrac{11}{3}x + \dfrac{10}{3} = 0$

$3\left(x^2 + \dfrac{11}{3}x + \dfrac{10}{3}\right) = 0 \cdot 3$

$3x^2 + 11x + 10 = 0$

59. $(x - r_1)(x - r_2) = 0$

$\left(x - \dfrac{1}{2}\right)\left(x - \dfrac{1}{3}\right) = 0$

$x^2 - \dfrac{5}{6}x + \dfrac{1}{6} = 0$

$6\left(x^2 - \dfrac{5}{6}x + \dfrac{1}{6}\right) = 0 \cdot 6$

$6x^2 - 5x + 1 = 0$

61. $(x - r_1)(x - r_2) = 0$

$\left(x - \dfrac{6}{5}\right)\left(x - \left(-\dfrac{1}{2}\right)\right) = 0$

$\left(x - \dfrac{6}{5}\right)\left(x + \dfrac{1}{2}\right) = 0$

$x^2 - \dfrac{7}{10}x - \dfrac{3}{5} = 0$

$10\left(x^2 - \dfrac{7}{10}x - \dfrac{3}{5}\right) = 0 \cdot 10$

$10x^2 - 7x - 6 = 0$

63. $(x - r_1)(x - r_2) = 0$

$\left(x - \left(-\dfrac{1}{4}\right)\right)\left(x - \left(-\dfrac{1}{2}\right)\right) = 0$

$\left(x + \dfrac{1}{4}\right)\left(x + \dfrac{1}{2}\right) = 0$

$x^2 + \dfrac{3}{4}x + \dfrac{1}{8} = 0$

$8\left(x^2 + \dfrac{3}{4}x + \dfrac{1}{8}\right) = 0 \cdot 8$

$8x^2 + 6x + 1 = 0$

65. $c = 0$

Objective C Exercises

67. $y^2 = 49$

$\sqrt{y^2} - \sqrt{49}$

$y = \pm\sqrt{49} = \pm 7$

The solutions are -7 and 7.

69. $z^2 = -4$

$\sqrt{z^2} = \sqrt{-4}$

$z = \pm\sqrt{-4} = \pm 2i$

The solutions are $-2i$ and $2i$.

71. $s^2 - 4 = 0$

$s^2 = 4$

$\sqrt{s^2} = \sqrt{4}$

$s = \pm\sqrt{4} = \pm 2$

The solutions are -2 and 2.

73. $4x^2 - 81 = 0$

$4x^2 = 81$

$x^2 = \dfrac{81}{4}$

$\sqrt{x^2} = \sqrt{\dfrac{81}{4}}$

$x = \pm\sqrt{\dfrac{81}{4}} = \pm\dfrac{9}{2}$

The solutions are $-\dfrac{9}{2}$ and $\dfrac{9}{2}$.

75. $y^2 + 49 = 0$

$y^2 = -49$

$\sqrt{y^2} = \sqrt{-49}$

$x = \pm\sqrt{-49} = \pm 7i$

The solutions are $-7i$ and $7i$.

77. $v^2 - 48 = 0$

$v^2 = 48$

$\sqrt{v^2} = \sqrt{48}$

$v = \pm\sqrt{48} = \pm 4\sqrt{3}$

The solutions are $-4\sqrt{3}$ and $4\sqrt{3}$.

79. $z^2 + 18 = 0$

$z^2 = -18$

$\sqrt{z^2} = \sqrt{-18}$

$z = \pm\sqrt{-18} = \pm 3i\sqrt{2}$

The solutions are $-3i\sqrt{2}$ and $3i\sqrt{2}$.

81. $(x-1)^2 = 36$

$\sqrt{(x-1)^2} = \sqrt{36}$

$x - 1 = \pm\sqrt{36} = \pm 6$

$x - 1 = 6 \quad x - 1 = -6$

$x = 7 \qquad x = -5$

The solutions are -5 and 7.

83. $5(z+2)^2 = 125$

$(z+2)^2 = 25$

$\sqrt{(z+2)^2} = \sqrt{25}$

$z + 2 = \pm\sqrt{25} = \pm 5$

$z + 2 = 5 \quad z + 2 = -5$

$z = 3 \qquad z = -7$

The solutions are -7 and 3.

85. $\left(v - \dfrac{1}{2}\right)^2 = \dfrac{1}{4}$

$\sqrt{\left(v - \dfrac{1}{2}\right)^2} = \sqrt{\dfrac{1}{4}}$

$v - \dfrac{1}{2} = \pm\sqrt{\dfrac{1}{4}} = \pm\dfrac{1}{2}$

$v - \dfrac{1}{2} = \dfrac{1}{2} \quad v - \dfrac{1}{2} = -\dfrac{1}{2}$

$v = 1 \qquad v = 0$

The solutions are 1 and 0.

87. $(x+5)^2 - 6 = 0$

$(x+5)^2 = 6$

$\sqrt{(x+5)^2} = \sqrt{6}$

$x+5 = \pm\sqrt{6}$

$x+5 = \sqrt{6}$ $x+5 = -\sqrt{6}$

$x = -5 + \sqrt{6}$ $x = -5 - \sqrt{6}$

The solutions are $-5 - \sqrt{6}$ and $-5 + \sqrt{6}$.

89. $(v-3)^2 + 45 = 0$

$(v-3)^2 = -45$

$\sqrt{(v-3)^2} = \sqrt{-45}$

$v-3 = \pm\sqrt{-45} = \pm 3i\sqrt{5}$

$v-3 = 3i\sqrt{5}$ $v-3 = -3i\sqrt{5}$

$v = 3 + 3i\sqrt{5}$ $v = 3 - 3i\sqrt{5}$

The solutions are $3 - 3i\sqrt{5}$ and $3 + 3i\sqrt{5}$.

91. $\left(u + \dfrac{2}{3}\right)^2 - 18 = 0$

$\left(u + \dfrac{2}{3}\right)^2 = 18$

$\sqrt{\left(u + \dfrac{2}{3}\right)^2} = \sqrt{18}$

$u + \dfrac{2}{3} = \pm\sqrt{18} = \pm 3\sqrt{2}$

$u + \dfrac{2}{3} = 3\sqrt{2}$ $u + \dfrac{2}{3} = -3\sqrt{2}$

$u = -\dfrac{2}{3} + 3\sqrt{2}$ $u = -\dfrac{2}{3} - 3\sqrt{2}$

$u = -\dfrac{2 + 9\sqrt{2}}{3}$ $u = -\dfrac{2 - 9\sqrt{2}}{3}$

The solutions are $-\dfrac{2 + 9\sqrt{2}}{3}$ and

$-\dfrac{2 - 9\sqrt{2}}{3}$.

93. $(x-a)^2 = -b$

Two complex solutions

95. $(x-a)^2 = 0$

Two equal real solutions

Critical Thinking

97. $(x-r_1)(x-r_2) = 0$

$(x - \sqrt{2})[x - (-\sqrt{2})] = 0$

$(x - \sqrt{2})(x + \sqrt{2}) = 0$

$x^2 - 2 = 0$

99. $(x - r_1)(x - r_2) = 0$

$(x-i)[x - (-i)] = 0$

$(x-i)(x+i) = 0$

$x^2 + 1 = 0$

101. $(x - r_1)(x - r_2) = 0$

$[x - (3 - \sqrt{2})][x - (3 + \sqrt{2})] = 0$

$x^2 - 6x + 7 = 0$

103. $(x - r_1)(x - r_2) = 0$

$[x - (5 - i)][x - (5 + i)] = 0$

$x^2 - 10x + 26 = 0$

105. $x^2 = \sqrt{7}$

$\sqrt{x^2} = \sqrt{\sqrt{7}}$

$x = \pm\sqrt[4]{7}$

The solutions are $-\sqrt[4]{7}$ and $\sqrt[4]{7}$.

107. $x^2 - \sqrt[3]{2} = 0$

$x^2 = \sqrt[3]{2}$

$\sqrt{x^2} = \sqrt{\sqrt[3]{2}}$

$x = \pm\sqrt[6]{2}$

The solutions are $-\sqrt[6]{2}$ and $\sqrt[6]{2}$.

Projects or Group Activities

109.   No

$b = -(1 + 5) = -6$

$c = (1)(5) = 5$

111. Yes
$$b = -(-1 - 2\sqrt{3}) + (-1 + 2\sqrt{3}) = -(-2) = 2$$
$$c = (-1 - 2\sqrt{3})(-1 + 2\sqrt{3}) = -11$$

113. Yes
$$b = -(2 - i\sqrt{3}) + (2 + i\sqrt{3}) = -(4) = -4$$
$$c = (2 - i\sqrt{3})(2 + i\sqrt{3}) = 7$$

Section 8.2

Concept Check

1. a) No
   b) Yes
   c) Yes
   d) No

3. Yes

Objective A Exercises

5. $x^2 - 4x - 5 = 0$
$$x^2 - 4x = 5$$
$$x^2 - 4x + 4 = 5 + 4$$
$$(x - 2)^2 = 9$$
$$\sqrt{(x - 2)^2} = \sqrt{9}$$
$$x - 2 = \pm\sqrt{9} = \pm 3$$
$$x - 2 = 3 \quad x - 2 = -3$$
$$x = 5 \qquad x = -1$$
The solutions are −1 and 5.

7. $z^2 - 6z + 9 = 0$
$$z^2 - 6z = -9$$
$$z^2 - 6z + 9 = -9 + 9$$
$$(z - 3)^2 = 0$$
$$\sqrt{(z - 3)^2} = \sqrt{0}$$
$$z - 3 = 0$$
$$z = 3$$
The solution is 3.

9. $r^2 + 4r - 7 = 0$
$$r^2 + 4r = 7$$
$$r^2 + 4r + 4 = 7 + 4$$
$$(r + 2)^2 = 11$$
$$\sqrt{(r + 2)^2} = \sqrt{11}$$
$$r + 2 = \pm\sqrt{11}$$
$$r + 2 = \sqrt{11} \quad r + 2 = -\sqrt{11}$$
$$r = -2 + \sqrt{11} \qquad r = -2 - \sqrt{11}$$
The solutions are $-2 - \sqrt{11}$ and $-2 + \sqrt{11}$

11. $x^2 - 6x + 7 = 0$
$$x^2 - 6x = -7$$
$$x^2 - 6x + 9 = -7 + 9$$
$$(x - 3)^2 = 2$$
$$\sqrt{(x - 3)^2} = \sqrt{2}$$
$$x - 3 = \pm\sqrt{2}$$
$$x - 3 = \sqrt{2} \quad x - 3 = -\sqrt{2}$$
$$x = 3 + \sqrt{2} \qquad x = 3 - \sqrt{2}$$
The solutions are $3 - \sqrt{2}$ and $3 + \sqrt{2}$.

13. $p^2 - 3p + 1 = 0$

$p^2 - 3p = -1$

$p^2 - 3p + \dfrac{9}{4} = -1 + \dfrac{9}{4}$

$\left(p - \dfrac{3}{2}\right)^2 = \dfrac{5}{4}$

$\sqrt{\left(p - \dfrac{3}{2}\right)^2} = \sqrt{\dfrac{5}{4}}$

$p - \dfrac{3}{2} = \pm \dfrac{\sqrt{5}}{2}$

$p - \dfrac{3}{2} = \dfrac{\sqrt{5}}{2} \qquad p - \dfrac{3}{2} = -\dfrac{\sqrt{5}}{2}$

$p = \dfrac{3}{2} + \dfrac{\sqrt{5}}{2} \qquad p = \dfrac{3}{2} - \dfrac{\sqrt{5}}{2}$

The solutions are $\dfrac{3 + \sqrt{5}}{2}$ and $\dfrac{3 - \sqrt{5}}{2}$.

15. $y^2 - 6y = 4$

$y^2 - 6y + 9 = 4 + 9$

$(y - 3)^2 = 13$

$\sqrt{(y - 3)^2} = \sqrt{13}$

$y - 3 = \pm\sqrt{13}$

$y - 3 = \sqrt{13} \qquad y - 3 = -\sqrt{13}$

$y = 3 + \sqrt{13} \qquad y = 3 - \sqrt{13}$

The solutions are $3 - \sqrt{13}$ and $3 + \sqrt{13}$.

17. $z^2 = z + 4$

$z^2 - z = 4$

$z^2 - z + \dfrac{1}{4} = 4 + \dfrac{1}{4}$

$\left(z - \dfrac{1}{2}\right)^2 = \dfrac{17}{4}$

$\sqrt{\left(z - \dfrac{1}{2}\right)^2} = \sqrt{\dfrac{17}{4}}$

$z - \dfrac{1}{2} = \pm\dfrac{\sqrt{17}}{2}$

$z - \dfrac{1}{2} = \dfrac{\sqrt{17}}{2} \qquad z - \dfrac{1}{2} = -\dfrac{\sqrt{17}}{2}$

$z = \dfrac{1}{2} + \dfrac{\sqrt{17}}{2} \qquad z = \dfrac{1}{2} - \dfrac{\sqrt{17}}{2}$

The solutions are $\dfrac{1 + \sqrt{17}}{2}$ and $\dfrac{1 - \sqrt{17}}{2}$.

19. $z^2 - 2z + 2 = 0$

$z^2 - 2z = -2$

$z^2 - 2z + 1 = -2 + 1$

$(z - 1)^2 = -1$

$\sqrt{(z - 1)^2} = \sqrt{-1}$

$z - 1 = \pm i$

$z - 1 = i \qquad z - 1 = -i$

$z = 1 + i \qquad z = 1 - i$

The solutions are $1 + i$ and $1 - i$.

21. $v^2 = 4v - 13$

$v^2 - 4v = -13$

$v^2 - 4v + 4 = -13 + 4$

$(v - 2)^2 = -9$

$\sqrt{(v - 2)^2} = \sqrt{-9}$

$v - 2 = \pm 3i$

$v - 2 = 3i \qquad v - 2 = -3i$

$v = 2 + 3i \qquad v = 2 - 3i$

The solutions are $2 + 3i$ and $2 - 3i$.

23. $p^2 + 6p = -13$

$p^2 + 6p + 9 = -13 + 9$

$(p + 3)^2 = -4$

$\sqrt{(p + 3)^2} = \sqrt{-4}$

$p + 3 = \pm 2i$

$p + 3 = 2i \qquad p + 3 = -2i$

$p = -3 + 2i \qquad p = -3 - 2i$

The solutions are $-3 - 2i$ and $-3 + 2i$.

25. $2s^2 = 4s + 5$

$2s^2 - 4s = 5$

$\dfrac{1}{2}\left(2s^2 - 4s\right) = \dfrac{1}{2}(5)$

$s^2 - 2s = \dfrac{5}{2}$

$s^2 - 2s + 1 = \dfrac{5}{2} + 1$

$(s - 1)^2 = \dfrac{7}{2}$

$\sqrt{(s - 1)^2} = \sqrt{\dfrac{7}{2}}$

$s - 1 = \pm\sqrt{\dfrac{7}{2}} = \pm\dfrac{\sqrt{14}}{2}$

$s - 1 = \dfrac{\sqrt{14}}{2} \qquad s - 1 = -\dfrac{\sqrt{14}}{2}$

$s = 1 + \dfrac{\sqrt{14}}{2} \qquad s = 1 - \dfrac{\sqrt{14}}{2}$

The solutions are $\dfrac{2 - \sqrt{14}}{2}$ and $\dfrac{2 + \sqrt{14}}{2}$.

27. $4x^2 - 4x + 5 = 0$

$4x^2 - 4x = -5$

$\dfrac{1}{4}\left(4x^2 - 4x\right) = \dfrac{1}{4}(-5)$

$x^2 - x = -\dfrac{5}{4}$

$x^2 - x + \dfrac{1}{4} = -\dfrac{5}{4} + \dfrac{1}{4}$

$\left(x - \dfrac{1}{2}\right)^2 = -1$

$\sqrt{\left(x - \dfrac{1}{2}\right)^2} = \sqrt{-1}$

$x - \dfrac{1}{2} = \pm i$

$x - \dfrac{1}{2} = i \qquad x - \dfrac{1}{2} = -i$

$x = \dfrac{1}{2} + i \qquad x = \dfrac{1}{2} - i$

The solutions are $\dfrac{1}{2} - i$ and $\dfrac{1}{2} + i$.

29. $9x^2 - 6x + 2 = 0$

$9x^2 - 6x = -2$

$\dfrac{1}{9}\left(9x^2 - 6x\right) = \dfrac{1}{9}(-2)$

$x^2 - \dfrac{2}{3}x = -\dfrac{2}{9}$

$x^2 - \dfrac{2}{3}x + \dfrac{1}{9} = -\dfrac{2}{9} + \dfrac{1}{9}$

$\left(x - \dfrac{1}{3}\right)^2 = -\dfrac{1}{9}$

$\sqrt{\left(x - \dfrac{1}{3}\right)^2} = \sqrt{-\dfrac{1}{9}}$

$x - \dfrac{1}{3} = \pm\dfrac{1}{3}i$

$x - \dfrac{1}{3} = \dfrac{1}{3}i \qquad x - \dfrac{1}{3} = -\dfrac{1}{3}i$

$x = \dfrac{1}{3} + \dfrac{1}{3}i \qquad x = \dfrac{1}{3} - \dfrac{1}{3}i$

The solutions are $\dfrac{1}{3} - \dfrac{1}{3}i$ and $\dfrac{1}{3} + \dfrac{1}{3}i$.

31. $y-2=(y-3)(y+2)$

$y-2=y^2-y-6$

$y^2-2y=4$

$y^2-2y+1=4+1$

$(y-1)^2=5$

$\sqrt{(y-1)^2}=\sqrt{5}$

$y-1=\pm\sqrt{5}$

$y-1=\sqrt{5} \quad y-1=-\sqrt{5}$

$y=1+\sqrt{5} \quad y=1-\sqrt{5}$

The solutions are $1-\sqrt{5}$ and $1+\sqrt{5}$.

33. $6t-2=(2t-3)(t-1)$

$6t-2=2t^2-5t+3$

$2t^2-11t=-5$

$\frac{1}{2}(2t^2-11t)=\frac{1}{2}(-5)$

$t^2-\frac{11}{2}t=-\frac{5}{2}$

$t^2-\frac{11}{2}t+\frac{121}{16}=-\frac{5}{2}+\frac{121}{16}$

$\left(t-\frac{11}{4}\right)^2=\frac{81}{16}$

$\sqrt{\left(t-\frac{11}{4}\right)^2}=\sqrt{\frac{81}{16}}$

$t-\frac{11}{4}=\pm\frac{9}{4}$

$t-\frac{11}{4}=\frac{9}{4} \quad t-\frac{11}{4}=-\frac{9}{4}$

$t=\frac{20}{4}=5 \quad t=\frac{2}{4}=\frac{1}{2}$

The solutions are $\frac{1}{2}$ and 5.

35. $(x-4)(x+1)=x-3$

$x^2-3x-4=x-3$

$x^2-4x=1$

$x^2-4x+4=1+4$

$(x-2)^2=5$

$\sqrt{(x-2)^2}=\sqrt{5}$

$x-2=\pm\sqrt{5}$

$x-2=\sqrt{5} \quad x-2=-\sqrt{5}$

$x=2+\sqrt{5} \quad x=2-\sqrt{5}$

The solutions are $2+\sqrt{5}$ and $2-\sqrt{5}$.

37. $z^2+2z=4$

$z^2+2z+1=4+1$

$(z+1)^2=5$

$\sqrt{(z+1)^2}=\sqrt{5}$

$z+1=\pm\sqrt{5}$

$z+1=\sqrt{5} \quad z+1=-\sqrt{5}$

$z=\sqrt{5}-1 \quad z=-\sqrt{5}-1$

$z=1.236 \quad z=-3.236$

The solutions are -3.326 and 1.236.

39. $2x^2=4x-1$

$2x^2-4x=-1$

$\frac{1}{2}(2x^2-4x)=\frac{1}{2}(-1)$

$x^2-2x=-\frac{1}{2}$

$x^2-2x+1=-\frac{1}{2}+1$

$(x-1)^2=\frac{1}{2}$

$\sqrt{(x-1)^2}=\sqrt{\frac{1}{2}}$

$$x - 1 = \pm\sqrt{\frac{1}{2}}$$

$$x - 1 = \sqrt{\frac{1}{2}} \qquad x - 1 = -\sqrt{\frac{1}{2}}$$

$$x = \sqrt{\frac{1}{2}} + 1 \qquad x = -\sqrt{\frac{1}{2}} + 1$$

$$x = 1.707 \qquad x = 0.293$$

The solutions are 0.293 and 1.707.

41. $c \le 4$

Objective B Exercises

43. The quadratic formula:

$$x = \frac{-b \pm \sqrt{b^2 - 4ac}}{2a}$$

a is the coefficient of x^2; b is the coefficient of x, and c is the constant term in the quadratic equation $ax^2 + bx + c, a \ne 0$.

45. $x^2 - 3x - 10 = 0$

$a = 1, b = -3, c = -10$

$$x = \frac{-b \pm \sqrt{b^2 - 4ac}}{2a}$$

$$x = \frac{-(-3) \pm \sqrt{(-3)^2 - 4(1)(-10)}}{2(1)}$$

$$x = \frac{3 \pm \sqrt{9 + 40}}{2} = \frac{3 \pm \sqrt{49}}{2}$$

$$x = \frac{3 \pm 7}{2}$$

$$x = \frac{3 + 7}{2} \qquad x = \frac{3 - 7}{2}$$

$$x = \frac{10}{2} = 5 \qquad x = -\frac{4}{2} = -2$$

The solutions are -2 and 5.

47. $x^2 - 8x + 9 = 0$

$a = 1, b = -8, c = 9$

$$x = \frac{-b \pm \sqrt{b^2 - 4ac}}{2a}$$

$$x = \frac{-(-8) \pm \sqrt{(-8)^2 - 4(1)(9)}}{2(1)}$$

$$x = \frac{8 \pm \sqrt{64 - 36}}{2} = \frac{8 \pm \sqrt{28}}{2}$$

$$x = \frac{8 \pm 2\sqrt{7}}{2} = 4 \pm \sqrt{7}$$

$$x = 4 + \sqrt{7} \qquad x = 4 - \sqrt{7}$$

The solutions are $4 + \sqrt{7}$ and $4 - \sqrt{7}$.

49. $v^2 = 6v + 19$

$v^2 - 6v - 19 = 0$

$a = 1, b = -6, c = -19$

$$x = \frac{-b \pm \sqrt{b^2 - 4ac}}{2a}$$

$$v = \frac{-(-6) \pm \sqrt{(-6)^2 - 4(1)(-19)}}{2(1)}$$

$$v = \frac{6 \pm \sqrt{36 + 76}}{2} = \frac{6 \pm \sqrt{112}}{2}$$

$$v = \frac{6 \pm 4\sqrt{7}}{2} = 3 \pm 2\sqrt{7}$$

$$v = 3 + 2\sqrt{7} \qquad v = 3 - 2\sqrt{7}$$

The solutions are $3 + 2\sqrt{7}$ and $3 - 2\sqrt{7}$.

51. $x^2 = 14x - 4$

$x^2 - 14x + 4 = 0$

$a = 1, b = -14, c = 4$

$x = \dfrac{-b \pm \sqrt{b^2 - 4ac}}{2a}$

$x = \dfrac{-(-14) \pm \sqrt{(-14)^2 - 4(1)(4)}}{2(1)}$

$x = \dfrac{14 \pm \sqrt{196 - 16}}{2} = \dfrac{14 \pm \sqrt{180}}{2}$

$x = \dfrac{14 \pm 6\sqrt{5}}{2}$

$x = \dfrac{14 + 6\sqrt{5}}{2} \qquad x = \dfrac{14 - 6\sqrt{5}}{2}$

$x = 7 + 3\sqrt{5} \qquad x = 7 - 3\sqrt{5}$

The solutions are $7 - 3\sqrt{5}$ and $7 + 3\sqrt{5}$.

53. $2z^2 - 2z - 1 = 0$

$a = 2, b = -2, c = -1$

$z = \dfrac{-b \pm \sqrt{b^2 - 4ac}}{2a}$

$z = \dfrac{-(-2) \pm \sqrt{(-2)^2 - 4(2)(-1)}}{2(2)}$

$z = \dfrac{2 \pm \sqrt{4 + 8}}{4} = \dfrac{2 \pm \sqrt{12}}{4}$

$z = \dfrac{2 \pm 2\sqrt{3}}{4} = \dfrac{1 \pm \sqrt{3}}{2}$

$z = \dfrac{1 + \sqrt{3}}{2} \qquad z = \dfrac{1 - \sqrt{3}}{2}$

The solutions are $\dfrac{1 - \sqrt{3}}{2}$ and $\dfrac{1 + \sqrt{3}}{2}$.

55. $4r^2 = 20r - 17$

$4r^2 - 20r + 17 = 0$

$a = 4, b = -20, c = 17$

$r = \dfrac{-b \pm \sqrt{b^2 - 4ac}}{2a}$

$r = \dfrac{-(-20) \pm \sqrt{(-20)^2 - 4(4)(17)}}{2(4)}$

$r = \dfrac{-(-20) \pm \sqrt{400 - 272}}{8} = \dfrac{20 \pm \sqrt{128}}{8}$

$r = \dfrac{20 \pm \sqrt{128}}{8} = \dfrac{20 \pm 8\sqrt{2}}{8} = \dfrac{5 \pm 2\sqrt{2}}{2}$

$r = \dfrac{5 + 2\sqrt{2}}{2} \qquad r = \dfrac{5 - 2\sqrt{2}}{2}$

The solutions are $\dfrac{5 - 2\sqrt{2}}{2}$ and $\dfrac{5 + 2\sqrt{2}}{2}$.

57. $z^2 + 2z + 2 = 0$

$a = 1, b = 2, c = 2$

$z = \dfrac{-b \pm \sqrt{b^2 - 4ac}}{2a}$

$z = \dfrac{-2 \pm \sqrt{2^2 - 4(1)(2)}}{2(1)}$

$z = \dfrac{-2 \pm \sqrt{4 - 8}}{2} = \dfrac{-2 \pm \sqrt{-4}}{2}$

$z = \dfrac{-2 \pm 2i}{2} = -1 \pm i$

$z = -1 + i \qquad z = -1 - i$

The solutions are $-1 - i$ and $-1 + i$.

59. $y^2 - 2y + 5 = 0$

$a = 1, b = -2, c = 5$

$$y = \frac{-b \pm \sqrt{b^2 - 4ac}}{2a}$$

$$y = \frac{-(-2) \pm \sqrt{(-2)^2 - 4(1)(5)}}{2(1)}$$

$$y = \frac{2 \pm \sqrt{4 - 20}}{2} = \frac{2 \pm \sqrt{-16}}{2}$$

$$y = \frac{2 \pm 4i}{2} = 1 \pm 2i$$

$y = 1 + 2i \qquad y = 1 - 2i$

The solutions are $1 - 2i$ and $1 + 2i$.

61. $s^2 - 4s + 13 = 0$

$a = 1, b = -4, c = 13$

$$s = \frac{-b \pm \sqrt{b^2 - 4ac}}{2a}$$

$$s = \frac{-(-4) \pm \sqrt{(-4)^2 - 4(1)(13)}}{2(1)}$$

$$s = \frac{4 \pm \sqrt{16 - 52}}{2} = \frac{4 \pm \sqrt{-36}}{2}$$

$$s = \frac{4 \pm 6i}{2} = 2 \pm 3i$$

$s = 2 + 3i \qquad s = 2 - 3i$

The solutions are $2 - 3i$ and $2 + 3i$.

63. $4x^2 - 4x + 33 = 0$

$a = 4, b = -4, c = 33$

$$x = \frac{-b \pm \sqrt{b^2 - 4ac}}{2a}$$

$$x = \frac{-(-4) \pm \sqrt{(-4)^2 - 4(4)(33)}}{2(4)}$$

$$x = \frac{4 \pm \sqrt{16 - 528}}{8} = \frac{4 \pm \sqrt{-512}}{8}$$

$$x = \frac{4 \pm 16i\sqrt{2}}{8} = \frac{1 \pm 4i\sqrt{2}}{2} = \frac{1}{2} \pm 2i\sqrt{2}$$

$x = \frac{1}{2} + 2i\sqrt{2} \qquad x = \frac{1}{2} - 2i\sqrt{2}$

The solutions are $\dfrac{1}{2} - 2i\sqrt{2}$ and $\dfrac{1}{2} + 2i\sqrt{2}$.

65. $9v^2 - 6v - 71 = 0$

$a = 9, b = -6, c = -71$

$$v = \frac{-b \pm \sqrt{b^2 - 4ac}}{2a}$$

$$v = \frac{-(-6) \pm \sqrt{(-6)^2 - 4(9)(-71)}}{2(9)}$$

$$v = \frac{6 \pm \sqrt{36 + 2556}}{18} = \frac{6 \pm \sqrt{2592}}{18}$$

$$v = \frac{6 \pm 36\sqrt{2}}{18} = \frac{1 \pm 6\sqrt{2}}{3}$$

$$v = \frac{1 + 6\sqrt{2}}{3} \qquad v = \frac{1 - 6\sqrt{2}}{3}$$

The solutions are $\dfrac{1 - 6\sqrt{2}}{3}$ and $\dfrac{1 + 6\sqrt{2}}{3}$.

67. $2w^2 - 2w - 5 = 0$

$a = 2, b = -2, c = -5$

$$w = \frac{-b \pm \sqrt{b^2 - 4ac}}{2a}$$

$$w = \frac{-(-2) \pm \sqrt{(-2)^2 - 4(2)(-5)}}{2(2)}$$

$$w = \frac{2 \pm \sqrt{4 + 40}}{4} = \frac{2 \pm \sqrt{44}}{4}$$

$$w = \frac{2 \pm 2\sqrt{11}}{4} = \frac{1 \pm \sqrt{11}}{2}$$

$$w = \frac{1 + \sqrt{11}}{2} \qquad w = \frac{1 - \sqrt{11}}{2}$$

The solutions are $\dfrac{1 - \sqrt{11}}{2}$ and $\dfrac{1 + \sqrt{11}}{2}$.

The solutions are $\dfrac{1}{2} - 2i\sqrt{2}$ and $\dfrac{1}{2} + 2i\sqrt{2}$.

69. $2x^2 + 4x - 6 = 0$

$a = 2, b = 4, c = -6$

$$x = \frac{-b \pm \sqrt{b^2 - 4ac}}{2a}$$

$$x = \frac{-(4) \pm \sqrt{(4)^2 - 4(2)(-6)}}{2(2)}$$

$$x = \frac{-4 \pm \sqrt{16 + 48}}{4} = \frac{-4 \pm \sqrt{64}}{4}$$

$$x = \frac{-4 \pm 8}{4}$$

$$x = \frac{4 \mid 8}{4} = 1 \qquad x = \frac{4 \quad 8}{4} = -3$$

The solutions are –3 and 1.

71. $2x^2 + x = (x-4)(x-2)$

$x^2 + 7x - 8 = 0$

$a = 1, b = 7 \ c = -8$

$$x = \frac{-b \pm \sqrt{b^2 - 4ac}}{2a}$$

$$x = \frac{-(7) \pm \sqrt{(7)^2 - 4(1)(-8)}}{2(1)}$$

$$x = \frac{-7 \pm \sqrt{49 + 32}}{2} = \frac{-7 \pm \sqrt{81}}{2}$$

$$x = \frac{-7 \pm 9}{2}$$

$$x = \frac{-7 + 9}{2} = 1 \qquad x = \frac{-7 - 9}{2} = -8$$

The solutions are 1 and –8.

73. $(2x+1)(x+2) = (x-4)(x+3)$

$x^2 + 6x + 14 = 0$

$a = 1, b = 6, c = 14$

$$x = \frac{-b \pm \sqrt{b^2 - 4ac}}{2a}$$

$$x = \frac{-(6) \pm \sqrt{(6)^2 - 4(1)(14)}}{2(1)}$$

$$x = \frac{-6 \pm \sqrt{36 - 56}}{2} = \frac{-6 \pm \sqrt{-20}}{2}$$

$$x = \frac{-6 \pm 2i\sqrt{5}}{2} = -3 \pm i\sqrt{5}$$

$$x = -3 + i\sqrt{5} \qquad x = -3 - i\sqrt{5}$$

The solutions are $-3 - i\sqrt{5}$ and $-3 + i\sqrt{5}$.

75. $2x^2 - x = (x+3)(x-2)$

$x^2 - 2x + 6 = 0$

$a = 1, b = -2, c = 6$

$$x = \frac{-b \pm \sqrt{b^2 - 4ac}}{2a}$$

$$x = \frac{-(-2) \pm \sqrt{(-2)^2 - 4(1)(6)}}{2(1)}$$

$$x = \frac{2 \pm \sqrt{4 - 24}}{2} = \frac{2 \pm \sqrt{-20}}{2}$$

$$x = \frac{2 \pm 2i\sqrt{5}}{2} = 1 \pm i\sqrt{5}$$

$$x = 1 + i\sqrt{5} \qquad x = 1 - i\sqrt{5}$$

The solutions are $1 - i\sqrt{5}$ and $1 + i\sqrt{5}$.

77. $5t^2 - 5t + 7 = (t-1)(t-2)$

$4t^2 - 2t + 5 = 0$

$a = 4, b = -2, c = 5$

$$t = \frac{-b \pm \sqrt{b^2 - 4ac}}{2a}$$

$$t = \frac{-(-2) \pm \sqrt{(-2)^2 - 4(4)(5)}}{2(4)}$$

$$t = \frac{2 \pm \sqrt{4 - 80}}{8} = \frac{2 \pm \sqrt{-76}}{8}$$

$$t = \frac{2 \pm 2i\sqrt{19}}{8} = \frac{1}{4} \pm \frac{i\sqrt{19}}{4}$$

$$t = \frac{1}{4} + \frac{i\sqrt{19}}{4} \qquad t = \frac{1}{4} - \frac{i\sqrt{19}}{4}$$

The solutions are $\dfrac{1}{4} + \dfrac{i\sqrt{19}}{4}$ and

$\dfrac{1}{4} - \dfrac{i\sqrt{19}}{4}$.

79. $p^2 - 8p + 3 = 0$

$a = 1, b = -8, c = 3$

$p = \dfrac{-b \pm \sqrt{b^2 - 4ac}}{2a}$

$p = \dfrac{-(-8) \pm \sqrt{(-8)^2 - 4(1)(3)}}{2(1)}$

$p = \dfrac{8 \pm \sqrt{64 - 12}}{2} = \dfrac{8 \pm \sqrt{52}}{2}$

$p = \dfrac{8 \pm 2\sqrt{13}}{2} = 4 \pm \sqrt{13}$

$p = 4 + \sqrt{13} \qquad p = 4 - \sqrt{13}$

The solutions are 0.394 and 7.606.

81. $w^2 + 4w = 1$

$w^2 + 4w - 1 = 0$

$a = 1, b = 4, c = -1$

$w = \dfrac{-b \pm \sqrt{b^2 - 4ac}}{2a}$

$w = \dfrac{-4 \pm \sqrt{4^2 - 4(1)(-1)}}{2(1)}$

$w = \dfrac{-4 \pm \sqrt{16 + 4}}{2} = \dfrac{-4 \pm \sqrt{20}}{2}$

$w = \dfrac{-4 \pm 2\sqrt{5}}{2} = -2 \pm \sqrt{5}$

$w = -2 + \sqrt{5} \qquad w = -2 - \sqrt{5}$

The solutions are -4.236 and 0.236.

83. $2y^2 = y + 5$

$2y^2 - y - 5 = 0$

$a = 2, b = -1, c = -5$

$y = \dfrac{-b \pm \sqrt{b^2 - 4ac}}{2a}$

$y = \dfrac{-(-1) \pm \sqrt{(-1)^2 - 4(2)(-5)}}{2(2)}$

$y = \dfrac{1 \pm \sqrt{1 + 40}}{4} = \dfrac{1 \pm \sqrt{41}}{4}$

$y = \dfrac{1 + \sqrt{41}}{4} \qquad y = \dfrac{1 - \sqrt{41}}{4}$

The solutions are -1.351 and 1.851.

85. $3y^2 + y + 1 = 0$

$a = 3, b = 1, c = 1$

$b^2 - 4ac = 1^2 - 4(3)(1)$

$= 1 - 12 = -11$

$-11 < 0$

Since the discriminant is less than zero, the equation has two complex number solutions.

87. $4x^2 + 20x + 25 = 0$

$a = 4, b = 20, c = 25$

$b^2 - 4ac = (20)^2 - 4(4)(25)$

$= 400 - 400 = 0$

Since the discriminant is equal to zero, the equation has two equal real number solutions.

89. $3w^2 + 3w - 2 = 0$

$a = 3, b = 3, c = -2$

$b^2 - 4ac = 3^2 - 4(3)(-2)$

$= 9 + 24 = 33$

$33 > 0$

Since the discriminant is greater than zero, the equation has two unequal real number solutions.

91. $\sqrt{4ac}$

Critical Thinking

93. $x^2 - 6x + p = 0$

$x^2 - 6x = -p$

$x^2 - 6x + 9 = -p + 9$

$(x - 3)^2 = -p + 9$

$\sqrt{(x - 3)^2} = \sqrt{9 - p}$

$x - 3 = \pm\sqrt{9 - p}$

$x = 3 \pm \sqrt{9 - p}$

x will have two real solutions if $9 - p > 0$.
Solving this inequality gives $p < 9$.
The values of p are $\{p \mid p < 9\}$.

95. $x^2 - 2x + p = 0$

$a = 1, b = -2, c = p$

$x = \dfrac{-b \pm \sqrt{b^2 - 4ac}}{2a}$

$x = \dfrac{-(-2) \pm \sqrt{(-2)^2 - 4(1)(p)}}{2(1)}$

$x = \dfrac{2 \pm \sqrt{4 - 4p}}{2}$

$x = 1 \pm \sqrt{1 - p}$

x will have two complex solutions when
$1 - p < 0$.
Solving the inequality gives $p > 1$.
The values of p are $(1, \infty)$.

97. $x^2 - 2ix + 15 = 0$

$a = 1, b = -2i, c = 15$

$x = \dfrac{-b \pm \sqrt{b^2 - 4ac}}{2a}$

$x = \dfrac{-(-2i) \pm \sqrt{(-2i)^2 - 4(1)(15)}}{2(1)}$

$x = \dfrac{2i \pm \sqrt{-4 - 60}}{2} = \dfrac{2i \pm \sqrt{-64}}{2}$

$x = \dfrac{2i \pm 8i}{2}$

$x = \dfrac{2i + 8i}{2} = 5i \qquad x = \dfrac{2i - 8i}{2} = -3i$

The solutions are $-3i$ and $5i$.

Projects or Group Activities

99. $h = -16t^2 + 70t + 4$

Find the time t it takes for the ball to hit the
ground. When the ball hits the ground h is
zero.

$0 = -16t^2 + 70t + 4$

$a = -16, b = 70, c = 4$

$x = \dfrac{-b \pm \sqrt{b^2 - 4ac}}{2a}$

$x = \dfrac{-(70) \pm \sqrt{(70)^2 - 4(-16)(4)}}{2(-16)}$

$x = \dfrac{-70 \pm \sqrt{5156}}{-32}$

$x = \dfrac{-70 + \sqrt{5156}}{-32} = -0.056$

$x = \dfrac{-70 - \sqrt{5156}}{-32} = 4.431$

The ball hits the ground 4.431 s after it is
struck by the batter.

$s = 44.5t = 44.5(4.431) = 197.2$ ft

No, the ball will not clear a fence 325 ft
from home plate. It will only have gone
197.2 ft when it hits the ground.

Section 8.3

Concept Check

1. (i), (ii), (iii), (iv) and (v)

Objective A Exercises

3. Yes

5. No

7. $x^4 - 13x^2 + 36 = 0$
$(x^2)^2 - 13(x^2) + 36 = 0$
$u^2 - 13u + 36 = 0$
$(u - 4)(u - 9) = 0$
$u - 4 = 0 \quad u - 9 = 0$
$u = 4 \qquad u = 9$
Replace u with x^2.
$x^2 = 4 \qquad x^2 = 9$
$\sqrt{x^2} = \sqrt{4} \quad \sqrt{x^2} = \sqrt{9}$
$x = \pm 2 \qquad x = \pm 3$
The solutions are -2, 2, -3 and 3.

9. $z^4 - 6z^2 + 8 = 0$
$(z^2)^2 - 6(z^2) + 8 = 0$
$u^2 - 6u + 8 = 0$
$(u - 4)(u - 2) = 0$
$u - 4 = 0 \quad u - 2 = 0$
$u = 4 \qquad u = 2$
Replace u with z^2.
$z^2 = 4 \qquad z^2 = 2$
$\sqrt{z^2} = \sqrt{4} \quad \sqrt{z^2} = \sqrt{2}$
$z = \pm 2 \qquad z = \pm\sqrt{2}$
The solutions are -2, 2, $-\sqrt{2}$ and $\sqrt{2}$.

11. $p - 3p^{1/2} + 2 = 0$
$(p^{1/2})^2 - 3(p^{1/2}) + 2 = 0$
$u^2 - 3u + 2 = 0$
$(u - 2)(u - 1) = 0$
$u - 2 = 0 \quad u - 1 = 0$
$u = 2 \qquad u = 1$
Replace u with $p^{1/2}$.
$p^{1/2} = 2 \qquad p^{1/2} = 1$
$\left(p^{1/2}\right)^2 = 2^2 \quad \left(p^{1/2}\right)^2 = 1^2$
$p = 4 \qquad p = 1$
The solutions are 1 and 4.

13. $x - x^{1/2} - 12 = 0$
$(x^{1/2})^2 - (x^{1/2}) - 12 = 0$
$u^2 - u - 12 = 0$
$(u - 4)(u + 3) = 0$
$u - 4 = 0 \quad u + 3 = 0$
$u = 4 \qquad u = -3$
Replace u with $x^{1/2}$.
$x^{1/2} = 4 \qquad x^{1/2} = -3$
$\left(x^{1/2}\right)^2 = 4^2 \quad \left(x^{1/2}\right)^2 = (-3)^2$
$x = 16 \qquad x = 9$
9 does not check as a solution. The solution is 16.

15. $z^4 + 3z^2 - 4 = 0$
$(z^2)^2 + 3(z^2) - 4 = 0$
$u^2 + 3u - 4 = 0$
$(u + 4)(u - 1) = 0$
$u + 4 = 0 \quad u - 1 = 0$
$u = -4 \qquad u = 1$
Replace u with z^2.

$z^2 = -4 \qquad z^2 = 1$

$\sqrt{z^2} = \sqrt{-4} \quad \sqrt{z^2} = \sqrt{1}$

$z = \pm 2i \qquad\qquad z = \pm 1$

The solutions are $-1, 1, -2i$ and $2i$.

17. $x^4 + 12x^2 - 64 = 0$

$(x^2)^2 + 12(x^2) - 64 = 0$

$u^2 + 12u - 64 = 0$

$(u + 16)(u - 4) = 0$

$u + 16 = 0 \quad u - 4 = 0$

$u = -16 \qquad u = 4$

Replace u with x^2.

$x^2 = -16 \qquad\quad x^2 = 4$

$\sqrt{x^2} = \sqrt{-16} \quad \sqrt{x^2} = \sqrt{4}$

$x = \pm 4i \qquad\qquad x = \pm 2$

The solutions are $-2, 2, -4i$ and $4i$.

19. $p + 2p^{1/2} - 24 = 0$

$(p^{1/2})^2 + 2(p^{1/2}) - 24 = 0$

$u^2 + 2u - 24 = 0$

$(u + 6)(u - 4) = 0$

$u + 6 = 0 \quad u - 4 = 0$

$u = -6 \qquad u = 4$

Replace u with $p^{1/2}$.

$p^{1/2} = -6 \qquad\quad p^{1/2} = 4$

$\left(p^{1/2}\right)^2 = (-6)^2 \quad \left(p^{1/2}\right)^2 = 4^2$

$p = 36 \qquad\qquad p = 16$

36 does not check as a solution. The solution is 16.

21. $y^{2/3} - 9y^{1/3} + 8 = 0$

$(y^{1/3})^2 - 9(y^{1/3}) + 8 = 0$

$u^2 - 9u + 8 = 0$

$(u - 8)(u - 1) = 0$

$u - 8 = 0 \quad u - 1 = 0$

$u = 8 \qquad u = 1$

Replace u with $y^{1/3}$.

$y^{1/3} = 8 \qquad\quad y^{1/3} = 1$

$(y^{1/3})^3 = 8^3 \quad (y^{1/3})^3 = 1^3$

$y = 512 \qquad\quad y = 1$

The solutions are 1 and 512.

23. $9w^4 - 13w^2 + 4 = 0$

$9(w^2)^2 - 13(w^2) + 4 = 0$

$9u^2 - 13u + 4 = 0$

$(9u - 4)(u - 1) = 0$

$9u - 4 = 0 \quad u - 1 = 0$

$9u = 4 \qquad u = 1$

$u = \dfrac{4}{9}$

Replace u with w^2.

$w^2 = \dfrac{4}{9} \qquad\quad w^2 = 1$

$\sqrt{w^2} = \sqrt{\dfrac{4}{9}} \quad \sqrt{w^2} = \sqrt{1}$

$w = \pm\dfrac{2}{3} \qquad\qquad w = \pm 1$

The solutions are $-1, 1, -\dfrac{2}{3}$ and $\dfrac{2}{3}$.

Objective B Exercises

25. Exercises 30, 31, 32, 36, 37, 38, 40, 41, 44

27. $\sqrt{x+1} + x = 5$

$\sqrt{x+1} = 5 - x$

$\left(\sqrt{x+1}\right)^2 = (5 - x)^2$

$x + 1 = 25 - 10x + x^2$

$0 = x^2 - 11x + 24$

$(x - 3)(x - 8) = 0$

$x - 3 = 0 \quad x - 8 = 0$

$x = 3 \qquad x = 8$

8 does not check as a solution. The solution is 3.

29. $x = \sqrt{x} + 6$

$x - 6 = \sqrt{x}$

$(x - 6)^2 = (\sqrt{x})^2$

$x^2 - 12x + 36 = x$

$x^2 - 13x + 36 = 0$

$(x - 9)(x - 4) = 0$

$x - 9 = 0 \quad x - 4 = 0$

$x = 9 \qquad x = 4$

4 does not check as a solution.
The solution is 9.

31. $\sqrt{3w + 3} = w + 1$

$\left(\sqrt{3w + 3}\right)^2 = (w + 1)^2$

$3w + 3 = w^2 + 2w + 1$

$0 = w^2 - w - 2$

$(w - 2)(w + 1) = 0$

$w - 2 = 0 \quad w + 1 = 0$

$w = 2 \qquad w = -1$

The solutions are −1 and 2.

33. $\sqrt{4y + 1} - y = 1$

$\sqrt{4y + 1} = 1 + y$

$\left(\sqrt{4y + 1}\right)^2 = (1 + y)^2$

$4y + 1 = 1 + 2y + y^2$

$0 = y^2 - 2y$

$y(y - 2) = 0$

$y = 0 \quad y - 2 = 0$

$\qquad\qquad y = 2$

The solutions are 0 and 2.

35. $\sqrt{10x + 5} - 2x = 1$

$\sqrt{10x + 5} = 1 + 2x$

$\left(\sqrt{10x + 5}\right)^2 = (1 + 2x)^2$

$10x + 5 = 1 + 4x + 4x^2$

$0 = 4x^2 - 6x - 4$

$2(2x + 1)(x - 2) = 0$

$2x + 1 = 0 \quad x - 2 = 0$

$2x = -1 \qquad x = 2$

$x = -\dfrac{1}{2}$

The solutions are $-\dfrac{1}{2}$ and 2.

37. $\sqrt{p + 11} = 1 - p$

$\left(\sqrt{p + 11}\right)^2 = (1 - p)^2$

$p + 11 = 1 - 2p + p^2$

$0 = p^2 - 3p - 10$

$(p - 5)(p + 2) = 0$

$p - 5 = 0 \quad p + 2 = 0$

$p = 5 \qquad p = -2$

5 does not check as a solution.
The solution is −2.

39. $\sqrt{x - 1} - \sqrt{x} = -1$

$\sqrt{x - 1} = \sqrt{x} - 1$

$\left(\sqrt{x - 1}\right)^2 = (\sqrt{x} - 1)^2$

$x - 1 = x - 2\sqrt{x} + 1$

$2\sqrt{x} = 2$

$\sqrt{x} = 1$

$\left(\sqrt{x}\right)^2 = 1^2$

$x = 1$

The solution is 1.

41. $\sqrt{2x-1} = 1 - \sqrt{x-1}$

$\left(\sqrt{2x-1}\right)^2 = (1-\sqrt{x-1})^2$

$2x-1 = 1 - 2\sqrt{x-1} + x - 1$

$2\sqrt{x-1} = -x+1$

$\left(2\sqrt{x-1}\right)^2 = (-x+1)^2$

$4(x-1) = x^2 - 2x + 1$

$4x - 4 = x^2 - 2x + 1$

$0 = x^2 - 6x + 5$

$(x-5)(x-1) = 0$

$x-5 = 0 \quad x-1 = 0$

$x = 5 \quad\quad x = 1$

5 does not check as a solution.
The solution is 1.

43. $\sqrt{t+3} + \sqrt{2t+7} = 1$

$\sqrt{t+3} = 1 - \sqrt{2t+7}$

$\left(\sqrt{t+3}\right)^2 = (1-\sqrt{2t+7})^2$

$t+3 = 1 - 2\sqrt{2t+7} + 2t + 7$

$2\sqrt{2t+7} = t+5$

$\left(2\sqrt{2t+7}\right)^2 = (t+5)^2$

$4(2t+7) = t^2 + 10t + 25$

$8t + 28 = t^2 + 10t + 25$

$0 = t^2 + 2t - 3$

$(t+3)(t-1) = 0$

$t+3 = 0 \quad t-1 = 0$

$t = -3 \quad\quad t = 1$

1 does not check as a solution.
The solution is −3.

Objective C exercises

45. $y + 2$

47. $x = \dfrac{10}{x-9}$

$(x-9)x = \left(\dfrac{10}{x-9}\right)(x-9)$

$x^2 - 9x = 10$

$x^2 - 9x - 10 = 0$

$(x-10)(x+1) = 0$

$x-10 = 0 \quad x+1 = 0$

$x = 10 \quad\quad x = -1$

The solutions are −1 and 10.

49. $\dfrac{y-1}{y+2} + y = 1$

$(y+2)\left(\dfrac{y-1}{y+2} + y\right) = 1(y+2)$

$y-1 + y(y+2) = y+2$

$y-1 + y^2 + 2y = y+2$

$y^2 + 2y - 3 = 0$

$(y+3)(y-1) = 0$

$y+3 = 0 \quad y-1 = 0$

$y = -3 \quad\quad y = 1$

The solutions are −3 and 1.

51. $\dfrac{3r+2}{r+2} - 2r = 1$

$(r+2)\left(\dfrac{3r+2}{r+2} - 2r\right) = 1(r+2)$

$3r+2 - 2r(r+2) = r+2$

$3r+2 - 2r^2 - 4r = r+2$

$-2r^2 - 2r = 0$

$-2r(r+1) = 0$

$-2r = 0 \quad r+1 = 0$

$r = 0 \quad\quad r = -1$

The solutions are −1 and 0.

53. $\dfrac{1}{x+2}+\dfrac{x}{x-2}=\dfrac{x+6}{x^2-4}$

$(x+2)(x-2)\left(\dfrac{1}{x+2}+\dfrac{x}{x-2}\right)=\left(\dfrac{x+6}{x^2-4}\right)(x+2)(x-2)$

$(x-2)+x(x+2)=x+6$

$x-2+x^2+2x=x+6$

$x^2+2x-8=0$

$(x+4)(x-2)=0$

$x+4=0 \quad x-2=0$

$x=-4 \qquad x=2$

The solution is –4.

55. $\dfrac{16}{z-2}+\dfrac{16}{z+2}=6$

$(z-2)(z+2)\left(\dfrac{16}{z-2}+\dfrac{16}{z+2}\right)=6(z-2)(z+2)$

$16(z+2)+16(z-2)=6(z^2-4)$

$16z+32+16z-32=6z^2-24$

$0=6z^2-32z-24$

$2(3z^2-16z-12)=0$

$2(3z+2)(z-6)=0$

$3z+2=0 \quad z-6=0$

$3z=-2 \qquad z=6$

$z=-\dfrac{2}{3}$

The solutions are $-\dfrac{2}{3}$ and 6.

57. $\dfrac{t}{t-2}+\dfrac{2}{t-1}=4$

$(t-2)(t-1)\left(\dfrac{t}{t-2}+\dfrac{2}{t-1}\right)=4(t-2)(t-1)$

$t(t-1)+2(t-2)=4(t^2-3t+2)$

$t^2-t+2t-4=4t^2-12t+8$

$0=3t^2-13t+12$

$(3t-4)(t-3)=0$

$3t-4=0 \quad t-3=0$

$3t=4 \qquad t=3$

$t=\dfrac{4}{3}$

The solutions are $\dfrac{4}{3}$ and 3.

Critical Thinking

59. $\left(\sqrt{x}+3\right)^2-4\sqrt{x}-17=0$

Let $u=\sqrt{x}+3$

$u^2-4(u-3)-17=0$

$u^2-4u+12-17=0$

$u^2-4u-5=0$

$(u-5)(u+1)=0$

$u-5=0 \quad u+1=0$

$u=5 \qquad u=-1$

Replace u with $\sqrt{x}+3$.

$\sqrt{x}+3=5 \quad \sqrt{x}+3=-1$

$\sqrt{x}=2 \qquad \sqrt{x}=-4$

$\left(\sqrt{x}\right)^2=2^2 \quad \left(\sqrt{x}\right)^2=(-4)^2$

$x=4 \qquad x=16$

The solution is 4.

Projects or Group Activities

61. a) $\{x \mid -\sqrt{29.7366} \le x \le \sqrt{29.7366}\}$

b)

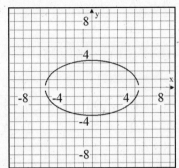

The $\pm$ symbol occurs in the equation so that the graph pictures the entire shape of the football.

c)

$$y = 3.3041\sqrt{1 - \frac{x^2}{29.7336}} = 3.3041\sqrt{1 - \frac{3^2}{29.7336}}$$

$$y = 2.7592 \text{ in}$$

Check Your Progress: Chapter 8

1. $3x^2 - 10x - 8 = 0$

$(3x + 2)(x - 4) = 0$

$3x + 2 = 0 \quad x - 4 = 0$

$3x = -2 \qquad x = 4$

$x = -\dfrac{2}{3}$

The solutions are $-\dfrac{2}{3}$ and 4.

2. $2x^2 + 3x = (x + 3)(x + 4)$

$x^2 - 4x - 12 = 0$

$(x - 6)(x + 2) = 0$

$x - 6 = 0 \quad x + 2 = 0$

$x = 6 \qquad x = -2$

The solutions are –2 and 6.

3. $(x - r_1)(x - r_2) = 0$

$(x - (-3))\left(x - \dfrac{3}{5}\right) = 0$

$(x + 3)\left(x - \dfrac{3}{5}\right) = 0$

$x^2 + \dfrac{12}{5}x - \dfrac{9}{5} = 0$

$5\left(x^2 + \dfrac{12}{5}x - \dfrac{9}{5}\right) = 0 \cdot 5$

$5x^2 + 12x - 9 = 0$

4. $(x + 3)^2 = 20$

$\sqrt{(x + 3)^2} = \sqrt{20}$

$x + 3 = \pm\sqrt{20} = \pm 2\sqrt{5}$

$x + 3 = 2\sqrt{5} \quad x + 3 = -2\sqrt{5}$

$x = -3 + 2\sqrt{5} \qquad x = -3 - 2\sqrt{5}$

The solutions are $-3 + 2\sqrt{5}$ and $-3 - 2\sqrt{5}$.

5. $(z - 4)^2 + 9 = 5$

$(z - 4)^2 = -4$

$\sqrt{(z - 4)^2} = \sqrt{-4}$

$z - 4 = \pm\sqrt{-4} = \pm 2i$

$z - 4 = 2i \quad z - 4 = -2i$

$z = 4 + 2i \qquad x = 4 - 2i$

The solutions are $4 + 2i$ and $4 - 2i$.

6. $x^2 + 2x = 49$

$x^2 + 2x + 1 = 49 + 1$

$(x + 1)^2 = 50$

$\sqrt{(x + 1)^2} = \sqrt{50}$

$x + 1 = \pm 5\sqrt{2}$

$x + 1 = 5\sqrt{2} \quad x + 1 = -5\sqrt{2}$

$x = -1 + 5\sqrt{2} \qquad x = -1 - 5\sqrt{2}$

The solutions are $-1 + 5\sqrt{2}$ and $-1 - 5\sqrt{2}$.

7. $4x^2 + 12x + 21 = 0$

$4x^2 + 12x = -21$

$\dfrac{1}{4}\left(4x^2 + 12x\right) = \dfrac{1}{4}(-21)$

$x^2 + 3x = -\dfrac{21}{4}$

$x^2 + 3x + \dfrac{9}{4} = -\dfrac{21}{4} + \dfrac{9}{4}$

$\left(x + \dfrac{3}{2}\right)^2 = -3$

$\sqrt{\left(x + \dfrac{3}{2}\right)^2} = \sqrt{-3}$

$x + \dfrac{3}{2} = \pm i\sqrt{3}$

$x + \dfrac{3}{2} = i\sqrt{3} \qquad x + \dfrac{3}{2} = -i\sqrt{3}$

$x = -\dfrac{3}{2} + i\sqrt{3} \qquad x = -\dfrac{3}{2} - i\sqrt{3}$

The solutions are $-\dfrac{3}{2} + i\sqrt{3}$ and

$-\dfrac{3}{2} - i\sqrt{3}$.

8. $4x^2 - 4x - 31 = 0$

$a = 4, b = -4, c = -31$

$x = \dfrac{-b \pm \sqrt{b^2 - 4ac}}{2a}$

$x = \dfrac{-(-4) \pm \sqrt{(-4)^2 - 4(4)(-31)}}{2(4)}$

$x = \dfrac{4 \pm \sqrt{16 + 496}}{8} = \dfrac{4 \pm \sqrt{512}}{8}$

$x = \dfrac{4 \pm 16\sqrt{2}}{8} = \dfrac{1 \pm 4\sqrt{2}}{2}$

$x = \dfrac{1 + 4\sqrt{2}}{2} \qquad x = \dfrac{1 - 4\sqrt{2}}{2}$

The solutions are $\dfrac{1 + 4\sqrt{2}}{2}$ and $\dfrac{1 - 4\sqrt{2}}{2}$.

9. $x^2 + 8x + 25 = 0$

$a = 1, b = 8, c = 25$

$x = \dfrac{-b \pm \sqrt{b^2 - 4ac}}{2a}$

$x = \dfrac{-(8) \pm \sqrt{(8)^2 - 4(1)(25)}}{2(1)}$

$x = \dfrac{-8 \pm \sqrt{64 - 100}}{2} = \dfrac{-8 \pm \sqrt{-36}}{2}$

$x = \dfrac{-8 \pm 6i}{2} = -4 \pm 3i$

$x = -4 + 3i \qquad x = -4 - 3i$

The solutions are $-4 + 3i$ and $-4 - 3i$.

10. $x^4 + 8x^2 - 20 = 0$

$(x^2)^2 + 8(x^2) - 20 = 0$

$u^2 + 8u - 20 = 0$

$(u + 10)(u - 2) = 0$

$u + 10 = 0 \quad u - 2 = 0$

$u = -10 \qquad u = 2$

Replace u with x^2.

$x^2 = -10 \qquad\qquad x^2 = 2$

$\sqrt{x^2} = \sqrt{-10} \qquad \sqrt{x^2} = \sqrt{2}$

$x = \pm i\sqrt{10} \qquad\qquad x = \pm\sqrt{2}$

The solutions are $i\sqrt{10}$, $-i\sqrt{10}$, $\sqrt{2}$ and $-\sqrt{2}$.

11. $\sqrt{2x + 1} - \sqrt{x + 1} = 2$

$\sqrt{2x + 1} = \sqrt{x + 1} + 2$

$\left(\sqrt{2x + 1}\right)^2 = \left(\sqrt{x + 1} + 2\right)^2$

$2x + 1 = x + 1 - 4\sqrt{x + 1} + 4$

$x - 4 = -4\sqrt{x + 1}$

$(x - 4)^2 = \left(-4\sqrt{x + 1}\right)^2$

$x^2 - 8x + 16 = 16x + 16$

$x^2 - 24x = 0$

$x(x - 24) = 0$

$x = 0 \qquad x - 24 = 0$

$\qquad\qquad x = 24$

The solution is 24.

12. $\dfrac{r}{r+1} - \dfrac{2}{r} = \dfrac{3}{10}$

$\left(10r(r+1)\right)\left(\dfrac{r}{r+1} - \dfrac{2}{r}\right) = \left(\dfrac{3}{10}\right)(10r(r+1))$

$10r^2 - 2(10r + 10) = 3r(r + 1)$

$10r^2 - 20r - 20 = 3r^2 + 3r$

$7r^2 - 23r - 20 = 0$

$(7r + 5)(r - 4) = 0$

$7r + 5 = 0 \quad r - 4 = 0$

$7r = -5 \qquad r = 4$

$r = -\dfrac{5}{7}$

The solution is $-\dfrac{5}{7}$ and 4.

Section 8.4

Concept Check

1. $\dfrac{1}{t}$

3. Down: $r + 2$
Up: $r - 2$

Objective A Exercises

5. Strategy: To find the maximum safe speed, substitute for d and solve for v.
Solution: $d = 0.04v^2 + 0.5v$

$60 = 0.04v^2 + 0.5v$

$0 = 0.04v^2 + 0.5v - 60$

$v = \dfrac{-b \pm \sqrt{b^2 - 4ac}}{2a}$

$v = \dfrac{-0.5 \pm \sqrt{(0.5)^2 - 4(0.04)(-60)}}{2(0.04)}$

$v = \dfrac{-0.5 \pm \sqrt{9.85}}{0.08}$

$v \approx 33$

$v \approx -45$

Since the speed cannot be a negative number the maximum speed is 33 mph.

7. Strategy: To find the time it takes for the projectile to return to Earth, substitute for s and v_0 and solve for t.

Solution: $s = v_0 t - 16t^2$

$0 = 200t - 16t^2$

$0 = 8t(25 - 2t)$

$8t = 0 \quad 25 - 2t = 0$

$t = 0 \qquad t = 12.5$

The projectile will take 12.5 s to return to Earth.

9. Strategy: Substitute the given value for H and V and solve for the length of the side. In a square base, $L = W$.

Solution: $V = LWH$
Let x represent the length of the side of the square base. $V = x^2 H$

$971,199 = x^2(31)$

$x^2 = 31,329$

$x = 177$

The length of a side of the square base is 177 m.

© 2014 Cengage Learning. All Rights Reserved. May not be scanned, copied or duplicated, or posted to a publicly accessible website, in whole or in part.

11. **Strategy:** Let t represent the time it takes the smaller pipe to fill the tank.
The time it takes the larger pipe to fill the tank is $t - 6$.

	Rate	Time	Part
Smaller pipe	$\dfrac{1}{t}$	4	$\dfrac{4}{t}$
Larger pipe	$\dfrac{1}{t-6}$	4	$\dfrac{4}{t-6}$

The sum of the parts of the task completed by each pipe equals 1.

Solution: $\dfrac{4}{t} + \dfrac{4}{t-6} = 1$

$t(t-6)\left(\dfrac{4}{t} + \dfrac{4}{t-6}\right) = 1t(t-6)$

$4(t-6) + 4t = t^2 - 6t$

$4t - 24 + 4t = t^2 - 6t$

$t^2 - 14t + 24 = 0$

$(t-12)(t-2) = 0$

$t - 12 = 0 \quad t - 2 = 0$

$t = 12 \qquad t = 2$

$t - 6 = 12 - 6 = 6$

$t - 6 = 2 - 6 = -4$

$t = 2$ is not possible since time cannot be a negative number. It will take the smaller pipe 12 min and the larger pipe 6 min.

13. **Strategy:** Let t represent the time it takes the faster computer working alone.

	Rate	Time	Part
Slower computer	$\dfrac{1}{t+4}$	3	$\dfrac{3}{t+4}$
Faster computer	$\dfrac{1}{t}$	1	$\dfrac{1}{t}$

The sum of the parts of the task completed by each computer equals 1.

Solution: $\dfrac{1}{t} + \dfrac{3}{t+4} = 1$

$t(t+4)\left(\dfrac{1}{t} + \dfrac{3}{t+4}\right) = 1t(t+4)$

$t + 4 + 3t = t^2 + 4t$

$4t + 4 = t^2 + 4t$

$t^2 - 4 = 0$

$(t+2)(t-2) = 0$

$t + 2 = 0 \quad t - 2 = 0$

$t = -2 \qquad t = 2$

Time cannot be a negative number. It will take the faster computer 2 h working alone.

15. Strategy: Let t represent the time it takes the experienced carpenter. The time it takes the apprentice is $t + 2$.

	Rate	Time	Part
Experienced carpenter	$\dfrac{1}{t}$	2	$\dfrac{2}{t}$
Apprentice carpenter	$\dfrac{1}{t+2}$	4	$\dfrac{4}{t+2}$

The sum of the parts of the task completed by each carpenter equals 1.

Solution: $\dfrac{2}{t} + \dfrac{4}{t+2} = 1$

$$t(t+2)\left(\frac{2}{t} + \frac{4}{t+2}\right) = 1t(t+2)$$

$$2(t+2) + 4t = t^2 + 2t$$

$$2t + 4 + 4t = t^2 + 2t$$

$$t^2 - 4t - 4 = 0$$

$$t = \frac{-b \pm \sqrt{b^2 - 4ac}}{2a}$$

$$t = \frac{-(-4) \pm \sqrt{(-4)^2 - 4(1)(-4)}}{2(1)}$$

$$t = \frac{4 \pm \sqrt{32}}{2} = 2 \pm 2\sqrt{2}$$

$$t \approx 4.8$$

$$t \approx -0.8$$

Time cannot be a negative number. It will take the apprentice carpenter $t + 2 \approx 6.8$ h working alone.

17. Strategy: Let r represent the rate of the wind.

	Distance	Rate	Time
With wind	4000	$1320 + r$	$\dfrac{4000}{1320 + r}$
Against wind	4000	$1320 - r$	$\dfrac{4000}{1320 - r}$

It took 0.5 h less time to make the return trip.

Solution: $\dfrac{4000}{1320 - r} - \dfrac{4000}{1320 + r} = 0.5$

$$(1320 - r)(1320 + r)\left(\frac{4000}{1320 - r} - \frac{4000}{1320 + r}\right) = 0.5(1320 - r)(1320 + r)$$

$$4000(1320 + r) - 4000(1320 - r) = 0.5(1{,}742{,}400 - r^2)$$

$$5{,}280{,}000 + 4000r - 5{,}280{,}000 + 4000r = 871{,}200 - 0.5r^2$$

$$8000r = 871{,}200 - 0.5r^2$$

$$0.5r^2 + 8000r - 871{,}200 = 0$$

$$r = \frac{-b \pm \sqrt{b^2 - 4ac}}{2a}$$

$$r = \frac{-(8000) \pm \sqrt{(8000)^2 - 4(0.5)(-871,200)}}{2(0.5)}$$

$$r = \frac{-8000 \pm \sqrt{65,742,400}}{1}$$

$$r \approx 108$$

$$r \cong -16,108$$

Since the rate cannot be a negative number. The rate of the wind was approximately 108 mph.

19. Strategy: Let r represent the rate of the jet stream.

	Distance	Rate	Time
With jet stream	3660	$630 + r$	$\dfrac{3660}{630+r}$
Against jet stream	3660	$630 - r$	$\dfrac{3660}{630-r}$

It took 1.75 h less time to make the trip flying with the jet stream.

Solution: $\dfrac{3660}{630 - r} - \dfrac{3660}{630 + r} = 1.75$

$$(630 + r)(630 - r)\left(\frac{3660}{630 - r} - \frac{3660}{630 + r}\right) = 1.75(630 + r)(630 - r)$$

$$3660(630 + r) - 3660(630 - r) = 1.75(396,900 - r^2)$$

$$2,305,800 + 3660r - 2,305,800 + 3660r = 694,575 - 1.75r^2$$

$$7320r = 694,575 - 1.75r^2$$

$$1.75r^2 + 7320r - 694,575 = 0$$

$$r = \frac{-b \pm \sqrt{b^2 - 4ac}}{2a}$$

$$r = \frac{-(7320) \pm \sqrt{(7320)^2 - 4(1.75)(-694,575)}}{2(1.75)}$$

$$r = \frac{-7320 \pm \sqrt{58,444,425}}{3.5}$$

$$r \approx 93$$

$$r \approx -4276$$

The rate cannot be a negative number. The rate of the jet stream is 93 mph.

21. Strategy: Let x represent the number of apartments.
The monthly rent is $1200 + 100x$.
The number of units rented is $100 - x$.

Solution:
$(1200 + 100x)(100 - x) = 153600$

$-100x^2 + 8800x + 120000 = 153600$

$100x^2 - 8800x - 33600 = 0$

$x^2 - 88x + 336 = 0$

$(x - 4)(x - 84) = 0$

$x - 4 = 0 \quad x - 84 = 0$

$x = 4 \qquad x = 84$

$100 - 4 = 96 \quad 1200 + 100(4) = 1600$
96 units at \$1600/month
$100 - 84 = 16 \quad 1200 + 100(84) = 9600$
16 units at \$9600/month

23. Strategy: Let x represent a side of the square base of the box.
The volume of the box is 49,000 cm³.

Solution: $V = LWH$
$49,000 = (x)(x)(10)$

$49,000 = 10x^2$

$x^2 = 4900$

$x = 70$
The side of the original square is 20 cm more than side x.
$x + 20 = 70 + 20 = 90$
The dimensions of the original square base are 90 cm by 90 cm.

25. Strategy: Let x represent the width of the rectangle.
The length of the rectangle is $40 - x$.
The area of the rectangle is 300 f².

Solution: $A = LW$
$300 = x(40 - x)$

$300 = 40x - x^2$

$x^2 - 40x + 300 = 0$

$(x - 10)(x - 30) = 0$

$x - 10 = 0 \quad x - 30 = 0$

$x = 10 \qquad x = 30$
$40 - x = 40 - 10 = 30$
$40 - x = 40 - 30 = 10$
The dimensions of the rectangle are 10 ft by 30 ft.

Critical Thinking

27. Strategy: Use the Pythagorean formula $a^2 + b^2 = c^2$ with $a = 1.5$, $b = 3.5$ and $c = x + 1.5$.

Solution: $a^2 + b^2 = c^2$
$(1.5)^2 + (3.5)^2 = (x + 1.5)^2$

$14.5 = (x + 1.5)^2$

$\sqrt{14.5} = \sqrt{(x + 1.5)^2}$

$\pm 3.8 \approx x + 1.5$

$3.8 = x + 1.5 \quad -3.18 = x + 1.5$

$x = 2.3 \qquad x = -5.3$
Distance cannot be a negative number.
The bottom of the scoop of ice cream is 2.3 in. from the bottom of the cone.

Section 8.5

Concept Check
1. It must be true that $x - 3 > 0$ and $x - 5 > 0$ or the $x - 3 < 0$ and $x - 5 < 0$. In other words, either both factors are positive or both factors are negative.

3. a) $x = 2$ True
   b) $x = -2$ False
   c) $x = -3$ False

5. a) $x = 2$ True
    b) $x = 3$ False
    c) $x = -1$ True

Objective A Exercises

7. $(x-4)(x+2) > 0$

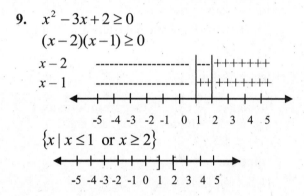

$\{x \mid x < -2 \ \cup \ x > 4\}$

9. $x^2 - 3x + 2 \geq 0$
     $(x-2)(x-1) \geq 0$

$\{x \mid x \leq 1 \ \text{or} \ x \geq 2\}$

11. $x^2 - x - 12 < 0$
     $(x-4)(x+3) < 0$

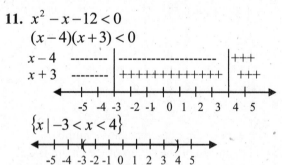

$\{x \mid -3 < x < 4\}$

13. $(x-1)(x+2)(x-3) < 0$

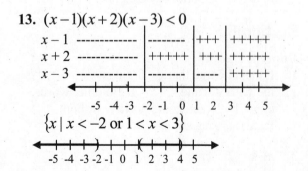

$\{x \mid x < -2 \ \text{or} \ 1 < x < 3\}$

15. $(x+4)(x-2)(x-1) \geq 0$

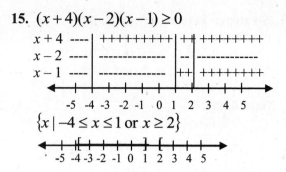

$\{x \mid -4 \leq x \leq 1 \ \text{or} \ x \geq 2\}$

17. $\dfrac{x-4}{x+2} > 0$

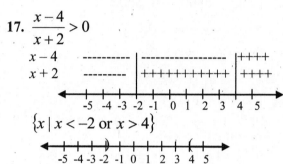

$\{x \mid x < -2 \ \text{or} \ x > 4\}$

19. $\dfrac{x-3}{x+1} \leq 0$

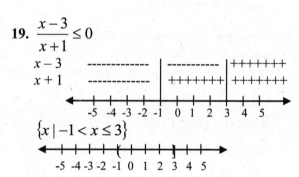

$\{x \mid -1 < x \leq 3\}$

21. $\dfrac{(x-1)(x+2)}{x-3} \leq 0$

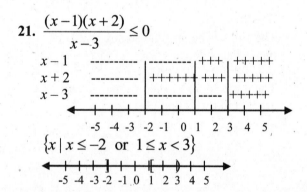

$\{x \mid x \leq -2 \ \text{or} \ 1 \leq x < 3\}$

23. $x^2 - 16 > 0$
$(x+4)(x-4) > 0$

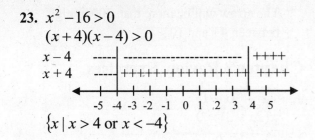

$\{x \mid x > 4 \text{ or } x < -4\}$

25. $x^2 - 9x \le 36$

$x^2 - 9x - 36 \le 0$

$(x-12)(x+3) \le 0$

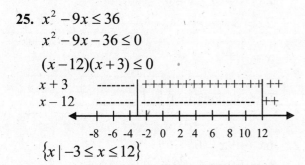

$\{x \mid -3 \le x \le 12\}$

27. $4x^2 - 8x + 3 < 0$
$(2x-3)(2x-1) < 0$

$2x - 1$ $\cdots\cdots\cdots$ $+++++++++$
$2x - 3$ $\cdots\cdots\cdots$ $+++++++$

-5 -4 -3 -2 -1 0 1 2 3 4 5

$\left\{x \mid \dfrac{1}{2} < x < \dfrac{3}{2}\right\}$

29. $\dfrac{3}{x-1} < 2$

$\dfrac{3}{x-1} - 2 < 0$

$\dfrac{3}{x-1} - \dfrac{2x-2}{x-1} < 0$

$\dfrac{-2x+5}{x-1} < 0$

$-2x + 5$ $++++++++++++$ $+++$ $\cdots\cdots$
$x - 1$ $\cdots\cdots\cdots\cdots$ $+++$ $++++++$

-5 -4 -3 -2 -1 0 1 2 3 4 5

$\left\{x \mid x < 1 \text{ or } x > \dfrac{5}{2}\right\}$

31. $\dfrac{x-2}{(x+1)(x-1)} \le 0$

$x - 2$ $\cdots\cdots\cdots\cdots$ $\cdots\cdots$ $\cdots$ $+++++++$
$x + 1$ $\cdots\cdots\cdots\cdots$ $++++$ $++$ $+++++++$
$x - 1$ $\cdots\cdots\cdots\cdots$ $\cdots\cdots$ $++$ $+++++++$

-5 -4 -3 -2 -1 0 1 2 3 4 5

$\{x \mid x < -1 \text{ or } 1 < x \le 2\}$

33. $\dfrac{x}{2x-1} \ge 1$

$\dfrac{x}{2x-1} - 1 \ge 0$

$\dfrac{x}{2x-1} - \dfrac{2x-1}{2x-1} \ge 0$

$\dfrac{-x+1}{2x-1} \ge 0$

$-x + 1$ $++++++++++$ $+$ $\cdots\cdots\cdots\cdots$
$2x - 1$ $\cdots\cdots\cdots\cdots$ $+$ $++++++++++$

-5 -4 -3 -2 -1 0 1 2 3 4 5

$\left\{x \mid \dfrac{1}{2} < x \le 1\right\}$

35. $\dfrac{3}{x-5} > \dfrac{1}{x+1}$

$\dfrac{3}{x-5} - \dfrac{1}{x+1} > 0$

$\dfrac{3(x+1)}{(x-5)(x+1)} - \dfrac{(x-5)}{(x-5)(x+1)} > 0$

$\dfrac{3x+3-x+5}{(x-5)(x+1)} > 0$

$\dfrac{2x+8}{(x-5)(x+1)} > 0$

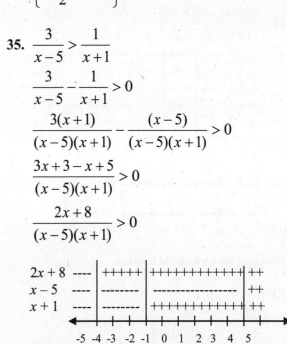

$\{x \mid x > 5 \text{ or } -4 < x < -1\}$

Critical Thinking

37. $(x-1)(x+3)(x-2)(x-4) \geq 0$

$x-1$	--------	------------	++	+++++	+++
$x+3$	--------	+++++++	++	+++++	+++
$x-2$	--------	------------	---	++++	+++
$x-4$	--------	------------	---	------	+++

-5 -4 -3 -2 -1 0 1 2 3 4 5

$\{x \mid x \leq -3 \text{ or } 1 \leq x \leq 2 \text{ or } x \geq 4\}$

-5 -4 -3 -2 -1 0 1 2 3 4 5

39. $(x^2 + 2x - 3)(x^2 + 3x + 2) \geq 0$

$(x-1)(x+3)(x+1)(x+2) \geq 0$

$x-1$	--------	---	---	------	+++++++++++
$x+3$	--------	++	++	+++	+++++++++++
$x+2$	--------	---	++	+++	+++++++++++
$x+1$	--------	---	---	+++	+++++++++++

-5 -4 -3 -2 -1 0 1 2 3 4 5

$\{x \mid x \leq -3 \text{ or } -2 \leq x \leq -1 \text{ or } x \geq 1\}$

-5 -4 -3 -2 -1 0 1 2 3 4 5

41. $\dfrac{x^2(3-x)(2x+1)}{(x+4)(x+2)} \geq 0$

x^2	+++	++++	++	+++++++	+++++
$3-x$	+++	+++	++	+++++++	--------
$2x+1$	----	--------	-----	+++++++	+++++
$x+4$	----	+++	+++	+++++++	+++++++
$x+2$	----	-----	+++	+++++++	+++++++

-5 -4 -3 -2 -1 0 1 2 3 4 5

$\left\{x \mid -4 < x < -2 \text{ or } -\dfrac{1}{2} \leq x \leq 3\right\}$

-5 -4 -3 -2 -1 0 1 2 3 4 5

Projects or Group Activities

43. $d = rt - 5t^2$

$70t - 5t^2 > 200$

$-5t^2 + 70t - 200 > 0$

$t^2 - 14t + 40 < 0$

$(t-4)(t-10) < 0$

$t-4 < 0 \qquad t-10 < 0$

$t < 4 \qquad\quad t < 10$

The arrow will be more than 200 m high between 4 s and 10 s.

Chapter 8 Review Exercises

1. $2x^2 - 3x = 0$

$x(2x-3) = 0$

$2x - 3 = 0 \quad x = 0$

$2x = 3$

$x = \dfrac{3}{2}$

The solutions are 0 and $\dfrac{3}{2}$.

2. $6x^2 + 9x = 6$

$6x^2 + 9x - 6 = 0$

$3(2x^2 + 3x - 2) = 0$

$3(2x-1)(x+2) = 0$

$2x - 1 = 0 \quad x + 2 = 0$

$2x = 1 \qquad\quad x = -2$

$x = \dfrac{1}{2}$

The solutions are -2 and $\dfrac{1}{2}$.

3. $x^2 = 48$

$\sqrt{x^2} = \sqrt{48}$

$x = \pm\sqrt{48} = \pm 4\sqrt{3}$

The solutions are $4\sqrt{3}$ and $-4\sqrt{3}$.

4. $\left(x + \dfrac{1}{2}\right)^2 + 4 = 0$

$\left(x + \dfrac{1}{2}\right)^2 = -4$

$\sqrt{\left(x + \dfrac{1}{2}\right)^2} = \sqrt{-4}$

$x + \dfrac{1}{2} = \pm\sqrt{-4} = \pm 2i$

$$x + \frac{1}{2} = 2i \quad x + \frac{1}{2} = -2i$$

$$x = 2i - \frac{1}{2} \quad x = -2i - \frac{1}{2}$$

The solutions are $2i - \frac{1}{2}$ and $-2i - \frac{1}{2}$.

5. $x^2 + 4x + 3 = 0$

$x^2 + 4x = -3$

$x^2 + 4x + 4 = -3 + 4$

$(x + 2)^2 = 1$

$\sqrt{(x + 2)^2} = \sqrt{1}$

$x + 2 = \pm 1$

$x + 2 = 1 \quad x + 2 = -1$

$x = -1 \quad\quad x = -3$

The solutions are -1 and -3.

6. $7x^2 - 14x + 3 = 0$

$7x^2 - 14x = -3$

$\frac{1}{7}(7x^2 - 14x) = \frac{1}{7}(-3)$

$x^2 - 2x = -\frac{3}{7}$

$x^2 - 2x + 1 = -\frac{3}{7} + 1$

$(x - 1)^2 = \frac{4}{7}$

$\sqrt{(x - 1)^2} = \sqrt{\frac{4}{7}}$

$x - 1 = \pm\sqrt{\frac{4}{7}} = \pm\frac{2\sqrt{7}}{7}$

$x - 1 = \frac{2\sqrt{7}}{7} \quad x - 1 = -\frac{2\sqrt{7}}{7}$

$x = 1 + \frac{2\sqrt{7}}{7} \quad x = 1 - \frac{2\sqrt{7}}{7}$

The solutions are $\frac{7 + 2\sqrt{7}}{7}$ and $\frac{7 - 2\sqrt{7}}{7}$.

7. $12x^2 - 25x + 12 = 0$

$a = 12, b = -25, c = 12$

$x = \frac{-b \pm \sqrt{b^2 - 4ac}}{2a}$

$x = \frac{-(-25) \pm \sqrt{(-25)^2 - 4(12)(12)}}{2(12)}$

$x = \frac{25 \pm \sqrt{625 - 576}}{24}$

$x = \frac{25 + 7}{24} = \frac{32}{24} = \frac{4}{3}$

$x = \frac{25 - 7}{24} = \frac{18}{24} = \frac{3}{4}$

$x = \frac{25 \pm \sqrt{49}}{24}$

$x = \frac{25 \pm 7}{24}$

The solutions are $\frac{4}{3}$ and $\frac{3}{4}$.

8. $x^2 - x + 8 = 0$

$a = 1, b = -1, c = 8$

$x = \frac{-b \pm \sqrt{b^2 - 4ac}}{2a}$

$x = \frac{-(-1) \pm \sqrt{(-1)^2 - 4(1)(8)}}{2(1)}$

$x = \frac{1 \pm \sqrt{1 - 32}}{2}$

$x = \frac{1 \pm \sqrt{-31}}{2}$

$x = \frac{1 \pm i\sqrt{31}}{2}$

The solutions are $\frac{1 + i\sqrt{31}}{2}$ and $\frac{1 - i\sqrt{31}}{2}$.

9. $(x - r_1)(x - r_2) = 0$

$(x - 0)(x - (-3)) = 0$

$x(x + 3) = 0$

$x^2 + 3x = 0$

10. $(x - r_1)(x - r_2) = 0$

$\left(x - \dfrac{3}{4}\right)\left(x - (-\dfrac{2}{3})\right) = 0$

$\left(x - \dfrac{3}{4}\right)\left(x + \dfrac{2}{3}\right) = 0$

$x^2 - \dfrac{1}{12}x - \dfrac{1}{2} = 0$

$12\left(x^2 - \dfrac{1}{12}x - \dfrac{1}{2}\right) = 0(12)$

$12x^2 - x - 6 = 0$

11. $x^2 - 2x + 8 = 0$

$x^2 - 2x = -8$

$x^2 - 2x + 1 = -8 + 1$

$(x - 1)^2 = -7$

$\sqrt{(x - 1)^2} = \sqrt{-7}$

$x - 1 = \pm\sqrt{-7} = \pm i\sqrt{7}$

$x = 1 \pm i\sqrt{7}$

The solutions are $1 + i\sqrt{7}$ and $1 - i\sqrt{7}$.

12. $(x - 2)(x + 3) = x - 10$

$x^2 + x - 6 = x - 10$

$x^2 = -4$

$\sqrt{x^2} = \sqrt{-4}$

$x = \pm\sqrt{-4}$

$x = \pm 2i$

The solutions are $2i$ and $-2i$.

13. $3x(x - 3) = 2x - 4$

$3x^2 - 9x = 2x - 4$

$3x^2 - 11x + 4 = 0$

$a = 3, b = -11, c = 4$

$x = \dfrac{-b \pm \sqrt{b^2 - 4ac}}{2a}$

$x = \dfrac{-(-11) \pm \sqrt{(-11)^2 - 4(3)(4)}}{2(3)}$

$x = \dfrac{11 \pm \sqrt{121 - 48}}{6}$

$x = \dfrac{11 \pm \sqrt{73}}{6}$

The solutions are $\dfrac{11 + \sqrt{73}}{6}$ and $\dfrac{11 - \sqrt{73}}{6}$.

14. $3x^2 - 5x + 3 = 0$

$a = 3, b = -5, c = 3$

$b^2 - 4ac$

$(-5)^2 - 4(3)(3) = 25 - 36 = -11$

$-11 > 0$

Since the discriminant is less than zero, the equation has two complex number solutions.

15. $(x + 3)(2x - 5) < 0$

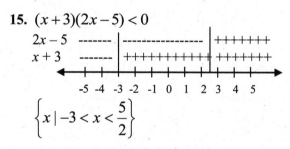

$\left\{x \,|\, -3 < x < \dfrac{5}{2}\right\}$

16. $(x - 2)(x + 4)(2x + 3) \le 0$

$\left\{x \,|\, x \le -4 \text{ 1.or } -\dfrac{3}{2} \le x \le 2\right\}$

17. $x^{2/3} + x^{1/3} - 12 = 0$

$\left(x^{1/3}\right)^2 + x^{1/3} - 12 = 0$

$u^2 + u - 12 = 0$

$(u + 4)(u - 3) = 0$

$u + 4 = 0 \quad u - 3 = 0$

$u = -4 \qquad u = 3$

Replace u with $x^{1/3}$.

$x^{1/3} = -4 \qquad x^{1/3} = 3$

$\left(x^{1/3}\right)^3 = (-4)^3 \quad \left(x^{1/3}\right)^3 = 3^3$

$x = -64 \qquad\qquad x = 27$

The solutions are -64 and 27.

18. $2(x-1) + 3\sqrt{x-1} - 2 = 0$

$2\left(\sqrt{x-1}\right)^2 + 3\sqrt{x-1} - 2 = 0$

$2u^2 + 3u - 2 = 0$

$(2u - 1)(u + 2) = 0$

$2u - 1 = 0 \qquad u + 2 = 0$

$2u = 1 \qquad\quad u = -2$

$u = \dfrac{1}{2}$

Replace u with $\sqrt{x-1}$.

$\sqrt{x-1} = \dfrac{1}{2} \qquad\qquad \sqrt{x-1} = 2$

$\left(\sqrt{x-1}\right)^2 = \left(\dfrac{1}{2}\right)^2 \quad \left(\sqrt{x-1}\right) = 2^2$

$x - 1 = \dfrac{1}{4} \qquad\qquad x - 1 = 4$

$x = \dfrac{5}{4} \qquad\qquad\quad x = 5$

5 does not check as a solution.

The solution is $\dfrac{5}{4}$.

19. $3x = \dfrac{9}{x-2}$

$3x(x-2) = \dfrac{9}{x-2}(x-2)$

$3x^2 - 6x = 9$

$3x^2 - 6x - 9 = 0$

$3(x^2 - 2x - 3) = 0$

$3(x-3)(x+1) = 0$

$x - 3 = 0 \quad x + 1 = 0$

$x = 3 \qquad x = -1$

The solutions are -1 and 3.

20. $\dfrac{3x+7}{x+2} + x = 3$

$\dfrac{3x+7}{x+2} = 3 - x$

$(x+2)\left(\dfrac{3x+7}{x+2}\right) = (3-x)(x+2)$

$3x + 7 = 3x + 6 - x^2 - 2x$

$x^2 + 2x + 1 = 0$

$(x+1)^2 = 0$

$\sqrt{(x+1)^2} = \sqrt{0}$

$x + 1 = 0$

$x = -1$

The solution is -1.

21. $\dfrac{x-2}{2x-3} \ge 0$

$$
\begin{array}{l}
x - 2 \quad \text{-------------} \ |\ \text{+++++++} \\
2x - 3 \quad \text{-------------} \ |\ \text{+++++++}
\end{array}
$$

-5 -4 -3 -2 -1 0 1 2 3 4 5

$\left\{ x \mid x < \dfrac{3}{2} \ \text{or} \ x \ge 2 \right\}$

-5 -4 -3 -2 -1 0 1 2 3 4 5

22. $\dfrac{(2x-1)(x+3)}{x-4} \le 0$

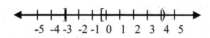

$$\left\{ x \mid x \le -3 \ \text{or} \ \frac{1}{2} \le x < 4 \right\}$$

23. $x = \sqrt{x} + 2$

$x - 2 = \sqrt{x}$

$(x-2)^2 = \left(\sqrt{x}\right)^2$

$x^2 - 4x + 4 = x$

$x^2 - 5x + 4 = 0$

$(x-4)(x-1) = 0$

$x - 4 = 0 \quad x - 1 = 0$

$x = 4 \qquad x = 1$

1 does not check as a solution.
The solution is 4.

24. $2x = \sqrt{5x + 24} + 3$

$2x - 3 = \sqrt{5x + 24}$

$(2x-3)^2 = \left(\sqrt{5x+24}\right)^2$

$4x^2 - 12x + 9 = 5x + 24$

$4x^2 - 17x - 15 = 0$

$(4x+3)(x-5) = 0$

$4x + 3 = 0 \quad x - 5 = 0$

$4x = -3 \qquad x = 5$

$x = -\dfrac{3}{4}$

$-\dfrac{3}{4}$ does not check as a solution.

The solution is 5.

25. $\dfrac{x-2}{2x+3} - \dfrac{x-4}{x} = 2$

$(2x+3)(x)\left(\dfrac{x-2}{2x+3} - \dfrac{x-4}{x}\right) = 2(2x+3)(x)$

$x(x-2) - (x-4)(2x+3) = 2x(2x+3)$

$x^2 - 2x - 2x^2 - 3x + 8x + 12 = 4x^2 + 6x$

$0 = 5x^2 + 3x - 12$

$a = 5, b = 3, c = -12$

$x = \dfrac{-b \pm \sqrt{b^2 - 4ac}}{2a}$

$x = \dfrac{-3 \pm \sqrt{3^2 - 4(5)(-12)}}{2(5)}$

$x = \dfrac{-3 \pm \sqrt{9 + 240}}{10}$

$x = \dfrac{-3 \pm \sqrt{249}}{10}$

The solutions are $\dfrac{-3 + \sqrt{249}}{10}$ and

$\dfrac{-3 - \sqrt{249}}{10}$.

26. $1 - \dfrac{x+4}{2-x} = \dfrac{x-3}{x+2}$

$(2-x)(x+2)\left(1 - \dfrac{x+4}{2-x}\right) = (2-x)(x+2)\dfrac{x-3}{x+2}$

$(2-x)(x+2) - (x+2)(x+4) = (2-x)(x-3)$

$4 - x^2 - x^2 - 6x - 8 = -x^2 + 5x - 6$

$x^2 + 11x - 2 = 0$

$a = 1, b = 11, c = -2$

$x = \dfrac{-b \pm \sqrt{b^2 - 4ac}}{2a}$

$$x = \frac{-11 \pm \sqrt{11^2 - 4(1)(-2)}}{2(1)}$$

$$x = \frac{-11 \pm \sqrt{121 + 8}}{2}$$

$$x = \frac{-11 \pm \sqrt{129}}{2}$$

The solutions are $\dfrac{-11 + \sqrt{129}}{2}$ and

$\dfrac{-11 - \sqrt{129}}{2}$.

27. $(x - r_1)(x - r_2) = 0$

$$\left(x - \frac{1}{3}\right)(x - (-3)) = 0$$

$$\left(x - \frac{1}{3}\right)(x + 3) = 0$$

$$x^2 + \frac{8}{3}x - 1 = 0$$

$$3\left(x^2 + \frac{8}{3}x - 1\right) = 0(3)$$

$$3x^2 + 8x - 3 = 0$$

28. $2x^2 + 9x = 5$

$2x^2 + 9x - 5 = 0$

$(2x - 1)(x + 5) = 0$

$2x - 1 = 0 \quad x + 5 = 0$

$2x = 1 \qquad x = -5$

$$x = \frac{1}{2}$$

The solutions are -5 and $\dfrac{1}{2}$.

29. $2(x + 1)^2 - 36 = 0$

$2(x + 1)^2 = 36$

$(x + 1)^2 = 18$

$(x + 1)^2 = \left(\sqrt{18}\right)^2$

$x + 1 = \pm\sqrt{18} = \pm 3\sqrt{2}$

$x = -1 \pm 3\sqrt{2}$

The solutions are $-1 + 3\sqrt{2}$ and $-1 - 3\sqrt{2}$.

30. $x^2 + 6x + 10 = 0$

$a = 1, b = 6, c = 10$

$$x = \frac{-b \pm \sqrt{b^2 - 4ac}}{2a}$$

$$x = \frac{-6 \pm \sqrt{6^2 - 4(1)(10)}}{2(1)}$$

$$x = \frac{-6 \pm \sqrt{36 - 40}}{2} = \frac{-6 \pm \sqrt{-4}}{2}$$

$$x = \frac{-6 \pm 2i}{2}$$

$x = -3 \pm i$

The solutions are $-3 + i$ and $-3 - i$.

31. $\dfrac{2}{x - 4} + 3 = \dfrac{x}{2x - 3}$

$$(2x - 3)(x - 4)\left(\frac{2}{x - 4} + 3\right) = (2x - 3)(x - 4)\frac{x}{2x - 3}$$

$2(2x - 3) + 3(x - 4)(2x - 3) = x(x - 4)$

$4x - 6 + 6x^2 - 33x + 36 = x^2 - 4x$

$5x^2 - 25x + 30 = 0$

$5(x^2 - 5x + 6) = 0$

$5(x - 3)(x - 2) = 0$

$x - 3 = 0 \quad x - 2 = 0$

$x = 3 \qquad x = 2$

The solutions are 2 and 3.

32. $x^4 - 28x^2 + 75 = 0$

$(x^2)^2 - 28x^2 + 75 = 0$

$u^2 - 28u + 75 = 0$

$(u - 25)(u - 3) = 0$

$u - 25 = 0 \quad u - 3 = 0$

$u = 25 \quad\quad u = 3$

Replace u with x^2.

$x^2 = 25 \quad\quad x^2 = 3$

$\sqrt{x^2} = \sqrt{25} \quad \sqrt{x^2} = \sqrt{3}$

$x = \pm 5 \quad\quad x = \pm\sqrt{3}$

The solutions are -5, 5, $\sqrt{3}$ and $-\sqrt{3}$.

33. $\sqrt{2x - 1} + \sqrt{2x} = 3$

$\sqrt{2x - 1} = 3 - \sqrt{2x}$

$(\sqrt{2x - 1})^2 = (3 - \sqrt{2x})^2$

$2x - 1 = 9 - 6\sqrt{2x} - 2x$

$-10 = -6\sqrt{2x}$

$5 = 3\sqrt{2x}$

$5^2 = (3\sqrt{2x})^2$

$25 = 18x$

$x = \dfrac{25}{18}$

The solution is $\dfrac{25}{18}$.

34. $2x^{2/3} + 3x^{1/3} - 2 = 0$

$2(x^{1/3})^2 + 3x^{1/3} - 2 = 0$

$2u^2 + 3u - 2 = 0$

$(2u - 1)(u + 2) = 0$

$2u - 1 = 0 \quad u + 2 = 0$

$2u = 1 \quad\quad u = -2$

$u = \dfrac{1}{2}$

Replace u with $x^{1/3}$.

$x^{1/3} = \dfrac{1}{2} \quad\quad\quad x^{1/3} = -2$

$(x^{1/3})^3 = \left(\dfrac{1}{2}\right)^3 \quad (x^{1/3})^3 = (-2)^3$

$x = \dfrac{1}{8} \quad\quad\quad\quad x = -8$

The solutions are -8 and $\dfrac{1}{8}$.

35. $\sqrt{3x - 2} + 4 = 3x$

$\sqrt{3x - 2} = 3x - 4$

$(\sqrt{3x - 2})^2 = (3x - 4)^2$

$3x - 2 = 9x^2 - 24x + 16$

$9x^2 - 27x + 18 = 0$

$9(x^2 - 3x + 2) = 0$

$(x - 2)(x - 1) = 0$

$x - 2 = 0 \quad x - 1 = 0$

$x = 2 \quad\quad x = 1$

1 does not check as a solution.

The solution is 2.

36. $x^2 - 10x + 7 = 0$

$x^2 - 10x = -7$

$x^2 - 10x + 25 = -7 + 25$

$(x - 5)^2 = 18$

$\sqrt{(x - 5)^2} = \sqrt{18}$

$x - 5 = \pm 3\sqrt{2}$

$x = 5 \pm 3\sqrt{2}$

The solutions are $5 + 3\sqrt{2}$ and $5 - 3\sqrt{2}$.

37. $\dfrac{2x}{x-4} + \dfrac{6}{x+1} = 11$

$(x-4)(x+1)\left(\dfrac{2x}{x-4} + \dfrac{6}{x+1}\right) = 11(x-4)(x+1)$

$2x(x+1) + 6(x-4) = 11(x-4)(x+1)$

$2x^2 + 2x + 6x - 24 = 11x^2 - 33x - 44$

$0 = 9x^2 - 41x - 20$

$0 = (9x+4)(x-5)$

$9x+4 = 0 \quad x-5 = 0$

$9x = -4 \quad\quad x = 5$

$x = -\dfrac{4}{9}$

The solutions are $-\dfrac{4}{9}$ and 5.

38. $9x^2 - 3x = 1$

$9x^2 - 3x - 1 = 0$

$a = 9,\, b = -3,\, c = -1$

$x = \dfrac{-b \pm \sqrt{b^2 - 4ac}}{2a}$

$x = \dfrac{-(-3) \pm \sqrt{(-3)^2 - 4(9)(-1)}}{2(9)}$

$x = \dfrac{3 \pm \sqrt{9+36}}{18} = \dfrac{3 \pm \sqrt{45}}{18}$

$x = \dfrac{3 \pm 3\sqrt{5}}{18}$

$x = \dfrac{1 \pm \sqrt{5}}{6}$

The solutions are $\dfrac{1+\sqrt{5}}{6}$ and $\dfrac{1-\sqrt{5}}{6}$.

39. $2x = 4 - 3\sqrt{x-1}$

$2x - 4 = 3\sqrt{x-1}$

$(2x-4)^2 = \left(3\sqrt{x-1}\right)^2$

$4x^2 - 16x + 16 = 9x - 9$

$4x^2 - 25x + 25 = 0$

$(4x-5)(x-5) = 0$

$4x-5 = 0 \quad x-5 = 0$

$4x = 5 \quad\quad x = 5$

$x = \dfrac{5}{4}$

$\dfrac{5}{4}$ does not check as a solution.

The solution is 5.

40. $1 - \dfrac{x+3}{3-x} = \dfrac{x-4}{x+3}$

$(3-x)(x+3)\left(1 - \dfrac{x+3}{3-x}\right) = (3-x)(x+3)\dfrac{x-4}{x+3}$

$(3-x)(x+3) - (x+3)(x+3) = (3-x)(x-4)$

$3x + 9 - x^2 - 3x - x^2 - 6x - 9 = -x^2 + 7x - 12$

$x^2 + 13x - 12 = 0$

$a = 1,\, b = 13,\, c = -12$

$x = \dfrac{-b \pm \sqrt{b^2 - 4ac}}{2a}$

$x = \dfrac{-13 \pm \sqrt{13^2 - 4(1)(-12)}}{2(1)}$

$x = \dfrac{-13 \pm \sqrt{169 + 48}}{2}$

$x = \dfrac{-13 \pm \sqrt{217}}{2}$

The solutions are $\dfrac{-13 + \sqrt{217}}{2}$ and

$-\dfrac{13 - \sqrt{217}}{2}$.

41. $2x^2 - 5x = 6$

$2x^2 - 5x - 6 = 0$

$a = 2, b = -5, c = -6$

$b^2 - 4ac = (-5)^2 - 4(2)(-6) = 73$

$73 > 0$

Since the discriminant is greater than zero the equation has two unequal real number solutions.

42. $x^2 - 3x \le 10$

$x^2 - 3x - 10 \le 0$

$(x - 5)(x + 2) \le 0$

The zeros are -2 and 5. The factors have opposite signs between the zeros. The solution set is $\{x \mid -2 \le x \le 5\}$.

43. Strategy: Let r represent the rate of the rowing in calm water.

	Distance	Rate	Time
With current	16	$r + 2$	$\dfrac{16}{r + 2}$
Against current	16	$r - 2$	$\dfrac{16}{r - 2}$

The total time traveled was 6 h.

Solution: $\dfrac{16}{r + 2} + \dfrac{16}{r - 2} = 6$

$(r - 2)(r + 2)\left(\dfrac{16}{r + 2} + \dfrac{16}{r - 2}\right) = 6(r - 2)(r + 2)$

$16(r - 2) + 16(r + 2) = 6r^2 - 24$

$16r - 32 + 16r + 32 = 6r^2 - 24$

$0 = 6r^2 - 32r - 24$

$0 = 2(3r^2 - 16r - 12)$

$0 = (3r + 2)(r - 6)$

$3r + 2 = 0 \quad r - 6 = 0$

$3r = -2 \qquad r = 6$

$r = -\dfrac{2}{3}$

The rate cannot be a negative number. The rowing rate in calm water is 6 mph.

44. Strategy: Let x represent the width of the rectangle.

The length of the rectangle is $2x + 2$.

The area of the rectangle is 60 cm^2.

Solution: $A = LW$

$60 = x(2x + 2)$

$60 = 2x^2 + 2x$

$0 = 2x^2 + 2x - 60$

$0 = 2(x^2 + x - 30)$

$0 = 2(x + 6)(x - 5)$

$x + 6 = 0 \quad x - 5 = 0$

$x = -6 \qquad x = 5$

The width cannot be a negative number.

$2x + 2 = 2(5) + 2 = 12$

The width is 5 cm.

The length is 12 cm.

45. Strategy: Let t represent the time it takes the new computer to print the payroll.

The time it takes the older computer to print the payroll is $t + 12$.

	Rate	Time	Part
New computer	$\dfrac{1}{t}$	8	$\dfrac{8}{t}$
Older computer	$\dfrac{1}{t + 12}$	8	$\dfrac{8}{t + 12}$

The sum of the parts of the task completed equals 1.

Solution: $\dfrac{8}{t} + \dfrac{8}{t + 12} = 1$

$t(t + 12)\left(\dfrac{8}{t} + \dfrac{8}{t + 12}\right) = 1t(t + 12)$

$8(t + 12) + 8t = t^2 + 12t$

$8t + 96 + 8t = t^2 + 12t$

$t^2 - 4t - 96 = 0$

$(t - 12)(t + 8) = 0$

$t - 12 = 0 \quad t + 8 = 0$

$t = 12 \qquad t = -8$

$t = -8$ is not possible since time cannot be a negative number. Working alone, the new computer can print the payroll in 12 min.

46. Strategy: Let r represent the rate of the first car.
The rate of the second car is $r + 10$.

	Distance	Rate	Time
1st car	200	r	$\dfrac{200}{r}$
2nd car	200	$r+10$	$\dfrac{200}{r+10}$

The second car's time is one hour less than the first car's time.

Solution: $\dfrac{200}{r+10} = \dfrac{200}{r} - 1$

$(r)(r+10)\left(\dfrac{200}{r+10}\right) = (r)(r+10)\left(\dfrac{200}{r} - 1\right)$

$200r = 200(r+10) - r(r+10)$

$200r = 200r + 2000 - r^2 - 10r$

$0 = r^2 - 10r - 2000$

$0 = (r - 40)(r + 50)$

$r - 40 = 0 \quad r + 50 = 0$

$r = 40 \qquad r = -50$

The rate cannot be a negative number.
The rate of the first car is 40 mph.
The rate of the second car is 50 mph.

Chapter 8 Test

1. $3x^2 + 10x = 8$

$3x^2 + 10x - 8 = 0$

$(3x - 2)(x + 4) = 0$

$(3x - 2)(x + 4) = 0$

$3x - 2 = 0 \quad x + 4 = 0$

$3x = 2 \qquad x = -4$

$x = \dfrac{2}{3}$

The solutions are -4 and $\dfrac{2}{3}$.

2. $6x^2 - 5x - 6 = 0$

$2x - 3 = 0 \quad 3x + 2 = 0$

$2x = 3 \qquad 3x = -2$

$x = \dfrac{3}{2} \qquad x = -\dfrac{2}{3}$

The solutions are $\dfrac{3}{2}$ and $-\dfrac{2}{3}$.

3. $(x - r_1)(x - r_2) = 0$

$(x - 3)(x - (-3)) = 0$

$(x - 3)(x + 3) = 0$

$x^2 - 9 = 0$

4. $(x - r_1)(x - r_2) = 0$

$\left(x - \dfrac{1}{2}\right)(x - (-4)) = 0$

$\left(x - \dfrac{1}{2}\right)(x + 4) = 0$

$x^2 + \dfrac{7}{2}x - 2 = 0$

$2\left(x^2 + \dfrac{7}{2}x - 2\right) = 0(2)$

$2x^2 + 7x - 4 = 0$

5. $3(x - 2)^2 - 24 = 0$

$3(x - 2)^2 = 24$

$(x - 2)^2 = 8$

$\sqrt{(x - 2)^2} = \sqrt{8}$

$x - 2 = \pm 2\sqrt{2}$

$x = 2 \pm 2\sqrt{2}$

The solutions are $2 + 2\sqrt{2}$ and $2 - 2\sqrt{2}$.

6. $x^2 - 6x - 2 = 0$

$x^2 - 6x = 2$

$x^2 - 6x + 9 = 2 + 9$

$(x - 3)^2 = 11$

$\sqrt{(x - 3)^2} = \sqrt{11}$

$x - 3 = \pm\sqrt{11}$

$x = 3 \pm \sqrt{11}$

The solutions are $3 + \sqrt{11}$ and $3 - \sqrt{11}$.

7. $3x^2 - 6x = 2$

$\frac{1}{3}(3x^2 - 6x) = 2\left(\frac{1}{3}\right)$

$x^2 - 2x = \frac{2}{3}$

$x^2 - 2x + 1 = \frac{2}{3} + 1$

$(x - 1)^2 = \frac{5}{3}$

$\sqrt{(x - 1)^2} = \sqrt{\frac{5}{3}}$

$x - 1 = \pm\frac{\sqrt{15}}{3}$

$x = 1 \pm \frac{\sqrt{15}}{3}$

$x = \frac{3 \pm \sqrt{15}}{3}$

The solutions are $\frac{3 + \sqrt{15}}{3}$ and $\frac{3 - \sqrt{15}}{3}$.

8. $2x^2 - 2x = 1$

$2x^2 - 2x - 1 = 0$

$a = 2, b = -2, c = -1$

$x = \frac{-b \pm \sqrt{b^2 - 4ac}}{2a}$

$x = \frac{-(-2) \pm \sqrt{(-2)^2 - 4(2)(-1)}}{2(2)}$

$x = \frac{2 \pm \sqrt{4 + 8}}{4} = \frac{2 \pm \sqrt{12}}{4}$

$x = \frac{2 \pm 2\sqrt{3}}{4}$

$x = \frac{1 \pm \sqrt{3}}{2}$

The solutions are $\frac{1 + \sqrt{3}}{2}$ and $\frac{1 - \sqrt{3}}{2}$.

9. $x^2 + 4x + 12 = 0$

$a = 1, b = 4, c = 12$

$x = \frac{-b \pm \sqrt{b^2 - 4ac}}{2a}$

$x = \frac{-4 \pm \sqrt{4^2 - 4(1)(12)}}{2(1)}$

$x = \frac{-4 \pm \sqrt{16 - 48}}{2} = \frac{-4 \pm \sqrt{-32}}{2}$

$x = \frac{-4 \pm 4i\sqrt{2}}{2}$

$x = -2 \pm 2i\sqrt{2}$

The solutions are $-2 + 2i\sqrt{2}$ and $-2 - 2i\sqrt{2}$.

10. $2x + 7x^{1/2} - 4 = 0$

$2\left(x^{1/2}\right)^2 + 7x^{1/2} - 4 = 0$

$2u^2 + 7u - 4 = 0$

$(2u - 1)(u + 4) = 0$

$2u - 1 = 0 \quad u + 4 = 0$

$2u = 1 \qquad u = -4$

$u = \dfrac{1}{2}$

Replace u with $x^{1/2}$.

$x^{1/2} = \dfrac{1}{2} \qquad x^{1/2} = -4$

$\left(x^{1/2}\right)^2 = \left(\dfrac{1}{2}\right)^2 \quad \left(x^{1/2}\right)^2 = (-4)^2$

$x = \dfrac{1}{4} \qquad\qquad x = 16$

16 does not check as a solution.

The solution is $\dfrac{1}{4}$.

11. $x^4 - 4x^2 + 3 = 0$

$\left(x^2\right)^2 - 4x^2 + 3 = 0$

$u^2 - 4u + 3 = 0$

$(u - 1)(u - 3) = 0$

$u - 1 = 0 \quad u - 3 = 0$

$u = 1 \qquad u = 3$

Replace u with x^2.

$x^2 = 1 \qquad\quad x^2 = 3$

$\sqrt{x^2} = \sqrt{1} \quad \sqrt{x^2} = \sqrt{3}$

$x = \pm 1 \qquad\quad x = \pm\sqrt{3}$

The solutions are -1, 1, $\sqrt{3}$ and $-\sqrt{3}$.

12. $\sqrt{2x + 1} + 5 = 2x$

$\sqrt{2x + 1} = 2x - 5$

$\left(\sqrt{2x + 1}\right)^2 = (2x - 5)^2$

$2x + 1 = 4x^2 - 20x + 25$

$4x^2 - 22x + 24 = 0$

$2(2x^2 - 11x + 12) = 0$

$2(2x - 3)(x - 4) = 0$

$2x - 3 = 0 \quad x - 4 = 0$

$2x = 3 \qquad\quad x = 4$

$x = \dfrac{3}{2}$

$\dfrac{3}{2}$ does not check as a solution.

The solution is 4.

13. $\sqrt{x - 2} = \sqrt{x} - 2$

$(\sqrt{x - 2})^2 = \left(\sqrt{x} - 2\right)^2$

$x - 2 = x - 4\sqrt{x} + 4$

$-6 = -4\sqrt{x}$

$3 = 2\sqrt{x}$

$3^2 = \left(2\sqrt{x}\right)^2$

$9 = 4x$

$x = \dfrac{9}{4}$

$\dfrac{9}{4}$ does not check as a solution.

There is no solution.

14. $\dfrac{2x}{x-3} + \dfrac{5}{x-1} = 1$

$(x-3)(x-1)\left(\dfrac{2x}{x-3} + \dfrac{5}{x-1}\right) = 1(x-3)(x-1)$

$2x(x-1) + 5(x-3) = 1(x-3)(x-1)$

$2x^2 - 2x + 5x - 15 = x^2 - 4x + 3$

$x^2 + 7x - 18 = 0$

$(x+9)(x-2) = 0$

$x+9 = 0 \quad x-2 = 0$

$x = -9 \qquad x = 2$

The solutions are -9 and 2.

15. $(x-2)(x+4)(x-4) < 0$

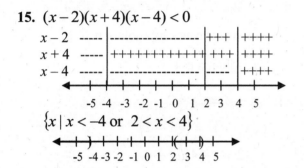

$\{x \mid x < -4 \text{ or } 2 < x < 4\}$

16. $\dfrac{(2x-3)}{x+4} \le 0$

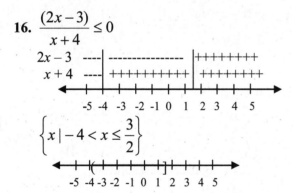

$\left\{x \mid -4 < x \le \dfrac{3}{2}\right\}$

17. $9x^2 + 24x = -16$

$9x^2 + 24x + 16 = 0$

$a = 9, b = 24, c = 16$

$b^2 - 4ac = (24)^2 - 4(9)(16) = 0$

Since the discriminant is equal to zero the equation has two equal real number solutions.

18. Strategy: To find the time when the ball hits the basket substitute 10 ft for h in the equation and solve for t.

Solution: $h = -16t^2 + 32t + 6.5$

$10 = -16t^2 + 32t + 6.5$

$0 = -16t^2 + 32t - 3.5$

$0 = 16t^2 - 32t + 3.5$

$a = 16, b = -32, c = 3.5$

$t = \dfrac{-b \pm \sqrt{b^2 - 4ac}}{2a}$

$t = \dfrac{-(-32) \pm \sqrt{(-32)^2 - 4(16)(3.5)}}{2(16)}$

$t = \dfrac{32 \pm \sqrt{800}}{32}$

$t = 1.88$

$t = 0.12$

We need to find the time it takes to reach the basket after the ball has reached its peak. This occurs 1.88 s after the ball has been released.

19. Strategy: Let t represent the time it takes Cora to stain a bookcase.
The time it takes Clive to stain the bookcase is $t + 6$.

	Rate	Time	Part
Cora	$\dfrac{1}{t}$	4	$\dfrac{4}{t}$
Clive	$\dfrac{1}{t+6}$	4	$\dfrac{4}{t+6}$

The sum of the parts of the task completed equals 1.

Solution: $\dfrac{4}{t} + \dfrac{4}{t+6} = 1$

$t(t+6)\left(\dfrac{4}{t} + \dfrac{4}{t+6}\right) = 1t(t+6)$

$4(t+6) + 4t = t^2 + 6t$

$4t + 24 + 4t = t^2 + 6t$

$t^2 - 2t - 24 = 0$

$(t-6)(t+4) = 0$

$t - 6 = 0 \quad t + 4 = 0$

$t = 6 \qquad t = -4$

$t = -4$ is not possible since time cannot be a negative number. Working alone, it will take Cora 6 h to stain the bookcase.

20. Strategy: Let r represent the rate of the canoe in calm water.

	Distance	Rate	Time
With current	6	$r + 2$	$\dfrac{6}{r+2}$
Against current	6	$r - 2$	$\dfrac{6}{r-2}$

The total time traveled was 4 h.

Solution: $\dfrac{6}{r+2} + \dfrac{6}{r-2} = 4$

$(r-2)(r+2)\left(\dfrac{6}{r+2} + \dfrac{6}{r-2}\right) = 4(r-2)(r+2)$

$6(r-2) + 6(r+2) = 4r^2 - 16$

$6r - 12 + 6r + 12 = 4r^2 - 16$

$0 = 4r^2 - 12r - 16$

$0 = 4(r^2 - 3r - 4)$

$0 = 4(r-4)(r+1)$

$r - 4 = 0 \quad r + 1 = 0$

$r = 4 \qquad r = -1$

The rate cannot be a negative number. The rate of the canoe in calm water is 4 mph.

Cumulative Review Exercises

1. $2a^2 - b^2 \div c^2$

$2(3)^2 - (-4)^2 \div (-2)^2 = 2(9) - 16 \div 4$

$= 18 - 16 \div 4 = 18 - 4$

$= 14$

2. $\dfrac{2x-3}{4} - \dfrac{x+4}{6} = \dfrac{3x-2}{8}$

$24\left(\dfrac{2x-3}{4} - \dfrac{x+4}{6}\right) = 24\left(\dfrac{3x-2}{8}\right)$

$6(2x-3) - 4(x+4) = 3(3x-2)$

$12x - 18 - 4x - 16 = 9x - 6$

$8x - 34 = 9x - 6$

$-x = 28$

$x = -28$

The solution is -28.

3. $(3, -4)$ and $(-1, 2)$

$m = \dfrac{y_2 - y_1}{x_2 - x_1} = \dfrac{2 - (-4)}{-1 - 3} = \dfrac{2+4}{-4} = \dfrac{6}{-4}$

$m = -\dfrac{3}{2}$

4. $x - y = 1$

$y = x - 1$

$m = 1$ and $(1, 2)$

$y - y_1 = m(x - x_1)$

$y - 2 = 1(x - 1)$

$y - 2 = x - 1$

$y = x + 1$

5. $-3x^3y + 6x^2y^2 - 9xy^3 = -3xy(x^2 - 2xy + 3y^2)$

6. $6x^2 - 7x - 20 = (2x - 5)(3x + 4)$

7. $x^2 + xy - 2x - 2y$

$= x(x + y) - 2(x + y)$

$= (x + y)(x - 2)$

8.
$$3x - 4 \overline{)3x^3 - 13x^2 + 0x + 10}$$
quotient: $x^2 - 3x - 4$

$$\underline{3x^3 - 4x^2}$$
$$-9x^2 + 0x$$
$$\underline{-9x^2 + 12x}$$
$$-12x + 10$$
$$\underline{-12x + 16}$$
$$-6$$

$$(3x^3 - 13x^2 + 10) \div (3x - 4) = x^2 - 3x - 4 + \frac{-6}{3x - 4}$$

9.
$$\frac{x^2 + 2x + 1}{8x^2 + 8x} \cdot \frac{4x^3 - 4x^2}{x^2 - 1}$$

$$= \frac{(x+1)(x+1)}{8x(x+1)} \cdot \frac{4x^2(x-1)}{(x+1)(x-1)}$$

$$= \frac{(x+1)(x+1) \cdot 4x^2(x-1)}{8x(x+1) \cdot (x+1)(x-1)}$$

$$= \frac{x}{2}$$

10. $(-2, 3)$ and $2, 5)$

$$d = \sqrt{(x_1 - x_2)^2 + (y_1 - y_2)^2}$$

$$d = \sqrt{(-2 - 2)^2 + (3 - 5)^2}$$

$$d = \sqrt{(-4)^2 + (-2)^2} = \sqrt{16 + 4} = \sqrt{20}$$

$$d = 2\sqrt{5}$$

The distance between the points is $2\sqrt{5}$.

11. $S = \dfrac{n}{2}(a + b)$

$$2S = n(a + b)$$

$$2S = na + nb$$

$$2S - na = nb$$

$$\frac{2S - na}{n} = b$$

12. $-2i(7 - 4i) = -14i + 8i^2 = -8 - 14i$

13. $a^{-1/2}(a^{1/2} - a^{3/2}) = a^0 - a^1$
$$= 1 - a$$

14.
$$\frac{\sqrt[3]{8x^4y^5}}{\sqrt[3]{16xy^6}} = \sqrt[3]{\frac{8x^4y^5}{16xy^6}} = \sqrt[3]{\frac{x^3}{2y}}$$

$$= \frac{\sqrt[3]{x^3}}{\sqrt[3]{2y}}$$

$$= \frac{x}{\sqrt[3]{2y}} \cdot \frac{\sqrt[3]{4y^2}}{\sqrt[3]{4y^2}} = \frac{x\sqrt[3]{4y^2}}{\sqrt[3]{8y^3}}$$

$$= \frac{x\sqrt[3]{4y^2}}{2y}$$

15. $\dfrac{x}{x+2} - \dfrac{4x}{x+3} = 1$

$$(x+2)(x+3)\left(\frac{x}{x+2} - \frac{4x}{x+3}\right) = 1(x+2)(x+3)$$

$$x(x+3) - 4x(x+2) = 1(x+2)(x+3)$$

$$x^2 + 3x - 4x^2 - 8x = x^2 + 5x + 6$$

$$0 = 4x^2 + 10x + 6$$

$$0 = 2(2x^2 + 5x + 3)$$

$$(2x + 3)(x + 1) = 0$$

$$2x + 3 = 0 \quad x + 1 = 0$$

$$2x = -3 \quad\quad x = -1$$

$$x = -\frac{3}{2}$$

The solutions are -1 and $-\dfrac{3}{2}$.

16. $\dfrac{x}{2x+3} - \dfrac{3}{4x^2-9} = \dfrac{x}{2x-3}$

$(2x+3)(2x-3)\left(\dfrac{x}{2x+3} - \dfrac{3}{(2x+3)(2x-3)}\right) = (2x+3)(2x-3)\dfrac{x}{2x-3}$

$x(2x-3)-3 = x(2x+3)$

$2x^2-3x-3 = 2x^2+3x$

$-3 = 6x$

$x = -\dfrac{1}{2}$

The solution is $-\dfrac{1}{2}$.

17. $x^4 - 6x^2 + 8 = 0$

$\left(x^2\right)^2 - 6x^2 + 8 = 0$

$u^2 - 6u + 8 = 0$

$(u-4)(u-2) = 0$

$u-4 = 0 \quad u-2 = 0$

$u = 4 \qquad u = 2$

Replace u with x^2.

$x^2 = 4 \qquad x^2 = 2$

$\sqrt{x^2} = \sqrt{4} \quad \sqrt{x^2} = \sqrt{2}$

$x = \pm 2 \qquad x = \pm\sqrt{2}$

The solutions are $-2, 2, \sqrt{2}$ and $-\sqrt{2}$.

18. $\sqrt{3x+1} - 1 = x$

$\sqrt{3x+1} = x+1$

$(\sqrt{3x+1})^2 = (x+1)^2$

$3x+1 = x^2+2x+1$

$0 = x^2 - x$

$0 = x(x-1)$

$x-1 = 0 \quad x = 0$

$x = 1$

The solutions are 1 and 0.

19. $|3x-2| < 8$

$-8 < 3x-2 < 8$

$-8+2 < 3x-2+2 < 8+2$

$-6 < 3x < 10$

$-2 < x < \dfrac{10}{3}$

$\left\{x \mid -2 < x < \dfrac{10}{3}\right\}$

20. $6x - 5y = 15$

$6x - 5(0) = 15$

$6x = 15$

$x = \dfrac{15}{6} = \dfrac{5}{2}$

The x-intercept is $\left(\dfrac{5}{2}, 0\right)$.

$6(0) - 5y = 15$

$-5y = 15$

$y = -3$

The y-intercept is $(0, -3)$.

21. Solve each inequality for y.

$x + y \le 3 \qquad 2x - y < 4$

$y \le 3 - x \qquad -y < 4 - 2x$

$\qquad\qquad\qquad y > 2x - 4$

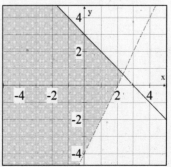

22. $x + y + z = 2$
$-x + 2y - 3z = -9$
$x - 2y - 2z = -1$

$$D = \begin{vmatrix} 1 & 1 & 1 \\ -1 & 2 & -3 \\ 1 & -2 & -2 \end{vmatrix} = -15$$

$$D_x = \begin{vmatrix} 2 & 1 & 1 \\ 4 & 2 & -2 \\ -1 & -2 & -2 \end{vmatrix} = -15$$

$$D_y = \begin{vmatrix} 1 & 2 & 1 \\ -1 & -9 & -3 \\ 1 & -1 & -2 \end{vmatrix} = 15$$

$$D_z = \begin{vmatrix} 1 & 1 & 2 \\ -1 & 2 & -9 \\ 1 & -2 & -1 \end{vmatrix} = -30$$

$$x = \frac{D_x}{D} = \frac{-15}{-15} = 1$$

$$y = \frac{D_y}{D} = \frac{15}{-15} = -1$$

$$z = \frac{D_z}{D} = \frac{-30}{-15} = 2$$

The solution is $(1, -1, 2)$

23. $f(x) = \dfrac{2x - 3}{x^2 - 1}$

$f(-2) = \dfrac{2(-2) - 3}{(-2)^2 - 1} = \dfrac{-4 - 3}{4 - 1} = -\dfrac{7}{3}$

24. $f(x) = \dfrac{x - 2}{x^2 - 2x - 15}$

$f(x) = \dfrac{x - 2}{(x - 5)(x + 3)}$

$x - 5 = 0 \quad x + 3 = 0$

$x = 5 \qquad x = -3$

$\{x \mid x \neq -3, 5\}$

25. $x^3 + x^2 - 6x < 0$

$x(x^2 + x - 6) < 0$

$x(x + 3)(x - 2) < 0$

$\{x \mid x < -3 \text{ or } 0 < x < 2\}$

26. $\dfrac{(x - 1)(x - 5)}{x + 3} \geq 0$

$\{x \mid -3 < x \leq 1 \text{ or } x \geq 5\}$

27. Strategy: Let P represent the length of the piston rod, T the tolerance and m the given length. Solve the absolute value inequality $|m - p| \leq T$ for m.

Solution: $|m - p| \leq T$

$$\left| m - 9\frac{3}{8} \right| \leq \frac{1}{64}$$

$$-\frac{1}{64} \leq m - 9\frac{3}{8} \leq \frac{1}{64}$$

$$-\frac{1}{64} + 9\frac{3}{8} \leq m - 9\frac{3}{8} + 9\frac{3}{8} \leq \frac{1}{64} + 9\frac{3}{8}$$

$$9\frac{23}{64} \leq m \leq 9\frac{25}{64}$$

The lower limit is $9\dfrac{23}{64}$ in.

The upper limit is $9\dfrac{25}{64}$ in.

28. **Strategy:** The base of the triangle is $x + 8$.
The height of the triangle is $2x - 4$.

Solution: $A = \dfrac{1}{2}bh$

$A = \dfrac{1}{2}(x+8)(2x-4)$

$A = \dfrac{1}{2}(2x^2 + 12x - 32)$

$A = (x^2 + 6x - 16) \text{ ft}^2$

29. $2x^2 + 4x + 3 = 0$
$a = 2, b = 4, c = 3$

$b^2 - 4ac = 4^2 - 4(2)(3) = 16 - 24 = -8$

$-8 < 0$

Since the discriminant is less than zero, the equation has two complex number solutions.

30. $(0, 250)$ and $(30, 0)$

$m = \dfrac{y_2 - y_1}{x_2 - x_1} = \dfrac{250 - 0}{0 - 30} = \dfrac{250}{-30} = -\dfrac{25}{3}$

$m = -\dfrac{25{,}000}{3}$

The building depreciates $\dfrac{\$25{,}000}{3}$, or about

$\$8333$, each year.

Chapter 9: Functions and Relations

Prep Test

1. $-\dfrac{b}{2a}$

$-\dfrac{(-4)}{2(2)} = -\dfrac{-4}{4} = -(-1) = 1$

2. $y = -x^2 + 2x + 1$
$y = -(-2)^2 + 2(-2) + 1 = -4 - 4 + 1 = -7$

3. $f(x) = x^2 - 3x + 2$
$f(-4) = (-4)^2 - 3(-4) + 2$
$f(-4) = 16 + 12 + 2$
$f(-4) = 30$

4. $p(r) = r^2 - 5$
$p(2 + h) = (2 + h)^2 - 5$
$\qquad\qquad = 4 + 4h + h^2 - 5$
$\qquad\qquad = h^2 + 4h - 1$

5. $0 = 3x^2 - 7x - 6$
$0 = (3x + 2)(x - 3)$

$0 = 3x + 2 \qquad 0 = x - 3$
$-2 = 3x \qquad\quad 3 = x$

$-\dfrac{2}{3} = x$

The solutions are $-\dfrac{2}{3}$ and 3.

6. $0 = x^2 - 4x + 1$
$a = 1 \quad b = -4 \quad c = 1$

$x = \dfrac{-b \pm \sqrt{b^2 - 4ac}}{2a}$

$x = \dfrac{-(-4) \pm \sqrt{(-4)^2 - 4(1)(1)}}{2(1)}$

$x = \dfrac{4 \pm \sqrt{16 - 4}}{2} = \dfrac{4 \pm \sqrt{12}}{2} = \dfrac{4 \pm 2\sqrt{3}}{2}$

$x = \dfrac{4}{2} \pm \dfrac{2\sqrt{3}}{2} = 2 \pm \sqrt{3}$

The solutions are $2 + \sqrt{3}$ and $2 - \sqrt{3}$.

7. $x = 2y + 4$
$2y + 4 = x$
$\qquad 2y = x - 4$

$\left(\dfrac{1}{2}\right)2y = \left(\dfrac{1}{2}\right)(x - 4)$

$\qquad y = \dfrac{1}{2}x - 2$

8. Domain: $\{-2, 3, 4, 6\}$
Range: $\{4, 5, 6\}$
Yes, the relation is a function.

9. Values for which $x - 8 = 0$ are excluded from the domain.
$x - 8 = 0$
$\qquad x = 8$

10.

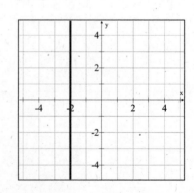

Section 9.1

Concept Check

1. (i) and (iii)

3. The axis of symmetry of the graph of a parabola is the vertical line that passes through the vertex of the parabola and is parallel to the y–axis..

5. Zero, one or two

7. $(-1,0)$ and $(3,0)$

9. When $a > 0$, the parabola opens up and the vertex of the parabola is the lowest point on the parabola, with the smallest y-coordinate. This point is called the minimum value of the function.
When $a < 0$, the parabola opens down and the vertex of the parabola is the highest point on the parabola, with the largest y-coordinate. This point is called the maximum value of the function.

Objective A Exercises

11. $y = x^2 - 2x - 4$

$$-\frac{b}{2a} = -\frac{-2}{2(1)} = -\frac{-2}{2} = -(-1) = 1$$

$$y = (1)^2 - 2(1) - 4 = -5$$

Vertex:
$(1, -5)$
Axis of symmetry:
$x = 1$

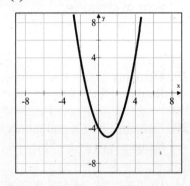

13. $y = -x^2 + 2x - 3$

$$-\frac{b}{2a} = -\frac{2}{2(-1)} = -\frac{2}{-2} = -(-1) = 1$$

$$y = -(1)^2 + 2(1) - 3 = -2$$

Vertex: $(1, -2)$
Axis of symmetry: $x = 1$

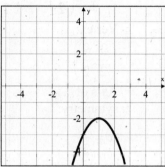

15. $f(x) = x^2 - x - 6$

$$-\frac{b}{2a} = -\frac{(-1)}{2(1)} = -\frac{-1}{2} = \frac{1}{2}$$

$$y = \left(\frac{1}{2}\right)^2 - \frac{1}{2} - 6 = -\frac{25}{4}$$

Vertex:
$\left(\frac{1}{2}, -\frac{25}{4}\right)$

Axis of symmetry:

$x = \dfrac{1}{2}$

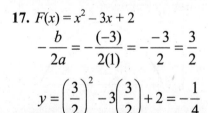

17. $F(x) = x^2 - 3x + 2$

$$-\frac{b}{2a} = -\frac{(-3)}{2(1)} = -\frac{-3}{2} = \frac{3}{2}$$

$$y = \left(\frac{3}{2}\right)^2 - 3\left(\frac{3}{2}\right) + 2 = -\frac{1}{4}$$

Vertex:
$\left(\frac{3}{2}, -\frac{1}{4}\right)$

Axis of symmetry:

$x = \dfrac{3}{2}$

19. $y = -2x^2 + 6x$

$$-\frac{b}{2a} = -\frac{6}{2(-2)} = -\frac{6}{-4} = \frac{3}{2}$$

$$y = -2\left(\frac{3}{2}\right)^2 + 6\left(\frac{3}{2}\right) = \frac{9}{2}$$

Vertex:

$\left(\dfrac{3}{2}, \dfrac{9}{2}\right)$

Axis of symmetry:

$x = \dfrac{3}{2}$

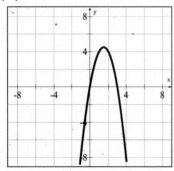

21. $y = -\dfrac{1}{4}x^2 - 1$

$$-\frac{b}{2a} = -\frac{(0)}{2\left(-\dfrac{1}{4}\right)} = -\frac{0}{-\dfrac{1}{2}} = 0$$

$$y = -\left(\frac{1}{4}\right)0^2 - 1 = -1$$

Vertex:

$(0, -1)$

Axis of symmetry:

$x = 0$

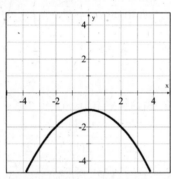

23. $P(x) = -\dfrac{1}{2}x^2 + 2x - 3$

$$-\frac{b}{2a} = -\frac{2}{2\left(-\dfrac{1}{2}\right)} = -\frac{2}{-1} = -(-2) = 2$$

$$y = \left(-\frac{1}{2}\right)2^2 + 2(2) - 3 = -1$$

Vertex:

$(2, -1)$

Axis of symmetry:

$x = 2$

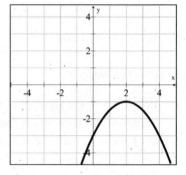

25. $y = -\dfrac{1}{2}x^2 + x - 3$

$$-\frac{b}{2a} = -\frac{1}{2\left(-\dfrac{1}{2}\right)} = -\frac{1}{-1} = -(-1) = 1$$

$$y = \left(-\frac{1}{2}\right)1^2 + 1 - 3 = -\frac{5}{2}$$

Vertex: $\left(1, -\dfrac{5}{2}\right)$

Axis of symmetry: $x = 1$

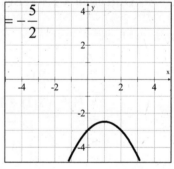

27. Domain: $\{x \mid x \in \text{real numbers}\}$
Range: $\{y \mid y \geq 2\}$

29. Domain: $\{x \mid x \in \text{real numbers}\}$
Range: $\{y \mid y \geq -5\}$

31. Domain: $\{x \mid x \in \text{real numbers}\}$
Range: $\{y \mid y \leq 0\}$

33. Domain: $\{x \mid x \in \text{real numbers}\}$
Range: $\{y \mid y \geq -7\}$

Objective B Exercises

35. $y = x^2 - 9$

$0 = x^2 - 9$

$0 = (x+3)(x-3)$

$x + 3 = 0 \qquad x - 3 = 0$

$\quad x = -3 \qquad \quad x = 3$

The x-intercepts are $(-3, 0)$ and $(3, 0)$.

37. $y = 3x^2 + 6x$

$0 = 3x^2 + 6x$

$0 = 3x(x+2)$

$3x = 0 \qquad x + 2 = 0$

$x = 0 \qquad \quad x = -2$

The x-intercepts are $(0, 0)$ and $(-2, 0)$.

39. $y = x^2 - 2x - 8$

$0 = x^2 - 2x - 8$

$0 = (x-4)(x+2)$

$x - 4 = 0 \qquad x + 2 = 0$

$\quad x = 4 \qquad \quad x = -2$

The x-intercepts are $(4, 0)$ and $(-2, 0)$.

41. $y = 2x^2 - 5x - 3$

$0 = 2x^2 - 5x - 3$

$0 = (x-3)(2x+1)$

$x - 3 = 0 \qquad 2x + 1 = 0$

$\quad x = 3 \qquad \quad 2x = -1$

$$x = -\frac{1}{2}$$

The x-intercepts are $(3, 0)$ and $\left(-\frac{1}{2}, 0\right)$.

43. $y = x^2 + 4x - 3$

$0 = x^2 + 4x - 3$

$a = 1 \quad b = 4 \quad c = -3$

$$x = \frac{-b \pm \sqrt{b^2 - 4ac}}{2a}$$

$$x = \frac{-4 \pm \sqrt{(4)^2 - 4(1)(-3)}}{2(1)}$$

$$x = \frac{-4 \pm \sqrt{16 + 12}}{2} = \frac{-4 \pm \sqrt{28}}{2} = \frac{-4 \pm 2\sqrt{7}}{2}$$

$x = -2 \pm \sqrt{7}$

The x-intercepts are $\left(-2 + \sqrt{7}, 0\right)$ and $\left(-2 - \sqrt{7}, 0\right)$.

45. $y = -x^2 - 4x - 5$

$0 = -x^2 - 4x - 5$

$a = -1 \quad b = -4 \quad c = -5$

$$x = \frac{-b \pm \sqrt{b^2 - 4ac}}{2a}$$

$$x = \frac{-(-4) \pm \sqrt{(-4)^2 - 4(-1)(-5)}}{2(-1)}$$

$$x = \frac{4 \pm \sqrt{16 - 20}}{-2} = \frac{4 \pm \sqrt{-4}}{-2} = \frac{4 \pm 2i}{-2}$$

$x = -2 \pm i$

There are no real solutions. The parabola has no x-intercepts.

47. $f(x) = x^2 - 4x - 5$

$x^2 - 4x - 5 = 0$

$(x-5)(x+1) = 0$

$x - 5 = 0 \qquad x + 1 = 0$

$\quad x = 5 \qquad \quad x = -1$

The zeros are 5 and -1.

49. $f(x) = 3x^2 - 2x - 8$

$3x^2 - 2x - 8 = 0$

$(3x + 4)(x - 2) = 0$

$3x + 4 = 0 \qquad x - 2 = 0$

$\qquad 3x = -4 \qquad\quad x = 2$

$\qquad x = -\dfrac{4}{3}$

The zeros are $-\dfrac{4}{3}$ and 2.

51. $h(x) = 4x^2 - 4x + 1$

$4x^2 - 4x + 1 = 0$

$(2x - 1)(2x - 1) = 0$

$2x - 1 = 0 \qquad 2x - 1 = 0$

$\quad 2x = 1 \qquad\qquad 2x = 1$

$\quad x = \dfrac{1}{2} \qquad\qquad x = \dfrac{1}{2}$

The zero is $\dfrac{1}{2}$.

53. $f(x) = -3x^2 + 4x$

$-3x^2 + 4x = 0$

$x(-3x + 4) = 0$

$-3x + 4 = 0 \qquad x = 0$

$\quad -3x = -4$

$\quad x = \dfrac{4}{3}$

The zeros are $\dfrac{4}{3}$ and 0.

55. $f(x) = -3x^2 + 12$

$-3x^2 + 12 = 0$

$-3(x^2 - 4) = 0$

$-3(x + 2)(x - 2) = 0$

$x + 2 = 0 \qquad x - 2 = 0$

$\quad x = -2 \qquad\quad x = 2$

The zeros are -2 and 2.

57. $f(x) = 2x^2 - 54$

$2x^2 - 54 = 0$

$2(x^2 - 27) = 0$

$x^2 - 27 = 0$

$x^2 = 27$

$x = \pm\sqrt{27} = \pm 3\sqrt{3}$

The zeros are $3\sqrt{3}$ and $-3\sqrt{3}$.

59. $f(x) = x^2 - 2x - 17$

$x^2 - 2x - 17 = 0$

$a = 1 \quad b = -2 \quad c = -17$

$x = \dfrac{-b \pm \sqrt{b^2 - 4ac}}{2a}$

$x = \dfrac{-(-2) \pm \sqrt{(-2)^2 - 4(1)(-17)}}{2(1)}$

$x = \dfrac{2 \pm \sqrt{4 + 68}}{2} = \dfrac{2 \pm \sqrt{72}}{2} = \dfrac{2 \pm 6\sqrt{2}}{2}$

$x = 1 \pm 3\sqrt{2}$

The zeros are $1 + 3\sqrt{2}$ and $1 - 3\sqrt{2}$.

61. $f(x) = x^2 + 4x + 5$

$x^2 + 4x + 5 = 0$

$a = 1 \quad b = 4 \quad c = 5$

$x = \dfrac{-b \pm \sqrt{b^2 - 4ac}}{2a}$

$x = \dfrac{-4 \pm \sqrt{(4)^2 - 4(1)(5)}}{2(1)}$

$x = \dfrac{-4 \pm \sqrt{16 - 20}}{2} = \dfrac{-4 \pm \sqrt{-4}}{2} = \dfrac{-4 \pm 2i}{2}$

$x = -2 \pm i$

The zeros are $-2 + i$ and $-2 - i$.

63. $f(x) = x^2 + 4x + 13$

$x^2 + 4x + 13 = 0$

$a = 1 \quad b = 4 \quad c = 13$

$x = \dfrac{-b \pm \sqrt{b^2 - 4ac}}{2a}$

$x = \dfrac{-4 \pm \sqrt{(4)^2 - 4(1)(13)}}{2(1)}$

$x = \dfrac{-4 \pm \sqrt{16 - 52}}{2} = \dfrac{-4 \pm \sqrt{-36}}{2} = \dfrac{-4 \pm 6i}{2}$

$x = -2 \pm 3i$

The zeros are $-2 + 3i$ and $-2 - 3i$.

65. $y = -x^2 - x + 3$

$a = -1 \quad b = -1 \quad c = 3$

$b^2 - 4ac$

$(-1)^2 - 4(-1)(3) = 1 + 12 = 13$

$13 > 0$

Since the discriminant is greater than zero, the parabola has two x-intercepts.

67. $y = x^2 - 10x + 25$

$a = 1 \quad b = -10 \quad c = 25$

$b^2 - 4ac$

$(-10)^2 - 4(1)(25) = 100 - 100 = 0$

Since the discriminant is equal to zero, the parabola has one x-intercept.

69. $y = -2x^2 + x - 1$

$a = -2 \quad b = 1 \quad c = -1$

$b^2 - 4ac$

$(1)^2 - 4(-2)(-1) = 1 - 8 = -7$

$-7 < 0$

Since the discriminant is less than zero, the parabola has no x-intercepts.

71. $y = 4x^2 - x - 2$

$a = 4 \quad b = -1 \quad c = -2$

$b^2 - 4ac$

$(-1)^2 - 4(4)(-2) = 1 + 32 = 33$

$33 > 0$

Since the discriminant is greater than zero, the parabola has two x-intercepts.

73. $y = 2x^2 + x + 4$

$a = 2 \quad b = 1 \quad c = 4$

$b^2 - 4ac$

$(1)^2 - 4(2)(4) = 1 - 32 = -31$

$-31 < 0$

Since the discriminant is less than zero, the parabola has no x-intercepts.

75. $y = 4x^2 + 2x - 5$

$a = 4 \quad b = 2 \quad c = -5$

$b^2 - 4ac$

$(2)^2 - 4(4)(-5) = 4 + 80 = 84$

$84 > 0$

Since the discriminant is greater than zero, the parabola has two x-intercepts.

77. a) $a > 0$

b) $a = 0$

c) $a < 0$

Objective C Exercises

79. $f(x) = 2x^2 + 4x$

$x = -\dfrac{b}{2a} = -\dfrac{4}{2(2)} = -1$

$f(x) = 2x^2 + 4x$

$f(-1) = 2(-1)^2 + 4(-1) = 2 - 4 = -2$

Since $a > 0$, the function has a minimum value. The minimum value of the function is -2.

81. $f(x) = -2x^2 + 4x - 5$

$$x = -\frac{b}{2a} = -\frac{4}{2(-2)} = 1$$

$$f(x) = -2x^2 + 4x - 5$$

$$f(1) = -2(1)^2 + 4(1) - 5 = -2 + 4 - 5 = -3$$

Since $a < 0$, the function has a maximum value. The maximum value of the function is -3.

83. $f(x) = -2x^2 - 3x$

$$x = -\frac{b}{2a} = -\frac{-3}{2(-2)} = -\frac{3}{4}$$

$$f(x) = -2x^2 - 3x$$

$$f\left(-\frac{3}{4}\right) = -2\left(-\frac{3}{4}\right)^2 - 3\left(-\frac{3}{4}\right) = -\frac{9}{8} + \frac{9}{4}$$

$$= \frac{9}{8}$$

Since $a < 0$, the function has a maximum value. The maximum value of the function is $\frac{9}{8}$.

85. $f(x) = 3x^2 + 3x - 2$

$$x = -\frac{b}{2a} = -\frac{3}{2(3)} = -\frac{1}{2}$$

$$f(x) = 3x^2 + 3x - 2$$

$$f\left(-\frac{1}{2}\right) = 3\left(-\frac{1}{2}\right)^2 + 3\left(-\frac{1}{2}\right) - 2 = \frac{3}{4} - \frac{3}{2} - 2$$

$$= -\frac{11}{4}$$

Since $a > 0$, the function has a minimum value. The minimum value of the function is $-\frac{11}{4}$.

87. $f(x) = -x^2 - x + 2$

$$x = -\frac{b}{2a} = -\frac{-1}{2(-1)} = -\frac{1}{2}$$

$$f(x) = -x^2 - x + 2$$

$$f\left(-\frac{1}{2}\right) = -\left(-\frac{1}{2}\right)^2 - \left(-\frac{1}{2}\right) + 2 = -\frac{1}{4} + \frac{1}{2} + 2$$

$$= \frac{9}{4}$$

Since $a < 0$, the function has a maximum value. The maximum value of the function is $\frac{9}{4}$.

89. $f(x) = 3x^2 + 5x + 2$

$$x = -\frac{b}{2a} = -\frac{5}{2(3)} = -\frac{5}{6}$$

$$f(x) = 3x^2 + 5x + 2$$

$$f\left(-\frac{5}{6}\right) = 3\left(-\frac{5}{6}\right)^2 + 5\left(-\frac{5}{6}\right) + 2 = \frac{25}{12} - \frac{25}{6} + 2$$

$$= -\frac{1}{12}$$

Since $a > 0$, the function has a minimum value. The minimum value of the function is $-\frac{1}{12}$.

Objective D Exercises

91. Strategy: To find the price that will give the maximum revenue, find the P-coordinate of the vertex.

Solution:

$$P = -\frac{b}{2a} = -\frac{125}{2\left(-\frac{1}{4}\right)} = 250$$

A price of $250 will give the maximum revenue.

93. Strategy: Let x represent one number. The other number is $20 - x$.
Their product is $x(20 - x)$.
To find one number, find the x-coordinate of the vertex. To find the second number, evaluate $20 - x$ at the x-coordinate of the vertex.

Solution:
$$x(20 - x) = 20x - x^2$$
$$x = -\frac{b}{2a} = -\frac{20}{2(-1)} = 10$$
$$20 - x = 20 - 10 = 10$$
The two numbers are 10 and 10.

95. Strategy: To find the time it takes the plane to reach its maximum height, find the t-coordinate of the vertex. To find the maximum height, evaluate the function at the t-coordinate of the vertex.

Solution:
$$t = -\frac{b}{2a} = -\frac{119}{2(-1.42)} \approx 42$$
$$h(t) = -1.42t^2 + 119t + 6000$$
$$h(42) = -1.42(42)^2 + 119(42) + 6000$$
$$= -2504.88 + 4998 + 6000 = 8493.12$$
The maximum height of the plane is about 8500 m.

97. To find the time at which the stream reaches its maximum height, find the t-coordinate of the vertex. To find the maximum height, evaluate the function at the t-coordinate of the vertex.

Solution:
$$t = -\frac{b}{2a} = -\frac{30}{2(-16)} = 0.9375$$
$$h(t) = -16t^2 + 30t$$
$$h(0.9375) = -16(0.9375)^2 + 30(0.9375)$$
$$= -14.0625 + 28.125 = 14.0625$$
The stream is at its maximum height in 0.9375 s.
The maximum height is 14.0625 ft.

99. Strategy: Let x represent the number of units. Find the maximum value of x in order to find the maximum revenue.
$$R(x) = (1200 + 100x)(100 - x)$$
$$R(x) = -100x^2 + 8800x + 120000$$

Solution:
$$x = -\frac{b}{2a} = -\frac{8800}{2(-100)} = 44$$
$$R(x) = -100x^2 + 8800x + 120000$$
$$R(44) = -100(44)^2 + 8800(44) + 120000$$
$$= -193600 + 387200 + 120000$$
$$= 313600$$
The maximum revenue of \$313,600 is reached when 44 units are rented.

101.   Strategy: Let x represent width of the rectangular corral. The length is $200 - 2x$.
The area is $x(200 - 2x)$.
To find the width, find the x-coordinate of the vertex. To find the length, evaluate $200 - 2x$ at the x-coordinate of the vertex.

Solution:
$$x(200 - 2x) = 200x - 2x^2$$
$$x = -\frac{b}{2a} = -\frac{200}{2(-2)} = 50$$
$$200 - 2x = 200 - 2(50) = 100$$
The width is 50 ft and the length is 100 ft.

103.   Strategy: Let x represent the width of the ball fields. The length is $\frac{2100 - 3x}{2}$.
The area is $x\left(\frac{2100 - 3x}{2}\right)$.
To find the width, find the x-coordinate of the vertex. To find the length, evaluate $\frac{2100 - 3x}{2}$ at the x-coordinate of the vertex.

Solution:

$$x\left(\frac{2100-3x}{2}\right) = x(1050-1.5x) = 1050x - 1.5x^2$$

$$x = -\frac{b}{2a} = -\frac{1050}{2(-1.5)} = 350$$

$$\frac{2100-3x}{2} = \frac{2100-1050}{2} = \frac{1050}{2} = 525$$

The dimensions are 350 ft by 525 ft.

Critical Thinking

105. To find the root, substitute $x = -2$ and $y = 0$ then $x = 3$ and $y = 0$ into $f(x) = mx^2 + nx + 1$.
Use these two equations to find the relationship between m and n.

Solution:

$$0 = m(-2)^2 + n(-2) + 1 \qquad 0 = m(3)^2 + n(3) + 1$$

$$0 = 4m - 2n + 1 \qquad\qquad 0 = 9m + 3n + 1$$

$$4m - 2n + 1 = 9m + 3n + 1$$

$$4m - 2n = 9m + 3n$$

$$-5n = 5m$$

$$n = -m$$

n and m are opposites.
Substituting $-n$ for m and $-m$ for n we get

$$f(x) = mx^2 + nx + 1 = (-n)x^2 + (-m)x + 1$$

$$= -nx^2 - mx + 1$$

$$= nx^2 + mx - 1 = g(x)$$

Therefore, since $f(x) = g(x)$, their roots are the same. $g(x)$ has roots -2 and 3.

Solution:

$$y = x^2 - 8x + k$$

$$x = -\frac{b}{2a} = -\frac{-8}{2(1)} = 4$$

$$x = 4 \quad y = 0$$

$$0 = (4)^2 - 8(4) + k$$

$$0 = 16 - 32 + k$$

$$0 = -16 + k$$

$$16 = k$$

Projects or Group Activities

107. Strategy: Let h be the height of the window, r is the radius of the semicircle, and $2r$ is the width of the window. Consider the perimeter and area of the rectangle added to the semicircle.

Solution:
$$P = (2h + 2r) + \pi r$$

$$50 = 2h + 2r + \pi r$$

$$h = \frac{50 - 2r - \pi r}{2} = 25 - r - \frac{1}{2}\pi r$$

$$A = (2rh) + \left(\frac{1}{2}\pi r^2\right)$$

$$A = 2r\left(25 - r - \frac{1}{2}\pi r\right) + \frac{1}{2}\pi r^2$$

$$= 50r - 2r^2 - \pi r^2 + \frac{1}{2}\pi r^2$$

$$= 50r - 3.57r^2$$

Find the maximum value for r.

$$r = -\frac{b}{2a} = -\frac{50}{2(-3.57)} \approx 7$$

$$h = 25 - r - \frac{1}{2}\pi r = 25 - 7 - \frac{1}{2}\pi(7) = 7$$

$$h \approx 7$$

Both r and h are approximately 7.

Section 9.2

Concept Check

1. Left

3. Up

5. $(-3, 5)$

Objective A Exercises

7.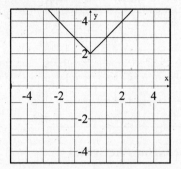

9.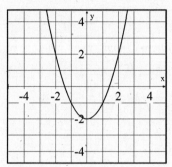

11.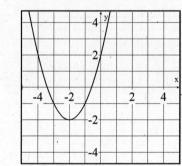

13.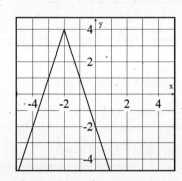

15.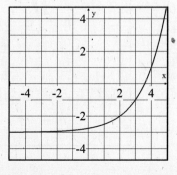

17.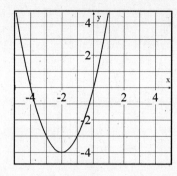

19.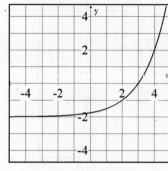

21.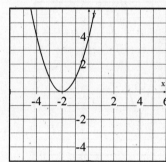

23. $(0, 9)$

Objective B Exercises

25.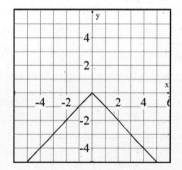

27.

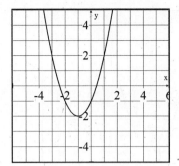

29.

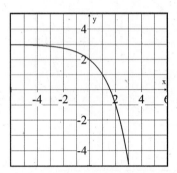

31.

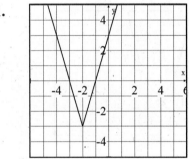

Critical Thinking

33.

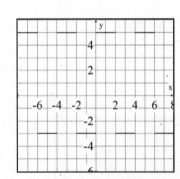

35.

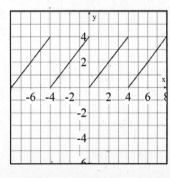

37.

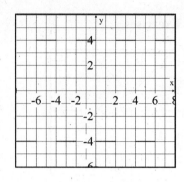

Projects or Group Activities

39.

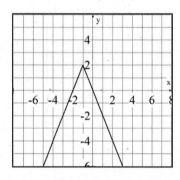

41.

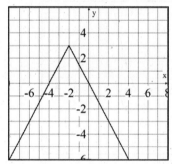

Check Your Progress: Chapter 9

1. $y = x^2 + 6x + 3$

$x = -\dfrac{b}{2a} = -\dfrac{6}{2(1)} = -3$

$y = (-3)^2 + 6(-3) + 3 = -6$

Vertex:
$(-3, -6)$
Axis of
symmetry:
$x = -3$

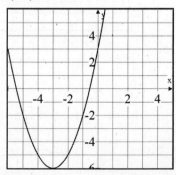

2. $f(x) = -\dfrac{1}{2}x^2 + 2x + 3$

$x = -\dfrac{b}{2a} = -\dfrac{2}{2\left(-\dfrac{1}{2}\right)} = -\dfrac{2}{-1} = 2$

$y = \left(-\dfrac{1}{2}\right)(2)^2 + 2(2) + 3 = 5$

Vertex: $(2,5)$
Axis of symmetry:
$x = 2$

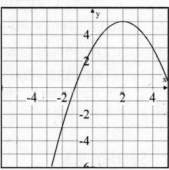

3. $y = 2x^2 + 5x - 3$

$0 = 2x^2 + 5x - 3$

$0 = (x+3)(2x-1)$

$x + 3 = 0 \qquad 2x - 1 = 0$

$\quad x = -3 \qquad\quad 2x = 1$

$\qquad\qquad\qquad x = \dfrac{1}{2}$

The x-intercepts are $(-3, 0)$ and $\left(\dfrac{1}{2}, 0\right)$.

4. $y = x^2 + 2x - 1$

$0 = x^2 + 2x - 1$

$a = 1 \quad b = 2 \quad c = -1$

$x = \dfrac{-b \pm \sqrt{b^2 - 4ac}}{2a}$

$x = \dfrac{-2 \pm \sqrt{(2)^2 - 4(1)(-1)}}{2(1)}$

$x = \dfrac{-2 \pm \sqrt{4+4}}{2} = \dfrac{-2 \pm \sqrt{8}}{2} = \dfrac{-2 \pm 2\sqrt{2}}{2}$

$x = -1 \pm \sqrt{2}$

The x-intercepts are $\left(-1 + \sqrt{2}, 0\right)$ and $\left(-1 - \sqrt{2}, 0\right)$.

5. $f(x) = x^2 + x - 30$

$x^2 + x - 30 = 0$

$(x + 6)(x - 5) = 0$

$x + 6 = 0 \quad x - 5 = 0$

$x = -6 \qquad x = 5$

The zeros are -6 and 5.

6. $f(x) = x^2 - 4x + 40$

$x^2 - 4x + 40 = 0$

$a = 1 \quad b = -4 \quad c = 40$

$x = \dfrac{-b \pm \sqrt{b^2 - 4ac}}{2a}$

$x = \dfrac{-(-4) \pm \sqrt{(-4)^2 - 4(1)(40)}}{2(1)}$

$x = \dfrac{4 \pm \sqrt{16 - 160}}{2} = \dfrac{4 \pm \sqrt{-144}}{2} = \dfrac{4 \pm 12i}{2}$

$x = 2 \pm 6i$

The zeros are $2 + 6i$ and $2 - 6i$.

7. $f(x) = x^2 - 4x - 32$

$x = -\dfrac{b}{2a} = -\dfrac{-4}{2(1)} = 2$

$f(x) = x^2 - 4x - 32$

$f(2) = (2)^2 - 4(2) - 32 = 4 - 8 - 32 = -36$

Since $a > 0$, the function has a minimum value. The minimum value of the function is -36.

8. $f(x) = x^2 - 4x + 8$

$x = -\dfrac{b}{2a} = -\dfrac{-4}{2(1)} = 2$

$f(x) = x^2 - 4x + 8$

$f(2) = (2)^2 - 4(2) + 8 = 4 - 8 + 8 = 4$

Since $a < 0$, the function has a maximum value. The maximum value of the function is 4.

9.

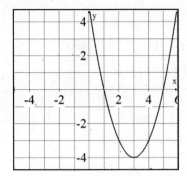

10.

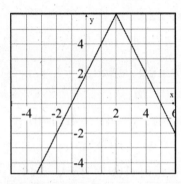

11.

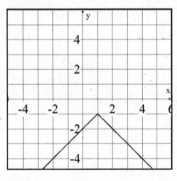

12.

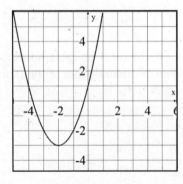

13. Strategy: Find the time at which the rock reaches its maximum height by finding the t-coordinate of the vertex. To find the maximum height, evaluate the function at the t-coordinate of the vertex.

Solution:

$$t = -\frac{b}{2a} = -\frac{64}{2(-16)} = 2$$

$$s(t) = -16t^2 + 64t + 76$$

$$s(2) = -16(2)^2 + 64(2) + 76$$

$$= -64 + 128 + 76 = 140$$

The maximum height of the rock is 140 ft.

Section 9.3

Concept Check

1. $f(x) = x^2 + 4 \qquad g(x) = \sqrt{x+4}$
   a) Yes
   b) Yes
   c) Yes
   d) No

3. a) No
   b) Yes

Objective A Exercises

5. $f(2) - g(2) = (2 \cdot 2^2 - 3) - (-2 \cdot 2 + 4)$
   $= (2 \cdot 4 - 3) - (-4 + 4)$
   $= (8 - 3) - (0)$
   $= 5$

7. $f(0) + g(0) = (2 \cdot 0^2 - 3) + (-2 \cdot 0 + 4)$
   $= (0 - 3) + (0 + 4)$
   $= -3 + 4$
   $= 1$

9. $(f \cdot g)(2) = f(2) \cdot g(2)$

$\qquad = (2 \cdot 2^2 - 3) \cdot (-2 \cdot 2 + 4)$

$\qquad = (2 \cdot 4 - 3) \cdot (-4 + 4)$

$\qquad = (8 - 3) \cdot (0)$

$\qquad = 0$

11. $\left(\dfrac{f}{g}\right)(4) = \dfrac{f(4)}{g(4)}$

$\qquad = \dfrac{2 \cdot (4)^2 - 3}{-2 \cdot (4) + 4}$

$\qquad = \dfrac{2 \cdot 16 - 3}{-8 + 4}$

$\qquad = \dfrac{29}{-4} = -\dfrac{29}{4}$

13. $f(1) + g(1) = (2 \cdot 1^2 + 3 \cdot 1 - 1) + (2 \cdot 1 - 4)$

$\qquad = (2 \cdot 1 + 3 - 1) + (2 \cdot 1 - 4)$

$\qquad = (2 + 3 - 1) + (2 - 4)$

$\qquad = 4 - 2$

$\qquad = 2$

15. $f(4) - g(4)$

$\qquad = (2 \cdot (4)^2 + 3 \cdot (4) - 1) - (2 \cdot (4) - 4)$

$\qquad = (2 \cdot 16 + 12 - 1) - (2 \cdot (4) - 4)$

$\qquad = (32 + 12 - 1) - (8 - 4)$

$\qquad = 43 - 4$

$\qquad = 39$

17. $(f \cdot g)(1) = f(1) \cdot g(1)$

$\qquad = (2 \cdot (1)^2 + 3 \cdot (1) - 1) \cdot (2 \cdot (1) - 4)$

$\qquad = (2 \cdot 1 + 3 - 1) \cdot (2 \cdot (1) - 4)$

$\qquad = (2 + 3 - 1) \cdot (2 - 4)$

$\qquad = 4 \cdot (-2)$

$\qquad = -8$

19. $\left(\dfrac{f}{g}\right)(2) = \dfrac{f(2)}{g(2)}$

$\qquad = \dfrac{2 \cdot (2)^2 + 3 \cdot (2) - 1}{2 \cdot (2) - 4}$

$\qquad = \dfrac{2 \cdot 4 + 6 - 1}{4 - 4}$

$\qquad = \dfrac{13}{0}$

Undefined

21. $f(2) - g(2)$

$\qquad = (2^2 + 3 \cdot (2) - 5) - (2^3 - 2 \cdot (2) + 3)$

$\qquad = (4 + 6 - 5) - (8 - 4 + 3)$

$\qquad = 5 - 7$

$\qquad = -2$

23. $\left(\dfrac{f}{g}\right)(-2) = \dfrac{f(-2)}{g(-2)}$

$\qquad = \dfrac{(-2)^2 + 3 \cdot (-2) - 5}{(-2)^3 - 2 \cdot (-2) + 3}$

$\qquad = \dfrac{4 - 6 - 5}{-8 + 4 + 3}$

$\qquad = \dfrac{-7}{-1}$

$\qquad = 7$

Objective B Exercises

25. $f(x) = 2x - 3 \qquad g(x) = 4x - 1$

$\qquad f(0) = 2(0) - 3 = 0 - 3 = -3$

$\qquad g(-3) = 4(-3) - 1 = -12 - 1 = -13$

$\qquad g[f(0)] = -13$

27. $f(x) = 2x - 3 \qquad g(x) = 4x - 1$

$\qquad f(-2) = 2(-2) - 3 = -4 - 3 = -7$

$\qquad g(-7) = 4(-7) - 1 = -28 - 1 = -29$

$\qquad g[f(-2)] = -29$

29. $f(x) = 2x - 3 \qquad g(x) = 4x - 1$

$g(2x - 3) = 4(2x - 3) - 1$

$= 8x - 12 - 1$

$= 8x - 13$

$g[f(x)] = 8x - 13$

31. $h(x) = 2x + 4 \qquad f(x) = \dfrac{1}{2}x + 2$

$h(0) = 2(0) + 4 = 0 + 4 = 4$

$f(4) = \dfrac{1}{2}(4) + 2 = 2 + 2 = 4$

$f[h(0)] = 4$

33. $h(x) = 2x + 4 \qquad f(x) = \dfrac{1}{2}x + 2$

$h(-1) = 2(-1) + 4 = -2 + 4 = 2$

$f(2) = \dfrac{1}{2}(2) + 2 = 1 + 2 = 3$

$f[h(-1)] = 3$

35. $h(x) = 2x + 4 \qquad f(x) = \dfrac{1}{2}x + 2$

$f(2x + 4) = \dfrac{1}{2}(2x + 4) + 2$

$= x + 2 + 2$

$= x + 4$

$f[h(x)] = x + 4$

37. $g(x) = x^2 + 3 \qquad h(x) = x - 2$

$g(0) = 0^2 + 3 = 3$

$h(3) = 3 - 2 = 1$

$h[g(0)] = 1$

39. $g(x) = x^2 + 3 \qquad h(x) = x - 2$

$g(-2) = (-2)^2 + 3 = 4 + 3 = 7$

$h(7) = 7 - 2 = 5$

$h[g(-2)] = 5$

41. $g(x) = x^2 + 3 \qquad h(x) = x - 2$

$h(x^2 + 3) = x^2 + 3 - 2$

$= x^2 + 1$

$h[g(x)] = x^2 + 1$

43. $f(x) = x^2 + x + 1 \qquad h(x) = 3x + 2$

$f(0) = (0)^2 + 0 + 1 == 1$

$h(1) = 3(1) + 2 = 3 + 2 = 5$

$h[f(0)] = 5$

45. $f(x) = x^2 + x + 1 \qquad h(x) = 3x + 2$

$f(-2) = (-2)^2 + (-2) + 1 = 4 - 2 + 1 = 3$

$h(3) = 3(3) + 2 = 9 + 2 = 11$

$h[f(-2)] = 11$

47. $f(x) = x^2 + x + 1 \qquad h(x) = 3x + 2$

$h(x^2 + x + 1) = 3(x^2 + x + 1) + 2$

$= 3x^2 + 3x + 3 + 2$

$= 3x^2 + 3x + 5$

$h[f(x)] = 3x^2 + 3x + 5$

49. $f(x) = x - 2 \qquad g(x) = x^3$

$g(-1) = (-1)^3 = -1$

$f(-1) = -1 - 2 = -3$

$f[g(-1)] = -3$

51. $f(x) = x - 2 \qquad g(x) = x^3$

$f(-1) = -1 - 2 = -3$

$g(-3) = (-3)^3 = -27$

$g[f(-1)] = -27$

53. $f(x) = x - 2 \qquad g(x) = x^3$

$g(x - 2) = (x - 2)^3$

$= (x - 2)(x - 2)(x - 2)$

$= x^3 - 6x^2 + 12x - 8$

$g[f(x)] = x^3 - 6x^2 + 12x - 8$

55. a) Strategy: Selling price equals the cost plus the mark up. If the cost is x and the mark up is 60%, then $S = x + 0.60x$.

If $M(x) = \dfrac{50x + 10,000}{x}$ is the cost per camera, then $(S \circ M)(x)$ is the selling price per camera. Find $(S \circ M)(x)$.

Solution:

$(S \circ M)(x) = S(M(x))$

$= M(x) + 0.60(M(x)) = 1.60(M(x))$

$= 1.60\left(\dfrac{50x + 10,000}{x}\right)$

$= \dfrac{80x + 16,000}{x}$

$S(M(x)) = 80 + \dfrac{16,000}{x}$

b)

$(S \circ M)(5000) = 80 + \dfrac{16000}{5000}$

$= 80 + 3.2$

$= \$83.20$

c) When 5000 digital cameras are manufactured, the camera store sells each camera for $83.20.

57. a) $I(n) = 12,500n \quad n(m) = 4m$

$(I \circ n)(m) = I(n(m))$

$= 12,500(4m)$

$= 50,000m$

b) $(I \circ n)(3) = 50,000(3) = \$150,000$

c) The garage's income from conversions done during a 3 month period is $150,000.

59. rebate: $r(p) = p - 1500$
discounted price: $d(p) = 0.90p$

a) If the dealer takes the rebate first and then the discount, we are finding
$d(r(p)) = 0.90(p - 1500) = 0.90p - 1350$.

b) If the dealer takes the discount first and then the rebate, we are finding
$r(d(p)) = 0.90p - 1500$.

c) As a buyer, you would prefer the dealer to use $r(d(p))$ since the cost would be less.

Critical Thinking

61. $f(1) = 2$ and $g(2) = 0$
$g[f(1)] = g(2) = 0$

63. $(f \circ g)(3) = f(g(3))$
$g(3) = 5$ and $f(5) = -2$
$(f \circ g)(3) = f(g(3)) = -2$

65. $g(0) = -4$ and $f(-4) = 7$
$f[g(0)] = f(-4) = 7$

67. $g(x) = x^2 - 1$
$g(3 + h) - g(3) = (3 + h)^2 - 1 - (3^2 - 1)$
$= 9 + 6h + h^2 - 1 - 8$
$g(3 + h) - g(3) = h^2 + 6h$

69. $g(x) = x^2 - 1$
$\dfrac{g(1 + h) - g(1)}{h} = \dfrac{(1 + h)^2 - 1 - (1^2 - 1)}{h}$
$= \dfrac{1 + 2h + h^2 - 1 - 0}{h}$
$= \dfrac{2h + h^2}{h} = 2 + h$
$\dfrac{g(1 + h) - g(1)}{h} = 2 + h$

71. $g(x) = x^2 - 1$
$\dfrac{g(a + h) - g(a)}{h} = \dfrac{(a + h)^2 - 1 - (a^2 - 1)}{h}$
$= \dfrac{a^2 + 2ah + h^2 - 1 - a^2 + 1}{h}$
$= \dfrac{2ah + h^2}{h} = 2a + h$
$\dfrac{g(a + h) - g(a)}{h} = 2a + h$

Projects or Group Activities

73. $f(x) = 2x$ $g(x) = 3x - 1$ $h(x) = x - 2$

$f(1) = 2(1) = 2$

$h(2) = 2 - 2 = 0$

$g(0) = 3(0) - 1 = 0 - 1 = -1$

$g(h[f(1)]) = -1$

75. $f(x) = 2x$ $g(x) = 3x - 1$ $h(x) = x - 2$

$g(0) = 3(0) - 1 = 0 - 1 = -1$

$h(-1) = -1 - 2 = -3$

$f(-3) = 2(-3) = -6$

$f(h[g(0)]) = -6$

77. $f(x) = 2x$ $g(x) = 3x - 1$ $h(x) = x - 2$

$h(x) = x - 2$

$f(x - 2) = 2(x - 2) = 2x - 4$

$g(2x - 4) = 3(2x - 4) - 1 = 6x - 13$

$g(f[h(x)]) = 6x - 13$

Section 9.4

Concept Check

1. A function is a 1-1 function if, for any a and b in the domain of f, $f(a) = f(b)$ implies that $a = b$.

3. a) Yes
   b) No

5. The inverse of a function f is the set of ordered pairs formed by reversing the coordinates of each ordered pair of f.

7. (ii)

Objective A Exercises

9. Yes, the graph represents a 1-1 function.

11. No, the graph is not a 1-1 function.

13. Yes, the graph represents a 1-1 function.

15. No, the graph is not a 1-1 function.

17. No, the graph is not a 1-1 function.

19. No, the graph is not a 1-1 function.

Objective B Exercises

21. inverse function: {(0, 1), (3, 2), (8, 3), (15, 4)}

23. There is no inverse because the numbers 5 and -5 would each be paired with two different values in the range.

25. inverse function:
   {(-2, 0), (5, -1), (3, 3), (6, -4)}

27. There is no inverse because the number 3 would be paired with three different values of the range.

29. $f(x) = 4x - 8$

$y = 4x - 8$

$x = 4y - 8$

$x + 8 = 4y$

$\dfrac{1}{4}x + 2 = y$

$f^{-1}(x) = \dfrac{1}{4}x + 2$

31. $f(x) = 2x + 4$

$y = 2x + 4$

$x = 2y + 4$

$x - 4 = 2y$

$\dfrac{1}{2}x - 2 = y$

$f^{-1}(x) = \dfrac{1}{2}x - 2$

33. $f(x) = \dfrac{1}{2}x - 1$

$y = \dfrac{1}{2}x - 1$

$x = \dfrac{1}{2}y - 1$

$x + 1 = \dfrac{1}{2}y$

$2x + 2 = y$

$f^{-1}(x) = 2x + 2$

35. $f(x) = -2x + 2$

$y = -2x + 2$

$x = -2y + 2$

$x - 2 = -2y$

$-\dfrac{1}{2}x + 1 = y$

$f^{-1}(x) = -\dfrac{1}{2}x + 1$

37. $f(x) = \dfrac{2}{3}x + 4$

$y = \dfrac{2}{3}x + 4$

$x = \dfrac{2}{3}y + 4$

$x - 4 = \dfrac{2}{3}y$

$\dfrac{3}{2}x - 6 = y$

$f^{-1}(x) = \dfrac{3}{2}x - 6$

39. $f(x) = -\dfrac{1}{3}x + 1$

$y = -\dfrac{1}{3}x + 1$

$x = -\dfrac{1}{3}y + 1$

$x - 1 = -\dfrac{1}{3}y$

$-3x + 3 = y$

$f^{-1}(x) = -3x + 3$

41. $f(x) = 2x - 5$

$y = 2x - 5$

$x = 2y - 5$

$x + 5 = 2y$

$\dfrac{1}{2}x + \dfrac{5}{2} = y$

$f^{-1}(x) = \dfrac{1}{2}x + \dfrac{5}{2}$

43. $f(x) = 5x - 2$

$y = 5x - 2$

$x = 5y - 2$

$x + 2 = 5y$

$\dfrac{1}{5}x + \dfrac{2}{5} = y$

$f^{-1}(x) = \dfrac{1}{5}x + \dfrac{2}{5}$

45. $f(x) = 6x - 3$

$y = 6x - 3$

$x = 6y - 3$

$x + 3 = 6y$

$\dfrac{1}{6}x + \dfrac{1}{2} = y$

$f^{-1}(x) = \dfrac{1}{6}x + \dfrac{1}{2}$

47. $f(x) = 3x - 5$

$y = 3x - 5$

$x = 3y - 5$

$x + 5 = 3y$

$\dfrac{1}{3}x + \dfrac{5}{3} = y$

$f^{-1}(x) = \dfrac{1}{3}x + \dfrac{5}{3}$

$f^{-1}(0) = \dfrac{1}{3}(0) + \dfrac{5}{3}$

$f^{-1}(0) = \dfrac{5}{3}$

49. $f(x) = 3x - 5$

$y = 3x - 5$

$x = 3y - 5$

$x + 5 = 3y$

$\dfrac{1}{3}x + \dfrac{5}{3} = y$

$f^{-1}(x) = \dfrac{1}{3}x + \dfrac{5}{3}$

$f^{-1}(4) = \dfrac{1}{3}(4) + \dfrac{5}{3}$

$f^{-1}(4) = \dfrac{9}{3} = 3$

51. Using the vertical-line test the graph is a function. Using the horizontal-line test, the graph is 1-1 and therefore does have an inverse.

53. $f(g(x)) = f\left(\dfrac{x}{4}\right) = 4\left(\dfrac{x}{4}\right) = x$

$g(f(x)) = g(4x) = \dfrac{4x}{4} = x$

Yes, the functions are inverses of each other.

55. $f(h(x)) = f\left(\dfrac{1}{3x}\right) = 3\left(\dfrac{1}{3x}\right) = \dfrac{1}{x}$

$h(f(x)) = h(3x) = \dfrac{1}{3(3x)} = \dfrac{1}{9x}$

No, the functions are not inverses of each other.

57. $f(g(x)) = f(3x + 2)$

$\quad = \dfrac{1}{3}(3x + 2) - \dfrac{2}{3} = x + \dfrac{2}{3} - \dfrac{2}{3} = x$

$g(f(x)) = g\left(\dfrac{1}{3}x - \dfrac{2}{3}\right)$

$\quad = 3\left(\dfrac{1}{3}x - \dfrac{2}{3}\right) + 2 = x - 2 + 2 = x$

Yes, the functions are inverses of each other.

59. $f(g(x)) = f(2x + 3)$

$\quad = \dfrac{1}{2}(2x + 3) - \dfrac{3}{2} = x + \dfrac{3}{2} - \dfrac{3}{2} = x$

$g(f(x)) = g\left(\dfrac{1}{2}x - \dfrac{3}{2}\right)$

$\quad = 2\left(\dfrac{1}{2}x - \dfrac{3}{2}\right) + 3 = x - 3 + 3 = x$

Yes, the functions are inverses of each other.

61. $f(x) = \dfrac{x}{16}$

$y = \dfrac{x}{16}$

$x = \dfrac{y}{16}$

$16x = y$

$f^{-1}(x) = 16x$

The inverse function converts pounds to ounces.

63. $f(x) = x + 30$

$y = x + 30$

$x = y + 30$

$x - 30 = y$

$f^{-1}(x) = x - 30$

The inverse function converts a dress size in France to a dress size in the United States.

65. $f(x) = 90x + 65$

$y = 90x + 65$

$x = 90y + 65$

$x - 65 = 90y$

$\dfrac{1}{90}x - \dfrac{13}{18} = y$

$f^{-1}(x) = \dfrac{1}{90}x - \dfrac{13}{18}$

The inverse function gives the training intensity percent for a given target heart rate.

67. $f(x) = 120.381x$

$y = 120.381x$

$x = 120.381y$

$\dfrac{x}{120.381} = y$

$0.008307x = y$

$f^{-1}(x) = 0.008307x$

Critical Thinking

69. Inverse of the function:

Grade	Score
A	90-100
B	80-89
C	70-79
D	60-69
F	0-59

No, the inverse of the grading scale is not a function because each grade is paired with more than one score.

71. A constant function is defined as $y = b$, where b is a constant value. The inverse of this function would be $x = a$, where a is a constant. This is not a function.

Projects or Group Activities

73.

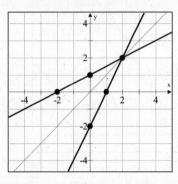

75.

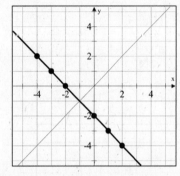

The inverse is the same graph.

77.

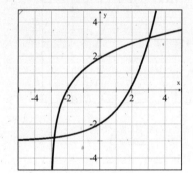

Chapter 9 Review Exercises

1.

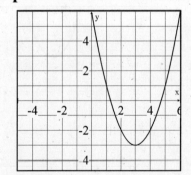

2. Yes, the graph is a function. It passes the vertical-line test.

3.

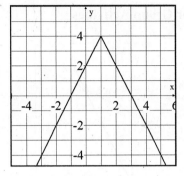

4.

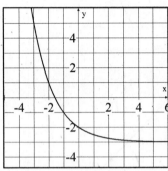

5.

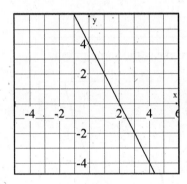

6. $f(x) = x^2 - 2x + 3$

$$-\frac{b}{2a} = -\frac{-2}{2(1)} = -\frac{-2}{2} = -(-1) = 1$$

$$y = 1^2 - 2(1) + 3 = 2$$

Vertex:
$(1, 2)$
Axis of symmetry:
$x = 1$

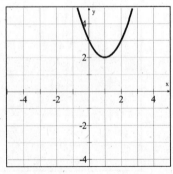

7. $y = -3x^2 + 4x + 6$

$a = -3 \quad b = 4 \quad c = 6$

$b^2 - 4ac$

$(4)^2 - 4(-3)(6) = 16 + 72 = 88$

$88 > 0$

Since the discriminant is greater than zero, the parabola has two x-intercepts.

8. $y = 2x^2 + x + 5$

$a = 2 \quad b = 1 \quad c = 5$

$b^2 - 4ac$

$(1)^2 - 4(2)(5) = 1 - 40 = -39$

$-39 < 0$

Since the discriminant is less than zero, the parabola has no x-intercepts.

9. $y = 3x^2 + 9x$

$0 = 3x^2 + 9x$

$0 = 3x(x + 3)$

$3x = 0 \quad x + 3 = 0$

$x = 0 \qquad x = -3$

The x-intercepts are $(0, 0)$ and $(-3, 0)$.

10. $f(x) = x^2 - 6x + 7$

$0 = x^2 - 6x + 7$

$a = 1 \quad b = -6 \quad c = 7$

$$x = \frac{-b \pm \sqrt{b^2 - 4ac}}{2a}$$

$$x = \frac{-(-6) \pm \sqrt{(-6)^2 - 4(1)(7)}}{2(1)}$$

$$x = \frac{6 \pm \sqrt{36 - 28}}{2} = \frac{6 \pm \sqrt{8}}{2} = \frac{6 \pm 2\sqrt{2}}{2}$$

$$x = 3 \pm \sqrt{2}$$

The x-intercepts are $(3 + \sqrt{2}, 0)$ and $(3 - \sqrt{2}, 0)$.

11. $f(x) = 2x^2 - 7x - 15$

$2x^2 - 7x - 15 = 0$

$(2x + 3)(x - 5) = 0$

$2x + 3 = 0 \qquad x - 5 = 0$

$2x = -3 \qquad\quad x = 5$

$x = -\dfrac{3}{2}$

The zeros are $-\dfrac{3}{2}$ and 5.

12. $f(x) = x^2 - 2x + 10$

$x^2 - 2x + 10 = 0$

$a = 1 \quad b = -2 \quad c = 10$

$x = \dfrac{-b \pm \sqrt{b^2 - 4ac}}{2a}$

$x = \dfrac{-(-2) \pm \sqrt{(-2)^2 - 4(1)(10)}}{2(1)}$

$x = \dfrac{2 \pm \sqrt{4 - 40}}{2} = \dfrac{2 \pm \sqrt{-36}}{2} = \dfrac{2 \pm 6i}{2}$

$x = 1 \pm 3i$

The zeros are $1 + 3i$ and $1 - 3i$.

13. $f(x) = -2x^2 + 4x + 1$

$x = -\dfrac{b}{2a} = -\dfrac{4}{2(-2)} = 1$

$f(x) = -2x^2 + 4x + 1$

$f(1) = -2(1)^2 + 4(1) + 1 = -2 + 4 + 1 = 3$

The maximum value of the function is 3.

14. $f(x) = x^2 - 7x + 8$

$x = -\dfrac{b}{2a} = -\dfrac{-7}{2(1)} = -\dfrac{7}{2}$

$f(x) = x^2 - 7x + 8$

$f\left(\dfrac{7}{2}\right) = \left(\dfrac{7}{2}\right)^2 - 7\left(\dfrac{7}{2}\right) + 8 = \dfrac{49}{4} - \dfrac{49}{2} + 8 = -\dfrac{17}{4}$

The minimum value of the function is $-\dfrac{17}{4}$.

15. $f(x) = x^2 + 4 \quad g(x) = 4x - 1$

$g(0) = 4(0) - 1 = -1$

$f(-1) = (-1)^2 + 4 = 1 + 4 = 5$

$f[g(0)] = 5$

16. $f(x) = 6x + 8 \quad g(x) = 4x + 2$

$f(-1) = 6(-1) + 8 = -6 + 8 = 2$

$g(2) = 4(2) + 2 = 8 + 2 = 10$

$g[f(-1)] = 10$

17. $f(x) = 3x^2 - 4 \quad g(x) = 2x + 1$

$f(g(x)) = f(2x + 1)$

$= 3(2x + 1)^2 - 4 = 3(2x + 1)(2x + 1) - 4$

$= 3(4x^2 + 4x + 1) - 4$

$= 12x^2 + 12x + 3 - 4 = 12x^2 + 12x - 1$

$f[g(x)] = 12x^2 + 12x - 1$

18. $f(x) = 2x^2 + x - 5 \quad g(x) = 3x - 1$

$g(f(x)) = g(2x^2 + x - 5)$

$= 3(2x^2 + x - 5) - 1$

$= 6x^2 + 3x - 15 - 1$

$= 6x^2 + 3x - 16$

$g[f(x)] = 6x^2 + 3x - 16$

19. $(f + g)(2) = f(2) + g(2)$

$= ((2)^2 + 2(2) - 3) + ((2)^2 - 2)$

$= (4 + 4 - 3) + (4 - 2)$

$= 5 + 2$

$= 7$

20. $(f - g)(-4) = f(-4) - g(-4)$

$= ((-4)^2 + 2(-4) - 3) - ((-4)^2 - 2)$

$= (16 - 8 - 3) - (16 - 2)$

$= 5 - 14$

$= -9$

21. $(f \cdot g)(-4) = f(-4) \cdot g(-4)$

$\quad = \left((-4)^2 + 2(-4) - 3\right) \cdot \left((-4)^2 - 2\right)$

$\quad = (16 - 8 - 3) \cdot (16 - 2)$

$\quad = 5 \cdot 14$

$\quad = 70$

22. $\left(\dfrac{f}{g}\right)(3) = \dfrac{f(3)}{g(3)}$

$\quad = \dfrac{(3)^2 + 2(3) - 3}{(3)^2 - 2}$

$\quad = \dfrac{9 + 6 - 3}{9 - 2}$

$\quad = \dfrac{12}{7}$

23. $f(x) = -6x + 4$

$\quad y = -6x + 4$

$\quad x = -6y + 4$

$\quad x - 4 = -6y$

$\quad -\dfrac{1}{6}x + \dfrac{2}{3} = y$

$\quad f^{-1}(x) = -\dfrac{1}{6}x + \dfrac{2}{3}$

24. $f(x) = \dfrac{2}{3}x - 12$

$\quad y = \dfrac{2}{3}x - 12$

$\quad x = \dfrac{2}{3}y - 12$

$\quad x + 12 = \dfrac{2}{3}y$

$\quad \dfrac{3}{2}x + 18 = y$

$\quad f^{-1}(x) = \dfrac{3}{2}x + 18$

25. $f(g(x)) = f(-4x + 5)$

$\quad = -\dfrac{1}{4}(-4x + 5) + \dfrac{5}{4}$

$\quad = x - \dfrac{5}{4} + \dfrac{5}{4} = x$

$\quad g(f(x)) = g\left(-\dfrac{1}{4}x + \dfrac{5}{4}\right)$

$\quad = -4\left(-\dfrac{1}{4}x + \dfrac{5}{4}\right) + 5$

$\quad = x - 5 + 5 = x$

Yes, the functions are inverses of each other.

26. $f(g(x)) = f(2x + 1)$

$\quad = \dfrac{1}{2}(2x + 1) = x + \dfrac{1}{2}$

$\quad g(f(x)) = g\left(\dfrac{1}{2}x\right)$

$\quad = 2\left(\dfrac{1}{2}x\right) + 1 = x + 1$

No, the functions are not inverses of each other.

27. $p(x) = 0.4x + 15$

$\quad p = 0.4x + 15$

$\quad x = 0.4p + 15$

$\quad x - 15 = 0.4p$

$\quad 2.5x - 37.5 = p$

$\quad p^{-1}(x) = 2.5x - 37.5$

The inverse function gives the diver's depth below the surface of the water for a given pressure on the diver.

28. Strategy: To find the number of gloves to make for a maximum profit, find the x-coordinate of the vertex. To find the maximum profit, evaluate the function at the x-coordinate of the vertex.

Solution:

$$x = -\frac{b}{2a} = -\frac{100}{2(-1)} = -\frac{100}{-2} = -(-50) = 50$$

$$P(x) = -x^2 + 100x + 2500$$

$$P(50) = -(50)^2 + 100(50) + 2500$$

$$= -2500 + 5000 + 2500 = 5000$$

The company should make 50 baseball gloves each month to maximize profit. The maximum profit is $5000.

29. Strategy: Let x represent width of the rectangle.
The length is $14 - x$.
The area is $x(14 - x)$.
To find the width, find the x-coordinate of the vertex. To find the length, evaluate $14 - x$ at the x-coordinate of the vertex.

Solution:

$$x(14 - x) = 14x - x^2$$

$$x = -\frac{b}{2a} = -\frac{14}{2(-1)} = -\frac{14}{-2} = -(-7) = 7$$

$$14 - x = 14 - 7 = 7$$

The dimensions are 7 ft by 7 ft.

Chapter 9 Test

1. $f(x) = x^2 - 6x + 4$

$$-\frac{b}{2a} = -\frac{-6}{2(1)} = -\frac{-6}{2} = -(-3) = 3$$

$$y = 3^2 - 6(3) + 4 = -5$$

Vertex:
$(3, -5)$
Axis of symmetry:
$x = 3$

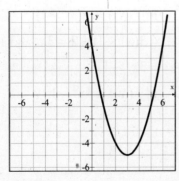

2.

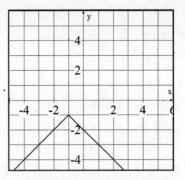

3.

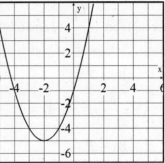

4.

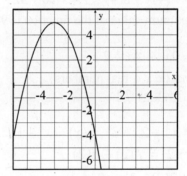

5. $y = 3x^2 - 4x + 6$

$a = 3 \quad b = -4 \quad c = 6$

$b^2 - 4ac$

$(-4)^2 - 4(3)(6) = 16 - 72 = -56$

$-56 < 0$

Since the discriminant is less than zero the parabola has no x-intercepts.

6. $y = 3x^2 - 7x - 6$

$0 = 3x^2 - 7x - 6$

$0 = (3x + 2)(x - 3)$

$3x + 2 = 0 \quad x - 3 = 0$

$3x = -2 \quad\quad x = 3$

$x = -\dfrac{2}{3}$

The x-intercepts are $(-\dfrac{2}{3}, 0)$ and $(3, 0)$.

7. $f(x) = -x^2 + 8x - 7$

$x = -\dfrac{b}{2a} = -\dfrac{8}{2(-1)} = -(-4) = 4$

$f(x) = -x^2 + 8x - 7$

$f(4) = -(4)^2 + 8(4) - 7 = -16 + 32 - 7 = 9$

The maximum value of the function is 9.

8. $f(x) = 2x^2 + 4x - 2$

Domain: $\{x \mid x \in \text{real numbers}\}$

Range: $\{y \mid y \geq -7\}$

9. $f(x) = x^2 + 2x - 3 \quad g(x) = x^3 - 1$

$(f - g)(2) = f(2) - g(2)$

$= (2^2 + 2(2) - 3) - (2^3 - 1)$

$= (4 + 4 - 3) - (8 - 1)$

$= 5 - 7$

$= -2$

10. $f(x) = x^3 + 1 \quad g(x) = 2x - 3$

$(f \cdot g)(-3) = f(-3) \cdot g(-3)$

$= ((-3)^3 + 1) \cdot (2(-3) - 3)$

$= (-27 + 1) \cdot (-6 - 3)$

$= (-26) \cdot (-9)$

$= 234$

11. $f(x) = 4x - 5 \quad g(x) = x^2 + 3x + 4$

$\left(\dfrac{f}{g}\right)(-2) = \dfrac{f(-2)}{g(-2)}$

$= \dfrac{4(-2) - 5}{(-2)^2 + 3(-2) + 4}$

$= \dfrac{-8 - 5}{4 - 6 + 4} = \dfrac{-13}{2}$

$= -\dfrac{13}{2}$

12. $f(x) = x^2 + 4 \quad g(x) = 2x^2 + 2x + 1$

$(f - g)(-4) = f(-4) - g(-4)$

$= ((-4)^2 + 4) - (2(-4)^2 + 2(-4) + 1)$

$= (16 + 4) - (32 - 8 + 1)$

$= 20 - 25$

$= -5$

13. $f(x) = 2x - 7 \quad g(x) = x^2 - 2x - 5$

$g(2) = 2^2 - 2(2) - 5 = 4 - 4 - 5 = -5$

$f(-5) = 2(-5) - 7 = -10 - 7 = -17$

$f[g(2)] = -17$

14. $f(x) = x^2 + 1 \quad g(x) = x^2 + x + 1$

$f(-2) = (-2)^2 + 1 = 4 + 1 = 5$

$g(5) = (5)^2 + 5 + 1 = 25 + 5 + 1 = 31$

$g[f(-2)] = 31$

15. $f(x) = x^2 - 1 \quad g(x) = 3x + 2$

$g(f(x)) = g(x^2 - 1)$

$= 3(x^2 - 1) + 2 = 3x^2 - 3 + 2$

$= 3x^2 - 1$

$g[f(x)] = 3x^2 - 1$

16. $f(x) = 2x^2 - 7 \quad g(x) = x - 1$

$f(g(x)) = f(x - 1)$

$= 2(x - 1)^2 - 7 = 2(x - 1)(x - 1) - 7$

$= 2(x^2 - 2x + 1) - 7$

$= 2x^2 - 4x + 2 - 7$

$= 2x^2 - 4x - 5$

$f[g(x)] = 2x^2 - 4x - 5$

17. There is no inverse because the numbers 4 and 5 would be paired with two different values of the range.

18. Inverse function: $\{(6, 2), (5, 3), (4, 4), (3, 5)\}$

19. $f(x) = 4x - 2$

$y = 4x - 2$

$x = 4y - 2$

$x + 2 = 4y$

$\dfrac{1}{4}x + \dfrac{1}{2} = y$

$f^{-1}(x) = \dfrac{1}{4}x + \dfrac{1}{2}$

20. $f(x) = \dfrac{1}{4}x - 4$

$y = \dfrac{1}{4}x - 4$

$x = \dfrac{1}{4}y - 4$

$x + 4 = \dfrac{1}{4}y$

$4x + 16 = y$

$f^{-1}(x) = 4x + 16$

21. $f(g(x)) = f(2x - 4)$

$= \dfrac{1}{2}(2x - 4) + 2 = x - 2 + 2 = x$

$g(f(x)) = g\left(\dfrac{1}{2}x + 2\right)$

$= 2\left(\dfrac{1}{2}x + 2\right) - 4 = x + 4 - 4 = x$

Yes, the functions are inverses of each other.

22. $f(g(x)) = f\left(\dfrac{3}{2}x + 3\right)$

$= \dfrac{2}{3}\left(\dfrac{3}{2}x + 3\right) - 2 = x + 2 - 2 = x$

$g(f(x)) = g\left(\dfrac{2}{3}x - 2\right)$

$= \dfrac{3}{2}\left(\dfrac{2}{3}x - 2\right) + 3 = x - 3 + 3 = x$

Yes, the functions are inverses of each other.

23. No, the graph is not a 1-1 function. It does not pass the horizontal-line test.

24. $C(x) = 1.25x + 5$

$C = 1.25x + 5$

$x = 1.25C + 5$

$x - 5 = 1.25C$

$0.8x - 4 = C$

$C^{-1}(x) = 0.8x - 4$

The inverse function gives the number of miles to a certain location for a given cost.

25. Strategy: To find the number of speakers for a minimum production cost, find the x-coordinate of the vertex. To find the minimum cost, evaluate the function at the x-coordinate of the vertex.

Solution:

$x = -\dfrac{b}{2a} = -\dfrac{-50}{2} = -\dfrac{-50}{2} = -(-25) = 25$

$C(x) = x^2 - 50x + 675$

$C(25) = (25)^2 - 50(25) + 675$

$= 625 - 1250 + 675 = 50$

The company should make 25 speakers each day to minimize production costs.
The minimum daily production cost is $50.

26. Strategy: Let x represent one number. The other number is $28 - x$.
Their product is $x(28 - x)$.
To find one number, find the x-coordinate of the vertex. To find the second number, evaluate $28 - x$ at the x-coordinate of the vertex.

Solution:

$$x(28 - x) = 28x - x^2$$

$$x = -\frac{b}{2a} = -\frac{28}{2(-1)} = -\frac{28}{-2} = -(-14) = 14$$

$$28 - x = 28 - 14 = 14$$

The two numbers are 14 and 14.

27. Strategy: Let x represent width of the rectangle. The length is $100 - x$.
The area is $x(100 - x)$.
To find the width, find the x-coordinate of the vertex. To find the length, evaluate $100 - x$ at the x-coordinate of the vertex.

Solution:

$$x(100 - x) = 100x - x^2$$

$$x = -\frac{b}{2a} = -\frac{100}{2(-1)} = -\frac{100}{-2} = -(-50) = 50$$

$$100 - x = 100 - 50 = 50$$

$$A = l \cdot w = 50 \cdot 50 = 2500$$

The dimensions of the rectangle are 50 cm by 50 cm. The area is 2500 cm^2.

Cumulative Review Exercises

1. $-3a + \left| \dfrac{3b - ab}{3b - c} \right|$

$$-3(2) + \left| \frac{3(2) - (2)(2)}{3(2) - (-2)} \right|$$

$$= -6 + \left| \frac{6 - 4}{6 + 2} \right| = -6 + \left| \frac{2}{8} \right| = -6 + \left| \frac{1}{4} \right|$$

$$= -6 + \frac{1}{4} = -\frac{23}{4}$$

2.

-5 -4 -3 -2 -1 0 1 2 3 4 5

3. $\dfrac{3x - 1}{6} - \dfrac{5 - x}{4} = \dfrac{5}{6}$

$$12\left(\frac{3x - 1}{6} - \frac{5 - x}{4} \right) = 12\left(\frac{5}{6} \right)$$

$$2(3x - 1) - 3(5 - x) = 10$$

$$6x - 2 - 15 + 3x = 10$$

$$9x - 17 = 10$$

$$9x = 27$$

$$x = 3$$

The solution is 3.

4. $4x - 2 < -10$ or $3x - 1 > 8$

$\quad 4x - 2 < -10 \qquad 3x - 1 > 8$

$\qquad 4x < -8 \qquad\quad 3x > 9$

$\qquad\quad x < -2 \qquad\qquad x > 3$

$\{x \mid x < -2\}$ or $\{x \mid x > 3\}$

$\{x \mid x < -2\} \cup \{x \mid x > 3\}$

5. $|8 - 2x| \geq 0$

$\quad 8 - 2x \leq 0 \qquad 8 - 2x \geq 0$

$\qquad -2x \leq -8 \qquad -2x \geq -8$

$\qquad\quad x \geq 4 \qquad\qquad x \leq 4$

$\{x \mid x \geq 4\}$ or $\{x \mid x \leq 4\}$

$\{x \mid x \geq 4\} \cup \{x \mid x \leq 4\}$

$\qquad = \{x \mid x \in \text{real numbers}\}$

6. $\left(\dfrac{3a^3 b}{2a} \right)^2 \left(\dfrac{a^2}{-3b^2} \right)^3 = \left(\dfrac{3a^2 b}{2} \right)^2 \left(\dfrac{a^2}{-3b^2} \right)^3$

$$= \left(\frac{3^2 a^4 b^2}{2^2} \right)\left(\frac{a^6}{(-3)^3 b^6} \right) = \frac{9a^{10} b^2}{4(-27)b^6}$$

$$= \frac{9a^{10}}{-108b^4} = -\frac{a^{10}}{12b^4}$$

7. $(x - 4)(2x^2 + 4x - 1)$

$$= x(2x^2 + 4x - 1) - 4(2x^2 + 4x - 1)$$

$$= 2x^3 + 4x^2 - x - 8x^2 - 16x + 4$$

$$= 2x^3 - 4x^2 - 17x + 4$$

8. $6x - 2y = -3$
$4x + y = 5$

$6x - 2y = -3$
$\underline{8x + 2y = 10}$
$14x \quad\;\; = 7$

$x = \dfrac{1}{2}$

$6\left(\dfrac{1}{2}\right) - 2y = -3$

$3 - 2y = -3$
$-2y = -6$
$y = 3$

The solution is $\left(\dfrac{1}{2}, 3\right)$.

9. $x^3y + x^2y^2 - 6xy^3 = xy(x^2 + xy - 6y^2)$
$= xy(x + 3y)(x - 2y)$

10. $(b + 2)(b - 5) = 2b + 14$
$b^2 - 3b - 10 = 2b + 14$
$b^2 - 5b - 24 = 0$
$(b - 8)(b + 3) = 0$
$b - 8 = 0 \quad\;\; b + 3 = 0$
$b = 8 \quad\qquad b = -3$
The solutions are -3 and 8.

11. $x^2 - 2x > 15$
$x^2 - 2x - 15 > 0$
$(x - 5)(x + 3) > 0$
$\{x \mid x < -3 \text{ or } x > 5\}$

12. $\dfrac{x^2 + 4x - 5}{2x^2 - 3x + 1} - \dfrac{x}{2x - 1}$

$= \dfrac{(x + 5)(x - 1)}{(2x - 1)(x - 1)} - \dfrac{x}{2x - 1}$

$= \dfrac{x + 5}{2x - 1} - \dfrac{x}{2x - 1} = \dfrac{x + 5 - x}{2x - 1}$

$= \dfrac{5}{2x - 1}$

13. $\dfrac{5}{x^2 + 7x + 12} = \dfrac{9}{x + 4} - \dfrac{2}{x + 3}$

$\dfrac{5}{(x + 4)(x + 3)} = \dfrac{9}{x + 4} \cdot \dfrac{x + 3}{x + 3} - \dfrac{2}{x + 3} \cdot \dfrac{x + 4}{x + 4}$

$\dfrac{5}{(x + 4)(x + 3)} = \dfrac{9x + 27}{(x + 4)(x + 3)} - \dfrac{2x + 8}{(x + 4)(x + 3)}$

$5 = (9x + 27) - (2x + 8)$

$5 = 7x + 19$

$-14 = 7x$

$-2 = x$
The solution is -2.

14. $\dfrac{4 - 6i}{2i} = \dfrac{4 - 6i}{2i} \cdot \dfrac{i}{i} = \dfrac{4i - 6i^2}{2i^2}$

$= \dfrac{4i + 6}{-2} = -3 - 2i$

15. $f(x) = \dfrac{1}{4}x^2$

$-\dfrac{b}{2a} = -\dfrac{0}{2\left(\dfrac{1}{4}\right)} = -\dfrac{0}{\dfrac{1}{4}} = 0$

$y = \left(\dfrac{1}{4}\right)0^2 = 0$

Vertex:
$(0, 0)$
Axis of
symmetry:
$x = 0$

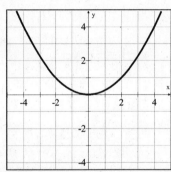

16. $3x - 4y \geq 8$
$-4y \geq -3x + 8$
$y \leq \dfrac{3}{4}x - 2$

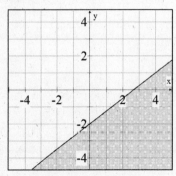

17. $m = \dfrac{y_2 - y_1}{x_2 - x_1} = \dfrac{-6-4}{2-(-3)} = \dfrac{-10}{5} = -2$

$y - y_1 = m(x - x_1)$

$y - 4 = -2(x - (-3))$

$y - 4 = -2(x + 3)$

$y - 4 = -2x - 6$

$y = -2x - 2$

18. The product of the slopes of perpendicular lines is -1.

$2x - 3y = 6$

$-3y = -2x + 6$

$y = \dfrac{2}{3}x - 2$

$m_1 \cdot m_2 = -1$

$\dfrac{2}{3} \cdot m_2 = -1$

$m_2 = -\dfrac{3}{2}$

$y - y_1 = m(x - x_1)$

$y - 1 = -\dfrac{3}{2}(x - (-3))$

$y - 1 = -\dfrac{3}{2}(x + 3)$

$y - 1 = -\dfrac{3}{2}x - \dfrac{9}{2}$

$y = -\dfrac{3}{2}x - \dfrac{7}{2}$

19. $3x^2 = 3x - 1$

$3x^2 - 3x + 1 = 0$

$a = 3 \quad b = -3 \quad c = 1$

$x = \dfrac{-b \pm \sqrt{b^2 - 4ac}}{2a}$

$x = \dfrac{-(-3) \pm \sqrt{(-3)^2 - 4(3)(1)}}{2(3)}$

$x = \dfrac{3 \pm \sqrt{9-12}}{6} = \dfrac{3 \pm \sqrt{-3}}{6} = \dfrac{3 \pm i\sqrt{3}}{6}$

$x = \dfrac{1}{2} \pm \dfrac{i\sqrt{3}}{6}$

The zeros are $\dfrac{1}{2} + \dfrac{i\sqrt{3}}{6}$ and $\dfrac{1}{2} - \dfrac{i\sqrt{3}}{6}$.

20. $\sqrt{8x+1} = 2x - 1$

$\left(\sqrt{8x+1}\right)^2 = (2x-1)^2$

$8x + 1 = 4x^2 - 4x + 1$

$0 = 4x^2 - 12x$

$0 = 4x(x - 3)$

$4x = 0 \qquad x - 3 = 0$

$x = 0 \qquad\quad x = 3$

Check both solutions in the original equation:

$\sqrt{8(0)+1} = 2(0) - 1$

$\sqrt{1} = -1$

$1 \neq -1$

$\sqrt{8(3)+1} = 2(3) - 1$

$\sqrt{25} = 5$

$5 = 5$

The solution is 3.

21. $f(x) = 2x^2 - 3$

$$x = -\frac{b}{2a} = -\frac{0}{2(2)} = -\frac{0}{4} = 0$$

$$f(x) = 2x^2 - 3$$

$$f(0) = 2(0)^2 - 3 = 0 - 3$$

$$= -3$$

Since $a > 0$, the function has a minimum value. The minimum value of the function is -3.

22. $f(x) = 3x - 4$

$$0 = 3x - 4$$

$$4 = 3x$$

$$\frac{4}{3} = x$$

The zero of the function is $\frac{4}{3}$

23. Yes

24. $\sqrt[3]{5x - 2} = 2$

$$\left(\sqrt[3]{5x - 2}\right)^3 = 2^3$$

$$5x - 2 = 8$$

$$5x = 10$$

$$x = 2$$

The solution is 2.

25. $g(x) = 3x - 5 \quad h(x) = \frac{1}{2}x + 4$

$$h(2) = \frac{1}{2}(2) + 4 = 1 + 4 = 5$$

$$g(5) = 3(5) - 5 = 15 - 5 = 10$$

$$g[h(2)] = 10$$

26. $f(x) = -3x + 9$

$$y = -3x + 9$$

$$x = -3y + 9$$

$$x - 9 = -3y$$

$$-\frac{1}{3}x + 3 = y$$

$$f^{-1}(x) = -\frac{1}{3}x + 3$$

27. Strategy: Let x represent the cost per pound of the mixture.

	Amount	Cost	Value
$4.50 tea	30	4.50	30(4.50)
$3.60 tea	45	3.60	45(3.60)
Mixture	75	x	75x

The sum of the values before mixing is equal to the value after mixing.

Solution:
$$30(4.50) + 45(3.60) = 75x$$
$$135 + 162 = 75x$$
$$297 = 75x$$
$$x = 3.96$$
The cost per pound of the mixture is $3.96.

28. Strategy: Let x represent the number of pounds of 80% copper alloy.

	Amount	Percent	Quantity
80%	x	0.80	0.80x
20%	50	0.20	0.20(50)
40%	$50 + x$	0.40	0.40(50 + x)

The sum of the quantities before mixing is equal to the quantity after mixing.

Solution:
$$0.80x + 0.20(50) = 0.40(50 + x)$$
$$0.80x + 10 = 20 + 0.40x$$
$$0.40x + 10 = 20$$
$$0.40x = 10$$
$$x = 25$$
25 lb of the 80% copper alloy must be used.

29. Strategy: Let x represent the additional amount of insecticide.
The total amount of insecticide is $x + 6$.
To find the additional amount of insecticide write and solve a proportion.

Solution: $\dfrac{6}{16} = \dfrac{x+6}{28}$.

$\dfrac{3}{8} = \dfrac{x+6}{28}$

$\dfrac{3}{8} \cdot 56 = \dfrac{x+6}{28} \cdot 56$

$21 = 2x + 12$

$9 = 2x$

$4.5 = x$

An additional 4.5 oz of insecticide are required.

30. Strategy: Let x represent the time it takes for the smaller pipe to fill the take.
The time it takes the larger pipe to fill the tank is $x - 8$.

	Rate	Time	Part
Smaller pipe	$\dfrac{1}{t}$	3	$\dfrac{3}{t}$
Larger pipe	$\dfrac{1}{t-8}$	3	$\dfrac{3}{t-8}$

The sum of the parts of the task completed must equal 1.

Solution:

$\dfrac{3}{t} + \dfrac{3}{t-8} = 1$

$t(t-8)\left(\dfrac{3}{t} + \dfrac{3}{t-8}\right) = 1(t(t-8))$

$3(t-8) + 3t = t^2 - 8t$

$3t - 24 + 3t = t^2 - 8t$

$6t - 24 = t^2 - 8t$

$0 = t^2 - 14t + 24$

$0 = (t-2)(t-12)$

$t - 2 = 0 \quad t - 12 = 0$

$\quad t = 2 \qquad t = 12$

The solution $t = 2$ is not possible since the time for the larger pipe would then be a negative number.
$t - 8 = 2 - 8 = -6$
It takes the larger pipe $t - 8 = 12 - 8 = 4$ min to fill the tank.

31. Strategy: Write the basic direct variation equation, replacing the variable with the given values. Solve for k.
Write the direct variation equation, replacing k with its value. Substitute 40 for f and solve for d.

Solution:

$d = kf \qquad d = \dfrac{3}{5}f$

$30 = k(50) \qquad = \dfrac{3}{5}(40)$

$\dfrac{3}{5} = k \qquad = 24$

A force of 40 lb will stretch the spring 24 in.

32. Strategy: Write the basic inverse variation equation, replacing the variable with the given values. Solve for k.
Write the inverse variation equation, replacing k with its value. Substitute 1.5 for L and solve for f.

Solution:

$f = \dfrac{k}{L} \qquad f = \dfrac{120}{L}$

$60 = \dfrac{k}{2} \qquad = \dfrac{120}{1.5}$

$120 = k \qquad = 80$

The frequency is 80 vibrations/min.

Chapter 10: Exponential and Logarithmic Functions

Prep Test

1. $3^{-2} = \dfrac{1}{3^2} = \dfrac{1}{9}$

2. $\left(\dfrac{1}{2}\right)^{-4} = \left(\dfrac{2}{1}\right)^4 = 2^4 = 16$

3. $\dfrac{1}{8} = \dfrac{1}{2^3} = 2^{-3}$

4. $f(x) = x^4 + x^3$
$f(-1) = (-1)^4 + (-1)^3 = 1 + (-1) = 0$
$f(3) = (3)^4 + (3)^3 = 81 + 27 = 108$

5. $3x + 7 = x - 5$
$2x + 7 = -5$
$2x = -12$
$x = -6$
The solution is -6.

6. $16 = x^2 - 6x$
$0 = x^2 - 6x - 16$
$0 = (x - 8)(x + 2)$
$x - 8 = 0 \quad x + 2 = 0$
$x = 8 \qquad x = -2$
The solutions are -2 and 8.

7. $A(1 + r)^n$
$5000(1 + 0.04)^6 = 5000(1.04)^6$
$= 6326.60$

8. $f(x) = x^2 - 1$
$x = -\dfrac{b}{2a} = \dfrac{0}{2(1)} = \dfrac{0}{2} = 0$
$f(0) = (0)^2 - 1 = -1$

Vertex:
$(0, -1)$.
Axis of
symmetry:
$x = 0$.

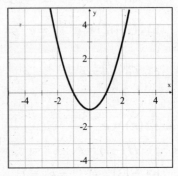

Section 10.1

Objective A Exercises

1. An exponential function with base b is defined by $f(x) = b^x$, $b > 0$, $b \neq 1$, and x is any real number.

3. $f(x) = b^x$, $b > 0$, $b \neq 1$
(iii) (v) and (vi) cannot be the base.

Objective A Exercises

5. $f(x) = 3^x$
a) $f(2) = 3^2 = 9$
b) $f(0) = 3^0 = 1$
c) $f(-2) = 3^{-2} = \dfrac{1}{3^2} = \dfrac{1}{9}$

7. $g(x) = 2^{x+1}$
a) $g(3) = 2^{3+1} = 2^4 = 16$
b) $g(1) = 2^{1+1} = 2^2 = 4$
c) $g(-3) = 2^{-3+1} = 2^{-2} = \dfrac{1}{2^2} = \dfrac{1}{4}$

9. a) $P(0) = \left(\dfrac{1}{2}\right)^{2(0)} = \left(\dfrac{1}{2}\right)^0 = 1$

b) $P\left(\dfrac{3}{2}\right) = \left(\dfrac{1}{2}\right)^{2(3/2)} = \left(\dfrac{1}{2}\right)^3 = \dfrac{1}{8}$

c) $P(-2) = \left(\dfrac{1}{2}\right)^{2(-2)} = \left(\dfrac{1}{2}\right)^{-4} = 2^4 = 16$

11. $G(x) = e^{x/2}$

   a) $G(4) = e^{4/2} = e^2 = 7.3891$

   b) $G(-2) = e^{-2/2} = e^{-1} = \dfrac{1}{e^1} = 0.3679$

   c) $G\left(\dfrac{1}{2}\right) = e^{(1/2)/2} = e^{1/4} = 1.2840$

13. $H(r) = e^{-r+3}$

   a) $H(-1) = e^{-(-1)+3} = e^4 = 54.5982$

   b) $H(3) = e^{-3+3} = e^0 = 1$

   c) $H(5) = e^{-5+3} = e^{-2} = \dfrac{1}{e^2} = 0.1353$

15. $F(x) = 2^{x^2}$

   a) $F(2) = 2^{2^2} = 2^4 = 16$

   b) $F(-2) = 2^{(-2)^2} = 2^4 = 16$

   c) $F\left(\dfrac{3}{4}\right) = 2^{(3/4)^2} = 2^{9/16} = 1.4768$

17. $f(x) = e^{-x^2/2}$

   a) $f(-2) = e^{-2^2/2} = e^{-2} = \dfrac{1}{e^2} = 0.1353$

   b) $f(2) = e^{-2^2/2} = e^{-2} = \dfrac{1}{e^2} = 0.1353$

   c) $f(-3) = e^{-3^2/2} = e^{-9/2} = \dfrac{1}{e^{9/2}} = 0.0111$

19. $f(a) > f(b)$

Objective B Exercises

21. $f(x) = 3^x$

x	y
0	1
-1	$\frac{1}{3}$
1	3

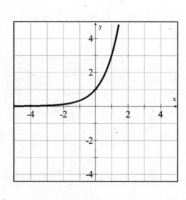

23. $f(x) = 2^{x+1}$

x	y
0	2
−1	1
1	4

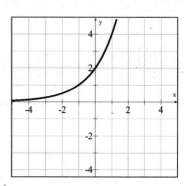

25. $f(x) = \left(\dfrac{1}{3}\right)^x$

x	y
0	1
−1	3
1	$\frac{1}{3}$

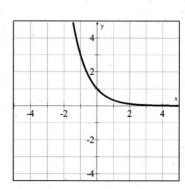

27. $f(x) = 2^{-x} + 1$

x	y
0	2
−1	3
1	$\frac{3}{2}$

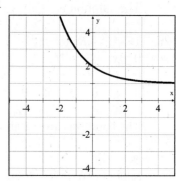

29. $f(x) = \left(\dfrac{1}{3}\right)^{-x}$

x	y
0	1
−1	$\frac{1}{3}$
1	3

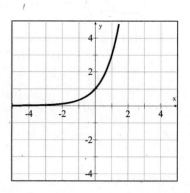

31. $f(x) = e^{x+1} - 1$

x	y
0	1.7
−1	0
1	6.4

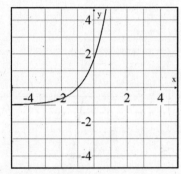

33. $v(t) = 32(1 - e^{-t})$

$v(4) = 32(1 - e^{-4}) \approx 31.41$

The speed of the object after 4 s is 31.41 ft/sec.

35. (i) and (iii) have the same graphs.
(ii) and (iv) have the same graphs.

Critical Thinking

37. $P(x) = \left(\sqrt{3}\right)^{x}$

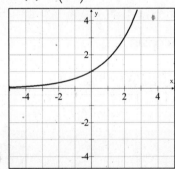

39. $f(x) = \pi^{x}$

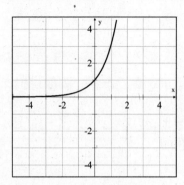

Projects or Group Activities

41.

n	$(1+n)^{1/n}$
0.01	2.704814
0.001	2.716924
0.0001	2.718146
0.00001	2.718268

As n decreases, $(1 + n)^{1/n}$ becomes closer to e.

Section 10.2

Concept Check

1. A common logarithm is a logarithm with a base of 10.

3. $\log_5 25 = 2$

5. $\log_4 \dfrac{1}{16} = -2$

7. $3^4 = 81$

9. $e^q = p$

11. False

13. True

15. True

Objective A Exercises

17. $\log_3 81 = x$

$3^x = 81$

$x = 4$

$\log_3 81 = 4$

21. $\log 100 = x$

$10^x = 100$

$x = 2$

$\log 100 = 2$

23. $\ln e^3 = x$

$3 \ln e = x$

$3(1) = x$

$x = 3$

$\ln e^3 = 3$

25. $\log_8 1 = x$

$8^x = 1$

$x = 0$

$\log_8 1 = 0$

27. $\log_5 625 = x$

$5^x = 625$

$x = 4$

$\log_5 625 = 4$

29. $\log_3 x = 2$

$3^2 = x$

$x = 9$

31. $\log_4 x = 3$

$4^3 = x$

$x = 64$

33. $\log_7 x = -1$

$7^{-1} = x$

$x = \dfrac{1}{7}$

35. $\log_6 x = 0$

$6^0 = x$

$x = 1$

37. $\log x = 2.5$

$10^{2.5} = x$

$x = 316.23$

39. $\log x = -1.75$

$10^{-1.75} = x$

$x = 0.02$

41. $\ln x = 2$

$e^2 = x$

$x = 7.39$

43. $\ln x = -\dfrac{1}{2}$

$e^{-1/2} = x$

$x = 0.61$

45. $x > 1$

Objective B Exercises

47. $\log_b(xy) = \log_b(x) + \log_b(y)$

49. False

51. $\log_{12} 1 = 0$

53. $\ln e = 1$

55. $\log_3 3^x = x$

57. $e^{\ln v} = v$

59. $2^{\log_2(x^2+1)} = x^2 + 1$

61. $\log_5 5^{x^2-x-1} = x^2 - x - 1$

63. $\log_8(xz) = \log_8 x + \log_8 z$

65. $\log_3 x^5 = 5\log_3 x$

67. $\log_b\left(\dfrac{r}{s}\right) = \log_b r - \log_b s$

69. $\log_3(x^2 y^6) = \log_3 x^2 + \log_3 y^6$
$= 2\log_3 x + 6\log_3 y$

71. $\log_7\left(\dfrac{u^3}{v^4}\right) = \log_7 u^3 - \log_7 v^4$
$= 3\log_7 u - 4\log_7 v$

73. $\log_2(rs)^2 = 2\log_2(rs) = 2[\log_2 r + \log_2 s]$

75. $\ln(x^2 yz) = \ln x^2 + \ln y + \ln z$
$= 2\ln x + \ln y + \ln z$

77. $\log_5\left(\dfrac{xy^2}{z^4}\right) = \log_5 xy^2 - \log_5 z^4$
$= \log_5 x + \log_5 y^2 - \log_5 z^4$
$= \log_5 x + 2\log_5 y - 4\log_5 z$

79. $\log_8\left(\dfrac{x^2}{yz^2}\right) = \log_8 x^2 - \log_8 yz^2$
$= \log_8 x^2 - (\log_8 y + \log_8 z^2)$
$= \log_8 x^2 - \log_8 y - \log_8 z^2$
$= 2\log_8 x - \log_8 y - 2\log_8 z$

81. $\log_4 \sqrt{x^3 y} = \log_4(x^3 y)^{1/2} = \dfrac{1}{2}\log_4(x^3 y)$
$= \dfrac{1}{2}[\log_4 x^3 + \log_4 y]$
$= \dfrac{1}{2}[3\log_4 x + \log_4 y]$
$= \dfrac{3}{2}\log_4 x + \dfrac{1}{2}\log_4 y$

83. $\log_7 \sqrt{\dfrac{x^3}{y}} = \log_7\left(\dfrac{x^3}{y}\right)^{1/2} = \dfrac{1}{2}\log_7 \dfrac{x^3}{y}$
$= \dfrac{1}{2}[\log_7 x^3 - \log_7 y]$
$= \dfrac{1}{2}[3\log_7 x - \log_7 y]$
$= \dfrac{3}{2}\log_7 x - \dfrac{1}{2}\log_7 y$

85. $\log_3\left(\dfrac{t}{\sqrt{x}}\right) = \log_3\left(\dfrac{t}{x^{1/2}}\right)$
$= \log_3 t - \log_3 x^{1/2}$
$= \log_3 t - \dfrac{1}{2}\log_3 x$

87. $\log_3 x^3 + \log_3 y^2 = \log_3(x^3 y^2)$

89. $\ln x^4 - \ln y^2 = \ln\left(\dfrac{x^4}{y^2}\right)$

91. $3\log_7 x = \log_7 x^3$

93. $3\ln x + 4\ln y = \ln x^3 + \ln y^4 = \ln(x^3 y^4)$

95. $2(\log_4 x + \log_4 y) = 2\log_4(xy)$
$= \log_4(xy)^2$
$= \log_4(x^2 y^2)$

97. $2\log_3 x - \log_3 y + 2\log_3 z$
$= \log_3 x^2 - \log_3 y + \log_3 z^2$
$= \log_3\left(\dfrac{x^2}{y}\right) + \log_3 z^2$
$= \log_3\left(\dfrac{x^2 z^2}{y}\right)$

99. $\ln x - (2\ln y + \ln z) = \ln x - (\ln y^2 + \ln z)$
$= \ln x - \ln(y^2 z)$
$= \ln\left(\dfrac{x}{y^2 z}\right)$

101. $\dfrac{1}{2}(\log_6 x - \log_6 y) = \dfrac{1}{2}\log_6\left(\dfrac{x}{y}\right)$

$= \log_6\left(\dfrac{x}{y}\right)^{1/2}$

$= \log_6\sqrt{\dfrac{x}{y}}$

103. $2(\log_4 s - 2\log_4 t + \log_4 r)$

$= 2(\log_4 s - \log_4 t^2 + \log_4 r)$

$= 2\left(\log_4 \dfrac{s}{t^2} + \log_4 r\right)$

$= 2\log_4\left(\dfrac{sr}{t^2}\right)$

$= \log_4\left(\dfrac{sr}{t^2}\right)^2$

$= \log_4 \dfrac{s^2 r^2}{t^4}$

105. $\ln x - 2(\ln y + \ln z)$

$= \ln x - 2\ln(yz)$

$= \ln x - \ln(yz)^2$

$= \ln\left(\dfrac{x}{(yz)^2}\right)$

$= \ln \dfrac{x}{y^2 z^2}$

107. $\dfrac{1}{2}(3\log_4 x - 2\log_4 y + \log_4 z)$

$= \dfrac{1}{2}(\log_4 x^3 - \log_4 y^2 + \log_4 z)$

$= \dfrac{1}{2}\left(\log_4\left(\dfrac{x^3}{y^2}\right) + \log_4 z\right)$

$= \log_4\left(\dfrac{x^3 z}{y^2}\right)^{1/2}$

$= \log_4\sqrt{\dfrac{x^3 z}{y^2}}$

109. $\dfrac{1}{2}\log_2 x - \dfrac{2}{3}\log_2 y + \dfrac{1}{2}\log_2 z$

$= \log_2 x^{1/2} - \log_2 y^{2/3} + \log_2 z^{1/2}$

$= \log_2\left(\dfrac{x^{1/2}}{y^{2/3}}\right) + \log_5 z^{1/2}$

$= \log_2\left(\dfrac{x^{1/2} z^{1/2}}{y^{2/3}}\right)$

$= \log_2\dfrac{\sqrt{xy}}{\sqrt[3]{y^2}}$

Objective C Exercises

111. $\log_8 6 = \dfrac{\log_{10} 6}{\log_{10} 8} = 0.8617$

113. $\log_5 30 = \dfrac{\log_{10} 30}{\log_{10} 5} = 2.1133$

115. $\log_3 0.5 = \dfrac{\log_{10} 0.5}{\log_{10} 3} = -0.6309$

117. $\log_7 1.7 = \dfrac{\log_{10} 1.7}{\log_{10} 7} = 0.2727$

119. $\log_5 15 = \dfrac{\log_{10} 15}{\log_{10} 5} = 1.6826$

121. $\log_{12} 120 = \dfrac{\log_{10} 120}{\log_{10} 12} = 1.9266$

123. $\log_4 2.55 = \dfrac{\log_{10} 2.55}{\log_{10} 4} = 0.6752$

125. $\log_5 67 = \dfrac{\log_{10} 67}{\log_{10} 5} = 2.6125$

127. $\log_5 x = \dfrac{\log x}{\log 5}$

131. $\log_2(\log_2 64) = x$

Let $y = \log_2(64)$

$y = \log_2(64) = 6$

$\log_2(6) = x$

$x \approx 2.58$

133. Because $x = 4$, $x - 5 = -1$. The logarithm of a negative number is undefined.

Critical Thinking

129. $\log_3(\log_3 x) = 2$

Let $y = \log_3(x)$

$\log_3(\log_3(x)) = 2$

$\log_3(y) = 2$

$3^2 = y$

$y = 9$

$y = \log_3(x)$

$9 = \log_3(x)$

$3^9 = x$

$x = 19{,}683$

Projects or Groups Activities

135. a) $D = -(p_1 \log_2 p_1 + p_2 \log_2 p_2 + p_3 \log_2 p_3 + p_4 \log_2 p_4 + p_5 \log_2 p_5)$

$D = -\left(\dfrac{1}{5}\log_2\dfrac{1}{5} + \dfrac{1}{5}\log_2\dfrac{1}{5} + \dfrac{1}{5}\log_2\dfrac{1}{5} + \dfrac{1}{5}\log_2\dfrac{1}{5} + \dfrac{1}{5}\log_2\dfrac{1}{5}\right)$

$D = -5\left(\dfrac{1}{5}\log_2 5\right) = \log_2 5 = 2.3219281$

b) $D = -(p_1 \log_2 p_1 + p_2 \log_2 p_2 + p_3 \log_2 p_3 + p_4 \log_2 p_4 + p_5 \log_2 p_5)$

$D = -\left(\dfrac{1}{8}\log_2\dfrac{1}{8} + \dfrac{3}{8}\log_2\dfrac{3}{8} + \dfrac{1}{16}\log_2\dfrac{1}{16} + \dfrac{1}{8}\log_2\dfrac{1}{8} + \dfrac{5}{16}\log_2\dfrac{5}{16}\right)$

$D = 2.055036$

Less diversity

c) $D = -(p_1 \log_2 p_1 + p_2 \log_2 p_2 + p_3 \log_2 p_3 + p_4 \log_2 p_4 + p_5 \log_2 p_5)$

$D = -\left(0\log_2 0 + \dfrac{1}{4}\log_2 \dfrac{1}{4} + 0\log_2 0 + 0\log_2 0 + \dfrac{3}{3}\log_2 \dfrac{3}{4}\right)$

$D = -\left(0 + \dfrac{1}{4}\log_2 \dfrac{1}{4} + 0 + 0 + \dfrac{3}{3}\log_2 \dfrac{3}{4}\right) = 0.82600$

Less diversity

d) $D = -(p_1 \log_2 p_1 + p_2 \log_2 p_2 + p_3 \log_2 p_3 + p_4 \log_2 p_4 + p_5 \log_2 p_5)$

$D = -(0\log_2 0 + 0\log_2 0 + 0\log_2 0 + 0\log_2 0 + 1\log_2 1)$

$D = -(0 + 0 + 0 + 0 + 0) = 0$

Because this system has only one species, there is no diversity in the system.

Section 10.3

Concept Check

1. They have the same graph.

3. $x = 2^{\frac{y}{3}}$

5. $x = e^{\frac{y-2}{3}}$

Objective A Exercises

7. $f(x) = \log_4 x$

$y = \log_4 x$ is equivalent to $x = 4^y$.

x	y
$\frac{1}{16}$	-2
$\frac{1}{4}$	-1
1	0
4	1

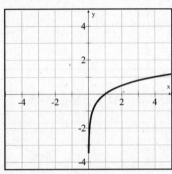

9. $f(x) = \log_3(2x - 1)$

$y = \log_3(2x - 1)$ is equivalent to

$2x - 1 = 3^y$ or $x = \frac{1}{2}(3^y + 1)$.

x	y
$\frac{2}{3}$	-1
1	0
2	1
5	2

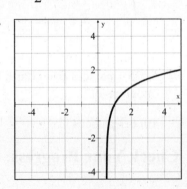

11. $f(x) = 3\log_2 x$

$y = 3\log_2 x$ is equivalent to $x = 2^{y/3}$.

x	y
$\frac{1}{2}$	-3
1	0
2	3
4	6

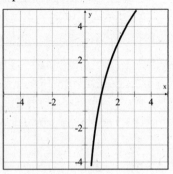

13. $f(x) = -\log_2 x$

$y = -\log_2 x$ is equivalent to $x = 2^{-y}$.

x	y
2	-1
1	0
$\frac{1}{2}$	1
$\frac{1}{4}$	2

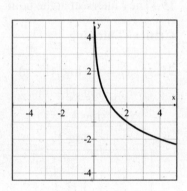

15. $f(x) = \log_2(x - 1)$

$y = \log_2(x - 1)$ is equivalent to $x - 1 = 2^y$

or $x = 2^y + 1$.

x	y
$\frac{3}{2}$	-1
2	0
3	1
5	2

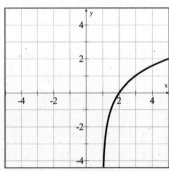

17. $f(x) = -\log_2(x-1)$

$y = -\log_2(x-1)$ is equivalent to

$x - 1 = 2^{-y}$ or $x = 2^{-y} + 1$.

x	y
3	-1
2	0
$\frac{3}{2}$	1
$\frac{5}{4}$	2

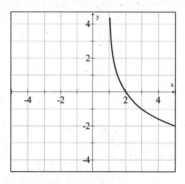

19. They intersect at the point $(1, 0)$.

Critical Thinking

21. $f(x) = x - \log_2(1-x)$

$y = x - \log_2(1-x)$

$y = x - \dfrac{\log(1-x)}{\log 2}$

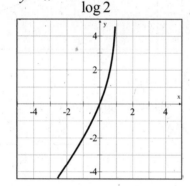

23. $f(x) = \dfrac{x}{2} - 2\log_2(x+1)$

$y = \dfrac{x}{2} - 2\log_2(x+1)$

$y = \dfrac{x}{2} - \log_2(x+1)^2$

$y = \dfrac{x}{2} - \dfrac{\log(x+1)^2}{\log 2}$

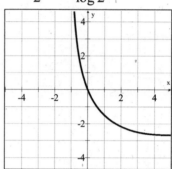

25. a) $M = 5\log s - 5$

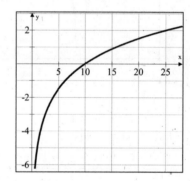

b) The point $(25.1, 2)$ means that a star that is 25.1 parsecs from Earth has a distance modulus of 2.

c) 6.3 parsecs

Check Your Progress: Chapter 10

1. $f(x) = 3^x$

$f(4) = 3^4 = 81$

2. $f(x) = 2^{x-5}$

$f(2) = 2^{2-5} = 2^{-3} = \dfrac{1}{2^3} = \dfrac{1}{8}$

3. $f(x) = 4^{2x+3}$

$f(-2) = 4^{2(-2)+3} = 4^{-1} = \dfrac{1}{4^1} = \dfrac{1}{4}$

4.

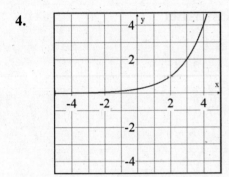

5.

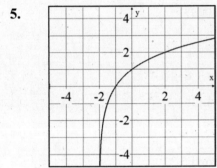

6.

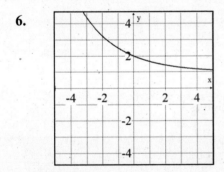

7. $\log_3 81 = x$

$3^x = 81$

$x = 4$

$\log_3 81 = 4$

8. $\log_4\left(\dfrac{1}{64}\right) = x$

$4^x = \dfrac{1}{64}$

$x = -3$

$\log_4\left(\dfrac{1}{64}\right) = -3$

9. $\log_5\left(\dfrac{1}{5}\right) = x$

$5^x - \dfrac{1}{5}$

$x = -1$

$\log_5\left(\dfrac{1}{5}\right) = -1$

10. $\log_7 7^{33} = x$

$7^x = 7^{33}$

$x = 33$

$\log_7 7^{33} = 33$

11. $\log_5 x = 4$

$5^4 = x$

$x = 625$

12. $\log_3 x = -3$

$3^{-3} = x$

$x = \dfrac{1}{27}$

13. $\log_7 x = 1$

$7^1 = x$

$x = 7$

14. $\log x = -4$

$10^{-4} = x$

$x = 0.0001$

15. $\log_7\left(x^2 y^5\right) = \log_7 x^2 + \log_7 y^5$
$= 2\log_7 x + 5\log_7 y$

16. $\log_8\left(\dfrac{x}{y^3}\right) = \log_8 x - \log_8 y^3$
$= \log_8 x - 3\log_8 y$

17. $\log_3\left(\dfrac{x^2}{\sqrt{yz}}\right) = \log_3 x^2 - \log_3 (yz)^{1/2}$

$= 2\log_3 x - \dfrac{1}{2}\log_3 (yz)$

$= 2\log_3 x - \dfrac{1}{2}\log_3 y - \dfrac{1}{2}\log_3 z$

18. $3\log_3 x - 4\log_3 y = \log_3 x^3 - \log_3 y^4$

$= \log_3\left(\dfrac{x^3}{y^4}\right)$

19. $\ln x - (4\ln y - 5\ln z) = \ln x - (\ln y^4 - \ln z^5)$

$= \ln x - \ln\dfrac{y^4}{z^5} = \ln\dfrac{x}{\dfrac{y^4}{z^5}} = \ln\dfrac{xz^5}{y^4}$

20. $\dfrac{1}{2}\left(\log x + \log y\right) = \dfrac{1}{2}\log(xy) = \log(xy)^{1/2}$

$= \log\sqrt{xy}$

21. $\log_3 12 = \dfrac{\log_{10} 12}{\log_{10} 3} \approx 2.2619$

22. $\log_5 0.1 = \dfrac{\log_{10} 0.1}{\log_{10} 5} \approx -1.4307$

23. $\log_7 5 = \dfrac{\log_{10} 5}{\log_{10} 7} \approx 0.8271$

Section 10.4

Concept Check

1. The 1-1 Property of Exponential Functions states that for $b > 0$, $b \neq 1$, if $b^x = b^y$, then $x = y$.

3. $x < 0$

Objective A Exercises

5. $5^{4x-1} = 5^{x-2}$
$4x - 1 = x - 2$
$3x - 1 = -2$
$3x = -1$
$x = -\dfrac{1}{3}$
The solution is $-\dfrac{1}{3}$.

7. $8^{x-4} = 8^{5x+8}$
$x - 4 = 5x + 8$
$-4 = 4x + 8$
$4x = -12$
$x = -3$
The solution is -3.

9. $9^x = 3^{x+1}$
$3^{2x} = 3^{x+1}$
$2x = x + 1$
$x = 1$
The solution is 1.

11. $8^{x+2} = 16^x$
$(2^3)^{x+2} = 2^{4x}$
$2^{3x+6} = 2^{4x}$
$3x + 6 = 4x$
$x = 6$
The solution is 6.

13. $16^{2-x} = 32^{2x}$

$(2^4)^{2-x} = (2^5)^{2x}$

$2^{8-4x} = 2^{10x}$

$8 - 4x = 10x$

$8 = 14x$

$x = \dfrac{8}{14} = \dfrac{4}{7}$

The solution is $\dfrac{4}{7}$.

15. $25^{3-x} = 125^{2x-1}$

$(5^2)^{3-x} = (5^3)^{2x-1}$

$5^{6-2x} = 5^{6x-3}$

$6 - 2x = 6x - 3$

$6 = 8x - 3$

$9 = 8x$

$x = \dfrac{9}{8}$

The solution is $\dfrac{9}{8}$.

17. $5^x = 6$

$\log 5^x = \log 6$

$x \log 5 = \log 6$

$x = \dfrac{\log 6}{\log 5}$

$x = 1.1133$

The solution is 1.1133.

19. $8^{x/4} = 0.4$

$\log 8^{x/4} = \log 0.4$

$\dfrac{x}{4} \log 8 = \log 0.4$

$\dfrac{x}{4} = \dfrac{\log 0.4}{\log 8}$

$x = 4 \cdot \dfrac{\log 0.4}{\log 8}$

$x = -1.7626$

The solution is -1.7626.

21. $2^{3x} = 5$

$\log 2^{3x} = \log 5$

$3x \log 2 = \log 5$

$3x = \dfrac{\log 5}{\log 2}$

$3x = 2.3219$

$x = 0.7740$

The solution is 0.7740.

23. $2^{-x} = 7$

$\log 2^{-x} = \log 7$

$-x \log 2 = \log 7$

$-x = \dfrac{\log 7}{\log 2}$

$-x = 2.8074$

$x = -2.8074$

The solution is -2.8074.

25. $2^{x-1} = 6$

$\log 2^{x-1} = \log 6$

$(x-1) \log 2 = \log 6$

$x - 1 = \dfrac{\log 6}{\log 2}$

$x = \dfrac{\log 6}{\log 2} + 1$

$x = 3.5850$

The solution is 3.5850.

27. $3^{2x-1} = 4$

$\log 3^{2x-1} = \log 4$

$(2x-1) \log 3 = \log 4$

$2x - 1 = \dfrac{\log 4}{\log 3}$

$2x - 1 = 1.2619$

$2x = 2.2619$

$x = 1.1309$

The solution is 1.1309.

29. $\left(\dfrac{1}{2}\right)^{x+1} = 3$

$\log\left(\dfrac{1}{2}\right)^{x+1} = \log 3$

$(x+1)\log\left(\dfrac{1}{2}\right) = \log 3$

$x + 1 = \dfrac{\log 3}{\log \dfrac{1}{2}}$

$x = \dfrac{\log 3}{\log \dfrac{1}{2}} - 1$

$x = -2.5850$

The solution is -2.5850.

31. $3 \cdot 2^x = 7$

$\log\left(3 \cdot 2^x\right) = \log 7$

$\log 3 + \log 2^x = \log 7$

$\log 3 + x \log 2 = \log 7$

$x \log 2 = \log 7 - \log 3$

$x = \dfrac{\log 7 - \log 3}{\log 2}$

$x = 1.2224$

The solution is 1.2224.

33. $7 = 10\left(\dfrac{1}{2}\right)^{x/8}$

$\log 7 = \log 10\left(\dfrac{1}{2}\right)^{x/8}$

$\log 7 = \log 10 + \log\left(\dfrac{1}{2}\right)^{x/8}$

$\log 7 = \log 10 + \dfrac{x}{8}\log\dfrac{1}{2}$

$\log 7 - \log 10 = \dfrac{x}{8}\log\dfrac{1}{2}$

$\dfrac{\log 7 - \log 10}{\log\dfrac{1}{2}} = \dfrac{x}{8}$

$0.5146 = \dfrac{x}{8}$

$4.1166 = x$

The solution is 4.1166.

35. $15 = 12(e)^{0.05x}$

$\ln 15 = \ln 12(e)^{0.05x}$

$\ln 15 = \ln 12 + \ln(e)^{0.05x}$

$\ln 15 = \ln 12 + 0.05x$

$\ln 15 - \ln 12 = 0.05x$

$0.2231 = 0.05x$

$4.4629 = x$

The solution is 4.4629.

Objective B Exercises

37. $\log x = \log(1 - x)$

$x = 1 - x$

$2x = 1$

$x = \dfrac{1}{2}$

The solution is $\dfrac{1}{2}$.

39. $\ln(3x + 2) = \ln(5x + 4)$

$3x + 2 = 5x + 4$

$-2x = 2$

$x = -1$

When we substitute $x = -1$ in either side of the equation, we get a logarithm of a negative number.
Because the logarithm of a negative number is not a real number, there is no solution.

41. $\log_2(8x) - \log_2(x^2 - 1) = \log_2 3$

$\log_2 \dfrac{8x}{x^x - 1} = \log_2 3$

$\dfrac{8x}{x^2 - 1} = 3$

$(x^2 - 1)\dfrac{8x}{x^2 - 1} = 3(x^2 - 1)$

$8x = 3x^2 - 3$

$0 = 3x^2 - 8x - 3$

$0 = (3x + 1)(x - 3)$

$3x + 1 = 0 \quad x - 3 = 0$

$\quad 3x = -1 \quad\quad x = 3$

$\quad\quad x = -\dfrac{1}{3}$

$-\dfrac{1}{3}$ does not check as a solution.

The solution is 3.

43. $\log_9 x + \log_9(2x - 3) = \log_9 2$

$\log_9(x(2x - 3)) = \log_9 2$

$x(2x - 3) = 2$

$2x^2 - 3x = 2$

$2x^2 - 3x - 2 = 0$

$(2x + 1)(x - 2) = 0$

$2x + 1 = 0 \quad x - 2 = 0$

$\quad 2x = -1 \quad\quad x = 2$

$\quad\quad x = -\dfrac{1}{2}$

$-\dfrac{1}{2}$ does not check as a solution.

The solution is 2.

45. $\log_2(2x - 3) = 3$

$2x - 3 = 2^3$

$2x - 3 = 8$

$2x = 11$

$x = \dfrac{11}{2}$

The solution is $\dfrac{11}{2}$.

47. $\ln(3x + 2) = 4$

$3x + 2 = e^4$

$3x + 2 = 54.5982$

$3x = 52.5982$

$x = 17.5327$

The solution is 17.5327.

49. $\log_2(x + 1) + \log_2(x + 3) = 3$

$\log_2((x + 1)(x + 3)) = 3$

$\log_2(x^2 + 4x + 3) = 3$

$x^2 + 4x + 3 = 2^3$

$x^2 + 4x + 3 = 8$

$x^2 + 4x - 5 = 0$

$(x + 5)(x - 1) = 0$

$x + 5 = 0 \quad x - 1 = 0$

$\quad x = -5 \quad\quad x = 1$

-5 does not check as a solution.

The solution is 1.

51. $\log_5(2x) - \log_5(x-1) = 1$

$\log_5 \dfrac{2x}{x-1} = 1$

$\dfrac{2x}{x-1} = 5^1$

$(x-1)\dfrac{2x}{x-1} = 5(x-1)$

$2x = 5x - 5$

$-3x = -5$

$x = \dfrac{-5}{-3} = \dfrac{5}{3}$

The solution is $\dfrac{5}{3}$.

53. $\log_8(6x) = \log_8 2 + \log_8(x-4)$

$\log_8(6x) = \log_8(2(x-4))$

$6x = 2x - 8$

$4x = -8$

$x = -2$

-2 does not check as a solution. The equation has no solution.

55. $x - 2 < x$ and therefore $\log(x-2) < \log x$. This means that $\log(x-2) - \log x < 0$ and could not equal the positive number 3.

Critical Thinking

57. $3^{x+1} = 2^{x-2}$

$\log 3^{x+1} = \log 2^{x-2}$

$(x+1)\log 3 = (x-2)\log 2$

$x + 1 = \dfrac{\log 2}{\log 3}(x-2)$

$x + 1 = 0.6309297536(x-2)$

$x + 1 = 0.6309297536x - 1.261859507$

$0.3690702464x = -2.261859507$

$x = -6.1285$

The solution is -6.1285.

59. $7^{2x-1} = 3^{2x+3}$

$\log 7^{2x-1} = \log 3^{2x+3}$

$(2x-1)\log 7 = (2x+3)\log 3$

$2x - 1 = \dfrac{\log 3}{\log 7}(2x+3)$

$2x - 1 = 0.5645750341(2x+3)$

$2x - 1 = 1.129150068x + 1.693725102$

$0.870849932x = 2.693725102$

$x = 3.0932$

The solution is 3.0935

61. a) $s = 312.5 \ln \dfrac{e^{0.32t} + e^{-0.32t}}{2}$

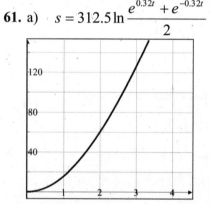

b) Use a graphing calculator to find t when $s = 100$.

$t \approx 2.64$

It will take 2.64s for the object to fall 100 ft.

Section 10.5

Concept Check

1. Compound interest is interest that is compounded not only on the original principal, but also on the interest already earned.

Objective A Exercises

3. **Strategy**: To find the value of the investment, use the compound interest formula.

$P = 1000,\ n = 8,\ i = \dfrac{8\%}{4} = \dfrac{0.08}{4} = 0.02$

Solution: $A = P(1+i)^n$

$A = 1000(1+0.02)^8$

$A = 1000(1.02)^8$

$A \approx 1171.66$

The value of the investment after 2 years is $1172.

5. **Strategy:** To find how many years it will take for the investment to be worth $15,000, solve the compound interest formula for n.

$A = 15,000$, $P = 5000$, $i = \dfrac{6\%}{12} = \dfrac{0.06}{12} = 0.005$

Solution: $A = P(1+i)^n$

$15000 = 5000(1+0.005)^n$

$3 = (1.005)^n$

$\log 3 = \log(1.005)^n$

$\log 3 = n \log 1.005$

$\dfrac{\log 3}{\log 1.005} = n$

$n \approx 220$

$\dfrac{n}{12} = \dfrac{220}{12} \approx 18$

The investment will be worth $15,000 in approximately 18 years.

7. a) **Strategy:** To find the technetium level, use the exponential decay formula.

$A_0 = 30$, $k = 6$, $t = 3$

Solution: $A = A_0(0.5)^{t/k}$

$A = 30(0.5)^{3/6}$

$A \approx 21.2$

The technetium level is 21.2 mg after 3 h.

b) **Strategy:** To find out how long it will take the technetium level to reach 20 mg, use the exponential decay formula.

$A_0 = 30$, $A = 20$, $k = 6$

Solution: $A = A_0(0.5)^{t/k}$

$20 = 30(0.5)^{t/6}$

$\dfrac{2}{3} = 0.5^{t/6}$

$\log \dfrac{2}{3} = \log 0.5^{t/6}$

$\log \dfrac{2}{3} = \dfrac{t}{6} \log 0.5$

$\dfrac{6 \log \dfrac{2}{3}}{\log 0.5} = t$

$t \approx 3.5$

The technetium level is 20 mg after 3.5 h.

9. **Strategy:** To find the half life, use the exponential decay formula.

$A_0 = 25$, $A = 18.95$, $t = 1$

Solution: $A = A_0(0.5)^{t/k}$

$18.95 = 25(0.5)^{1/k}$

$0.758 = 0.5^{1/k}$

$\log 0.758 = \log 0.5^{1/k}$

$\log 0.758 = \dfrac{1}{k} \log 0.5$

$k \log 0.758 = \log 0.5$

$k = \dfrac{\log 0.5}{\log 0.758}$

$k \approx 2.5$

The half life is 2.5 years.

11. **Strategy**: To determine the intensity of the earthquake, use the Richter scale equation.
$M = 8.9$

Solution: $M = \log \dfrac{I}{I_0}$

$8.9 = \log \dfrac{I}{I_0}$

$10^{8.9} = \dfrac{I}{I_0}$

$I = 10^{8.9} I_0$

$I \approx 794,328,235 I_0$

The intensity of the earthquake was $794,328,235 I_0$.

13. **Strategy**: To determine the how many times stronger the Honshu earthquake was, use the Richter scale equation.
$M_1 = 6.9 \quad M_2 = 6.4$

Solution: $M = \log \dfrac{I}{I_0} = \log I - \log I_0$

$6.9 = \log I_1 - \log I_0$

$6.4 = \log I_2 - \log I_0$

Subtract the equations.

$0.5 = \log I_1 - \log I_2$

$0.5 = \log \dfrac{I_1}{I_2}$

$10^{0.5} = \dfrac{I_1}{I_2}$

$I_1 = 10^{0.5} I_2$

$I_1 \approx 3.2 I_2$

The Honshu earthquake was 3.2 times stronger than the Quetta earthquake.

15. **Strategy**: To determine the magnitude of the earthquake for the seismogram given, use the given equation.
$A = 23 \quad t = 24$

Solution: $M = \log A + 3 \log 8t - 2.92$

$M = \log 23 + 3 \log 8(24) - 2.92$

$M = \log 23 + 3 \log 192 - 2.92$

$M \approx 5.3$

The magnitude of the earthquake was 5.3.

17. **Strategy**: To determine the magnitude of the earthquake for the seismogram given, use the given equation.
$A = 28 \quad t = 28$

Solution: $M = \log A + 3 \log 8t - 2.92$

$M = \log 28 + 3 \log 8(28) - 2.92$

$M = \log 28 + 3 \log 224 - 2.92$

$M \approx 5.6$

The magnitude of the earthquake was 5.6.

19. **Strategy**: To find the pH, replace H^+ with its given value and solve for pH.

Solution: $pH = -\log(H^+)$

$pH = -\log(3.98 \times 10^{-9})$

$pH \approx 8.4$

The pH of baking soda is 8.4.

21. **Strategy**: To find the hydrogen ion concentration, replace pH with its given value and solve for H^+.

Solution: $pH = -\log(H^+)$

$5.3 < -\log(H^+) < 6.6$

$-5.3 > \log(H^+) > -6.6$

$10^{-5.3} > H^+ > 10^{-6.6}$

$5 \times 10^{-6} > H^+ > 2.5 \times 10^{-7}$

The range of hydrogen ion concentration for peanuts is 2.5×10^{-7} to 5.0×10^{-6}

23. Strategy: To find the number of decibels, replace I with its given value in the equation and solve for D.

Solution: $D = 10(\log I + 16)$

$D = 10(\log(630) + 16)$

$D \approx 10(18.7993)$

$D = 187.993$

The blue whale sounds emit 188 decibels.

25. Strategy: To find the intensity, replace D with its given value in the equation and solve for I.

Solution: $D = 10(\log I + 16)$

$25 = 10(\log(I) + 16)$

$25 = 10\log I + 160$

$-135 = 10\log I$

$-13.5 = \log I$

$10^{-13.5} = I$

$I \approx 3.16 \times 10^{-14}$

The intensity is 3.16×10^{-14} watts/cm^2.

27. Strategy: To find the percent solve the equation for P.

$d = 0.005 \quad k = 20$

Solution: $\log P = -kd$

$\log P = -20(0.005)$

$\log P = -0.1$

$P = 10^{-0.1}$

$P \approx 0.7943$

79.4% of the light will pass through the glass.

29. Strategy: To find the thickness of copper needed, replace I and I_o with the given values then solve for x. $I = 0.25 \quad I_0 = 1$

Solution: $I = I_0 e^{-3.2x}$

$0.25 = e^{-3.2x}$

$\ln 0.25 = \ln e^{-3.2x}$

$-1.39 \approx -3.2x$

$x \approx 0.43$

The thickness of the copper is 0.4 cm.

31. a) decay
  b) growth
  c) decay
  d) growth

Critical Thinking

33. a) **Strategy:** To find the value of the investments after 3 years, use the given equation.

Solution: $A = A_0 e^{rt}$

$A = 5000e^{(0.06)(3)}$

$A = 5000e^{0.18}$

$A \approx 5986.09$

The value of the investment will be worth $5986.09.

b) **Strategy:** To find the interest rate needed to grow an investment from $1000 to $1250 in 2 years, use the given equation.

Solution: $A = A_0 e^{rt}$

$1250 = 1000e^{2r}$

$1.25 = e^{2r}$

$\ln 1.25 = \ln e^{2r}$

$0.2231 = 2r$

$r \approx 0.112$

The rate must be 11.2%.

Projects or Group Activities

35. a) Strategy: Evaluate the given function at $x = 375$ ft.

$$f(x) = \left(\frac{0.5774v + 155.3}{v}\right)x + 565.3\ln\left(\frac{v - 0.2747x}{v}\right) + 3.5$$

Solution:

$$f(375) = \left(\frac{0.5774(160) + 155.3}{160}\right)375 + 565.3\ln\left(\frac{160 - 0.2747(375)}{160}\right) + 3.5$$

$$f(375) = \left(\frac{247.684}{160}\right)375 + 565.3\ln\left(\frac{56.9875}{160}\right) + 3.5$$

$$f(375) \approx 580.5094 - 583.5829 + 3.5$$

$$f(375) \approx 0.43$$

The ball will hit 0.43 ft from the bottom of the fence.

b) Strategy: Increase the speed by 4% so that $v = 166.4$ ft/s.

$$f(375) = \left(\frac{0.5774(166.4) + 155.3}{166.4}\right)375 + 565.3\ln\left(\frac{166.4 - 0.2747(375)}{166.4}\right) + 3.5$$

$$f(375) = 566.5100 - 545.5868 + 3.5$$

$$f(375) = 24.4$$

The height of the ball is 24.4 feet so it will clear the 15-foot fence by approximately 9 ft.

c) Strategy: Determine the value for x for which $f(x)$ is greater than 15.

$$f(x) = \left(\frac{0.5774v + 155.3}{v}\right)x + 565.3\ln\left(\frac{v - 0.2747x}{v}\right) + 3.5$$

$$f(x) = \left(\frac{0.5774(166.4) + 155.3}{166.4}\right)x + 565.3\ln\left(\frac{166.4 - 0.2747x}{166.4}\right) + 3.5$$

$$x \approx 385$$

Use a graphing calculator to determine that $x = 385$ ft. The height of the ball will be 15.029 ft and will clear the fence.

Chapter 10 Review Exercises

1. $f(2) = e^{2-2} = e^0 = 1$

2. $5^2 = 25$

3. $f(x) = 3^{-x} + 2$

x	y
0	3
−1	5
1	$\frac{7}{3}$
2	$\frac{19}{9}$

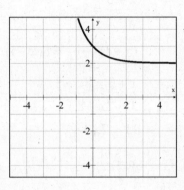

4. $f(x) = \log_3(x-1)$

$y = \log_3(x-1)$ is equivalent to

$x - 1 = 3^y$ or $x = 3^y + 1$.

x	y
$\frac{4}{3}$	−1
2	0
4	1

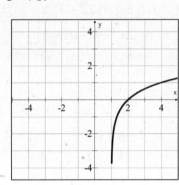

5. $\log_3 \sqrt[5]{x^2 y^4} = \log_3 (x^2 y^4)^{1/5} = \frac{1}{5} \log_3 (x^2 y^4)$

$= \frac{1}{5}[\log_3 x^2 + \log_3 y^4]$

$= \frac{1}{5}[2\log_3 x + 4\log_3 y]$

$= \frac{2}{5}\log_3 x + \frac{4}{5}\log_3 y$

6. $2\log_3 x - 5\log_3 y = \log_3 x^2 - \log_3 y^5$

$= \log_3\left(\frac{x^2}{y^5}\right)$

7. $27^{2x+4} = 81^{x-3}$

$(3^3)^{2x+4} = (3^4)^{x-3}$

$6x + 12 = 4x - 12$

$2x = -24$

$x = -12$

The solution is −12.

8. $\log_5 \dfrac{7x+2}{3x} = 1$

Rewrite in exponential form.

$5^1 = \dfrac{7x+2}{3x}$

$15x = 7x + 2$

$8x = 2$

$x = \dfrac{1}{4}$

The solution is $\dfrac{1}{4}$.

9. $\log_6 22 = \dfrac{\log_{10} 22}{\log_{10} 6} \approx 1.7251$

10. $\log_2 x = 5$

Rewrite in exponential form.

$2^5 = x$

$x = 32$

The solution is 32.

11. $\log_3(x+2) = 4$

Rewrite in exponential form.

$3^4 = x + 2$

$81 = x + 2$

$x = 79$

The solution is 79.

12. $\log_{10} x = 3$

Rewrite in exponential form.

$10^3 = x$

$1x = 1000$

The solution is 1000.

13. $\dfrac{1}{3}(\log_7 x + 4\log_7 y) = \dfrac{1}{3}(\log_7 x + \log_7 y^4)$

$\quad = \dfrac{1}{3}(\log_7(xy^4)) = \log_7(xy^4)^{1/3}$

$\quad = \log_7 \sqrt[3]{xy^4}$

14. $\log_8 \sqrt{\dfrac{x^5}{y^3}} = \log_8 \left(\dfrac{x^5}{y^3}\right)^{1/2}$

$\quad = \dfrac{1}{2}(\log_8 x^5 - \log_8 y^3)$

$\quad = \dfrac{5}{2}\log_8 x - \dfrac{3}{2}\log_8 y$

15. $\log_2 32 = 5$

16. $\log_3 1.6 = \dfrac{\log_{10} 1.6}{\log_{10} 3} \approx 0.4278$

17. $\quad 3^{x+2} = 5$

$\quad \log 3^{x+2} = \log 5$

$\quad (x+2)\log 3 = \log 5$

$\quad x+2 = \dfrac{\log 5}{\log 3}$

$\quad x = \dfrac{\log 5}{\log 3} - 2$

$\quad x = -0.535$

The solution is -0.535.

18. $f(-3) = \left(\dfrac{2}{3}\right)^{-3+2} = \left(\dfrac{2}{3}\right)^{-1} = \dfrac{3}{2}$

19. $\log_2(x+3) - \log_2(x-1) = 3$

$\quad \log_2 \dfrac{x+3}{x-1} = 3$

$\quad \dfrac{x+3}{x-1} = 2^3$

$\quad \dfrac{x+3}{x-1} = 8$

$\quad (x-1)\dfrac{x+3}{x-1} = 8(x-1)$

$\quad x+3 = 8x-8$

$\quad 3 = 7x-8$

$\quad 11 = 7x$

$\quad x = \dfrac{11}{7}$

The solution is $\dfrac{11}{7}$.

20. $\quad \log_3(2x+3) + \log_3(x-2) = 2$

$\quad \log_3((2x+3)(x-2)) = 2$

$\quad 2x^2 - x - 6 = 3^2$

$\quad 2x^2 - x - 6 = 9$

$\quad 2x^2 - x - 15 = 0$

$\quad (2x+5)(x-3) = 0$

$\quad 2x+5 = 0 \quad x-3 = 0$

$\quad 2x = -5 \qquad x = 3$

$\quad x = -\dfrac{5}{2}$

$-\dfrac{5}{2}$ does not check as a solution.

The solution is 3.

21. $f(x) = \left(\dfrac{2}{3}\right)^{x+1}$

x	y
-1	1
-2	$\sfrac{3}{2}$
0	$\sfrac{2}{3}$
1	$\sfrac{4}{9}$

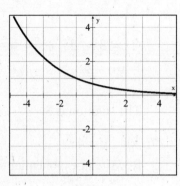

22. $f(x) = \log_2(2x-1)$

$y = \log_2(2x-1)$ is equivalent to

$2x - 1 = 2^y$ or $x = \dfrac{2^y + 1}{2}$

x	y
$\sfrac{3}{4}$	-1
1	0
$\sfrac{3}{2}$	1

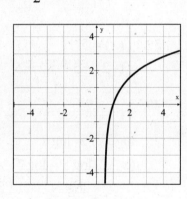

23. $\log_6 36 = x$

$6^x = 36$

$x = 2$

24. $\dfrac{1}{3}(\log_2 x - \log_2 y) = \dfrac{1}{3}\log_2\left(\dfrac{x}{y}\right)$

$= \log_2\left(\dfrac{x}{y}\right)^{1/3} = \log_2 \sqrt[3]{\dfrac{x}{y}}$

25. $9^{2x} = 3^{x+3}$

$(3^2)^{2x} = 3^{x+3}$

$3^{4x} = 3^{x+3}$

$4x = x + 3$

$3x = 3$

$x = 1$

The solution is 1.

26. $5 \cdot 3^{x/2} = 12$

$3^{x/2} = 2.4$

$\log 3^{x/2} = \log 2.4$

$\dfrac{x}{2}\log 3 = \log 2.4$

$\dfrac{x}{2} = \dfrac{\log 2.4}{\log 3}$

$\dfrac{x}{2} \approx 0.7969$

$x = 1.5938$

The solution is 1.5938.

27. $\log_5 x = -1$

Rewrite in exponential form.

$5^{-1} = x$

$\dfrac{1}{5} = x$

The solution is $\dfrac{1}{5}$.

28. $\log_3 81 = 4$

29. $\log x + \log(x-2) = \log 15$

$\log(x(x-2)) = \log 15$

$x(x-2) = 15$

$x^2 - 2x - 15 = 0$

$(x-5)(x+3) = 0$

$x - 5 = 0 \quad x + 3 = 0$

$x = 5 \qquad x = -3$

-3 does not check as a solution.

The solution is 5.

30. $\log_5 \sqrt[3]{x^2 y} = \log_3 (x^2 y)^{1/3} = \frac{1}{3}\log_3 (x^2 y)$

$= \frac{1}{3}[\log_3 x^2 + \log_3 y]$

$= \frac{1}{3}[2\log_3 x + \log_3 y]$

$= \frac{2}{3}\log_3 x + \frac{1}{3}\log_3 y$

31. $\quad 6e^{-2x} = 17$

$\ln 6e^{-2x} = \ln 17$

$\ln 6 + \ln e^{-2x} = \ln 17$

$-2x = \ln 17 - \ln 6$

$-2x = 1.0415$

$x = -0.5207$

The solution is -0.5207

32. $f(-3) = 7^{-3+2} = 7^{-1} = \frac{1}{7}$

33. $\log_2 16 = x$

$2^x = 16$

$x = 4$

34. $\log_6 x = \log_6 2 + \log_6 (2x-3)$

$\log_6 x = \log_6 (2(2x-3)$

$x = 2(2x-3)$

$x = 4x - 6$

$-3x = -6$

$x = 2$

The solution is 2.

35. $\log_2 5 = x$

$x = \frac{\log 5}{\log 2} \approx 2.3219$

36. $\quad 4^x = 8^{x-1}$

$(2^2)^x = (2^3)^{x-1}$

$2^{2x} = 2^{3x-3}$

$2x = 3x - 3$

$-x = -3$

$x = 3$

The solution is 3.

37. $\log_5 x = 4$

Rewrite in exponential form.

$x = 5^4 = 625$

38. $3\log_b x - 7\log_b y = \log_b x^3 - \log_b y^7$

$= \log_b \left(\frac{x^3}{y^7}\right)$

39. $f(x) = 5^{-x-1}$

$f(-2) = 5^{-(-2)-1} = 5^{2-1} = 5^1 = 5$

40. $5^{x-2} = 7$

$\log 5^{x-2} = \log 7$

$(x-2)\log 5 = \log 7$

$x - 2 = \frac{\log 7}{\log 5}$

$x = \frac{\log 7}{\log 5} + 2$

$x \approx 3.2091$

The solution is 3.2091.

41. Strategy: To find the value of the investment, use the compound interest formula.

$$P = 4000, \ n = 24, \ i = \frac{8\%}{12} = \frac{0.08}{12} = 0.00\overline{6}$$

Solution: $A = P(1+i)^n$

$A = 4000(1 + 0.00\overline{6})^{24}$

$A = 4000(1.00\overline{6})^{24}$

$A = 4691.55$

The value of the investment after 2 years is $4692.

42. Strategy: To determine the magnitude of the earthquake, use the Richter scale equation.

Solution: $M = \log \dfrac{I}{I_0}$

$M = \log \dfrac{1,584.893,192 I_0}{I_0}$

$M = \log 1584893192$

$M \cong 9.2$

The magnitude of the earthquake was 9.2.

43. Strategy: To find the half life use the exponential decay formula.

$A_0 = 25, \ A = 15, \ t = 20$

Solution: $A = A_0 (0.5)^{t/k}$

$15 = 25(0.5)^{20/k}$

$0.6 = 0.5^{20/k}$

$\log 0.6 = \log 0.5^{20/k}$

$\log 0.6 = \dfrac{20}{k} \log 0.5$

$k \log 0.6 = 20 \log 0.5$

$k = \dfrac{20 \log 0.5}{\log 0.6}$

$k \cong 27$

The half life is 27 days.

44. Strategy: To find the number of decibels replace I with its given value in the equation and solve for D.

Solution: $D = 10(\log I + 16)$

$D = 10(\log(5 \times 10^{-6}) + 16)$

$D \approx 10(10.6990)$

$D = 106.99$

The sound emitted from a busy street corner is 107 decibels.

Chapter 10 Test

1. $f(0) = \left(\dfrac{2}{3}\right)^0 = 1$

2. $f(-2) = 3^{-2+1} = 3^{-1} = \dfrac{1}{3}$

3. $f(x) = 2^x - 3$

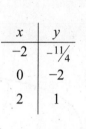

x	y
-2	$-11/4$
0	-2
2	1

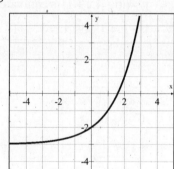

4. $f(x) = 2^x + 2$

x	y
-2	$9/4$
0	3
1	4

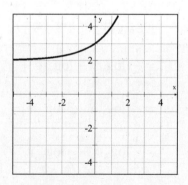

5. $\log_4 16 = x$

$4^x = 16$

$x = 2$

6. $\log_3 x = -2$

$x = 3^{-2} = \dfrac{1}{3^2} = \dfrac{1}{9}$

7. $f(x) = \log_2(2x)$

$y = \log_2(2x)$ is equivalent to

$2x = 2^y$ or $x = \dfrac{2^y}{2}$.

X	y
¼	−1
½	0
1	1
2	2

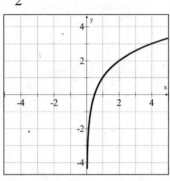

8. $f(x) = \log_3(x+1)$

$y = \log_3(x+1)$ is equivalent to

$x + 1 = 3^y$ or $x = 3^y - 1$.

x	y
−⅔	−1
0	0
2	1

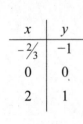
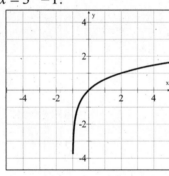

9. $\log_6 \sqrt{xy^3} = \log_6(xy^3)^{1/2} = \dfrac{1}{2}\log_3(xy^3)$

$= \dfrac{1}{2}[\log_6 x + \log_6 y^3]$

$= \dfrac{1}{2}[\log_6 x + 3\log_6 y]$

$= \dfrac{1}{2}\log_6 x + \dfrac{3}{2}\log_6 y$

10. $\dfrac{1}{2}(\log_3 x - \log_3 y) = \dfrac{1}{2}\log_3\left(\dfrac{x}{y}\right)$

$= \log_3\sqrt{\dfrac{x}{y}}$

11. $\ln\dfrac{x}{\sqrt{z}} = \ln x - \ln\sqrt{z} = \ln x - \ln z^{1/2}$

$= \ln x - \dfrac{1}{2}\ln z$

12. $3\ln x - \ln y - \dfrac{1}{2}\ln z = \ln x^3 - \ln y - \ln z^{1/2}$

$= \ln\dfrac{x^3}{y} - \ln z^{1/2} = \ln\dfrac{x^3}{yz^{1/2}}$

$= \ln\dfrac{x^3}{y\sqrt{z}}$

13. $3^{7x+1} = 3^{4x-5}$

$7x + 1 = 4x - 5$

$3x = -6$

$x = -2$

The solution is −2.

14. $8^x = 2^{x-6}$

$(2^3)^x = 2^{x-6}$

$3x = x - 6$

$2x = -6$

$x = -3$

The solution is −3.

15. $3^x = 17$

$\log 3^x = \log 17$

$x\log 3 = \log 17$

$x = \dfrac{\log 17}{\log 3}$

$x \approx 2.5789$

The solution is 2.5789.

16. $\log x + \log(x-4) = \log 12$

$\qquad \log(x(x-4)) = \log 12$

$\qquad x(x-4) = 12$

$\qquad x^2 - 4x - 12 = 0$

$\qquad (x-6)(x+2) = 0$

$x - 6 = 0 \quad x + 2 = 0$

$x = 6 \qquad x = -2$

-2 does not check as a solution.
The solution is 6.

17. $\log_6 x + \log_6(x-1) = 1$

$\qquad \log_6(x(x-1)) = 1$

$\qquad x(x-1) = 6^1$

$\qquad x^2 - x - 6 = 0$

$\qquad (x-3)(x+2) = 0$

$\qquad x - 3 = 0 \quad x + 2 = 0$

$\qquad x = 3 \qquad x = -2$

-2 does not check as a solution.
The solution is 3.

18. $\log_5 9 = x$

$x = \dfrac{\log 9}{\log 5} \approx 1.3652$

19. $\log_3 19 = x$

$x = \dfrac{\log 19}{\log 3} \approx 2.6801$

20. $\qquad 5^{2x-5} = 9$

$\log 5^{2x-5} = \log 9$

$(2x-5)\log 5 = \log 9$

$2x - 5 = \dfrac{\log 9}{\log 5}$

$2x - 5 \approx 1.3652$

$2x = 6.3652$

$x = 3.1826$

The solution is 3.1826.

21. $2e^{x/4} = 9$

$\ln 2e^{x/4} = \ln 9$

$\ln 2 + \ln e^{x/4} = \ln 9$

$\dfrac{x}{4} = \ln 9 - \ln 2$

$\dfrac{x}{4} \approx 1.5041$

$x = 6.0163$

The solution is 6.0163.

22. $\log_5(30x) - \log_5(x+1) = 2$

$\qquad \log_5 \dfrac{30x}{x+1} = 2$

$\qquad \dfrac{30x}{x+1} = 5^2$

$\qquad \dfrac{30x}{x+1} = 25$

$\qquad (x+1)\dfrac{30x}{x+1} = 25(x+1)$

$\qquad 30x = 25x + 25$

$\qquad 5x = 25$

$\qquad x = 5$

The solution is 5.

23. Strategy: To find the approximate age of the shard, use the given equation.

$A_0 = 250, \; A = 170$

Solution: $A = A_0(0.5)^{t/5570}$

$170 = 250(0.5)^{t/5570}$

$0.68 = 0.5^{t/5570}$

$\log 0.68 = \log 0.5^{t/5570}$

$\log 0.68 = \dfrac{t}{5570}\log 0.5$

$\dfrac{5570 \log 0.68}{\log 0.5} = t$

$t \approx 3099$

The shard is approximately 3099 years old.

24. Strategy: To find the intensity, replace D with its given value in the equation and solve for I.

Solution: $D = 10(\log I + 16)$

$75 = 10(\log(I) + 16)$

$75 = 10 \log I + 160$

$-85 = 10 \log I$

$-8.5 = \log I$

$10^{-8.5} = I$

$I \approx 3.16 \times 10^{-9}$

The intensity is 3.16×10^{-9} watts/cm^2.

25. Strategy: To find out the half life of a radioactive material use the exponential decay formula.

$A_0 = 10, A = 9, t = 5$

Solution: $A = A_0(0.5)^{t/k}$

$9 = 10(0.5)^{5/k}$

$0.9 = 0.5^{5/k}$

$\log 0.9 = \log 0.5^{5/k}$

$\log 0.9 = \dfrac{5}{k} \log 0.5$

$k = \dfrac{5 \log 0.5}{\log 0.9}$

$k \approx 33$

The half life is 33 h.

Cumulative Review Exercises

1. $4 - 2[x - 3(2 - 3x) - 4x] = 2x$

$\qquad 4 - 2[x - 6 + 9x - 4x] = 2x$

$\qquad\qquad 4 - 2[6x - 6] = 2x$

$\qquad\qquad 4 - 12x + 12 = 2x$

$\qquad\qquad\qquad 16 = 14x$

$\qquad\qquad\qquad\qquad x = \dfrac{16}{14} = \dfrac{8}{7}$

The solution is $\dfrac{8}{7}$.

2. $2x - y = 5$

$\quad -y = -2x + 5$

$y = 2x - 5$

$m = 2$ and $(2, -2)$

$y - y_1 = m(x - x_1)$

$y - (-2) = 2(x - 2)$

$y + 2 = 2x - 4$

$y = 2x - 6$

3. $4x^4 + 7x^2 + 3 = (4x^2 + 3)(x^2 + 1)$

4. $\dfrac{1 - \dfrac{5}{x} + \dfrac{6}{x^2}}{1 + \dfrac{1}{x} - \dfrac{6}{x^2}} = \dfrac{1 - \dfrac{5}{x} + \dfrac{6}{x^2}}{1 + \dfrac{1}{x} - \dfrac{6}{x^2}} \cdot \dfrac{x^2}{x^2}$

$\qquad = \dfrac{x^2 - 5x + 6}{x^2 + x - 6} = \dfrac{(x-2)(x-3)}{(x-2)(x+3)}$

$\qquad = \dfrac{x-3}{x+3}$

5. $\dfrac{\sqrt{xy}}{\sqrt{x} - \sqrt{y}} = \dfrac{\sqrt{xy}}{\sqrt{x} - \sqrt{y}} \cdot \dfrac{\sqrt{x} + \sqrt{y}}{\sqrt{x} + \sqrt{y}}$

$\qquad = \dfrac{\sqrt{x^2 y} + \sqrt{xy^2}}{\sqrt{x^2} - \sqrt{y^2}}$

$\qquad = \dfrac{x\sqrt{y} - y\sqrt{x}}{x - y}$

6. $x^2 - 4x - 6 = 0$

$x^2 - 4x + 4 = 6 + 4$

$(x - 2)^2 = 10$

$\sqrt{(x-2)^2} = \sqrt{10}$

$x - 2 = \pm\sqrt{10}$

$x = 2 \pm \sqrt{10}$

The solutions are $2 + \sqrt{10}$ and $2 - \sqrt{10}$.

7. $(x - r_1)(x - r_2) = 0$

$\left(x - \dfrac{1}{3}\right)(x - (-3)) = 0$

$\left(x - \dfrac{1}{3}\right)(x + 3) = 0$

$x^2 + \dfrac{8}{3}x - 1 = 0$

$3x^2 + 8x - 3 = 0$

8.
$2x - y < 3 \qquad\quad x + y < 1$
$\quad -y < -2x + 3 \qquad y < -x + 1$
$\qquad y > 2x - 3$

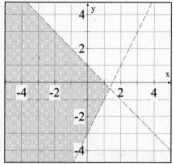

9. (1) $3x - y + z = 3$
  (2) $x + y + 4z = 7$
  (3 $3x - 2y + 3z = 8$
Eliminate y. Add equations (1) and (2).
$3x - y + z = 3$
$\ x + y + 4z = 7$
(4) $4x + 5z = 10$
Multiply equation (2) by 2 and add to
equation (3).
$\ 2(x + y + 4z) = 2(7)$
$\ 3x - 2y + 3z = 8$

$2x + 2y + 8z = 14$
$3x - 2y + 3z = 8$
(5) $5x + 11z = 22$
Multiply equation (4) by 5 and equation (5)
by -4, and add.
$5(4x + 5z) = 5(10)$
$-4(5x + 11z) = -4(22)$

$20x + 25z = 50$
$-20x - 44z = -88$
$\qquad -19z = -38$
$\qquad\quad z = 2$

Replace z with 2 in equation (4).
$4x + 5(2) = 10$
$\quad 4x = 0$
$\quad\ x = 0$
Replace x with 0 and z with 2 in equation
(1).
$3(0) - y + 2 = 3$
$\quad -y + 2 = 3$
$\qquad -y = 1$
$\qquad\ y = -1$
The solution is $(0, -1, 2)$.

10. $\dfrac{x-4}{2-x} - \dfrac{1-6x}{2x^2 - 7x + 6}$

$= \dfrac{x-4}{2-x} - \dfrac{1-6x}{(2x-3)(x-2)}$

$= \dfrac{x-4}{2-x} + \dfrac{1-6x}{(2x-3)(2-x)}$

$= \dfrac{(x-4)}{(2-x)} \cdot \dfrac{(2x-3)}{(2x-3)} + \dfrac{1-6x}{(2x-3)(2-x)}$

$= \dfrac{2x^2 - 11x + 12 + 1 - 6x}{(2-x)(2x-3)} = \dfrac{2x^2 - 17x + 13}{(2-x)(2x-3)}$

$= -\dfrac{2x^2 - 17x + 13}{(x-2)(2x-3)}$

11. $x^2 + 4x - 5 \le 0$
$(x + 5)(x - 1) \le 0$
$\{x \mid -5 \le x \le 1\}$

12. $|2x - 5| \le 3$
$\quad -3 \le 2x - 5 \le 3$
$-3 + 5 \le 2x - 5 + 5 \le 3 + 5$
$\quad\ 2 \le 2x \le 8$
$\quad\ 1 \le x \le 4$
$\{x \mid 1 \le x \le 4\}$

13. $f(x) = \left(\dfrac{1}{2}\right)^x + 1$

x	y
-1	3
0	2
2	$\tfrac{5}{4}$

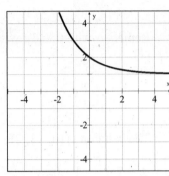

14. $f(x) = \log_2 x - 1$

$y + 1 = \log_2 x$

$x = 2^{y+1}$

x	y
1	-1
2	0
4	1

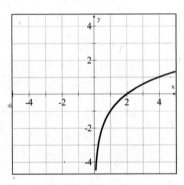

15. $f(-3) = 2^{-(-3)-1} = 2^2 = 4$

16. $\log_5 x = 3$

$x = 5^3 = 125$

17. $3\log_b x - 5\log_b y = \log_b x^3 - \log_b y^5$

$= \log_b \dfrac{x^3}{y^5}$

18. $\log_3 7 = x$

$x = \dfrac{\log 7}{\log 3} \approx 1.7712$

19. $4^{5x-2} = 4^{3x+2}$

$5x - 2 = 3x + 2$

$2x = 4$

$x = 2$

The solution is 2.

20. $\log x + \log(2x + 3) = \log 2$

$\log(x(2x + 3)) = \log 2$

$x(2x + 3) = 2$

$2x^2 + 3x - 2 = 0$

$(2x - 1)(x + 2) = 0$

$2x - 1 = 0 \quad x + 2 = 0$

$2x = 1 \qquad x = -2$

$x = \dfrac{1}{2}$

-2 does not check as a solution.

The solution is $\dfrac{1}{2}$.

21. Strategy: Let c represent the number of checks. To find the number of checks, write and solve an inequality.

Solution: $5.00 + 0.02c > 2.00 + 0.08c$

$5 > 2 + 0.06c$

$3 > 0.06c$

$50 > c$

The customer can write at most 49 checks.

22. Strategy: Let x represent the cost per pound of the mixture.

	Amount	Cost	Value
$4.00	16	4.00	16(4.00)
$2.50	24	2.50	24(2.50)
Mixture	40	x	$40x$

The sum of the values before mixing is equal to the value after mixing.

Solution:

$$16(4.00) + 24(2.50) = 40x$$
$$64 + 60 = 40x$$
$$124 = 40x$$
$$x = 3.1$$

The cost per pound of the mixture is $3.10.

23. Strategy: Let x represent the rate of the wind.

	Distance	Rate	Time
With wind	1000	$225 + x$	$\dfrac{1000}{225 + x}$
Against wind	800	$225 - x$	$\dfrac{800}{225 \ x}$

The flying time with the wind is the same as the flying time against the wind.

Solution:

$$\frac{1000}{225 + x} = \frac{800}{225 - x}$$

$$(225 + x)(225 - x)\frac{1000}{225 + x} = \frac{800}{225 - x}(225 + x)(225 - x)$$

$$(225 - x)1000 = 800(225 + x)$$

$$225000 - 1000x = 180{,}000 + 800x$$

$$45{,}000 = 1800x$$

$$25 = x$$

The rate of the wind is 25 mph.

24. Strategy: Write the basic direct variation equation replacing the variable with the given values. Solve for k.
Write the direct variation equation replacing k with its value. Substitute 34 for f and solve for d.

Solution:

$d = kf$	$d = 0.3f$
$6 = k(20)$	$= 0.3(34)$
$0.3 = k$	$= 10.2$

The string will stretch 10.2 in.

25. Strategy: Let x represent the cost of redwood. The cost of fir is y.

First purchase:

	Amount	Cost	Value
Redwood	80	x	$80x$
Fir	140	y	$140y$

Second purchase:

	Amount	Cost	Value
Redwood	140	x	$140x$
Fir	100	y	$100y$

The total cost of the first purchase is $67.
The total cost of the second purchase is $81.

Solution:

$$80x + 140y = 67$$
$$140x + 100y = 81$$

$$-5(80x + 140y) = -5(67)$$
$$7(140x + 100y) = 7(81)$$

$$-400x - 700y = -335$$
$$980x + 700y = 567$$

$$580x = 232$$
$$x = 0.40$$

$$80(0.40) + 140y = 67$$
$$32 + 140y = 67$$
$$140y = 35$$
$$y = 0.25$$

The cost of the redwood is $.40 per foot.
The cost of the fir is $.25 per foot.

26. Strategy: To find how many years it will take for the investment to double in value, use the compound interest formula.

$$A = 10{,}000, \ P = 5000, \ i = \frac{7\%}{2} = \frac{0.07}{2} = 0.035$$

Solution: $A = P(1 + i)^n$

$$10000 = 5000(1 + 0.035)^n$$

$$2 = (1.035)^n$$

$$\log 2 = \log(1.035)^n$$

$$\log 2 = n \log 1.035$$

$$\frac{\log 2}{\log 1.035} = n$$

$$n \approx 20$$

$$\frac{n}{2} = \frac{20}{2} = 10$$

The investment will take 10 years to double in value.

Chapter 11: Conic Sections

Prep Test

1. $d = \sqrt{(x_1 - x_2)^2 + (y_1 - y_2)^2}$

$d = \sqrt{(-2 - 4)^2 + (3 - (-1))^2}$

$= \sqrt{(-6)^2 + (4)^2}$

$= \sqrt{36 + 16}$

$= \sqrt{52} = 7.21$

2. $x^2 - 8x + \left[\dfrac{1}{2}(-8)\right]^2 = x^2 - 8x + (-4)^2$

$= x^2 - 8x + 16$

$= (x - 4)^2$

3. $\dfrac{x^2}{16} + \dfrac{y^2}{9} = 1$

When $y = 3$:

$\dfrac{x^2}{16} + \dfrac{(3)^2}{9} = 1$

$\dfrac{x^2}{16} + 1 = 1$

$\dfrac{x^2}{16} = 0$

$x^2 = 0$

$x = 0$

When $y = 0$:

$\dfrac{x^2}{16} + \dfrac{(0)^2}{9} = 1$

$\dfrac{x^2}{16} + 0 = 1$

$\dfrac{x^2}{16} = 1$

$x^2 = 16$

$x = \pm 4$

4. $7x + 4y = 3$

$\qquad y = x - 2$

$7x + 4(x - 2) = 3$

$7x + 4x - 8 = 3$

$\qquad 11x = 11$

$\qquad\quad x = 1$

$y = x - 2 = 1 - 2 = -1$

The solution is $(1, -1)$.

5. (1) $4x - y = 9$

(2) $2x + 3y = -13$

Eliminate y.

$3(4x - y) = 3(9)$

$2x + 3y = -13$

$12x - 3y = 27$

$\ \ 2x + 3y = -13$

$\qquad 14x = 14$

$\qquad\ \ x = 1$

Replace x with 1 in equation (1).

$4(1) - y = 9$

$\quad 4 - y = 9$

$\qquad -y = 5$

$\qquad\ \ y = -5$

The solution is $(1, -5)$.

6. $y = x^2 - 4x + 2$

$x = -\dfrac{b}{2a} = -\dfrac{-4}{2(1)} = -(-2) = 2$

$y = (2)^2 - 4(2) + 2 = 4 - 8 + 2 = -2$

The axis of symmetry is $x = 2$.

The vertex is $(2, -2)$.

7. $f(x) = -2x^2 + 4x$

x	y
-1	-6
0	0
1	2
2	0
3	-6

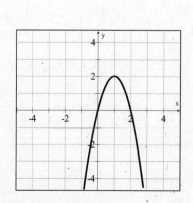

8. $x + 2y \leq 4$

$2y \leq -x + 4$

$y \leq -\dfrac{1}{2}x + 2$

$x - y \leq 2$

$-y \leq -x + 2$

$y \geq x - 2$

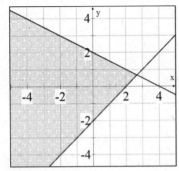

Section 11.1

Concept Check

1. a) The axis of symmetry is a vertical line.
   b) Since $a > 0$, the parabola opens up.

3. a) The axis of symmetry is a horizontal line.
   b) Since $a > 0$, the parabola opens right.

5. a) The axis of symmetry is a horizontal line.
   b) Since $a < 0$, the parabola opens left.

Objective A Exercises

7. $x = y^2 - 3y - 4$

$y = -\dfrac{b}{2a} = -\dfrac{-3}{2(1)} = -\left(-\dfrac{3}{2}\right) = \dfrac{3}{2}$

$x = \left(\dfrac{3}{2}\right)^2 - 3\left(\dfrac{3}{2}\right) - 4 = \dfrac{9}{4} - \dfrac{9}{2} - 4 = -\dfrac{25}{4}$

The axis of symmetry is $y = \dfrac{3}{2}$.

The vertex is $\left(-\dfrac{25}{4}, \dfrac{3}{2}\right)$.

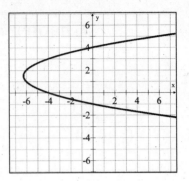

9. $x = -\dfrac{1}{2}y^2 + 2y - 3$

$y = -\dfrac{b}{2a} = -\dfrac{2}{2\left(-\dfrac{1}{2}\right)} = -\dfrac{2}{-1} = 2$

$x = \left(-\dfrac{1}{2}\right)(2)^2 + 2(2) - 3 = -2 + 4 - 3 = -1$

The axis of symmetry is $y = 2$.
The vertex is $(-1, 2)$.

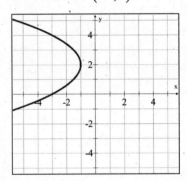

11. $y=\frac{1}{2}x^2+x-3$

$x=-\frac{b}{2a}=-\frac{1}{2\left(\frac{1}{2}\right)}=-\frac{1}{1}=-1$

$y=\frac{1}{2}(-1)^2+(-1)-3=\frac{1}{2}-1-3=-\frac{7}{2}$

The axis of symmetry is $x=-1$.

The vertex is $\left(-1,-\frac{7}{2}\right)$.

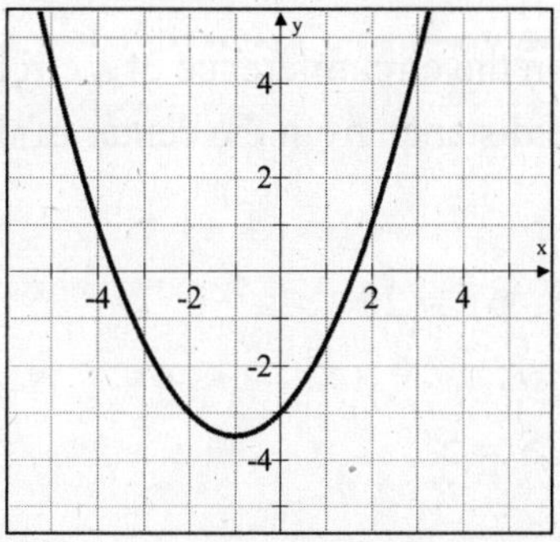

13. $x=y^2-y-6$

$y=-\frac{b}{2a}=-\frac{-1}{2(1)}=-(-\frac{3}{2})=\frac{1}{2}$

$x=\left(\frac{1}{2}\right)^2-\left(\frac{1}{2}\right)-6=\frac{1}{4}-\frac{1}{2}-6=-\frac{25}{4}$

The axis of symmetry is $y=\frac{1}{2}$.

The vertex is $\left(-\frac{25}{4},\frac{1}{2}\right)$.

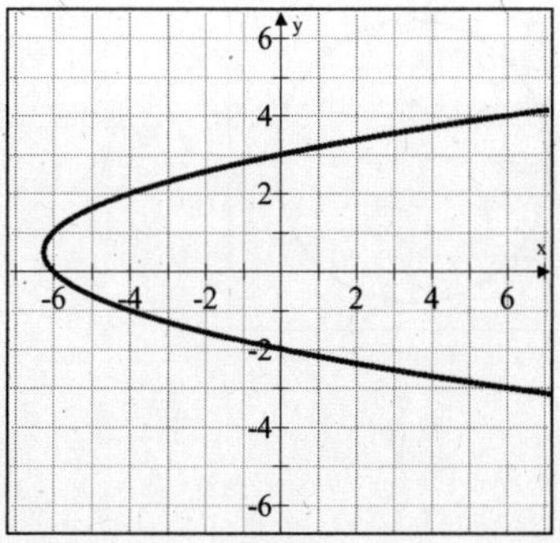

15. $y=2x^2+4x-5$

$x=-\frac{b}{2a}=-\frac{4}{2(2)}=-1$

$y=2(-1)^2+4(-1)-5=2-4-5=-7$

The axis of symmetry is $x=-1$.
The vertex is $(-1,-7)$.

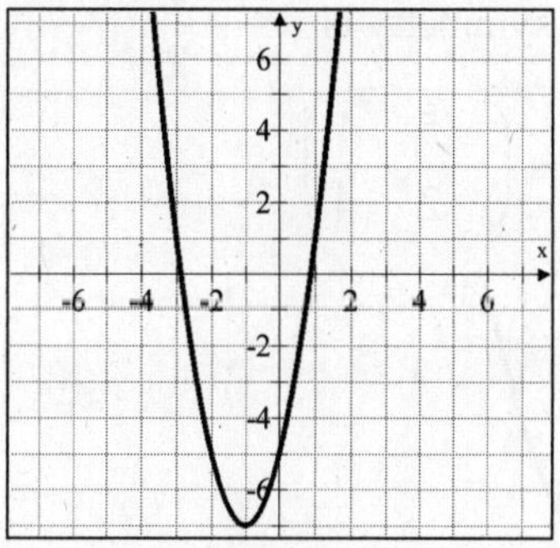

17. $y=2x^2-x-3$

$x=-\frac{b}{2a}=-\frac{-1}{2(2)}=\frac{1}{4}$

$y=2\left(\frac{1}{4}\right)^2-\frac{1}{4}-3=\frac{1}{8}-\frac{1}{4}-3=-\frac{25}{8}$

The axis of symmetry is $x=\frac{1}{4}$.

The vertex is $\left(\frac{1}{4},-\frac{25}{8}\right)$.

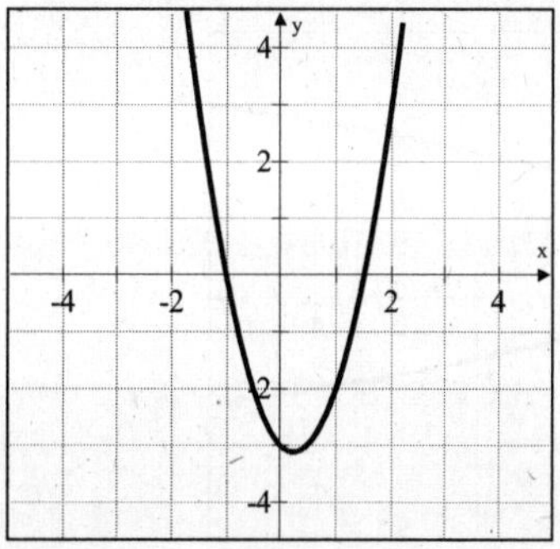

19. $y = x^2 + 5x + 6$

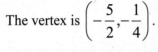

$$y = \left(-\frac{5}{2}\right)^2 + 5\left(-\frac{5}{2}\right) + 6 = \frac{25}{4} - \frac{25}{2} + 6 = -\frac{1}{4}$$

The axis of symmetry is $x = -\dfrac{5}{2}$.

The vertex is $\left(-\dfrac{5}{2}, -\dfrac{1}{4}\right)$.

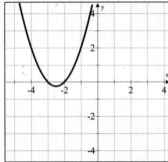

21. $x = y^2 - 2y - 5$

$$y = -\frac{b}{2a} = -\frac{-2}{2(1)} = -(-1) = 1$$

$$x = (1)^2 - 2(1) - 5 = 1 - 2 - 5 = -6$$

The axis of symmetry is $y = 1$.
The vertex is $(-6, 1)$.

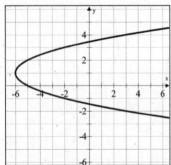

Critical Thinking

23. $p = \dfrac{1}{4a}; \quad a = 2$

$$p = \frac{1}{4(2)} = \frac{1}{8}$$

The focus is $\left(0, \dfrac{1}{8}\right)$.

Section 11.2

Concept Check

1. All points on the circumference of a circle are the same distance from the center of the circle.

3. $(x - 4)^2 + (y - 6)^2 = 25$
Center: $(4, 6)$
Radius: $\sqrt{25} = 5$

5. $x^2 + (y - 1)^2 = 5$
Center: $(0, 1)$
Radius: $\sqrt{5}$

Objective A Exercises

7. $(x - 2)^2 + (y + 2)^2 = 9$
Center: $(2, -2)$
Radius: $\sqrt{9} = 3$

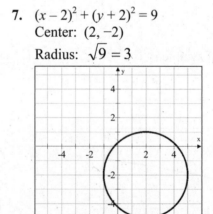

9. $(x + 3)^2 + (y - 1)^2 = 25$

Center: $(-3, 1)$

Radius: $\sqrt{25} = 5$

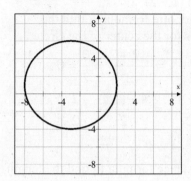

11. $(x + 2)^2 + (y + 2)^2 = 4$

Center: $(-2, -2)$

Radius: $\sqrt{4} = 2$

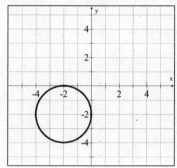

13. $(x - h)^2 + (y - k)^2 = r^2$

Center: $(2, -1)$

Radius: $r = 2$

$(x - 2)^2 + (y - (-1))^2 = 2^2$

$(x - 2)^2 + (y + 1)^2 = 4$

15. $(x - h)^2 + (y - k)^2 = r^2$

Center: $(0, -4)$

Radius: $r = \sqrt{7}$

$(x - 0)^2 + (y - (-4))^2 = (\sqrt{7})^2$

$x^2 + (y + 4)^2 = 7$

17. $(x - h)^2 + (y - k)^2 = r^2$

Center: $(-3, 0)$

Radius: $r = 9$

$(x - (-3))^2 + (y - 0)^2 = (9)^2$

$(x + 3)^2 + y^2 = 81$

19. $(1, 2)$ and $(-1, 1)$

Use the distance formula to find the length of the radius:

$d = \sqrt{(x_1 - x_2)^2 + (y_1 - y_2)^2}$

$= \sqrt{(1 - (-1))^2 + (2 - 1)^2}$

$= \sqrt{2^2 + 1^2} = \sqrt{4 + 1} = \sqrt{5}$

$(x - h)^2 + (y - k)^2 = r^2$

Center: $(-1, 1)$

Radius: $r = \sqrt{5}$

$(x - (-1))^2 + (y - 1)^2 = (\sqrt{5})^2$

$(x + 1)^2 + (y - 1)^2 - 5$

21. $(-3, -4)$ and $(-1, 0)$

Use the distance formula to find the length of the radius:

$d = \sqrt{(x_1 - x_2)^2 + (y_1 - y_2)^2}$

$= \sqrt{(-3 - (-1))^2 + (-4 - 0)^2}$

$= \sqrt{(-2)^2 + (-4)^2} = \sqrt{4 + 16} = \sqrt{20}$

$(x - h)^2 + (y - k)^2 = r^2$

Center: $(-1, 0)$

Radius: $r = \sqrt{20}$

$(x - (-1))^2 + (y - 0)^2 = (\sqrt{20})^2$

$(x + 1)^2 + y^2 = 20$

23. $(1, -5)$ and $(0, 3)$

Use the distance formula to find the length of the radius:

$d = \sqrt{(x_1 - x_2)^2 + (y_1 - y_2)^2}$

$= \sqrt{(1 - 0)^2 + (-5 - 3)^2}$

$= \sqrt{1^2 + (-8)^2} = \sqrt{1 + 64} = \sqrt{65}$

$(x - h)^2 + (y - k)^2 = r^2$

Center: $(0, 3)$

Radius: $r = \sqrt{65}$

$(x - 0)^2 + (y - 3)^2 = (\sqrt{65})^2$

$x^2 + (y - 3)^2 = 65$

25. $x^2 + y^2 = r^2$

Objective B Exercises

27.
$$x^2 + y^2 - 2x + 4y - 20 = 0$$
$$(x^2 - 2x) + (y^2 + 4y) = 20$$
$$(x^2 - 2x + 1) + (y^2 + 4y + 4) = 20 + 1 + 4$$
$$(x - 1)^2 + (y + 2)^2 = 25$$

29.
$$x^2 + y^2 + 6x + 8y + 9 = 0$$
$$(x^2 + 6x) + (y^2 + 8y) = -9$$
$$(x^2 + 6x + 9) + (y^2 + 8y + 16) = -9 + 9 + 16$$
$$(x + 3)^2 + (y + 4)^2 = 16$$

31.
$$x^2 + y^2 - 10x - 8y - 9 = 0$$
$$(x^2 - 10x) + (y^2 - 8y) = 9$$
$$(x^2 - 10x + 25) + (y^2 - 8y + 16) = 9 + 25 + 16$$
$$(x - 5)^2 + (y - 4)^2 = 50$$

33.
$$x^2 + y^2 - 12x + 12 = 0$$
$$(x^2 - 12x) + y^2 = -12$$
$$(x^2 - 12x + 36) + y^2 = -12 + 36$$
$$(x - 6)^2 + y^2 = 24$$

35.
$$x^2 + y^2 + 14y + 13 = 0$$
$$x^2 + (y^2 + 14y) = -13$$
$$x^2 + (y^2 + 14y + 49) = -13 + 49$$
$$x^2 + (y + 7)^2 = 36$$

37.
$$x^2 + y^2 - 4x - 7y - 5 = 0$$
$$(x^2 - 4x) + (y^2 - 7y) = 5$$
$$(x^2 - 4x + 4) + (y^2 - 7y + \frac{49}{4}) = 5 + 4 + \frac{49}{4}$$
$$(x - 2)^2 + (y - \frac{7}{2})^2 = \frac{85}{4}$$

39.
$$x^2 + y^2 + 5x + 5y - 8 = 0$$
$$(x^2 + 5x) + (y^2 + 5y) = 8$$
$$(x^2 + 5x + \frac{25}{4}) + (y^2 + 5y + \frac{25}{4}) = 8 + \frac{25}{4} + \frac{25}{4}$$
$$\left(x + \frac{5}{2}\right)^2 + \left(y + \frac{5}{2}\right)^2 = \frac{41}{2}$$

Applying the Concepts

41. $(-2, 4)$ and $(2, -2)$
Use the midpoint formula to find the coordinates for the center of the circle.
$$x_m = \frac{x_1 + x_2}{2} = \frac{-2 + 2}{2} = \frac{0}{2} = 0$$
$$y_m = \frac{y_1 + y_2}{2} = \frac{4 + (-2)}{2} = \frac{2}{2} = 1$$
Center: $(0, 1)$

Use the distance formula to find the length of the diameter:
$$d = \sqrt{(x_1 - x_2)^2 + (y_1 - y_2)^2}$$
$$= \sqrt{(-2 - 2)^2 + (4 - (-2))^2}$$
$$= \sqrt{(-4)^2 + (6)^2} = \sqrt{16 + 36} = \sqrt{52}$$
Radius: $\dfrac{\sqrt{52}}{2} = \dfrac{2\sqrt{13}}{2} = \sqrt{13}$
$$(x - h)^2 + (y - k)^2 = r^2$$
$$(x - 0)^2 + (y - 1)^2 = (\sqrt{13})^2$$
$$x^2 + (y - 1)^2 = 13$$

Projects or Group Activities

43.
$$r^2 + 6^2 = 12^2$$
$$r^2 + 36 = 144$$
$$r^2 = 108$$
$$r = \sqrt{108} = \sqrt{36 \cdot 3} = 6\sqrt{3}$$
The radius of the circle will be $6\sqrt{3}$.

Section 11.3

Concept Check

1. a) The graph of an ellipse is not a function. It does not pass the vertical line test.
b) The graph of a hyperbola is not a function. It does not pass the vertical line test.

Objective A Exercises

3. *x*-intercepts: (2, 0) and (−2, 0)
   y-intercepts: (0, 3) and (0, −3)

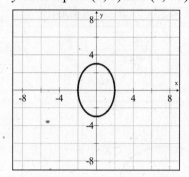

5. *x*-intercepts: (5, 0) and (−5, 0)
   y-intercepts: (0, 3) and (0, −3)

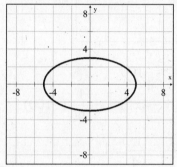

7. *x*-intercepts: (6, 0) and (−6, 0)
   y-intercepts: (0, 4) and (0, −4)

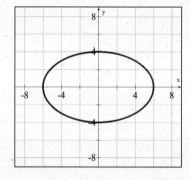

9. *x*-intercepts: (4, 0) and (−4, 0)
   y-intercepts: (0, 7) and (0, −7)

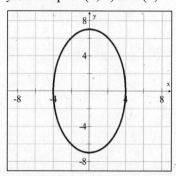

11. *x*-intercepts: (2, 0) and (−2, 0)
    y-intercepts: (0, 5) and (0, −5)

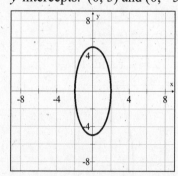

13. The graph of the ellipse gets rounder.

Objective B Exercises

15. Axis of symmetry: *x*-axis
    Vertices: (5, 0) and (−5, 0)

    Asymptotes: $y = \dfrac{2}{5}x$ and $y = -\dfrac{2}{5}x$

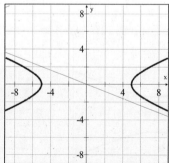

17. Axis of symmetry: y-axis

Vertices: $(0, 4)$ and $(0, -4)$

Asymptotes: $y = \dfrac{4}{5}x$ and $y = -\dfrac{4}{5}x$

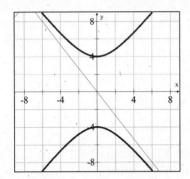

19. Axis of symmetry: x-axis

Vertices: $(3, 0)$ and $(-3, 0)$

Asymptotes: $y = \dfrac{7}{3}x$ and $y = -\dfrac{7}{3}x$

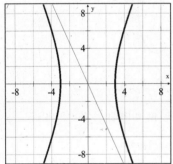

21. Axis of symmetry: y-axis

Vertices: $(0, 2)$ and $(0, -2)$

Asymptotes: $y = \dfrac{1}{2}x$ and $y = -\dfrac{1}{2}x$

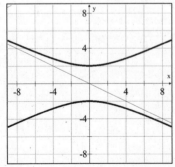

23. The vertices are on the x-axis.

Critical Thinking

25.
$$16x^2 + 25y^2 = 400$$
$$\frac{1}{400}(16x^2 + 25y^2) = \frac{1}{400}(400)$$
$$\frac{x^2}{25} + \frac{y^2}{16} = 1$$

ellipse

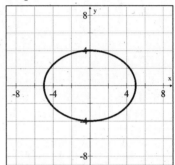

27. $25y^2 - 4x^2 = -100$
$$-\frac{1}{100}(25y^2 - 4x^2) = -\frac{1}{100}(-100)$$
$$\frac{x^2}{25} - \frac{y^2}{4} = 1$$

hyperbola

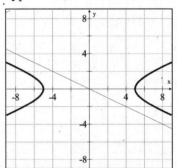

29. $4y^2 - x^2 = 36$

$$\frac{1}{36}(4y^2 - x^2) = \frac{1}{36}(36)$$

$$\frac{y^2}{9} - \frac{x^2}{36} = 1$$

hyperbola

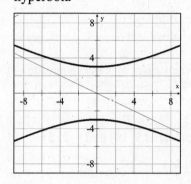

Check Your Progress: Chapter 11

1. $y = x^2 + 4x - 2$

$$x = -\frac{b}{2a} = -\frac{4}{2(1)} = -2$$

$$y = (-2)^2 + 4(-2) - 2 = 4 - 8 - 2 = -6$$

The axis of symmetry is $x = -2$.

The vertex is $(-2, -6)$.

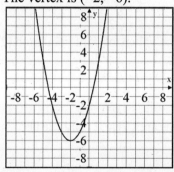

2. $x = -y^2 - 2y + 3$

$$y = -\frac{b}{2a} = -\frac{-2}{2(-1)} = -1$$

$$x = -(-1)^2 - 2(-1) + 3 = -1 + 2 + 3 = 4$$

The axis of symmetry is $y = -1$.

The vertex is $(3, -1)$.

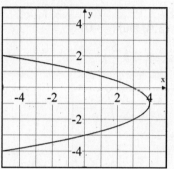

3. $(x - h)^2 + (y - k)^2 = r^2$

Center: $(2, -5)$

Radius: $r = 3$

$(x - 2)^2 + (y - (-5))^2 = 3^2$

$(x - 2)^2 + (y + 5)^2 = 9$

4. $(-1, 3)$ and $(2, 5)$

Use the distance formula to find the length of the radius:

$$d = \sqrt{(x_1 - x_2)^2 + (y_1 - y_2)^2}$$

$$= \sqrt{(-1 - (2))^2 + (3 - 5)^2}$$

$$= \sqrt{(-3)^2 + (-2)^2} = \sqrt{9 + 4} = \sqrt{13}$$

$(x - h)^2 + (y - k)^2 = r^2$

Center: $(2, 5)$

Radius: $r = \sqrt{13}$

$(x - 2)^2 + (y - 5)^2 = \left(\sqrt{13}\right)^2$

$(x - 2)^2 + (y - 5)^2 = 13$

5.
$$x^2 - 6x + y^2 - 7y - 2 = 0$$
$$(x^2 - 6x) + (y^2 - 7y) = 2$$
$$(x^2 - 6x + 9) + \left(y^2 - 7y + \frac{49}{4}\right) = 2 + 9 + \frac{49}{4}$$
$$(x - 3)^2 + \left(y - \frac{7}{2}\right)^2 = \frac{93}{4}$$

6. Intercepts: $(3,0)$ and $(-3,0)$

Asymptotes: $y = \pm\dfrac{5}{3}x$

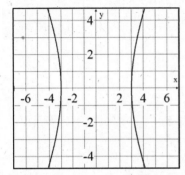

7. Intercepts: $(3,0)$, $(-3,0)$, $(0,5)$ and $(0,-5)$

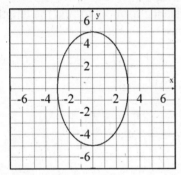

Section 11.4

Concept Check

1. If a system of equations contains the equation of a line and the equation of a hyperbola it can have 0, 1 or 2 solutions.

3. If a system of equations contains the equation of a parabola and the equation of a ellipse it can have 0, 1, 2, 3 or 4 solutions.

Objective A Exercises

5. (1) $y = x^2 - x - 1$
(2) $y = 2x + 9$
Use the substitution method:
$$y = x^2 - x - 1$$
$$2x + 9 = x^2 - x - 1$$
$$0 = x^2 - 3x - 10$$
$$0 = (x - 5)(x + 2)$$
$$x - 5 = 0 \qquad x + 2 = 0$$
$$x = 5 \qquad\quad x = -2$$
Substitute the values of x into equation (2).
$$y = 2(5) + 9 \qquad y = 2(-2) + 9$$
$$y = 10 + 9 \qquad\quad y = -4 + 9$$
$$y = 19 \qquad\qquad y = 5$$
The solutions are $(5, 19)$ and $(-2, 5)$.

7. (1) $y^2 = -x + 3$
(2) $x - y = 1$
Solve equation (2) for x.
$$x - y = 1$$
$$x = y + 1$$
Use the substitution method:
$$y^2 = -x + 3$$
$$y^2 = -(y + 1) + 3$$
$$y^2 = -y - 1 + 3$$
$$y^2 = -y + 2$$
$$y^2 + y - 2 = 0$$
$$(y - 1)(y + 2) = 0$$
$$y - 1 = 0 \qquad y + 2 = 0$$
$$y = 1 \qquad\quad y = -2$$
Substitute the values of y into equation (2).
$$x - y = 1 \qquad\qquad x - y = 1$$
$$x - 1 = 1 \qquad\quad x - (-2) = 1$$
$$x = 2 \qquad\qquad x + 2 = 1$$
$$\qquad\qquad\qquad\qquad x = -1$$
The solutions are $(-1, -2)$ and $(2, 1)$.

9. (1) $y^2 = 2x$
(2) $x + 2y = -2$
Solve equation (2) for x.
$$x + 2y = -2$$
$$x = -2y - 2$$
Use the substitution method:
$$y^2 = 2x$$
$$y^2 = 2(-2y - 2)$$
$$y^2 = -4y - 4$$
$$y^2 + 4y + 4 = 0$$
$$(y + 2)(y + 2) = 0$$
$$y + 2 = 0 \qquad y + 2 = 0$$

$y = -2$ $y = -2$

Substitute the value of y into equation (2).

$x + 2y = -2$

$x + 2(-2) = -2$

$x - 4 = -2$

$x = 2$

The solution is $(2, -2)$.

11. (1) $x^2 + 2y^2 = 12$

(2) $2x - y = 2$

Solve equation (2) for y.

$2x - y = 2$

$2x = y + 2$

$y = 2x - 2$

Use the substitution method:

$$x^2 + 2y^2 = 12$$
$$x^2 + 2(2x - 2)^2 = 12$$
$$x^2 + 2(4x^2 - 8x + 4) = 12$$
$$x^2 + 8x^2 - 16x + 8 = 12$$
$$9x^2 - 16x - 4 = 0$$
$$(x - 2)(9x + 2) = 0$$
$$x - 2 = 0 \quad 9x + 2 = 0$$
$$x = 2 \quad\quad 9x = -2$$
$$x = -\frac{2}{9}$$

Substitute the values of x into equation (2).

$2x - y = 2$ $2x - y = 2$

$2(2) - y = 2$ $2(-\dfrac{2}{9}) - y = 2$

$4 - y = 2$ $-\dfrac{4}{9} - y = 2$

$-y = -2$ $-y = \dfrac{22}{9}$

$y = 2$ $y = -\dfrac{22}{9}$

The solutions are $(2, 2)$ and $\left(-\dfrac{2}{9}, -\dfrac{22}{9}\right)$.

13. (1) $x^2 + y^2 = 13$

(2) $x + y = 5$

Solve equation (2) for y.

$x + y = 5$

$y = -x + 5$

Use the substitution method:

$$x^2 + y^2 = 13$$
$$x^2 + (-x + 5)^2 = 13$$
$$x^2 + x^2 - 10x + 25 = 13$$
$$2x^2 - 10x + 12 = 0$$

$$2(x^2 - 5x + 6) = 0$$
$$(x - 3)(x - 2) = 0$$
$$x - 3 = 0 \quad x - 2 = 0$$
$$x = 3 \quad\quad x = 2$$

Substitute the values of x into equation (2).

$3 + y = 5$ $2 + y = 5$

$y = 2$ $y = 3$

The solutions are $(3, 2)$ and $(2, 3)$.

15. (1) $4x^2 + y^2 = 12$

(2) $y = 4x^2$

Use the substitution method:

$$4x^2 + y^2 = 12$$
$$4x^2 + (4x^2)^2 = 12$$
$$4x^2 + 16x^4 = 12$$
$$16x^4 + 4x^2 - 12 = 0$$
$$4(x^4 + x^2 - 3) = 0$$
$$4(4x^2 - 3)(x^2 + 1) = 0$$
$$4x^2 - 3 = 0 \quad x^2 + 1 = 0$$
$$4x^2 = 3 \quad\quad x^2 = -1$$
$$x^2 = \frac{3}{4} \quad\quad x = \pm\sqrt{-1}$$
$$x = \pm\frac{\sqrt{3}}{2}$$

Substitute the real number values of x into equation (2).

$y = 4x^2$

$y = 4\left(\dfrac{\sqrt{3}}{2}\right)^2$ $y = 4\left(-\dfrac{\sqrt{3}}{2}\right)^2$

$y = 4\left(\dfrac{3}{4}\right)$ $y = 4\left(\dfrac{3}{4}\right)$

$y = 3$ $y = 3$

The solutions are $\left(\dfrac{\sqrt{3}}{2}, 3\right)$ and $\left(-\dfrac{\sqrt{3}}{2}, 3\right)$.

17. (1) $y = x^2 - 2x - 3$
(2) $y = x - 6$
Use the substitution method:
$$y = x^2 - 2x - 3$$
$$x - 6 = x^2 - 2x - 3$$
$$0 = x^2 - 3x + 3$$
$$x = \frac{-b \pm \sqrt{b^2 - 4ac}}{2a}$$
$$= \frac{-(-3) \pm \sqrt{(-3)^2 - 4(1)(3)}}{2(1)}$$
$$= \frac{3 \pm \sqrt{9 - 12}}{2} = \frac{3 \pm \sqrt{-3}}{2}$$
The system of equations has no real number solution.

19. (1) $3x^2 - y^2 = -1$
(2) $x^2 + 4y^2 = 17$
Use the addition method. Multiply equation (1) by 4.
$$12x^2 - 4y^2 = -4$$
$$x^2 + 4y^2 = 17$$
$$13x^2 = 13$$
$$x^2 = 1$$
$$x = \pm\sqrt{1} = \pm 1$$
Substitute the values of x into equation (2).

$x^2 + 4y^2 = 17$	$x^2 + 4y^2 = 17$
$1^2 + 4y^2 = 17$	$(-1)^2 + 4y^2 = 17$
$1 + 4y^2 = 16$	$1 + 4y^2 = 17$
$4y^2 = 16$	$4y^2 = 16$
$y^2 = 4$	$y^2 = 4$
$y = \pm\sqrt{4} = \pm 2$	$y = \pm\sqrt{4} = \pm 2$

The solutions are $(1, 2)$, $(1, -2)$, $(-1, 2)$ and $(-1, -2)$.

21. (1) $2x^2 + 3y^2 = 30$
(2) $x^2 + y^2 = 13$
Use the addition method. Multiply equation (2) by -2.
$$2x^2 + 3y^2 = 30$$
$$-2x^2 - 2y^2 = -26$$
$$y^2 = 4$$
$$y = \pm\sqrt{4} = \pm 2$$
Substitute the values of y into equation (2).

$x^2 + y^2 = 13$	$x^2 + y^2 = 13$
$x^2 + 2^2 = 13$	$x^2 + (-2)^2 = 13$
$x^2 + 4 = 13$	$x^2 + 4 = 13$

$x^2 = 9$	$x^2 = 9$
$x = \pm\sqrt{9} = \pm 3$	$x = \pm\sqrt{9} = \pm 3$

The solutions are $(3, 2)$, $(-3, 2)$, $(3, -2)$ and $(-3, -2)$.

23. (1) $y = 2x^2 - x + 1$
(2) $y = x^2 - x + 5$
Use the substitution method:
$$y = 2x^2 - x + 1$$
$$x^2 - x + 5 = 2x^2 - x + 1$$
$$0 = x^2 - 4$$
$$0 = (x - 2)(x + 2)$$
$$x - 2 = 0 \qquad x + 2 = 0$$
$$x = 2 \qquad \quad x = -2$$
Substitute the values of x into equation (2).

$y = x^2 - x + 5$	$y = x^2 - x + 5$
$y = 2^2 - 2 + 5$	$y = (-2)^2 - (-2) + 5$
$y = 4 - 2 + 5$	$y = 4 + 2 + 5$
$y = 7$	$y = 11$

The solutions are $(2, 7)$ and $(-2, 11)$.

25. (1) $2x^2 + 3y^2 = 24$
(2) $x^2 - y^2 = 7$
Use the addition method. Multiply equation (2) by -2.
$$2x^2 + 3y^2 = 24$$
$$-2x^2 + 2y^2 = -14$$
$$5y^2 = 10$$
$$y^2 = 2$$
$$y = \pm\sqrt{2}$$
Substitute the values of y into equation (2).

$x^2 - y^2 = 7$	$x^2 - y^2 = 7$
$x^2 - \left(\sqrt{2}\right)^2 = 7$	$x^2 - \left(-\sqrt{2}\right)^2 = 7$
$x^2 - 2 = 7$	$x^2 - 2 = 7$
$x^2 = 9$	$x^2 = 9$
$x = \pm\sqrt{9} = \pm 3$	$x = \pm\sqrt{9} = \pm 3$

The solutions are $(3, \sqrt{2})$, $(-3, \sqrt{2})$, $(3, -\sqrt{2})$ and $(-3, -\sqrt{2})$.

27. (1) $x^2 + y^2 = 36$
(2) $4x^2 + 9y^2 = 36$
Use the addition method. Multiply equation (1) by -4.
$$-4x^2 - 4y^2 = -144$$
$$4x^2 + 9y^2 = 36$$
$$5y^2 = -108$$

$$y^2 = -\frac{108}{5}$$

$$y = \pm\sqrt{-\frac{108}{5}}$$

The system of equations has no real number solution.

29. (1) $11x^2 - 2y^2 = 4$
(2) $3x^2 + y^2 = 15$
Use the addition method. Multiply equation (2) by 2.

$$11x^2 - 2y^2 = 4$$
$$6x^2 + 2y^2 = 30$$
$$17x^2 = 34$$
$$x^2 = 2$$
$$x = \pm\sqrt{2}$$

Substitute the values of x into equation (2).

$3x^2 + y^2 = 15$	$3x^2 + y^2 = 15$
$3\left(\sqrt{2}\right)^2 + y^2 = 15$	$3\left(-\sqrt{2}\right)^2 + y^2 = 15$
$6 + y^2 = 15$	$6 + y^2 = 15$
$y^2 = 9$	$y^2 = 9$
$y = \pm\sqrt{9} = \pm 3$	$y = \pm\sqrt{9} = \pm 3$

The solutions are $\left(3, \sqrt{2}\right)$, $\left(-3, \sqrt{2}\right)$, $\left(3, -\sqrt{2}\right)$ and $\left(-3, -\sqrt{2}\right)$.

31. (1) $2x^2 - y^2 = 7$
(2) $2x - y = 5$
Solve equation (2) for y.

$$2x - y = 5$$
$$-y = -2x + 5$$
$$y = 2x - 5$$

Use the substitution method:

$$2x^2 - y^2 = 7$$
$$2x^2 - (2x - 5)^2 = 7$$
$$2x^2 - (4x^2 - 20x + 25) = 7$$
$$2x^2 - 4x^2 + 20x - 25 = 7$$
$$-2x^2 + 20x - 32 = 0$$
$$-2(x^2 - 10x + 16) = 0$$
$$-2(x - 8)(x - 2) = 0$$

$x - 8 = 0$	$x - 2 = 0$
$x = 8$	$x = 2$

Substitute the values of x into equation (2).

$2x - y = 5$	$2x - y = 5$
$2(8) - y = 5$	$2(2) - y = 5$
$-y = -11$	$-y = 1$
$y = 11$	$y = -1$

The solutions are $(8, 11)$ and $(2, -1)$.

33. (1) $y = 3x^2 + x - 4$
(2) $y = 3x^2 - 8x + 5$
Use the substitution method:

$$y = 3x^2 + x - 4$$
$$3x^2 - 8x + 5 = 3x^2 + x - 4$$
$$0 = 9x - 9$$
$$9 = 9x$$
$$x = 1$$

Substitute the value of x into equation (1).

$$y = 3x^2 + x - 4$$
$$y = 3(1)^2 + 1 - 4$$
$$y = 3 + 1 - 4$$
$$y = 0$$

The solution is $(1, 0)$.

35. (1) $x = y + 3$
(2) $x^2 + y^2 = 5$
Use the substitution method:

$$(y + 3)^2 + y^2 = 5$$
$$y^2 + 6y + 9 + y^2 = 5$$
$$2y^2 + 6y + 4 = 0$$
$$2(y^2 + 3y + 2) = 0$$
$$2(y + 2)(y + 1) = 0$$

$y + 2 = 0$	$y + 1 = 0$
$y = -2$	$y = -1$

Substitute the values of y into equation (1).

$x = y + 3$	$x = y + 3$
$x = -2 + 3$	$x = -1 + 3$
$x = 1$	$x = 2$

The solutions are $(1, -2)$ and $(2, -1)$.

37. (1) $y = x^2 + 4x + 4$
(2) $x + 2y = 4$
Use the substitution method:

$$x + 2y = 4$$
$$x + 2(x^2 + 4x + 4) = 4$$
$$x + 2x^2 + 8x + 8 = 4$$
$$2x^2 + 9x + 4 = 0$$
$$(2x + 1)(x + 4) = 0$$

$2x + 1 = 0$	$x + 4 = 0$
$2x = -1$	$x = -4$
$x = -\dfrac{1}{2}$	

Substitute the values of x into equation (2).

$x + 2y = 4$ $\qquad$ $x + 2y = 4$

$-\dfrac{1}{2} + 2y = 4$ $\qquad$ $-4 + 2y = 4$

$2y = \dfrac{9}{2}$ $\qquad$ $2y = 8$

$y = \dfrac{9}{4}$ $\qquad$ $y = 4$

The solutions are $\left(-\dfrac{1}{2}, \dfrac{9}{4} \right)$ and $(-4, 4)$.

Critical Thinking

39. $y = 2^x$
$\quad$ $x + y = 3$

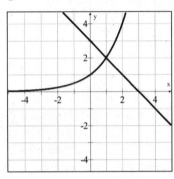

The solution is $(1, 2)$.

41. $y = \log_4 x$
$\quad$ $x + y = 4$

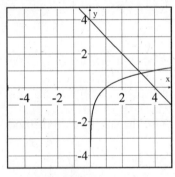

The solution is $(3.168, 0.832)$.

Projects or Group Activities

43. No. A graph of these two equations shows that they do not intersect. Therefore, the meteorite will not hit the earth.

Section 11.5

Concept Check

1. A solid curve is used for the boundaries of inequalities that use $\le$ or $\ge$, and a dashed curve is used for the boundaries of inequalities that use $<$ or $>$.

Objective A Exercises

3. $y \le x^2 - 4x + 3$

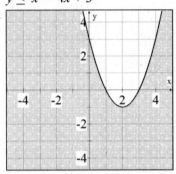

5. $(x - 1)^2 + (y + 2)^2 \ge 9$

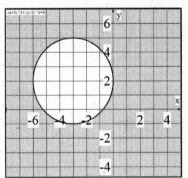

7. $\dfrac{x^2}{16} + \dfrac{y^2}{25} < 1$

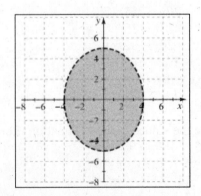

9. $\dfrac{x^2}{25} - \dfrac{y^2}{9} \le 1$

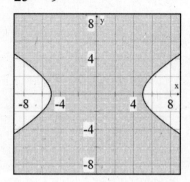

11. $\dfrac{x^2}{4} + \dfrac{y^2}{16} \ge 1$

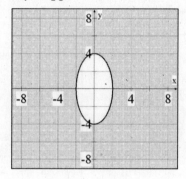

13. $\dfrac{y^2}{9} - \dfrac{x^2}{16} \le 1$

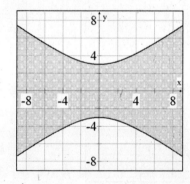

15. $\dfrac{x^2}{9} + \dfrac{y^2}{1} \le 1$

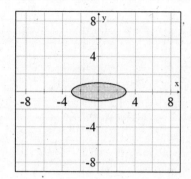

17. $(x-1)^2 + (y+3)^2 \le 25$

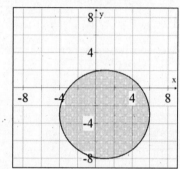

19. $\dfrac{y^2}{25} - \dfrac{x^2}{4} \le 1$

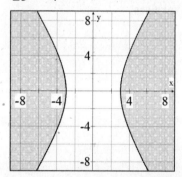

21. $\dfrac{x^2}{25} + \dfrac{y^2}{9} \leq 1$

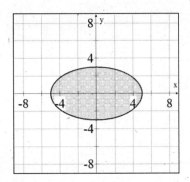

27. $x^2 + y^2 < 16$
$y > x + 1$

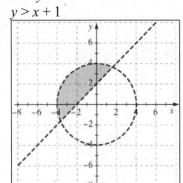

23. $\dfrac{x^2}{36} + \dfrac{y^2}{4} \leq 1$

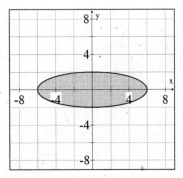

29. $\dfrac{x^2}{4} + \dfrac{y^2}{16} \leq 1$
$y \leq -\dfrac{1}{2}x + 2$

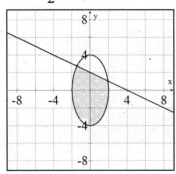

Objective B Exercises

25. $y \leq (x-2)^2$
$y + x > 4$

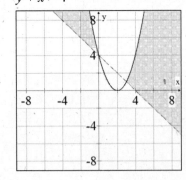

31. $x \geq y^2 - 3y + 2$
$y \geq 2x - 2$

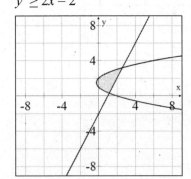

33. $x^2 + y^2 < 25$

$\dfrac{x^2}{9} + \dfrac{y^2}{36} < 1$

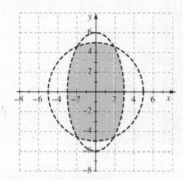

35. $x^2 + y^2 > 4$
$x^2 + y^2 < 25$

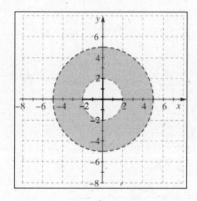

37. The solution set would be the empty set.

Critical Thinking

39. $y > x^2 - 3$
$y < x + 3$
$x \leq 0$

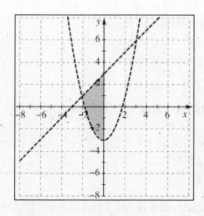

41. $x^2 + y^2 < 3$
$x > y^2 - 1$
$y \geq 0$

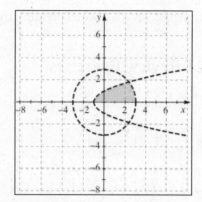

43. $\dfrac{x^2}{16} + \dfrac{y^2}{4} \leq 1$

$x^2 + y^2 \leq 4$

$x \geq 0$
$y \leq 0$

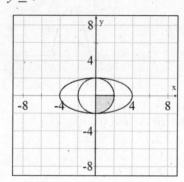

45. $y > 2^x$
$x + y < 4$

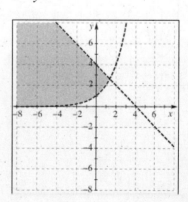

Projects or Group Activities

47. Because x is a variable, its value can be positive or negative. If x is negative, the inequality sign is reversed when dividing by x.

Chapter 11 Review Exercises

1. $y = -2x^2 + x - 2$

$$x = -\frac{b}{2a} = -\frac{1}{2(-2)} = -\frac{1}{-4} = \frac{1}{4}$$

$$y = -2\left(\frac{1}{4}\right)^2 + \frac{1}{4} - 2 = -\frac{1}{8} + \frac{1}{4} - 2 = -\frac{15}{8}$$

The axis of symmetry is $x = \dfrac{1}{4}$.

The vertex is $\left(\dfrac{1}{4}, -\dfrac{15}{8}\right)$.

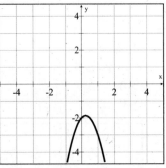

2. x-intercepts: $(1, 0)$ and $(-1, 0)$

y-intercepts: $(0, 3)$ and $(0, -3)$

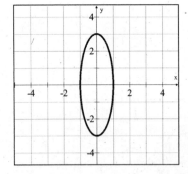

3. $\dfrac{x^2}{9} - \dfrac{y^2}{16} < 1$

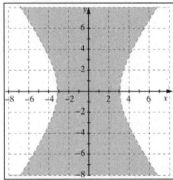

4. $(x + 3)^2 + (y + 1)^2 = 1$

Center: $(-3, -1)$

Radius: 1

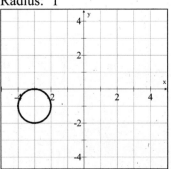

5. $y = x^2 - 4x + 8$

$$x = -\frac{b}{2a} = -\frac{-4}{2(1)} = -\frac{-4}{2} = 2$$

$$y = (2)^2 - 4(2) + 8 = 4 - 8 + 8 = 4$$

The axis of symmetry is $x = 2$.

The vertex is $(2, 4)$.

6. (1) $y^2 = 2x^2 - 3x + 6$

(2) $y^2 = 2x^2 + 5x - 2$

Use the substitution method.

$$y^2 = 2x^2 - 3x + 6$$
$$2x^2 + 5x - 2 = 2x^2 - 3x + 6$$
$$8x = 8$$
$$x = 1$$

Substitute the value of x into equation (1).

$$y^2 = 2x^2 - 3x + 6$$
$$y^2 = 2(1)^2 - 3(1) + 6$$
$$= 2 - 3 + 6 = 5$$
$$y = \pm\sqrt{5}$$

The solutions are $\left(1, \sqrt{5}\right)$ and $\left(1, -\sqrt{5}\right)$.

7. $\dfrac{x^2}{25} + \dfrac{y^2}{16} \le 1$

$\dfrac{y^2}{4} - \dfrac{x^2}{4} \ge 1$

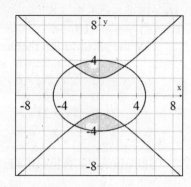

8. Axis of symmetry: x-axis

Vertices: $(5, 0)$ and $(-5, 0)$

Asymptotes: $y = \dfrac{1}{5}x$ and $y = -\dfrac{1}{5}x$

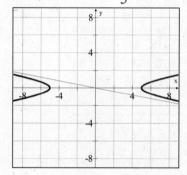

9. $(2, -1)$ and $(-1, 2)$

Use the distance formula to find the length of the radius:

$d = \sqrt{(x_1 - x_2)^2 + (y_1 - y_2)^2}$

$= \sqrt{(2 - (-1))^2 + (-1 - 2)^2}$

$= \sqrt{3^2 + (-3)^2} = \sqrt{9 + 9} = \sqrt{18}$

$(x - h)^2 + (y - k)^2 = r^2$

Center: $(-1, 2)$

Radius: $r = \sqrt{18}$

$(x - (-1))^2 + (y - 2)^2 = (\sqrt{18})^2$

$(x + 1)^2 + (y - 2)^2 = 18$

10. $y = -x^2 + 7x - 8$

$x = -\dfrac{b}{2a} = -\dfrac{7}{2(-1)} = -\dfrac{7}{-2} = \dfrac{7}{2}$

$y = -\left(\dfrac{7}{2}\right)^2 + 7\left(\dfrac{7}{2}\right) - 8 = -\dfrac{49}{4} + \dfrac{49}{2} - 8 = \dfrac{17}{4}$

The axis of symmetry is $x = \dfrac{7}{2}$.

The vertex is $\left(\dfrac{7}{2}, \dfrac{17}{4}\right)$.

11. (1) $x = 2y^2 - 3y + 1$

(2) $3x - 2y = 0$

Solve equation (2) for x.

$3x - 2y = 0$

$3x = 2y$

$x = \dfrac{2}{3}y$

Use the substitution method:

$x = 2y^2 - 3y + 1$

$\dfrac{2}{3}y = 2y^2 - 3y + 1$

$2y = 6y^2 - 9y + 3$

$0 = 6y^2 - 11y + 3$

$0 = (2y - 3)(3y - 1)$

$2y - 3 = 0 \qquad 3y - 1 = 0$

$2y = 3 \qquad\quad 3y = 1$

$y = \dfrac{3}{2} \qquad\quad y = \dfrac{1}{3}$

Substitute the values of y into equation (2).

$3x - 2y = 0 \qquad\qquad 3x - 2y = 0$

$3x - 2\left(\dfrac{3}{2}\right) = 0 \qquad 3x - 2\left(\dfrac{1}{3}\right) = 0$

$3x - 3 = 0 \qquad\qquad 3x - \left(\dfrac{2}{3}\right) = 0$

$3x = 3 \qquad\qquad\quad 3x = \dfrac{2}{3}$

$x = 1 \qquad\qquad\qquad x = \dfrac{2}{9}$

The solutions are $\left(1, \dfrac{3}{2}\right)$ and $\left(\dfrac{2}{9}, \dfrac{1}{3}\right)$.

12. $(4, 6)$ and $(0, -3)$

Use the distance formula to find the length of the radius:

$$d = \sqrt{(x_1 - x_2)^2 + (y_1 - y_2)^2}$$
$$= \sqrt{(4 - 0)^2 + (6 - (-3))^2}$$
$$= \sqrt{4^2 + 9^2} = \sqrt{16 + 81} = \sqrt{97}$$

$(x - h)^2 + (y - k)^2 = r^2$

Center: $(0, -3)$

Radius: $r = \sqrt{97}$

$(x - 0)^2 + (y - (-3))^2 = (\sqrt{97})^2$

$x^2 + (y + 3)^2 = 97$

13. Center: $(-1, 5)$

Radius: 6

$(x - h)^2 + (y - k)^2 = r^2$

$(x - (-1))^2 + (y - 5)^2 = 6^2$

$(x + 1)^2 + (y - 5)^2 = 36$

14. (1) $2x^2 + y^2 = 19$

(2) $3x^2 - y^2 = 6$

Use the addition method.

$$2x^2 + y^2 = 19$$
$$3x^2 - y^2 = 6$$
$$5x^2 = 25$$
$$x^2 = 5$$
$$x = \pm\sqrt{5}$$

Substitute the values of x into equation (1).

$2x^2 + y^2 = 19$ $2x^2 + y^2 = 19$

$2(\sqrt{5})^2 + y^2 = 19$ $2(-\sqrt{5})^2 + y^2 = 19$

$10 + y^2 = 19$ $10 + y^2 = 19$

$y^2 = 9$ $y^2 = 9$

$y = \pm\sqrt{9} = \pm 3$ $y = \pm\sqrt{9} = \pm 3$

The solutions are $(\sqrt{5}, 3)$, $(-\sqrt{5}, 3)$, $(\sqrt{5}, -3)$ and $(-\sqrt{5}, -3)$.

15. $x^2 + y^2 + 4x - 2y = 4$

$(x^2 + 4x) + (y^2 - 2y) = 4$

$x^2 + 4x + 4 + y^2 - 2y + 1 = 4 + 4 + 1$

$(x + 2)^2 + (y - 1)^2 = 9$

16. (1) $y = x^2 + 5x - 6$

(2) $y = x - 10$

Use the substitution method:

$$y = x^2 + 5x - 6$$
$$x - 10 = x^2 + 5x - 6$$
$$0 = x^2 + 4x + 4$$
$$0 = (x + 2)^2$$

$x + 2 = 0$ $x + 2 = 0$

$x = -2$ $x = -2$

Substitute the value of x into equation (2).

$y = x - 10$

$y = -2 - 10$

$y = -12$

The solution is $(-2, -12)$.

17. $\dfrac{x^2}{16} + \dfrac{y^2}{4} > 1$

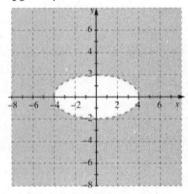

18. Axis of symmetry: y-axis

Vertices: $(0, 4)$ and $(0, -4)$

Asymptotes: $y = \dfrac{4}{3}x$ and $y = -\dfrac{4}{3}x$

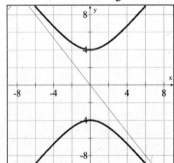

19. $(x - 2)^2 + (y + 1)^2 \leq 16$

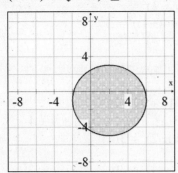

20. $x^2 + (y - 2)^2 = 9$
Center: $(0, 2)$
Radius: 3

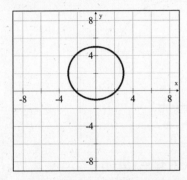

21. x-intercepts: $(5, 0)$ and $(-5, 0)$
y-intercepts: $(0, 3)$ and $(0, -3)$

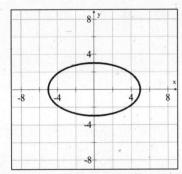

22. $y \geq -x^2 - 2x + 3$

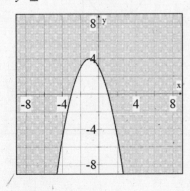

23. $\dfrac{x^2}{16} + \dfrac{y^2}{4} < 1$
$x^2 + y^2 > 9$

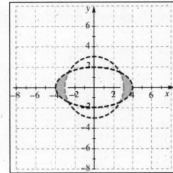

24. $y \geq x^2 - 4x + 2$

$y \leq \dfrac{1}{3}x - 1$

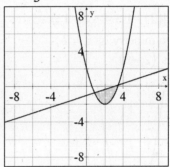

25. $x = 2y^2 - 6y + 5$

The axis of symmetry is $y = \dfrac{3}{2}$.

The vertex is $\left(\dfrac{1}{2}, \dfrac{3}{2}\right)$.

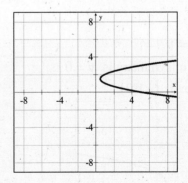

26. $\dfrac{x^2}{9} + \dfrac{y^2}{1} \geq 1$

$\dfrac{x^2}{4} - \dfrac{y^2}{1} \leq 1$

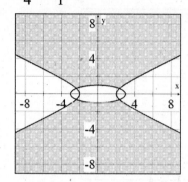

27. $y = x^2 - 4x - 1$

$x = -\dfrac{b}{2a} = -\dfrac{-4}{2(1)} = -\dfrac{-4}{2} = -(-2) = 2$

$y = (2)^2 - 4(2) - 1 = 4 - 8 - 1 = -5$

The axis of symmetry is $x = 2$.

The vertex is $(2, -5)$.

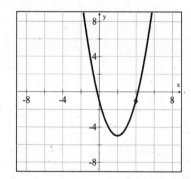

28. $x = y^2 - 1$

$y = -\dfrac{b}{2a} = -\dfrac{0}{2(1)} = 0$

$x = (0)^2 - 1 = -1$

The axis of symmetry is $y = 0$.

The vertex is $(-1, 0)$.

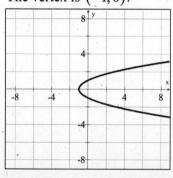

29. Center: $(3, -4)$

Radius: 5

$(x - h)^2 + (y - k)^2 = r^2$

$(x - 3)^2 + (y - (-4))^2 = 5^2$

$(x - 3)^2 + (y + 4)^2 = 25$

30. $x^2 + y^2 - 6x + 4y - 23 = 0$

$x^2 + y^2 - 6x + 4y = 23$

$(x^2 - 6x) + (y^2 + 4y) = 23$

$x^2 - 6x + 9 + y^2 + 4y + 4 = 23 + 9 + 4$

$(x - 3)^2 + (y + 2)^2 = 36$

31. $(x + 1)^2 + (y - 4)^2 = 36$

Center: $(-1, 4)$

Radius: 6

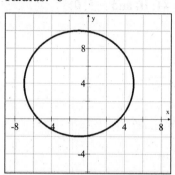

32. $x^2 + y^2 - 8x + 4y + 16 = 0$

$x^2 + y^2 - 8x + 4y = -16$

$(x^2 - 8x) + (y^2 + 4y) = -16$

$x^2 - 8x + 16 + y^2 + 4y + 4 = -16 + 16 + 4$

$(x - 4)^2 + (y + 2)^2 = 4$

Center: $(4, -2)$

Radius: 2

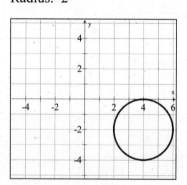

33. x-intercepts: $(6, 0)$ and $(-6, 0)$
y-intercepts: $(0, 4)$ and $(0, -4)$

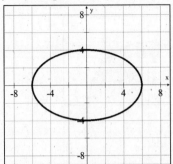

34. x-intercepts: $(1, 0)$ and $(-1, 0)$
y-intercepts: $(0, 5)$ and $(0, -5)$

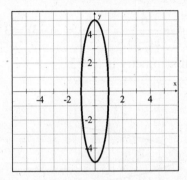

35. Axis of symmetry: x-axis
Vertices: $(3, 0)$ and $(-3, 0)$
Asymptotes: $y = 2x$ and $y = -2x$

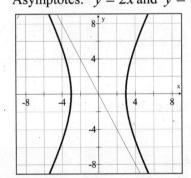

36. Axis of symmetry: y-axis
Vertices: $(0, 6)$ and $(0, -6)$
Asymptotes: $y = 3x$ and $y = -3x$

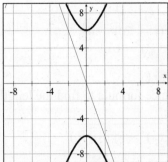

37. (1) $y = x^2 - 3x - 4$
(2) $y = x + 1$
Use the substitution method:
$$y = x^2 - 3x - 4$$
$$x + 1 = x^2 - 3x - 4$$
$$0 = x^2 - 4x - 5$$
$$0 = (x + 1)(x - 5)$$
$$x + 1 = 0 \qquad x - 5 = 0$$
$$x = -1 \qquad\quad x = 5$$
Substitute the value of x into equation (2).
$$y = x + 1 \qquad\qquad y = x + 1$$
$$y = -1 + 1 \qquad\quad y = 5 + 1$$
$$y = 0 \qquad\qquad\quad y = 6$$
The solutions are $(-1, 0)$ and $(5, 6)$.

38. $x \leq y^2 + 2y - 3$

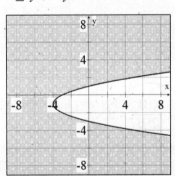

Chapter 11 Test

1. Center: $(-3, -3)$
Radius: 4
$(x - h)^2 + (y - k)^2 = r^2$
$(x - (-3))^2 + (y - (-3))^2 = 4^2$
$(x + 3)^2 + (y + 3)^2 = 16$

2. $(2, 5)$ and $(-2, 1)$
Use the distance formula to find the length of the radius:

$d = \sqrt{(x_1 - x_2)^2 + (y_1 - y_2)^2}$

$= \sqrt{(2 - (-2))^2 + (5 - 1)^2}$

$= \sqrt{4^2 + 4^2} = \sqrt{16 + 16} = \sqrt{32}$

$(x - h)^2 + (y - k)^2 = r^2$
Center: $(-2, 1)$
Radius: $r = \sqrt{32}$
$(x - (-2))^2 + (y - 1)^2 = (\sqrt{32})^2$
$(x + 2)^2 + (y - 1)^2 = 32$

3. Axis of symmetry: y-axis
Vertices: $(0, 5)$ and $(0, -5)$
Asymptotes: $y = \dfrac{5}{4}x$ and $y = -\dfrac{5}{4}x$

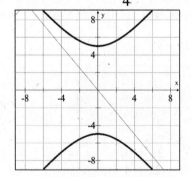

4. $x^2 + y^2 < 36$
$x + y > 4$

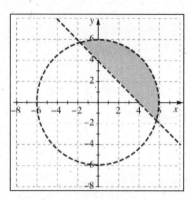

5. $y = -x^2 + 6x - 5$

$x = -\dfrac{b}{2a} = -\dfrac{6}{2(-1)} = -\dfrac{6}{-2} = -(-3) = 3$

The axis of symmetry is $x = 3$.

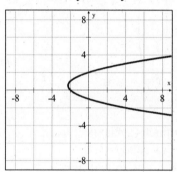

6. (1) $x^2 - y^2 = 24$
(2) $2x^2 + 5y^2 = 55$
Use the addition method. Multiply equation (1) by -2.
$-2x^2 + 2y^2 = -48$
$2x^2 + 5y^2 = 55$
$7y^2 = 7$
$y^2 = 1$
$y = \pm 1$

Substitute the values of y into equation (1).

$\begin{array}{ll} x^2 - y^2 = 24 & x^2 - y^2 = 24 \\ x^2 - (1)^2 = 24 & x^2 - (-1)^2 = 24 \\ x^2 - 1 = 24 & x^2 - 1 = 24 \\ x^2 = 25 & x^2 = 25 \\ x = \pm\sqrt{25} = \pm 5 & x = \pm\sqrt{25} = \pm 5 \end{array}$

The solutions are $(5, 1)$, $(-5, 1)$, $(5, -1)$ and $(-5, -1)$.

7. $y = -x^2 + 3x - 2$

$$x = -\frac{b}{2a} = -\frac{3}{2(-1)} = -\frac{3}{-2} = \frac{3}{2}$$

$$y = -\left(\frac{3}{2}\right)^2 + 3\left(\frac{3}{2}\right) - 2 = -\frac{9}{4} + \frac{9}{2} - 2 = \frac{1}{4}$$

The vertex is $\left(\dfrac{3}{2}, \dfrac{1}{4}\right)$.

8. $x = y^2 - y - 2$

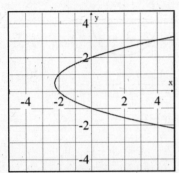

9. x-intercepts: $(4, 0)$ and $(-4, 0)$
y-intercepts: $(0, 2)$ and $(0, -2)$

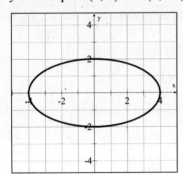

10. $\dfrac{x^2}{25} + \dfrac{y^2}{4} \le 1$

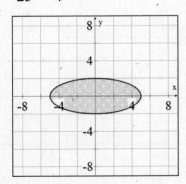

11. Center: $(-2, 4)$
Radius: 3
$(x - h)^2 + (y - k)^2 = r^2$
$(x - (-2))^2 + (y - 4)^2 = 3^2$
$(x + 2)^2 + (y - 4)^2 = 9$

12. (1) $x = 3y^2 + 2y - 4$
(2) $x = y^2 - 5y$
Use the substitution method:
$$x = 3y^2 + 2y - 4$$
$$y^2 - 5y = 3y^2 + 2y - 4$$
$$0 = 2y^2 + 7y - 4$$
$$0 = (y + 4)(2y - 1)$$
$$y + 4 = 0 \qquad 2y - 1 = 0$$
$$y = -4 \qquad y = \frac{1}{2}$$

Substitute the values of y into equation (2).
$x = y^2 - 5y \qquad\qquad x = y^2 - 5y$

$x = (-4)^2 - 5(-4) \qquad x = \left(\frac{1}{2}\right)^2 - 5\left(\frac{1}{2}\right)$

$x = 36 \qquad\qquad x = -\dfrac{9}{4}$

The solutions are $\left(-\dfrac{9}{4}, \dfrac{1}{2}\right)$ and $(36, -4)$.

13. (1) $x^2 + 2y^2 = 4$
(2) $x + y = 2$
Solve equation (2) for y.
$x + y = 2$
$y = -x + 2$
Use the substitution method:
$$x^2 + 2y^2 = 4$$
$$x^2 + 2(-x + 2)^2 = 4$$
$$x^2 + 2(x^2 - 4x + 4) = 4$$
$$x^2 + 2x^2 - 8x + 8 = 4$$
$$3x^2 - 8x + 4 = 0$$
$$(3x - 2)(x - 2) = 0$$
$$3x - 2 = 0 \qquad x - 2 = 0$$
$$x = \frac{2}{3} \qquad\qquad x = 2$$

Substitute the values of x into equation (2).

$x + y = 2$ $\qquad$ $x + y = 2$

$\dfrac{2}{3} + y = 2$ $\qquad$ $2 + y = 2$

$y = \dfrac{4}{3}$ $\qquad\qquad$ $y = 0$

The solutions are $\left(\dfrac{2}{3}, \dfrac{4}{3} \right)$ and $(2, 0)$.

14. $(2, 4)$ and $(-1, -3)$

Use the distance formula to find the length of the radius:

$d = \sqrt{(x_1 - x_2)^2 + (y_1 - y_2)^2}$

$\quad = \sqrt{(2 - (-1))^2 + (4 - (-3))^2}$

$\quad = \sqrt{3^2 + 7^2} = \sqrt{9 + 49} = \sqrt{58}$

$(x - h)^2 + (y - k)^2 = r^2$

Center: $(-1, -3)$

Radius: $r = \sqrt{58}$

$(x - (-1))^2 + (y - (-3))^2 = (\sqrt{58})^2$

$(x + 1)^2 + (y + 3)^2 = 58$

15. $\dfrac{x^2}{25} - \dfrac{y^2}{16} > 1$

$x^2 + y^2 < 3$

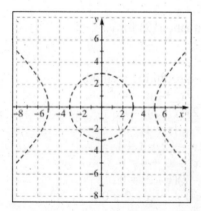

The solution sets of these inequalities do not intersect, so the system has no real number solution.

16. $x^2 + y^2 - 4x + 2y + 1 = 0$

$x^2 + y^2 - 4x + 2y = -1$

$(x^2 - 4x) + (y^2 + 2y) = -1$

$x^2 - 4x + 4 + y^2 + 2y + 1 = -1 + 4 + 1$

$(x - 2)^2 + (y + 1)^2 = 4$

Center: $(2, -1)$

Radius: 2

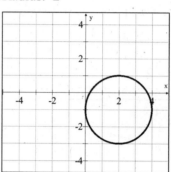

17. $y = \dfrac{1}{2}x^2 + x - 4$

$x = -\dfrac{b}{2a} = -\dfrac{1}{2\left(\dfrac{1}{2}\right)} = -\dfrac{1}{1} = -1$

$y = \left(\dfrac{1}{2}\right)(-1)^2 + (-1) - 4 = \dfrac{1}{2} - 1 - 4 = -\dfrac{9}{2}$

The axis of symmetry is $x = -1$.

The vertex is $\left(-1, -\dfrac{9}{2} \right)$.

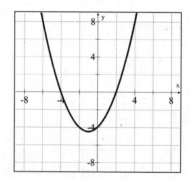

18. $(x-2)^2 + (y+1)^2 = 9$
Center: $(2, -1)$
Radius: 3

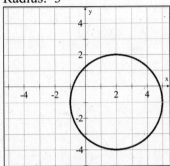

19. Axis of symmetry: x-axis
Vertices: $(3, 0)$ and $(-3, 0)$

Asymptotes: $y = \dfrac{2}{3}x$ and $y = -\dfrac{2}{3}x$

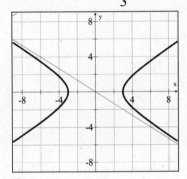

20. $\dfrac{x^2}{16} - \dfrac{y^2}{25} < 1$

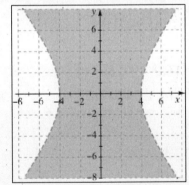

Cumulative Review Exercises

1. $\{x \mid x < 4\} \cap \{x \mid x > 2\} = \{x \mid 2 < x < 4\}$

$$\xleftarrow{\quad} \overset{}{\underset{-5\ -4\ -3\ -2\ -1\ 0\ 1\ 2\ 3\ 4\ 5}{\rule{0pt}{0pt}}} \xrightarrow{\quad}$$

2. $\dfrac{5x-2}{3} - \dfrac{1-x}{5} = \dfrac{x+4}{10}$

$30\left(\dfrac{5x-2}{3} - \dfrac{1-x}{5}\right) = 30\left(\dfrac{x+4}{10}\right)$

$10(5x-2) - 6(1-x) = 3(x+4)$

$50x - 20 - 6 + 6x = 3x + 12$

$56x - 26 = 3x + 12$

$53x = 38$

$x = \dfrac{38}{53}$

The solution is $\dfrac{38}{53}$.

3. $4 + |3x+2| < 6$

$|3x+2| < 2$

$-2 < 3x + 2 < 2$

$-4 < 3x < 0$

$-\dfrac{4}{3} < x < 0$

$\left\{x \mid -\dfrac{4}{3} < x < 0\right\}$

4. $m = -\dfrac{3}{2}$ and $(2, -3)$

$y - y_1 = m(x - x_1)$

$y - (-3) = -\dfrac{3}{2}(x - 2)$

$y + 3 = -\dfrac{3}{2}x + 3$

$y = -\dfrac{3}{2}x$

5. The product of the slope of two perpendicular lines is -1.

$m_1 \cdot m_2 = -1$

$m_1 \cdot (-1) = -1$

$m_1 = 1$

$y - y_1 = m(x - x_1)$

$y - (-2) = 1(x - 4)$

$y + 2 = x - 4$

$y = x - 6$

6. $3a^2b(4a - 3b^2) = 12a^3b - 9a^2b^3$

7. $(x-1)^3 - y^3$
$= [(x-1) - y][(x-1)^2 + (x-1)y + y^2]$
$= (x - 1 - y)(x^2 - 2x + 1 + xy - y + y^2)$
$= (x - y - 1)(x^2 - 2x + 1 + xy - y + y^2)$

8. $\dfrac{3x - 2}{x + 4} \le 1$

$\dfrac{3x - 2}{x + 4} - 1 \le 0$

$\dfrac{3x - 2}{x + 4} - \dfrac{x + 4}{x + 4} \le 0$

$\dfrac{2x - 6}{x + 4} \le 0$

$\{x \mid -4 < x \le 3\}$

9. $\dfrac{ax - bx}{ax + ay - bx - by} = \dfrac{x(a - b)}{a(x + y) - b(x + y)}$

$= \dfrac{x(a - b)}{(x + y)(a - b)} = \dfrac{x}{x + y}$

10. $\dfrac{x - 4}{3x - 2} - \dfrac{1 + x}{3x^2 + x - 2} = \dfrac{x - 4}{3x - 2} - \dfrac{1 + x}{(3x - 2)(x + 1)}$

$= \dfrac{x - 4}{3x - 2} \cdot \dfrac{x + 1}{x + 1} - \dfrac{1 + x}{(3x - 2)(x + 1)}$

$= \dfrac{x^2 - 3x - 4}{(3x - 2)(x + 1)} - \dfrac{1 + x}{(3x - 2)(x + 1)}$

$= \dfrac{x^2 - 4x - 5}{(3x - 2)(x + 1)} = \dfrac{(x - 5)(x + 1)}{(3x - 2)(x + 1)}$

$= \dfrac{x - 5}{3x - 2}$

11. $\dfrac{6x}{2x - 3} - \dfrac{1}{2x - 3} = 7$

$(2x - 3)\left(\dfrac{6x}{2x - 3} - \dfrac{1}{2x - 3}\right) = 7(2x - 3)$

$6x - 1 = 14x - 21$

$20 = 8x$

$\dfrac{5}{2} = x$

The solution is $\dfrac{5}{2}$.

12. $5x + 2y > 10$
$2y > -5x + 10$
$y > -\dfrac{5}{2}x + 5$

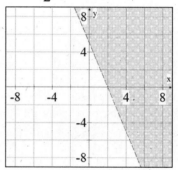

13. $\left(\dfrac{12a^2b^2}{a^{-3}b^{-4}}\right)^{-1}\left(\dfrac{ab}{4^{-1}a^{-2}b^4}\right)^2 = (12a^5b^6)^{-1}\left(\dfrac{4a^3}{b^3}\right)^2$

$= (12^{-1}a^{-5}b^{-6})\left(\dfrac{4^2a^6}{b^6}\right)$

$= \dfrac{1}{12a^5b^6} \cdot \dfrac{16a^6}{b^6} = \dfrac{16a^6}{12a^5b^{12}}$

$= \dfrac{4a}{3b^{12}}$

14. $2x^{3/4}$

15. $(3 + 5i)(4 - 6i) = 12 - 18i + 20i - 30i^2$
$= 12 + 2i - 30(-1)$
$= 12 + 2i + 30$
$= 42 + 2i$

16. $2x^2 + 2x - 3 = 0$

$$x = \frac{-b \pm \sqrt{b^2 - 4ac}}{2a}$$

$$= \frac{-2 \pm \sqrt{2^2 - 4(2)(-3)}}{2(2)}$$

$$\frac{-2 \pm \sqrt{4 + 24}}{4} = \frac{-2 \pm \sqrt{28}}{4} = \frac{-2 \pm 2\sqrt{7}}{4}$$

$$= \frac{-1 \pm \sqrt{7}}{2}$$

The solutions are $\dfrac{-1 + \sqrt{7}}{2}$ and $\dfrac{-1 - \sqrt{7}}{2}$.

17. (1) $x^2 + y^2 = 20$
(2) $x^2 - y^2 = 12$
Use the addition method.
$x^2 + y^2 = 20$
$x^2 - y^2 = 12$
$\quad 2x^2 = 32$
$\quad\ x^2 = 16$
$\quad\ \ x = \pm 4$
Substitute the values of x into equation (1).
$x^2 + y^2 = 20 \qquad x^2 + y^2 = 20$
$(4)^2 + y^2 = 20 \qquad (-4)^2 + y^2 = 20$
$y^2 = 4 \qquad\qquad y^2 = 4$
$y = \pm 2 \qquad\qquad y = \pm 2$
The solutions are $(4, 2)$, $(-4, 2)$,
$(4, -2)$ and $(-4, -2)$.

18. $x - \sqrt{2x - 3} = 3$

$\quad -\sqrt{2x - 3} = -x + 3$

$\quad\ \ \sqrt{2x - 3} = x - 3$

$\quad \left(\sqrt{2x - 3}\right)^2 = (x - 3)^2$

$\quad\ \ 2x - 3 = x^2 - 6x + 9$

$\quad\ \ \ 0 = x^2 - 8x + 12$

$\quad\ \ \ 0 = (x - 6)(x - 2)$

$x - 6 = 0 \qquad x - 2 = 0$

$\quad x = 6 \qquad\quad x = 2$

The solution is 6.

19. $f(-3) = -(-3)^2 + 3(-3) - 2$
$\qquad\quad = -9 - 9 - 2$
$\qquad\quad = -20$

20. $f(x) = 4x + 8$
$y = 4x + 8$
$x = 4y + 8$
$x - 8 = 4y$
$y = \dfrac{1}{4}x - 2$
$f^{-1}(x) = \dfrac{1}{4}x - 2$

21. $f(x) = -2x^2 + 4x - 2$

$$x = -\frac{b}{2a} = -\frac{4}{2(-2)} = -\frac{4}{-4} = -(-1) = 1$$

$f(1) = -2(1)^2 + 4(1) - 2 = -2 + 4 - 2 = 0$
The maximum value of the function is 0.

22. **Strategy**: Let x represent one number. The other number is $40 - x$. To find the first number, find the x-coordinate of the vertex of the given function.
Solution: $f(x) = -x^2 + 40x$

$$x = -\frac{b}{2a} = -\frac{40}{2(-1)} = -\frac{40}{-2} = -(-20) = 20$$

$40 - x = 40 - 20 = 20$
The maximum product of the two numbers whose sum is 40 is 400.

23. $(2, 4)$ and $(-1, 0)$
$$d = \sqrt{(x_1 - x_2)^2 + (y_1 - y_2)^2}$$
$$= \sqrt{(2 - (-1))^2 + (4 - 0)^2}$$
$$= \sqrt{3^2 + 4^2} = \sqrt{9 + 16} = \sqrt{25}$$
$$= 5$$

24. (3, 1) and (−1, 2)

Use the distance formula to find the length of the radius:

$$d = \sqrt{(x_1 - x_2)^2 + (y_1 - y_2)^2}$$
$$= \sqrt{(3 - (-1))^2 + (1 - 2)^2}$$
$$= \sqrt{4^2 + (-1)^2} = \sqrt{16 + 1} = \sqrt{17}$$

$(x - h)^2 + (y - k)^2 = r^2$

Center: (−1, 2)

Radius: $r = \sqrt{17}$

$(x - (-1))^2 + (y - 2)^2 = (\sqrt{17})^2$

$(x + 1)^2 + (y - 2)^2 = 17$

25. $x = y^2 - 2y + 3$

$$y = -\frac{b}{2a} = -\frac{-2}{2(1)} = -(-1) = 1$$

$$x = (1)^2 - 2(1) + 3 = 1 - 2 + 3 = 2$$

The axis of symmetry is $y = 1$.

The vertex is (2, 1).

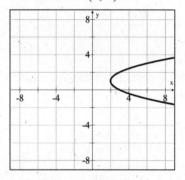

26. x-intercepts: (5, 0) and (−5, 0)

y-intercepts: (0, 2) and (0, −2)

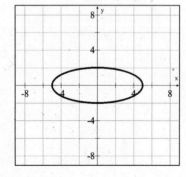

27. Axis of symmetry: y-axis

Vertices: (0, 2) and (0, −2)

Asymptotes: $y = \dfrac{2}{5}x$ and $y = -\dfrac{2}{5}x$

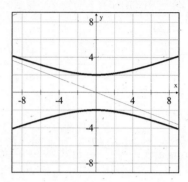

28. $(x - 1)^2 + y^2 \leq 25$

$y^2 < x$

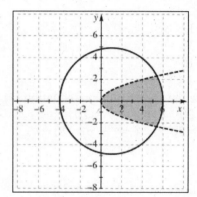

29. Strategy: Let x represent the number of adult tickets.

The number of children's tickets is $192 - x$.

	Amount	Cost	Value
Adult	x	12.00	$12x$
Children	$192 - x$	4.50	$4.5(192 - x)$

The sum of the values of each type of ticket sold equals the total value of all of the tickets sold ($1479).

Solution:

$$12x + 4.5(192 - x) = 1479$$
$$12x + 864 - 4.5x = 1479$$
$$7.5x + 864 = 1479$$
$$7.5x = 615$$
$$x = 82$$

There were 82 adult tickets sold.

30. Strategy: Let x represent the rate of the motorcycle.
The rate of the car is $x - 12$.

	Distance	Rate	Time
Motorcycle	180	x	$\dfrac{180}{x}$
Car	144	$x - 12$	$\dfrac{144}{x-12}$

The time traveled by the motorcycle is the same as the time traveled by the car.

Solution:
$$\frac{180}{x} = \frac{144}{x-12}$$
$$(x)(x-12)\frac{180}{x} = \frac{144}{x-12}(x)(x-12)$$
$$(x-12)180 = 144(x)$$
$$180x - 2160 = 144x$$
$$36x = 2160$$
$$x = 60$$
The rate of the motorcycle is 60 mph.

31. Strategy: Let x represent the rowing rate of the crew in calm water.

	Distance	Rate	Time
With current	12	$x + 1.5$	$\dfrac{12}{x+1.5}$
Against current	12	$x - 1.5$	$\dfrac{12}{x-1.5}$

The sum of the time traveling upriver and the time traveling downriver equals the total time traveled (6 h).

Solution:
$$\frac{12}{x+1.5} + \frac{12}{x-1.5} = 6$$

$$(x+1.5)(x-1.5)\left(\frac{12}{x+1.5} + \frac{12}{x-1.5}\right) = 6(x+1.5)(x-1.5)$$

$$12(x-1.5) + 12(x+1.5) = 6(x+1.5)(x-1.5)$$
$$12x - 18 + 12x + 18 = 6x^2 - 13.5$$
$$24x = 6x^2 - 13.5$$
$$0 = 6x^2 - 24x - 13.5$$

$$0 = 12x^2 - 48x - 27$$
$$0 = 3(4x^2 - 16x - 9)$$
$$0 = 3(2x+1)(2x-9)$$
$$2x+1 = 0 \qquad 2x-9 = 0$$
$$x = -\frac{1}{2} \qquad\qquad x = \frac{9}{2}$$

The rowing rate of the crew is 4.5 mph.

32. Strategy: Write the basic inverse variation equation replacing the variable with the given values. Solve for k.
Write the inverse variation equation replacing k with its value. Substitute 60 for t and solve for v.

Solution:
$$v = \frac{k}{t} \qquad\qquad v = \frac{1080}{t}$$
$$30 = \frac{k}{36} \qquad\qquad = \frac{1080}{60}$$
$$1080 = k \qquad\qquad = 18$$

The gear will make 18 revolutions/min.

Chapter 12: Sequences and Series

Prep Test

1. $[3(1) - 2] + [3(2) - 2] + [3(3) - 2]$
$= [3 - 2] + [6 - 2] + [9 - 2]$
$= 1 + 4 + 7$
$= 12$

2. $f(n) = \dfrac{n}{n+2}$

$f(6) = \dfrac{6}{6+2} = \dfrac{6}{8} = \dfrac{3}{4}$

3. $a_1 + (n-1)d$
$2 + (5-1)4 = 2 + (4)(4)$
$= 2 + 16 = 18$

4. $a_1 r^{n-1}$

$(-3)(-2)^{6-1} = (-3)(-2)^5$
$= (-3)(-32) = 96$

5. $\dfrac{a_1(1 - r^n)}{1 - r}$

$\dfrac{(-2)(1 - (-4)^5)}{1 - (-4)} = \dfrac{(-2)(1 - (-1024))}{1 + 4}$

$= \dfrac{(-2)(1025)}{5} = \dfrac{-2050}{5}$

$= -410$

6. $\dfrac{\dfrac{4}{10}}{1 - \dfrac{1}{10}} = \dfrac{\dfrac{4}{10}}{\dfrac{9}{10}} = \dfrac{4}{10} \div \dfrac{9}{10}$

$= \dfrac{4}{10} \cdot \dfrac{10}{9} = \dfrac{4}{9}$

7. $(x + y)(x + y) = x^2 + 2xy + y^2$

8. $(x + y)^3 = (x + y)^2(x + y)$
$= (x^2 + 2xy + y^2)(x + y)$
$= x(x^2 + 2xy + y^2) + y(x^2 + 2xy + y^2)$
$= x^3 + 2x^2y + xy^2 + x^2y + 2xy^2 + y^3$
$= x^3 + 3x^2y + 3xy^2 + y^3$

Section 12.1

Objective A Exercises

1. A sequence is an ordered list of numbers.

3. 8

Objective A Exercises

5. $a_n = n + 1$
$a_1 = 1 + 1 = 2$ The first term is 2.
$a_2 = 2 + 1 = 3$ The second term is 3.
$a_3 = 3 + 1 = 4$ The third term is 4.
$a_4 = 4 + 1 = 5$ The fourth term is 5.

7. $a_n = 2n + 1$
$a_1 = 2(1) + 1 = 3$ The first term is 3.
$a_2 = 2(2) + 1 = 5$ The second term is 5.
$a_3 = 2(3) + 1 = 7$ The third term is 7.
$a_4 = 2(4) + 1 = 9$ The fourth term is 9.

9. $a_n = 2 - 2n$
$a_1 = 2 - 2(1) = 0$ The first term is 0.
$a_2 = 2 - 2(2) = -2$ The second term is -2.
$a_3 = 2 - 2(3) = -4$ The third term is -4.
$a_4 = 2 - 2(4) = -6$ The fourth term is -6.

11. $a_n = 2^n$
$a_1 = 2^1 = 2$ The first term is 2.
$a_2 = 2^2 = 4$ The second term is 4.
$a_3 = 2^3 = 8$ The third term is 8.
$a_4 = 2^4 = 16$ The fourth term is 16.

13. $a_n = n^2 + 1$
$a_1 = 1^2 + 1 = 2$ The first term is 2.
$a_2 = 2^2 + 1 = 5$ The second term is 5.
$a_3 = 3^2 + 1 = 10$ The third term is 10.
$a_4 = 4^2 + 1 = 17$ The fourth term is 17.

15. $a_n = n^2 - \dfrac{1}{n}$

$a_1 = 1^2 - \dfrac{1}{1} = 0$ The first term is 0.

$a_2 = 2^2 - \dfrac{1}{2} = \dfrac{7}{2}$ The second term is $\dfrac{7}{2}$.

$$a_3 = 3^2 - \frac{1}{3} = \frac{26}{3} \quad \text{The third term is } \frac{26}{3}.$$

$$a_4 = 4^2 - \frac{1}{4} = \frac{63}{4} \quad \text{The fourth term is } \frac{63}{4}.$$

17. $a_n = 3n + 4$
$a_{12} = 3(12) + 4 = 40$
The twelfth term is 40.

19. $a_n = n(n-1)$
$a_{11} = 11(11-1) = 110$
The eleventh term is 110.

21. $u_n = (-1)^{n-1} n^2$
$a_{15} = (-1)^{14}(15)^2 = 225$
The fifteenth term is 225.

23. $a_n = \left(\frac{1}{2}\right)^n$

$$a_8 = \left(\frac{1}{2}\right)^8 = \frac{1}{256}$$

The eighth term is $\frac{1}{256}$.

25. $a_n = (n+2)(n+3)$
$a_{17} = (17+2)(17+3) = (19)(20) = 380$
The seventeenth term is 380.

27. $a_n = \dfrac{(-1)^{2n-1}}{n^2}$

$$a_6 = \frac{(-1)^{11}}{6^2} = -\frac{1}{36}$$

The sixth term is $-\dfrac{1}{36}$.

29. $a_n = n^2$

Objective B Exercises

31. $\displaystyle\sum_{n=1}^{5}(2n+3) = (2 \cdot 1 + 3) + (2 \cdot 2 + 3) + (2 \cdot 3 + 3) + (2 \cdot 4 + 3) + (2 \cdot 5 + 3)$
$= 5 + 7 + 9 + 11 + 13 = 45$

33. $\displaystyle\sum_{i=1}^{4}(2i) = (2 \cdot 1) + (2 \cdot 2) + (2 \cdot 3) + (2 \cdot 4) = 2 + 4 + 6 + 8 = 20$

35. $\displaystyle\sum_{i=1}^{6} i^2 = 1^2 + 2^2 + 3^2 + 4^2 + 5^2 + 6^2 = 1 + 4 + 9 + 16 + 25 + 36 = 91$

37. $\displaystyle\sum_{n=1}^{6}(-1)^n = (-1)^1 + (-1)^2 + (-1)^3 + (-1)^4 + (-1)^5 + (-1)^6$
$= -1 + 1 - 1 + 1 - 1 + 1 = 0$

39. $\displaystyle\sum_{i=3}^{6} i^3 = 3^3 + 4^3 + 5^3 + 6^3 = 27 + 64 + 125 + 216 = 432$

41. $\displaystyle\sum_{n=3}^{5} \frac{(-1)^{n-1}}{n-2} = \frac{(-1)^2}{3-2} + \frac{(-1)^3}{4-2} + \frac{(-1)^4}{5-2} = \frac{1}{1} + \frac{-1}{2} + \frac{1}{3}$

$\displaystyle = \frac{6-3+2}{6} = \frac{5}{6}$

43. $\displaystyle\sum_{i=1}^{n} i^2$

Critical Thinking

45. $a_n = na_{n-1},\, n > 1$

$a_1 = 1$

$a_2 = 2a_1 = 2(1) = 2$

$a_3 = 3a_2 = 3(2) = 6$

$a_4 = 4a_3 = 4(6) = 24$

Projects or Group Activities

47. a) $a_n - a_{n-1} = 0.04(P - a_{n-1})$

$a_n - a_{n-1} = 0.04(5000 - a_{n-1})$

$a_n - a_{n-1} = 200 - 0.04a_{n-1}$

$a_n = 0.96a_{n-1} + 200$

b) $a_0 = 150$

$a_1 = 0.96a_0 + 200 = 344$

$a_2 = 0.96a_1 + 200 = 0.96(344) + 200 = 530$

$a_3 = 0.96a_2 + 200 = 0.96(530) + 200 = 709$

$a_4 = 0.96a_3 + 200 = 0.96(709) + 200 = 881$

There will be 881 people infected after 4 days.

Section 12.2

Concept Check

1. An arithmetic sequence is one in which the difference between any two consecutive terms is constant.

3. a) Yes

b) Yes

c) No

d) No

e) yes

f) No

Objective A Exercises

5. $d = a_2 - a_1 = 11 - 1 = 10$

$a_n = a_1 + (n-1)d$

$a_{15} = 1 + (15-1)(10) = 1 + 14(10) = 1 + 140$

$a_{15} = 141$

7. $d = a_2 - a_1 = -2 - (-6) = 4$

$a_n = a_1 + (n-1)d$

$a_{15} = -6 + (15-1)(4) = -6 + 14(4) = -6 + 56$

$a_{15} = 50$

9. $d = a_2 - a_1 = \dfrac{5}{2} - 2 = \dfrac{1}{2}$

$a_n = a_1 + (n-1)d$

$a_{31} = 2 + (31-1)\left(\dfrac{1}{2}\right) = 2 + 30\left(\dfrac{1}{2}\right) = 2 + 15$

$a_{31} = 17$

11. $d = a_2 - a_1 = -\dfrac{5}{2} - (-4) = \dfrac{3}{2}$

$a_n = a_1 + (n-1)d$

$a_{12} = -4 + (12-1)\left(\dfrac{3}{2}\right) = -4 + 11\left(\dfrac{3}{2}\right) = -4 + \dfrac{33}{2}$

$a_{12} = \dfrac{25}{2}$

13. $d = a_2 - a_1 = 5 - 8 = -3$

$a_n = a_1 + (n-1)d$

$a_{40} = 8 + (40-1)(-3) = 8 + 39(-3) = 8 - 117$

$a_{40} = -109$

15. $d = a_2 - a_1 = 4 - 1 = 3$

$a_n = a_1 + (n-1)d$

$a_n = 1 + (n-1)(3) = 1 + 3n - 3$

$a_n = 3n - 2$

17. $d = a_2 - a_1 = 0 - 3 = -3$

$a_n = a_1 + (n-1)d$

$a_n = 3 + (n-1)(-3) = 3 - 3n + 3$

$a_n = -3n + 6$

19. $d = a_2 - a_1 = 4.5 - 7 = -2.5$

$a_n = a_1 + (n-1)d$

$a_n = 7 + (n-1)(-2.5) = 7 - 2.5n + 2.5$

$a_n = -2.5n + 9.5$

21. $d = a_2 - a_1 = 8 - 3 = 5$

$a_n = a_1 + (n-1)d$

$98 = 3 + (n-1)(5)$

$98 = 3 + 5n - 5$

$98 = 5n - 2$

$100 = 5n$

$20 = n$

There are 20 terms in the sequence.

23. $d = a_2 - a_1 = -3 - 1 = -4$

$a_n = a_1 + (n-1)d$

$-75 = 1 + (n-1)(-4)$

$-75 = 1 - 4n + 4$

$-75 = -4n + 5$

$-80 = -4n$

$20 = n$

There are 20 terms in the sequence.

25. $d = a_2 - a_1 = \dfrac{13}{3} - \dfrac{7}{3} = \dfrac{6}{3} = 2$

$a_n = a_1 + (n-1)d$

$\dfrac{79}{3} = \dfrac{7}{3} + (n-1)2$

$\dfrac{79}{3} = \dfrac{7}{3} + 2n - 2$

$\dfrac{79}{3} = 2n + \dfrac{1}{3}$

$\dfrac{78}{3} = 2n$

$26 = 2n$

$13 = n$

There are 13 terms in the sequence.

27. $d = a_2 - a_1 = 2 - 3.5 = -1.5$

$a_n = a_1 + (n-1)d$

$-25 = 3.5 + (n-1)(-1.5)$

$-25 = 3.5 - 1.5n + 1.5$

$-25 = -1.5n + 5$

$-30 = -1.5n$

$20 = n$

There are 20 terms in the sequence.

Objective B Exercises

29. $d = a_2 - a_1 = 3 - 1 = 2$

$a_n = a_1 + (n-1)d$

$a_{50} = 1 + (50-1)2 = 1 + 49(2) = 99$

$S_n = \dfrac{n}{2}(a_1 + a_n)$

$S_{50} = \dfrac{50}{2}(1 + 99) = 25(100) = 2500$

31. $d = a_2 - a_1 = 18 - 20 = -2$

$a_n = a_1 + (n-1)d$

$a_{40} = 20 + (40-1)(-2) = 20 + 39(-2) = -58$

$S_n = \dfrac{n}{2}(a_1 + a_n)$

$S_{40} = \dfrac{40}{2}(20 - 58) = 20(-38) = -760$

33. $d = a_2 - a_1 = 1 - \dfrac{1}{2} = \dfrac{1}{2}$

$a_n = a_1 + (n-1)d$

$a_{27} = \dfrac{1}{2} + (27-1)\dfrac{1}{2} = \dfrac{1}{2} + (26)\dfrac{1}{2}$

$a_{27} = \dfrac{1}{2} + 13 = \dfrac{27}{2}$

$S_n = \dfrac{n}{2}(a_1 + a_n)$

$S_{27} = \dfrac{27}{2}\left(\dfrac{1}{2} + \dfrac{27}{2}\right) = \dfrac{27}{2}\left(\dfrac{28}{2}\right) = 189$

35. $a_i = 3i - 1$

$a_1 = 3(1) - 1 = 2$

$a_{15} = 3(15) - 1 = 44$

$S_i = \dfrac{i}{2}(a_1 + a_i)$

$S_{15} = \dfrac{15}{2}(2 + 44) = \left(\dfrac{15}{2}\right)(46) = 345$

37. $a_n = \dfrac{1}{2}n + 1$

$a_1 = \dfrac{1}{2}(1) + 1 = \dfrac{3}{2}$

$a_{17} = \dfrac{1}{2}(17) + 1 = \dfrac{19}{2}$

$S_n = \dfrac{n}{2}(a_1 + a_n)$

$S_{17} = \dfrac{17}{2}\left(\dfrac{3}{2} + \dfrac{19}{2}\right) = \left(\dfrac{17}{2}\right)(11) = \dfrac{187}{2}$

39. $a_i = 4 - 2i$

$a_1 = 4 - 2(1) = 2$

$a_{15} = 4 - 2(15) = -26$

$S_i = \dfrac{i}{2}(a_1 + a_i)$

$S_{15} = \dfrac{15}{2}(2 + (-26)) = \left(\dfrac{15}{2}\right)(-24) = -180$

41. $d = a_2 - a_1 = 3 - 1 = 2$

$a_n = a_1 + (n-1)d$

$a_n = 1 + (n-1)2 = 1 + 2n - 2 = 2n - 1$

$S_n = \dfrac{n}{2}(a_1 + a_n)$

$S_n = \dfrac{n}{2}(1 + 2n - 1) = \dfrac{n}{2}(2n) = n^2$

Objective C Exercises

43. Strategy: Write the arithmetic sequence.
Find the common difference, d.
Use the formula for the nth Term of an
Arithmetic Sequence to find the number of
rows.
Use the Formula for the Sum of n Terms of
an Arithmetic Sequence.

Solution: 2, 4, 5, 6….
$$d = a_2 - a_1 = 5 - 4 = 1$$
$$a_n = a_1 + (n-1)d$$
$$a_{200} = 4 + (200 - 1)(1)$$
$$a_{200} = 4 + 199 = 203$$
$$S_n = \frac{n}{2}(a_1 + a_n)$$
$$S_{200} = \frac{200}{2}(4 + 203) = 100(207) = 20700$$
There are 203 stitches in the 200th row.
There are 20,700 stitches in all 200 rows.

45. Strategy: Write the arithmetic sequence.
Find the common difference, d.
Use the formula for the nth Term of an
Arithmetic Sequence to find the number of
terms in the sequence.

Solution: 10, 15, 20, …, 60
$$d = a_2 - a_1 = 15 - 10 = 5$$
$$a_n = a_1 + (n-1)d$$
$$60 = 10 + (n-1)(5)$$
$$60 = 10 + 5n - 5$$
$$60 = 5n + 5$$
$$55 = 5n$$
$$11 = n$$
In 11 weeks, the person will walk 60 min
per day.

47. $a_n = 200 - 0.208n$

49. No

Critical Thinking

51. $S_n = \dfrac{n}{2}(a_1 + a_n)$

$$36 = \frac{n}{2}(-9 + 21)$$
$$36 = \frac{n}{2}(12)$$
$$36 = 6n$$
$$n = 6$$
$$a_n = a_1 + (n-1)d$$
$$21 = -9 + (6-1)d$$
$$21 = -9 + 5d$$
$$30 = 5d$$
$$6 = d$$

Check Your Progress: Chapter 12

1. $a_n = 2n - 1$
$$a_1 = 2(1) - 1 = 1$$
$$a_2 = 2(2) - 1 = 3$$
$$a_3 = 2(3) - 1 = 5$$
$$a_4 = 2(4) - 1 = 7$$

2. $a_n = n^2$
$$a_1 = 1^2 = 1$$
$$a_2 = 2^2 = 4$$
$$a_3 = 3^2 = 9$$
$$a_4 = 4^2 = 16$$

3. $a_n = \dfrac{(-1)^n}{2^n}$

$$a_9 = \frac{(-1)^9}{2^9} = -\frac{1}{512}$$

4. $a_n = \dfrac{1}{n(n+1)}$

$$a_8 = \frac{1}{8(8+1)} = \frac{1}{72}$$

5. $\displaystyle\sum_{i=1}^{5}(i^2+2i)$

$=(1^2+2)+(2^2+4)+(3^2+6)+(4^2+8)+(5^2+10)$

$=3+8+15+24+35=85$

6. $\displaystyle\sum_{i=2}^{6}(3i-1)$

$=(3(2)-1)+(3(3)-1)+(3(4)-1)+(3(5)-1)+(3(6)-1)$

$=5+8+11+14+17=55$

7. $a_n=a_1+(n-1)d$

$a_{25}=3+(25-1)4=3+24(4)=99$

8. $a_n=a_1+(n-1)d$

$65=2+(n-1)3$

$65=2+3n-3$

$66=3n$

$22=n$

9. $a_n=a_1+(n-1)d$

$a_n=2+(n-1)5$

$a_n=2+5n-5$

$a_n=5n-3$

10. $a_n=a_1+(n-1)d$

$a_{25}=2+(25-1)4$

$a_{25}=2+(24)4=98$

$S_n=\dfrac{n}{2}(a_1+a_n)$

$S_{25}=\dfrac{25}{2}(2+98)=\dfrac{25}{2}(100)=1250$

11. $a_i=5i-1$

$a_1=5(1)-1=4$

$a_{40}=5(40)-1=199$

$S_i=\dfrac{i}{2}(a_1+a_i)$

$S_{40}=\dfrac{40}{2}(4+199)=\left(\dfrac{40}{2}\right)(203)=4060$

12. $a_i=6i-5$

$a_1=6(1)-5=1$

$a_{25}=6(25)-5=145$

$S_i=\dfrac{i}{2}(a_1+a_i)$

$S_{25}=\dfrac{25}{2}(1+145)=\left(\dfrac{25}{2}\right)(146)=1825$

Section 12.3

Concept Check

1. An arithmetic sequence is one in which the difference between any two consecutive terms is constant.
A geometric sequence is one in which each successive term in the sequence is the same nonzero constant multiple of the preceding term.

3. a) No
b) Yes
c) Yes
d) No
e) No
f) Yes

Objective A Exercises

5. $r=\dfrac{a_2}{a_1}=\dfrac{8}{2}=4$

$a_n=a_1 r^{n-1}$

$a_9=2(4)^8=2(65,536)$

$a_9=131,072$

7. $r=\dfrac{a_2}{a_1}=\dfrac{-4}{6}=-\dfrac{2}{3}$

$a_n=a_1 r^{n-1}$

$a_7=6\left(-\dfrac{2}{3}\right)^6=6\left(\dfrac{64}{729}\right)$

$a_7=\dfrac{128}{243}$

9. $r = \dfrac{a_2}{a_1} = \dfrac{\frac{1}{8}}{-\frac{1}{16}} = -2$

$a_n = a_1 r^{n-1}$

$a_{10} = -\dfrac{1}{16}(-2)^9 = -\dfrac{1}{16}(-512)$

$a_{10} = 32$

11. $a_n = a_1 r^{n-1}$

$a_4 = 9r^3$

$\dfrac{8}{3} = 9r^3$

$\dfrac{8}{27} = r^3$

$\dfrac{2}{3} = r$

$a_n = a_1 r^{n-1}$

$a_2 = 9\left(\dfrac{2}{3}\right)^1 = 6$

$a_3 = 9\left(\dfrac{2}{3}\right)^2 = 9\left(\dfrac{4}{9}\right) = 4$

13. $a_n = a_1 r^{n-1}$

$a_4 = 3r^3$

$-\dfrac{8}{9} = 3r^3$

$-\dfrac{8}{27} = r^3$

$-\dfrac{2}{3} = r$

$a_n = a_1 r^{n-1}$

$a_2 = 3\left(-\dfrac{2}{3}\right)^1 = -2$

$a_3 = 3\left(-\dfrac{2}{3}\right)^2 = 3\left(\dfrac{4}{9}\right) = \dfrac{4}{3}$

15. $a_n = a_1 r^{n-1}$

$a_4 = -3r^3$

$192 = -3r^3$

$-64 = r^3$

$-4 = r$

$a_n = a_1 r^{n-1}$

$a_2 = -3(-4)^1 = 12$

$a_3 = -3(-4)^2 = -3(16) = -48$

Objective B Exercises

17. $r = \dfrac{a_2}{a_1} = \dfrac{6}{2} = 3$

$S_n = \dfrac{a_1(1-r^n)}{1-r}$

$S_7 = \dfrac{2(1-3^7)}{1-3} = \dfrac{2(1-2187)}{-2} = \dfrac{2(-2186)}{-2}$

$S_7 = 2186$

19. $r = \dfrac{a_2}{a_1} = \dfrac{-2}{3} = -\dfrac{2}{3}$

$S_n = \dfrac{a_1(1-r^n)}{1-r}$

$S_5 = \dfrac{3\left(1-\left(-\dfrac{2}{3}\right)^5\right)}{1-\left(-\dfrac{2}{3}\right)} = \dfrac{3\left(1+\left(\dfrac{32}{243}\right)\right)}{\dfrac{5}{3}} = \dfrac{3\left(\dfrac{275}{243}\right)}{\dfrac{5}{3}}$

$S_5 = \dfrac{\dfrac{275}{81}}{\dfrac{5}{3}} = \dfrac{275}{81} \cdot \dfrac{3}{5} = \dfrac{55}{27}$

21. $r = \dfrac{a_2}{a_1} = \dfrac{9}{12} = \dfrac{3}{4}$

$S_n = \dfrac{a_1(1-r^n)}{1-r}$

$S_5 = \dfrac{12\left(1-\left(\dfrac{3}{4}\right)^5\right)}{1-\left(\dfrac{3}{4}\right)} = \dfrac{12\left(1-\left(\dfrac{243}{1024}\right)\right)}{\dfrac{1}{4}} = \dfrac{12\left(\dfrac{781}{1024}\right)}{\dfrac{1}{4}}$

$S_5 = \dfrac{\dfrac{2343}{256}}{\dfrac{1}{4}} = \dfrac{2343}{256} \cdot \dfrac{4}{1} = \dfrac{2343}{64}$

23. $a_n = 2^i$

$a_1 = 2^1 = 2$

$a_2 = 2^2 = 4$

$r = \dfrac{a_2}{a_1} = \dfrac{4}{2} = 2$

$S_n = \dfrac{a_1(1-r^n)}{1-r}$

$S_5 = \dfrac{2(1-2^5)}{1-2} = \dfrac{2(1-32)}{-1} = \dfrac{2(-31)}{-1}$

$S_5 = 62$

25. $a_i = \left(\dfrac{1}{3}\right)^i$

$a_1 = \left(\dfrac{1}{3}\right)^1 = \dfrac{1}{3}$

$a_2 = \left(\dfrac{1}{3}\right)^2 = \dfrac{1}{9}$

$r = \dfrac{a_2}{a_1} = \dfrac{\dfrac{1}{9}}{\dfrac{1}{3}} = \dfrac{1}{3}$

$S_n = \dfrac{a_1(1-r^n)}{1-r}$

$S_5 = \dfrac{\dfrac{1}{3}\left(1-\left(\dfrac{1}{3}\right)^5\right)}{1-\dfrac{1}{3}} = \dfrac{\dfrac{1}{3}\left(1-\dfrac{1}{243}\right)}{\dfrac{2}{3}} = \dfrac{\dfrac{1}{3}\left(\dfrac{242}{243}\right)}{\dfrac{2}{3}}$

$S_6 = \dfrac{1}{3} \cdot \dfrac{242}{243} \cdot \dfrac{3}{2} = \dfrac{121}{243}$

27. Positive

Objective C Exercises

29. $r = \dfrac{a_2}{a_1} = \dfrac{2}{3}$

$S = \dfrac{a_1}{1-r} = \dfrac{3}{1-\dfrac{2}{3}} = \dfrac{3}{\dfrac{1}{3}}$

$S = 9$

31. $r = \dfrac{a_2}{a_1} = \dfrac{\dfrac{7}{100}}{\dfrac{7}{10}} = \dfrac{1}{10}$

$S = \dfrac{a_1}{1-r} = \dfrac{\dfrac{7}{10}}{1-\dfrac{1}{10}} = \dfrac{\dfrac{7}{10}}{\dfrac{9}{10}} = \dfrac{7}{9}$

33. $0.88\overline{8} = 0.8 + 0.08 + 0.008 + \dots$

$$0.88\overline{8} = \frac{8}{10} + \frac{8}{100} + \frac{8}{1000} + \dots$$

$$S = \frac{a_1}{1-r} = \frac{\frac{8}{10}}{1 - \frac{1}{10}} = \frac{\frac{8}{10}}{\frac{9}{10}} = \frac{8}{9}$$

The equivalent fraction is $\frac{8}{9}$.

35. $0.22\overline{2} = 0.2 + 0.02 + 0.002 + \dots$

$$0.22\overline{2} = \frac{2}{10} + \frac{2}{100} + \frac{2}{1000} + \dots$$

$$S = \frac{a_1}{1-r} = \frac{\frac{2}{10}}{1 - \frac{1}{10}} = \frac{\frac{2}{10}}{\frac{9}{10}} = \frac{2}{9}$$

The equivalent fraction is $\frac{2}{9}$.

37. $0.45\overline{45} = 0.45 + 0.0045 + 0.000045 + \dots$

$$0.45\overline{45} = \frac{45}{100} + \frac{45}{10,000} + \frac{45}{1,000,000} + \dots$$

$$S = \frac{a_1}{1-r} = \frac{\frac{45}{100}}{1 - \frac{1}{100}} = \frac{\frac{45}{100}}{\frac{99}{100}} = \frac{45}{99} = \frac{5}{11}$$

The equivalent fraction is $\frac{5}{11}$.

39. $0.16\overline{6} = 0.1 + 0.06 + 0.006 + \dots$

$$0.16\overline{6} = \frac{1}{10} + \frac{6}{100} + \frac{6}{1000} + \dots$$

$$S = \frac{a_1}{1-r} = \frac{\frac{6}{100}}{1 - \frac{1}{10}} = \frac{\frac{6}{100}}{\frac{9}{10}} = \frac{6}{90} = \frac{1}{15}$$

$$0.16\overline{6} = \frac{1}{10} + \frac{1}{15} = \frac{5}{30} = \frac{1}{6}$$

The equivalent fraction is $\frac{1}{6}$.

41. Yes

43. No

Objective D Exercises

45. Strategy: To find the height of the ball on the sixth bounce, use the formula for the nth Term of a Geometric Sequence.

Solution: $n = 6$, $a_1 = 75\%$ of $10 = 7.5$

$$r = 75\% = \frac{3}{4}$$

$$a_n = a_1 r^{n-1}$$

$$a_6 = 7.5\left(\frac{3}{4}\right)^5$$

$$a_6 = 7.5\left(\frac{243}{1024}\right) = 1.8$$

The ball bounces to a height of 1.8 ft on the sixth bounce.

47. Strategy: Determine when a_n is first less than 1.

Solution: $200(0.5)^{\frac{n}{30}} < 1$

$$(0.5)^{\frac{n}{30}} < 0.005$$

$$\log_{0.5}(0.005) < \frac{n}{30}$$

$$30\log_{0.5}(0.005) < n$$

$$n = 230 \text{ years.}$$

49. Strategy: To find the total salary for the second job, use the formula for the Sum of a Finite Geometric Sequence.

Solution: $a_1 = 0.01$, $a_2 = 0.02$, $n = 30$

$$r = \frac{a_2}{a_1} = \frac{0.02}{0.01} = 2$$

$$S_{30} = \frac{0.01(1 - 2^{30})}{1 - 2} = 10,737,418$$

The second job pays more because it pays over $10 million after 30 days.

Critical Thinking

51. $a_n = e^n$ and $b_n = \ln a_n$

$b_n = \ln a_n = \log e^n$

$b_1 = \ln e$

$b_2 = \ln e^2 = 2\ln e$

$b_3 = \ln e^3 = 3\ln e$

$b_4 = \ln e^4 = 4\ln e$

$d = b_2 - b_1 = 2\ln e - \ln e = \ln e = 1$

$d = b_3 - b_2 = 3\ln e - 2\ln e = \ln e = 1$

$d = b_4 - b_3 = 4\ln e - 3\ln e = \ln e = 1$

$d = b_n - b_{n-1} = n\ln e - (n-1)\ln e = \ln e = 1$

The common difference is 1.

53. $f(n) = ab^n$

$f(1) = ab$

$f(2) = ab^2$

$f(3) = ab^3$

$f(4) = ab^4$

$r = \dfrac{ab^2}{ab} = b$

$r = \dfrac{ab^3}{ab^2} = b$

$r = \dfrac{ab^4}{ab^3} = b$

$r = \dfrac{ab^{n+1}}{ab^n} = b$

The common ratio is b.

Projects or Group Activities

55. $a_n = 3^{n-1}$

Section 12.4

Concept Check

1. $n!$ is the product of the first n consecutive natural numbers.
$n! = n(n-1)(n-2)\ldots3\cdot2\cdot1$

3. The degree of each term is n.

Objective A Exercises

5. $3! = 3 \cdot 2 \cdot 1 = 6$

7. $8! = 8 \cdot 7 \cdot 6 \cdot 5 \cdot 4 \cdot 3 \cdot 2 \cdot 1 = 40{,}320$

9. $0! = 1$

11. $\dfrac{5!}{2!3!} = \dfrac{5 \cdot 4 \cdot 3 \cdot 2 \cdot 1}{(2\cdot1)(3\cdot2\cdot1)} = 10$

13. $\dfrac{6!}{6!0!} = \dfrac{6\cdot5\cdot4\cdot3\cdot2\cdot1}{(6\cdot5\cdot4\cdot3\cdot2\cdot1)(1)} = 1$

15. $\dfrac{9!}{6!3!} = \dfrac{9 \cdot 8 \cdot 7 \cdot 6 \cdot 5 \cdot 4 \cdot 3 \cdot 2 \cdot 1}{(6 \cdot 5 \cdot 4 \cdot 3 \cdot 2 \cdot 1)(3 \cdot 2 \cdot 1)} = 84$

17. $\dbinom{7}{2} = \dfrac{7!}{(7-2)!2!} = \dfrac{7!}{5!2!} = \dfrac{7 \cdot 6 \cdot 5 \cdot 4 \cdot 3 \cdot 2 \cdot 1}{(5 \cdot 4 \cdot 3 \cdot 2 \cdot 1)(2 \cdot 1)} = 21$

19. $\dbinom{9}{0} = \dfrac{9!}{(9-0)!0!} = \dfrac{9!}{9!0!} = \dfrac{9 \cdot 8 \cdot 7 \cdot 6 \cdot 5 \cdot 4 \cdot 3 \cdot 2 \cdot 1}{(9 \cdot 8 \cdot 7 \cdot 6 \cdot 5 \cdot 4 \cdot 3 \cdot 2 \cdot 1)(1)} = 1$

21. $\dbinom{6}{3} = \dfrac{6!}{(6-3)!3!} = \dfrac{6!}{3!3!} = \dfrac{6 \cdot 5 \cdot 4 \cdot 3 \cdot 2 \cdot 1}{(3 \cdot 2 \cdot 1)(3 \cdot 2 \cdot 1)} = 20$

23. $\dbinom{11}{1} = \dfrac{11!}{(11-1)!1!} = \dfrac{11!}{10!1!} = \dfrac{11 \cdot 10 \cdot 9 \cdot 8 \cdot 7 \cdot 6 \cdot 5 \cdot 4 \cdot 3 \cdot 2 \cdot 1}{(10 \cdot 9 \cdot 8 \cdot 7 \cdot 6 \cdot 5 \cdot 4 \cdot 3 \cdot 2 \cdot 1)(1)} = 11$

25. $(x+y)^4 = \dbinom{4}{0}x^4 + \dbinom{4}{1}x^3y + \dbinom{4}{2}x^2y^2 + \dbinom{4}{3}xy^3 + \dbinom{4}{4}y^4 = x^4 + 4x^3y + 6x^2y^2 + 4xy^3 + y^4$

27. $(x-y)^5 = \dbinom{5}{0}x^5 + \dbinom{5}{1}x^4(-y) + \dbinom{5}{2}x^3(-y)^2 + \dbinom{5}{3}x^2(-y)^3 + \dbinom{5}{4}x(-y)^4 + \dbinom{5}{5}(-y)^5$

$= x^5 - 5x^4y + 10x^3y^2 - 10x^2y^3 + 5xy^4 - y^5$

29. $(2m+1)^4 = \dbinom{4}{0}(2m)^4 + \dbinom{4}{1}(2m)^3(1) + \dbinom{4}{2}(2m)^2(1)^2 + \dbinom{4}{3}(2m)(1)^3 + \dbinom{4}{4}(1)^4$

$= 16m^4 + 4(8m^3) + 6(4m^2) + 4(2m) + 1$

$= 16m^4 + 32m^3 + 24m^2 + 8m + 1$

31. $(2r-3)^5 = \dbinom{5}{0}(2r)^5 + \dbinom{5}{1}(2r)^4(-3) + \dbinom{5}{2}(2r)^3(-3)^2 + \dbinom{5}{3}(2r)^2(-3)^3 + \dbinom{5}{4}(2r)(-3)^4 + \dbinom{5}{5}(-3)^5$

$= 32r^5 + 5(16r^4)(-3) + 10(8r^3)(9) + 10(4r^2)(-27) + 5(2r)(81) - 243$

$= 32r^5 - 240r^4 + 720r^3 - 1080r^2 + 810r - 243$

33. $(a+b)^{10} = \dbinom{10}{0}a^{10} + \dbinom{10}{1}a^9b + \dbinom{10}{2}a^8b^2 + \ldots$

$= a^{10} + 10a^9b + 45a^8b^2 + \ldots$

35. $(2x+y)^8 = \dbinom{8}{0}(2x)^8 + \dbinom{8}{1}(2x)^7(y) + \dbinom{8}{2}(2x)^6(y)^2 + \ldots$

$= 256x^8 + 8(128x^7)(y) + 28(64x^6)y^2 + \ldots$

$= 256x^8 + 1024x^7y + 1792x^6y^2 + \ldots$

37. $(4x-3y)^8 = \binom{8}{0}(4x)^8 + \binom{8}{1}(4x)^7(-3y) + \binom{8}{2}(4x)^6(-3y)^2 + \ldots$

$\qquad = 65{,}536x^8 + 8(16{,}384x^7)(-3y) + 28(4096x^6)(9y^2) + \ldots$

$\qquad = 65{,}536x^8 - 393{,}216x^7y + 1{,}032{,}192x^6y^2 + \ldots$

39. $\left(x+\dfrac{1}{x}\right)^7 = \binom{7}{0}x^7 + \binom{7}{1}x^6\left(\dfrac{1}{x}\right) + \binom{7}{2}x^5\left(\dfrac{1}{x}\right)^2 + \ldots$

$\qquad = x^7 + 7x^6\left(\dfrac{1}{x}\right) + 21x^5\left(\dfrac{1}{x}\right)^2 + \ldots$

$\qquad = x^7 + 7x^5 + 21x^3 + \ldots$

41. $n=7, a=2x, b=-1, r=4$

$\qquad \binom{7}{4-1}(2x)^{7-4+1}(-1)^{4-1} = \binom{7}{3}(2x)^4(-1)^3 = 35(16x^4)(-1) = -560x^4$

43. $n=6, a=x^2, b=-y^2, r=2$

$\qquad \binom{6}{2-1}(x^2)^{6-2+1}(-y^2)^{2-1} = \binom{6}{1}(x^2)^5(-y^2) = -6x^{10}y^2$

45. $n=5, a=n, b=\dfrac{1}{n}, r=2$

$\qquad \binom{5}{2-1}n^{5-2+1}\left(\dfrac{1}{n}\right)^{2-1} = \binom{5}{1}n^4\left(\dfrac{1}{n}\right) = 5n^4\left(\dfrac{1}{n}\right) = 5n^3$

47. False

Critical Thinking

49. $\dfrac{n!}{(n-1)!} = \dfrac{n(n-1)(n-2)...3\cdot2\cdot1}{(n-1)(n-2)...3\cdot2\cdot1} = n$

51. $\dbinom{n}{r} = \dfrac{n!}{r!(n-r)!}$

$\dbinom{n}{n-r} = \dfrac{n!}{(n-r)![n-(n-r)]!}$

$= \dfrac{n!}{(n-r)!r!}$

Projects or Group Activities

53. $-4-4i$

55. $41+38i$

57. $-\dfrac{1}{4}$

59. i

Chapter 12 Review Exercises

1. Neither

2. $a_n = 3n-2$
$a_{10} = 3(10) - 2 = 30 - 2 = 28$
The tenth term is 28.

3. $0.\overline{7} = 0.7 + 0.07 + 0.007 + ...$

$0.\overline{7} = \dfrac{7}{10} + \dfrac{7}{100} + \dfrac{7}{1000} + ...$

$S = \dfrac{a_1}{1-r} = \dfrac{\dfrac{7}{10}}{1-\dfrac{1}{10}} = \dfrac{\dfrac{7}{10}}{\dfrac{9}{10}} = \dfrac{7}{9}$

$0.\overline{7} = \dfrac{7}{9}$

The equivalent fraction is $\dfrac{7}{9}$.

4. $r = \dfrac{a_2}{a_1} = \dfrac{\dfrac{3}{100}}{\dfrac{3}{10}} = \dfrac{1}{10}$

$S = \dfrac{a_1}{1-r} = \dfrac{\dfrac{3}{10}}{1-\dfrac{1}{10}} = \dfrac{\dfrac{3}{10}}{\dfrac{9}{10}} = \dfrac{1}{3}$

5. $\displaystyle\sum_{i=1}^{30} 4i - 1 = 3,7,11,15...119$

$S_n = \dfrac{30}{2}(3+119) = 15(122) = 1830$

6. $\displaystyle\sum_{n=1}^{5}(3n-2) = (3\cdot1-2) + (3\cdot2-2)$
$+ (3\cdot3-2) + (3\cdot4-2) + (3\cdot5-2)$
$= 1+4+7+10+13 = 35$

7. $0.23\overline{23} = 0.23 + 0.0023 + 0.000023 + ...$

$0.23\overline{23} = \dfrac{23}{100} + \dfrac{23}{10,000} + \dfrac{23}{1,000,000} + ...$

$S = \dfrac{a_1}{1-r} = \dfrac{\dfrac{23}{100}}{1-\dfrac{1}{100}} = \dfrac{\dfrac{23}{100}}{\dfrac{99}{100}} = \dfrac{23}{99}$

The equivalent fraction is $\dfrac{23}{99}$.

8. $\dbinom{9}{3} = \dfrac{9!}{(9-3)!3!} = \dfrac{9!}{6!3!}$

$= \dfrac{9\cdot8\cdot7\cdot6\cdot5\cdot4\cdot3\cdot2\cdot1}{(6\cdot5\cdot4\cdot3\cdot2\cdot1)(3\cdot2\cdot1)} = 84$

9. $\displaystyle\sum_{n=1}^{5} 2(3)^n = 6, 18, 54...$

$r = \dfrac{a_2}{a_1} = \dfrac{18}{6} = 3$

$$S_n = \frac{a_1(1 - r^n)}{1 - r}$$

$$S_5 = \frac{6(1 - 3^5)}{1 - 3} = \frac{6(1 - 243)}{-2} = \frac{6(-242)}{-2}$$

$$S_5 = 726$$

10. $d = a_2 - a_1 = 2 - 8 = -6$

$a_n = a_1 + (n - 1)d$

$-118 = 8 + (n - 1)(-6)$

$-118 = 8 - 6n + 6$

$-118 = -6n + 14$

$-132 = -6n$

$22 = n$

There are 22 terms in the sequence.

11. $r = \dfrac{a_2}{a_1} = \dfrac{\frac{1}{4}}{\frac{1}{2}} = \dfrac{1}{2}$

$$S_n = \frac{a_1(1 - r^n)}{1 - r}$$

$$S_8 = \frac{\frac{1}{2}\left(1 - \left(\frac{1}{2}\right)^8\right)}{1 - \frac{1}{2}} = \frac{\frac{1}{2}\left(1 - \frac{1}{256}\right)}{\frac{1}{2}} = \frac{255}{256}$$

$$S_8 = 0.996$$

12. $n = 8, a = 3x, b = -y, r = 5$

$$\binom{8}{5 - 1}(3x)^{8 - 5 + 1}(-y)^{5 - 1} = \binom{8}{4}(3x)^4(-y)^4$$

$$= 70(81x^4)y^4 = 5670x^4y^4$$

13. $a_n = \dfrac{(-1)^{2n - 1} n}{n^2 + 2}$

$$a_5 = \frac{(-1)^9(5)}{5^2 + 2} = -\frac{5}{27}$$

The fifth term is $-\dfrac{5}{27}$.

14. $r = \dfrac{a_2}{a_1} = \dfrac{5\sqrt{5}}{5} = \sqrt{5}$

$$S_n = \frac{a_1(1 - r^n)}{1 - r}$$

$$S_7 = \frac{5(1 - (\sqrt{5})^7)}{1 - \sqrt{5}} = \frac{5(1 - 125\sqrt{5})}{1 - \sqrt{5}}$$

$$= \frac{5 - 625\sqrt{5}}{1 - \sqrt{5}} = \frac{5 - 625\sqrt{5}}{1 - \sqrt{5}} \cdot \frac{1 + \sqrt{5}}{1 + \sqrt{5}}$$

$$= \frac{5 + 5\sqrt{5} - 625\sqrt{5} - 3125}{1 - 5} = \frac{-620\sqrt{5} - 3120}{-4}$$

$$S_7 = 1127$$

15. $d = a_2 - a_1 = -2 - (-7) = -2 + 7 = 5$

$a_n = a_1 + (n - 1)d$

$a_n = -7 + (n - 1)(5)$

$\quad = -7 + 5n - 5$

$\quad = 5n - 12$

16. $n = 11, a = x, b = -2y, r = 8$

$$\binom{11}{8 - 1}(x)^{11 - 8 + 1}(-2y)^{8 - 1} = \binom{11}{7}x^4(-2y)^7$$

$$= 330x^4(-128y^7) = -42{,}240x^4y^7$$

17. $r = \dfrac{a_2}{a_1} = \dfrac{\sqrt{3}}{1} = \sqrt{3}$

$a_n = a_1 r^{n - 1}$

$a_{12} = 1(\sqrt{3})^{11} = 243\sqrt{3}$

18. $d = a_2 - a_1 = 13 - 11 = 2$

$a_n = a_1 + (n - 1)d$

$a_{40} = 11 + (40 - 1)(2) = 11 + (39)(2)$

$a_{40} = 89$

$$S_n = \frac{n}{2}(a_1 + a_n)$$

$$S_{40} = \frac{40}{2}(11 + 89) = 20(100) = 2000$$

19. $\dbinom{12}{9} = \dfrac{12!}{(12-9)!\,9!} = \dfrac{12!}{3!\,9!}$

$= \dfrac{12\cdot 11\cdot 10\cdot 9\cdot 8\cdot 7\cdot 6\cdot 5\cdot 4\cdot 3\cdot 2\cdot 1}{(3\cdot 2\cdot 1)(9\cdot 8\cdot 7\cdot 6\cdot 5\cdot 4\cdot 3\cdot 2\cdot 1)}$

$= 220$

20. $\displaystyle\sum_{n=1}^{4} \dfrac{(-1)^{n-1}n}{n+1} = \dfrac{(-1)^{0}(1)}{1+1} + \dfrac{(-1)^{1}(2)}{2+1} + \dfrac{(-1)^{2}(3)}{3+1} + \dfrac{(-1)^{3}(4)}{4+1} = \dfrac{1}{2} + \left(-\dfrac{2}{3}\right) + \dfrac{3}{4} + \left(-\dfrac{4}{5}\right) = -\dfrac{13}{60}$

21. $0.\overline{36} = 0.36 + 0.0036 + 0.000036\ldots$

$0.\overline{36} = \dfrac{36}{100} + \dfrac{36}{10000} + \ldots$

$S = \dfrac{a_1}{1-r} = \dfrac{\dfrac{36}{100}}{1 - \dfrac{1}{100}} = \dfrac{\dfrac{36}{100}}{\dfrac{99}{100}} = \dfrac{36}{99} = \dfrac{4}{11}$

$0.\overline{36} = \dfrac{4}{11}$

The equivalent fraction is $\dfrac{4}{11}$.

22. $r = \dfrac{a_2}{a_1} = \dfrac{1}{3}$

$a_n = a_1 r^{n-1}$

$a_8 = 3\left(\dfrac{1}{3}\right)^{7} = \dfrac{1}{729}$

23. $(x - 3y^2)^5 = \dbinom{5}{0}x^5 + \dbinom{5}{1}x^4(-3y^2) + \dbinom{5}{2}x^3(-3y^2)^2 + \dbinom{5}{3}x^2(-3y^2)^3 + \dbinom{5}{4}x(-3y^2)^4 + \dbinom{5}{5}(-3y^2)^5$

$= x^5 + (5x^4)(-3y^2) + 10x^3(9y^4) + 10x^2(-27y^6) + 5x(81y^8) - 243y^{10}$

$= x^5 - 15x^4y^2 + 90x^3y^4 - 270x^2y^6 + 405xy^8 - 243y^{10}$

24. $r = \dfrac{a_2}{a_1} = -\dfrac{1}{4}$

$S = \dfrac{a_1}{1-r} = \dfrac{4}{1-\left(-\dfrac{1}{4}\right)} = \dfrac{4}{\dfrac{5}{4}} = \dfrac{16}{5}$

25. $r = \dfrac{a_2}{a_1} = \dfrac{\dfrac{4}{3}}{2} = \dfrac{2}{3}$

$S = \dfrac{a_1}{1-r} = \dfrac{2}{1-\left(\dfrac{2}{3}\right)} = \dfrac{2}{\dfrac{1}{3}} = 6$

26. $\dfrac{12!}{5!\,8!} = \dfrac{12\cdot11\cdot10\cdot9\cdot8\cdot7\cdot6\cdot5\cdot4\cdot3\cdot2\cdot1}{(5\cdot4\cdot3\cdot2\cdot1)(8\cdot7\cdot6\cdot5\cdot4\cdot3\cdot2\cdot1)}$

$= 99$

27. $d = a_2 - a_1 = 5 - 1 = 4$

$a_n = a_1 + (n-1)d$

$a_{20} = 1 + (20-1)(4) = 1 + (19)(4)$

$a_{20} = 77$

28. $\displaystyle\sum_{n=0}^{4}\dfrac{1}{n!} = \dfrac{1}{0!} + \dfrac{1}{1!} + \dfrac{1}{2!} + \dfrac{1}{3!} + \dfrac{1}{4!} = 2\dfrac{17}{24} = \dfrac{65}{24}$

29. $d = a_2 - a_1 = -7 - (-3) = -4$

$a_n = a_1 + (n-1)d$

$-59 = -3 + (n-1)(-4)$

$-59 = -3 - 4n + 4$

$-59 = -4n + 1$

$-60 = -4n$

$15 = n$

There are 15 terms in the sequence.

30. $r = \dfrac{a_2}{a_1} = \dfrac{5\sqrt{3}}{5} = \sqrt{3}$

$a_n = a_1 r^{n-1}$

$a_7 = 5\left(\sqrt{3}\right)^6 = 135$

31. $r = \dfrac{a_2}{a_1} = \dfrac{2}{3}$

$S = \dfrac{a_1}{1-r} = \dfrac{3}{1-\left(\dfrac{2}{3}\right)} = \dfrac{3}{\dfrac{1}{3}} = 9$

32. $a_n = \dfrac{10}{n+3}$

$a_{17} = \dfrac{10}{17+3} = \dfrac{10}{20} = \dfrac{1}{2}$

33. $d = a_2 - a_1 = -4 - (-10) = 6$

$a_n = a_1 + (n-1)d$

$a_{10} = -10 + (10-1)(6) = -10 + (9)(6)$

$a_{10} = 44$

34. $d = a_2 - a_1 = -12 - (-19) = 7$

$a_n = a_1 + (n-1)d$

$a_{18} = -19 + (18-1)(7) = -19 + (17)(7)$

$a_{18} = 100$

$S_n = \dfrac{n}{2}(a_1 + a_n)$

$S_{18} = \dfrac{18}{2}(-19 + 100) = 9(81) = 729$

35. $r = \dfrac{a_2}{a_1} = \dfrac{-16}{8} = -2$

$S_n = \dfrac{a_1(1 - r^n)}{1-r}$

$S_6 = \dfrac{8(1-(-2)^6)}{1-(-2)} = \dfrac{8(1-64)}{3}$

$= \dfrac{8(-63)}{3} = 8(-21)$

$S_6 = -168$

36. $\dfrac{7!}{5!\,2!} = \dfrac{7\cdot6\cdot5\cdot4\cdot3\cdot2\cdot1}{(5\cdot4\cdot3\cdot2\cdot1)(2\cdot1)} = 21$

37. $n = 9, a = 3x, b = y, r = 7$

$$\binom{9}{7-1}(3x)^{9-7+1}(y)^{7-1} = \binom{9}{6}(3x)^3 y^6$$

$$= 84(27x^3)y^6 = 2268x^3 y^6$$

38. $\sum_{n=1}^{5}(4n-3) = 1 + 5 + 9 + 13 + 17 = 45$

39. $a_n = \dfrac{n+2}{2n}$

$$a_8 = \dfrac{8+2}{2(8)} = \dfrac{10}{16} = \dfrac{5}{8}$$

40. $d = a_2 - a_1 = 0 - (-4) = 4$

$$a_n = a_1 + (n-1)d$$

$$a_n = -4 + (n-1)(4) = -4 + 4n - 4$$

$$a_n = 4n - 8$$

41. $r = \dfrac{a_2}{a_1} = \dfrac{-2}{10} = -\dfrac{1}{5}$

$$a_n = a_1 r^{n-1}$$

$$a_5 = 10\left(-\dfrac{1}{5}\right)^4 = 10\left(\dfrac{1}{625}\right) = \dfrac{10}{625} = \dfrac{2}{125}$$

42. $0.36\overline{6} = 0.3 + 0.06 + 0.006 + \ldots$

$$0.36\overline{6} = \dfrac{3}{10} + \dfrac{6}{100} + \dfrac{6}{1000} + \ldots$$

$$S = \dfrac{a_1}{1-r} = \dfrac{\dfrac{6}{100}}{1 - \dfrac{1}{10}} = \dfrac{\dfrac{6}{100}}{\dfrac{9}{10}} = \dfrac{6}{90} = \dfrac{2}{30}$$

$$0.36\overline{6} = \dfrac{3}{10} + \dfrac{2}{30} = \dfrac{11}{30}$$

The equivalent fraction is $\dfrac{11}{30}$.

43. $d = a_2 - a_1 = -32 - (-37) = 5$

$$a_n = a_1 + (n-1)d$$

$$a_{37} = -37 + (37-1)(5) = -37 + (36)(5)$$

$$a_{37} = -37 + 180$$

$$a_{37} = 143$$

44. $r = \dfrac{a_2}{a_1} = \dfrac{4}{1} = 4$

$$S_n = \dfrac{a_1(1-r^n)}{1-r}$$

$$S_5 = \dfrac{1(1-4^5)}{1-4} = \dfrac{1(1-1024)}{-3}$$

$$S_5 = \dfrac{-1023}{-3} = 341$$

45. Strategy: Write the arithmetic sequence. Find the common difference, d. Use the formula for the nth Term of an Arithmetic Sequence to find the number of terms in the sequence.

Solution: $d = 3, a_1 = 15, a_n = 60$

$$a_n = a_1 + (n-1)d$$

$$60 = 15 + (n-1)(3)$$

$$60 = 15 + 3n - 3$$

$$60 = 3n + 12$$

$$48 = 3n$$

$$16 = n$$

In 16 weeks, the person will walk 60 min per day.

46. Strategy: Use the formula for the nth Term of a Geometric Sequence to find the temperature after 8 h.

Solution:

$$n = 8, a_1 = 0.95(102) = 96.9, r = 0.95$$

$$a_n = a_1 r^{n-1}$$

$$a_8 = 96.9(0.95)^7 = 67.7$$

The temperature is $67.7°$ F after 8 h.

47. Strategy: Write the arithmetic sequence. Find the common difference, d. Use the formula for the nth Term of an Arithmetic Sequence to find the salary after 9 months. Use the Formula for the Sum of n Terms of an Arithmetic Sequence.

Solution: $2400, $2480, $2560, \ldots$

$d = a_2 - a_1 = 2480 - 2400 = 80$

$a_n = a_1 + (n-1)d$

$a_9 = 2400 + (9-1)(80) = 2400 + (8)(80)$

$a_9 = 3040$

$S_n = \dfrac{n}{2}(a_1 + a_n)$

$S_9 = \dfrac{9}{2}(2400 + 3040) = 24{,}480$

The total salary for the nine-month period is $24,480.

48. Strategy: To find the amount of radioactive material at the beginning of the seventh hour use the Formula for the nth term of a Geometric Sequence.

Solution: $n = 7, a_1 = 200, r = \dfrac{1}{2}$

$a_n = a_1 r^{n-1}$

$a_7 = 200\left(\dfrac{1}{2}\right)^6 = \dfrac{200}{64} = 3.125$

There will be 3.125 mg of radioactive material in the sample at the beginning of the seventh hour.

Chapter 12 Test

1. $\displaystyle\sum_{n=1}^{4}(3n+1) = 4 + 7 + 10 + 13 = 34$

2. $\dbinom{9}{6} = \dfrac{9!}{(9-6)!\,6!} = \dfrac{9!}{3!\,6!}$

$\quad = \dfrac{9 \cdot 8 \cdot 7 \cdot 6 \cdot 5 \cdot 4 \cdot 3 \cdot 2 \cdot 1}{(3 \cdot 2 \cdot 1)(6 \cdot 5 \cdot 4 \cdot 3 \cdot 2 \cdot 1)} = 84$

3. $r = \dfrac{a_2}{a_1} = \dfrac{4\sqrt{2}}{4} = \sqrt{2}$

$a_n = a_1 r^{n-1}$

$a_7 = 4\left(\sqrt{2}\right)^6 = 32$

4. $a_n = \dfrac{8}{n+2}$

$a_{14} = \dfrac{8}{14+2} = \dfrac{8}{16} = \dfrac{1}{2}$

5. $r = \dfrac{a_2}{a_1}$

$r = \dfrac{\frac{1}{8}}{-\frac{1}{4}} = -\dfrac{1}{2}$

$a_2 = 2\left(-\dfrac{1}{2}\right) = -1$

6. $d = a_2 - a_1 = -19 - (-25) = 6$

$a_n = a_1 + (n-1)d$

$a_{18} = -25 + (18-1)(6) = -25 + (17)(6)$

$a_{18} = 77$

$S_n = \dfrac{n}{2}(a_1 + a_n)$

$S_{18} = \dfrac{18}{2}(-25 + 77) = 9(52) = 468$

7. $r = \dfrac{a_2}{a_1} = \dfrac{3}{4}$

$S_n = \dfrac{a_1}{1-r}$

$S = \dfrac{4}{1-\frac{3}{4}} = \dfrac{4}{\frac{1}{4}} = 16$

8. $d = a_2 - a_1 = -8 - (-5) = -3$

$a_n = a_1 + (n-1)d$

$-50 = -5 + (n-1)(-3)$

$-50 = -5 - 3n + 3$

$-50 = -3n - 2$

$-48 = -3n$

$16 = n$

9. $0.2\overline{33} = 0.2 + 0.03 + 0.003 + \ldots$

$0.2\overline{33} = \dfrac{2}{10} + \dfrac{3}{100} + \dfrac{3}{1000} + \ldots$

$S = \dfrac{a_1}{1-r} = \dfrac{\dfrac{3}{100}}{1 - \dfrac{1}{10}} = \dfrac{\dfrac{3}{100}}{\dfrac{9}{10}} = \dfrac{3}{90} = \dfrac{1}{30}$

$0.2\overline{33} = \dfrac{2}{10} + \dfrac{1}{30} = \dfrac{7}{30}$

The equivalent fraction is $\dfrac{7}{30}$.

10. $\dfrac{8!}{4!\,4!} = \dfrac{8 \cdot 7 \cdot 6 \cdot 5 \cdot 4 \cdot 3 \cdot 2 \cdot 1}{(4 \cdot 3 \cdot 2 \cdot 1)(4 \cdot 3 \cdot 2 \cdot 1)} = 70$

11. $r = \dfrac{a_2}{a_1} = \dfrac{2}{6} = \dfrac{1}{3}$

$a_n = a_1 r^{n-1}$

$a_5 = 6\left(\dfrac{1}{3}\right)^4 = \dfrac{6}{81} = \dfrac{2}{27}$

12. $n = 7,\ a = x,\ b = -2y,\ r = 4$

$\binom{7}{4-1}(x)^{7-4+1}(-2y)^{4-1} = \binom{7}{3}x^4(-2y)^3$

$= 35x^4(-8y^3) = -280x^4y^3$

13. $d = a_2 - a_1 = -16 - (-13) = -3$

$a_n = a_1 + (n-1)d$

$a_{35} = -13 + (35-1)(-3) = -13 + (34)(-3)$

$a_{35} = -115$

14. $d = a_2 - a_1 = 9 - 12 = -3$

$a_n = a_1 + (n-1)d$

$a_n = 12 + (n-1)(-3) = 12 - 3n + 3$

$a_n = -3n + 15$

15. $r = \dfrac{a_2}{a_1} = \dfrac{12}{-6} = -2$

$S_n = \dfrac{a_1(1 - r^n)}{1 - r}$

$S_5 = \dfrac{-6(1 - (-2)^5)}{1 - (-2)} = \dfrac{-6(1 + 32)}{3}$

$= \dfrac{-6(33)}{3} = -2(-33)$

$S_5 = -66$

16. $a_n = \dfrac{n+1}{n}$

$a_6 = \dfrac{6+1}{6} = \dfrac{7}{6}$

$a_7 = \dfrac{7+1}{7} = \dfrac{8}{7}$

17. $r = \dfrac{a_2}{a_1} = \dfrac{\dfrac{3}{2}}{1} = \dfrac{3}{2}$

$S_n = \dfrac{a_1(1 - r^n)}{1 - r}$

$S_6 = \dfrac{1\left(1 - \left(\dfrac{3}{2}\right)^6\right)}{1 - \dfrac{3}{2}} = \dfrac{1 - \dfrac{729}{64}}{-\dfrac{1}{2}} = \dfrac{665}{32}$

18. $d = a_2 - a_1 = 12 - 5 = 7$

$a_n = a_1 + (n-1)d$

$a_{21} = 5 + (21-1)(7) = 5 + (20)(7)$

$a_{21} = 145$

$S_n = \dfrac{n}{2}(a_1 + a_n)$

$S_{21} = \dfrac{21}{2}(5 + 145) = 1575$

19. Strategy: Write the arithmetic sequence. Find the common difference, d. Use the formula for the nth Term of an Arithmetic Sequence to find the number skeins in stock in October.

Solution: 7500, 6950, 6400, ...

$d = a_2 - a_1 = 6950 - 7500 = -550$

$a_n = a_1 + (n-1)d$

$a_{10} = 7500 + (10-1)(-550) = 75000 + (9)(-550)$

$a_{10} = 2550$

The inventory after the October 1st shipment was 2550 skeins.

20. Strategy: To find the amount of radioactive material at the beginning of the fifth day, use the Formula for the nth term of a Geometric Sequence.

Solution: $n = 5, a_1 = 320, r = \dfrac{1}{2}$

$a_n = a_1 r^{n-1}$

$a_5 = 320\left(\dfrac{1}{2}\right)^4 = \dfrac{320}{16} = 20$

There will be 20 mg of radioactive material in the sample at the beginning of the fifth day.

Cumulative Review Exercises

1. $3x - 2y = -4$

$-2y = -3x - 4$

$y = \dfrac{3}{2}x + 2$

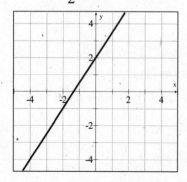

2. $2x^6 + 16 = 2(x^6 + 8)$
$= 2((x^2)^3 + 2^3)$
$= 2(x^2 + 2)(x^4 - 2x^2 + 4)$

3. $\dfrac{4x^2}{x^2 + x - 2} - \dfrac{3x - 2}{x + 2}$

$= \dfrac{4x^2}{(x+2)(x-1)} - \dfrac{3x-2}{x+2} \cdot \dfrac{x-1}{x-1}$

$= \dfrac{4x^2 - (3x^2 - 5x + 2)}{(x+2)(x-1)}$

$= \dfrac{x^2 + 5x - 2}{(x+2)(x-1)}$

4. $f(-2) = 2(-2)^2 - 3(-2) = 2(4) + 6 = 14$

5. $\sqrt{2y}(\sqrt{8xy} - \sqrt{y}) = \sqrt{16xy^2} - \sqrt{2y^2}$

$= \sqrt{16y^2 x} - \sqrt{y^2(2)}$

$= 4y\sqrt{x} - y\sqrt{2}$

6. $2x^2 - x + 7 = 0$

$a = 2 \quad b = -1 \quad c = 7$

$x = \dfrac{-b \pm \sqrt{b^2 - 4ac}}{2a}$

$x = \dfrac{-(-1) \pm \sqrt{(-1)^2 - 4(2)(7)}}{2(2)}$

$x = \dfrac{1 \pm \sqrt{1 - 56}}{4} = \dfrac{1 \pm \sqrt{-55}}{4}$

$x = \dfrac{1}{4} \pm \dfrac{\sqrt{55}}{4}i$

The solutions are $\dfrac{1}{4} + \dfrac{\sqrt{55}}{4}i$ and $\dfrac{1}{4} - \dfrac{\sqrt{55}}{4}i$.

7.
$$5 - \sqrt{x} = \sqrt{x+5}$$
$$\left(5 - \sqrt{x}\right)^2 = \left(\sqrt{x+5}\right)^2$$
$$25 - 10\sqrt{x} + x = x + 5$$
$$-10\sqrt{x} = -20$$
$$\sqrt{x} = 2$$
$$\left(\sqrt{x}\right)^2 = 2^2$$
$$x = 4$$
The solution is 4.

8. $(4, 2)$ and $(-1, -1)$
$$d = \sqrt{(x_1 - x_2)^2 + (y_1 - y_2)^2}$$
$$= \sqrt{(4 - (-1))^2 + (2 - (-1))^2}$$
$$= \sqrt{(5)^2 + 3^2} = \sqrt{25 + 9} = \sqrt{34}$$
Center: $(-1, -1)$
Radius: $r = \sqrt{34}$
$$(x + 1)^2 + (y + 1)^2 = 34$$

9. (1) $3x - 3y = 2$
(2) $6x - 4y = 5$
Eliminate x. Multiply equation (1) by -2.
$$-6x + 6y = -4$$
$$6x - 4y = 5$$
$$2y = 1$$
$$y = \frac{1}{2}$$
Substitute $\frac{1}{2}$ for y in equation (2).
$$6x - (4)\frac{1}{2} = 5$$
$$6x - 2 = 5$$
$$6x = 7$$
$$x = \frac{7}{6}$$
The solution is $\left(\frac{7}{6}, \frac{1}{2}\right)$.

10. $\begin{vmatrix} -3 & 1 \\ 4 & 2 \end{vmatrix} = -3(2) - 4(1) = -6 - 4 = -10$

11. $2x - 1 > 3$ or $1 - 3x > 7$
$\qquad 2x > 4 \qquad\qquad -3x > 6$
$\qquad\ x > 2 \qquad\qquad\ x < -2$
$\{x \mid x > 2\}$ or $\{x \mid x < -2\}$
$\{x \mid x < -2\} \cup \{x > 2\}$

12. $2x - 3y < 9$
$$-3y < -2x + 9$$
$$y > \frac{2}{3}x - 3$$

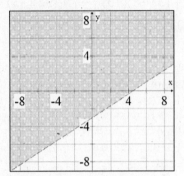

13. $\log_5 \sqrt{\dfrac{x}{y}} = \dfrac{1}{2}[\log_5 x - \log_5 y]$
$$= \frac{1}{2}\log_5 x - \frac{1}{2}\log_5 y$$

14. $4^x = 8^{x-1}$
$$(2^2)^x = (2^3)^{x-1}$$
$$2x = 3(x-1)$$
$$2x = 3x - 3$$
$$3 = x$$
The solution is 3.

15. $a_n = n(n - 1)$
$a_5 = 5(5 - 1) = 5(4) = 20$
$a_6 = 6(6 - 1) = 6(5) = 30$

16. $\displaystyle\sum_{n=1}^{7}(-1)^{n-1}(n+2) = (-1)^0(1+2)+(-1)^1(2+2)+(-1)^2(3+2)+(-1)^3(4+2)+(-1)^4(5+2)+(-1)^5(6+2)+(-1)^6(7+2)$

$$= 1(3)+(-1)(4)+1(5)+(-1)(6)+1(7)+(-1)(8)+1(9)$$
$$= 3-4+5-6+7-8+9 = 6$$

17. $d = a_2 - a_1 = -10-(-7) = -3$

$a_n = a_1 + (n-1)d$

$a_{33} = -7+(33-1)(-3) = -7+(32)(-3)$

$a_{35} = -103$

18. $r = \dfrac{a_2}{a_1} = -\dfrac{2}{3}$

$S_n = \dfrac{a_1}{1-r}$

$S = \dfrac{3}{1-\left(-\dfrac{2}{3}\right)} = \dfrac{3}{\dfrac{5}{3}} = \dfrac{9}{5}$

19. $0.4\overline{6} = 0.4+0.06+0.006+...$

$0.4\overline{6} = \dfrac{4}{10}+\dfrac{6}{100}+\dfrac{6}{1000}+...$

$S = \dfrac{a_1}{1-r} = \dfrac{\dfrac{6}{100}}{1-\dfrac{1}{10}} = \dfrac{\dfrac{6}{100}}{\dfrac{9}{10}} = \dfrac{1}{15}$

$0.4\overline{6} = \dfrac{4}{10}+\dfrac{1}{15} = \dfrac{14}{30} = \dfrac{7}{15}$

The equivalent fraction is $\dfrac{7}{15}$.

20. $n = 6,\ a = 2x,\ b = y,\ r = 6$

$\dbinom{6}{6-1}(2x)^{6-6+1}(y)^{6-1} = \dbinom{6}{5}(2x)^1(y)^5$

$= 6(2x)y^5 = 12xy^5$

21. Strategy: Let x represent the amount of water.

	Amount	Percent	Value
Water	x	0	$0x$
8%	200	0.08	0.08(200)
5%	$200+x$	0.05	$0.05(200+x)$

The sum of the values before mixing is equal to the value after mixing.

Solution:
$0x + 0.08(200) = 0.05(200+x)$
$16 = 10+0.05x$
$6 = 0.05x$
$x = 120$
120 oz of water must be added.

22. Strategy: Let x represent the time required for the older computer.
The time required for the new computer is $x-16$.

	Rate	Time	Part
Older computer	$\dfrac{1}{x}$	15	$\dfrac{15}{x}$
New computer	$\dfrac{1}{x-16}$	15	$\dfrac{15}{x-16}$

The sum of the parts of the task completed must equal 1.

Solution:
$$\dfrac{15}{x}+\dfrac{15}{x-16} = 1$$
$$x(x-16)\left(\dfrac{15}{x}+\dfrac{15}{x-16}\right) = 1(x)(x-16)$$
$$15(x-16)+15x = x^2-16x$$
$$15x-240+15x = x^2-16x$$

$0 = x^2 - 46x + 240$

$0 = (x - 40)(x - 6)$

$x - 40 = 0 \quad x - 6 = 0$

$\quad x = 40 \qquad x = 6$

$x = 6$ does not check as a solution since $x - 16 = 6 - 16 = -10$; $x = 40$, $40 - 16 = 24$ The new computer takes 24 minutes to complete the payroll while the older computer takes 40 minutes.

23. Strategy: Let x represent the rate of the boat in calm water.
The rate of the current is y.

	Rate	Time	Distance
With current	$x+y$	2	$2(x+y)$
Against current	$x-y$	3	$3(x-y)$

The distance traveled with the current is 15 mi. The distance traveled against the current is 15 mi.

Solution:

$2(x + y) = 15$

$3(x - y) = 15$

$\dfrac{1}{2} \cdot 2(x + y) = 15 \cdot \dfrac{1}{2}$

$\dfrac{1}{3} \cdot 3(x - y) = 15 \cdot \dfrac{1}{3}$

$x + y = 7.5$

$x - y = 5$

$\quad 2x = 12.5$

$\quad\quad x = 6.25$

$x + y = 7.5$

$6.25 + y = 7.5$

$y = 1.25$

The boat was traveling at 6.25 mph and the current was moving at 1.25 mph.

24. Strategy: To find the half-life use the exponential decay formula.
$A_0 = 80$, $A = 55$, $t = 30$

Solution: $A = A_0(0.5)^{t/k}$

$55 = 80(0.5)^{30/k}$

$0.6875 = 0.5^{30/k}$

$\log 0.6875 = \log 0.5^{30/k}$

$\log 0.6875 = \dfrac{30}{k} \log 0.5$

$k \log 0.6875 = 30 \log 0.5$

$k = \dfrac{30 \log 0.5}{\log 0.6875} = 55.49$

The half-life is 55 days.

25. Strategy: Write the arithmetic sequence. Find the common difference, d.
Use the formula for the nth Term of an Arithmetic Sequence to find the number of rows.
Use the Formula for the Sum of n Terms of an Arithmetic Sequence.

Solution: 62, 74, 86 …

$d = a_2 - a_1 = 74 - 62 = 12$

$a_n = a_1 + (n - 1)d$

$a_{12} = 62 + (12 - 1)(12) = 62 + (11)(12)$

$a_{12} = 194$

$S_n = \dfrac{n}{2}(a_1 + a_n)$

$S_{12} = \dfrac{12}{2}(62 + 194) = 6(256) = 1536$

There are 1536 seats in the theater.

26. Strategy: To find the height of the ball on the fifth bounce, use the formula for the nth Term of a Geometric Sequence.

Solution: $n = 5$, $a_1 = 80\%$ of $8 = 6.4$
$r = 80\% = 0.8$

$a_n = a_1 r^{n-1}$

$a_5 = 6.4(0.8)^4$

$a_5 = 6.4(0.4096) = 2.62144$

The ball bounces to a height of 2.6 ft on the fifth bounce.

Final Exam

1. $12 - 8[3 - (-2)]^2 \div 5 - 3$

$= 12 - 8(3 + 2)^2 \div 5 - 3$

$= 12 - 8(5)^2 \div 5 - 3$

$= 12 - 8(25) \div 5 - 3$

$= 12 - 200 \div 5 - 3$

$= 12 - 40 - 3$

$= -31$

2. $\dfrac{a^2 - b^2}{a - b}$

$\dfrac{3^2 - (-4)^2}{3 - (-4)} = \dfrac{9 - 16}{3 + 4} = \dfrac{-7}{7}$

$= -1$

3. $f(x) = 3x - 7 \quad g(x) = x^2 - 4x$

$(f \circ g)(3) = f(g(3))$

$g(3) = 3^3 - 4(3) = 9 - 12 = -3$

$f(-3) = 3(-3) - 7 = -9 - 7 = -16$

$(f \circ g)(3) = -16$

4. $\dfrac{3}{4}x - 2 = 4$

$\dfrac{3}{4}x = 6$

$\dfrac{4}{3} \cdot \dfrac{3}{4}x = 6 \cdot \dfrac{4}{3}$

$x = 8$

The solution is 8.

5. $\dfrac{2 - 4x}{3} - \dfrac{x - 6}{12} = \dfrac{5x - 2}{6}$

$12\left(\dfrac{2 - 4x}{3} - \dfrac{x - 6}{12}\right) = 12\left(\dfrac{5x - 2}{6}\right)$

$4(2 - 4x) - (x - 6) = 2(5x - 2)$

$8 - 16x - x + 6 = 10x - 4$

$-17x + 14 = 10x - 4$

$-27x = -18$

$x = \dfrac{18}{27} = \dfrac{2}{3}$

The solution is $\dfrac{2}{3}$.

6. $8 - |5 - 3x| = 1$

$-|5 - 3x| = -7$

$|5 - 3x| = 7$

$5 - 3x = 7 \quad 5 - 3x = -7$

$-3x = 2 \qquad -3x = -12$

$x = -\dfrac{2}{3} \qquad x = 4$

The solutions are $-\dfrac{2}{3}$ and 4.

7.
$$2x - 3y = 9$$
$$2x - 3(0) = 9$$
$$2x = 9$$
$$x = \frac{9}{2}$$

The x-intercept is $\left(\frac{9}{2}, 0\right)$.

$$2(0) - 3y = 9$$
$$-3y = 9$$
$$y = -3$$

The y-intercept is $(0, -3)$.

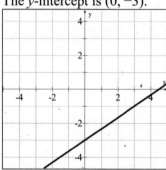

8. $(3, -2)$ and $(1, 4)$

$$m = \frac{y_2 - y_1}{x_2 - x_1} = \frac{4 - (-2)}{1 - 3} = \frac{4 + 2}{-2} = \frac{6}{-2}$$
$$m = -3$$
$$y - y_1 = m(x - x_1)$$
$$y - 4 = -3(x - 1)$$
$$y - 4 = -3x + 3$$
$$y = -3x + 7$$

9.
$$3x - 2y = 6$$
$$-2y = -3x + 6$$
$$y = \frac{3}{2}x - 3$$
$$m_1 = \frac{3}{2}$$
$$m_1 \cdot m_2 = -1$$
$$\frac{3}{2}m_2 = -1$$
$$m_2 = -\frac{2}{3} \quad (-2, 1)$$
$$y - y_1 = m(x - x_1)$$
$$y - 1 = -\frac{2}{3}(x - (-2))$$
$$y - 1 = -\frac{2}{3}(x + 2)$$
$$y - 1 = -\frac{2}{3}x - \frac{4}{3}$$
$$y = -\frac{2}{3}x - \frac{1}{3}$$

10. $2a[5 - a(2 - 3a) - 2a] + 3a^2$
$$= 2a[5 - 2a + 3a^2 - 2a] + 3a^2$$
$$= 2a[5 - 4a + 3a^2] + 3a^2$$
$$= 10a - 8a^2 + 6a^3 + 3a^2$$
$$= 6a^3 - 5a^2 + 10a$$

11. $8 - x^3y^3 = 2^3 - (xy)^3$
$$= (2 - xy)(4 + 2xy + x^2y^2)$$

12. $x - y - x^3 + x^2y = x - y - x^2(x - y)$
$$= 1(x - y) - x^2(x - y)$$
$$= (x - y)(1 - x^2)$$
$$= (x - y)(1 - x)(1 + x)$$

13.
$$\begin{array}{r} x^2 - 2x - 3 \\ 2x-3\overline{)2x^3 - 7x^2 + 0x + 4} \\ \underline{2x^3 - 3x^2} \\ -4x^2 + 0x \\ \underline{-4x^2 + 6x} \\ -6x + 4 \\ \underline{-6x + 9} \\ -5 \end{array}$$

$$(2x^3 - 7x^2 + 4) \div (2x - 3) = x^2 - 2x - 3 + \frac{-5}{2x-3}$$

14. $\dfrac{x^2 - 3x}{2x^2 - 3x - 5} \div \dfrac{4x - 12}{4x^2 - 4}$

$= \dfrac{x^2 - 3x}{2x^2 - 3x - 5} \cdot \dfrac{4x^2 - 4}{4x - 12}$

$= \dfrac{x(x-3)}{(2x-5)(x+1)} \cdot \dfrac{4(x+1)(x-1)}{4(x-3)}$

$= \dfrac{x(x-3) \cdot 4(x+1)(x-1)}{(2x-5)(x+1) \cdot 4(x-3)}$

$= \dfrac{x(x-1)}{2x-5}$

15. The LCM is $(x-3)(x+2)$.

$\dfrac{x-2}{x+2} - \dfrac{x+3}{x-3} = \dfrac{x-2}{x+2} \cdot \dfrac{x-3}{x-3} - \dfrac{x+3}{x-3} \cdot \dfrac{x+2}{x+2}$

$= \dfrac{(x-2)(x-3) - (x+3)(x+2)}{(x-3)(x+2)}$

$= \dfrac{x^2 - 5x + 6 - x^2 - 5x - 6}{(x-3)(x+2)}$

$= \dfrac{-10x}{(x-3)(x+2)}$

16. The LCM is $x(x+4)$.

$\dfrac{\dfrac{3}{x} + \dfrac{1}{x+4}}{\dfrac{1}{x} + \dfrac{3}{x+4}} = \dfrac{\dfrac{3}{x} + \dfrac{1}{x+4}}{\dfrac{1}{x} + \dfrac{3}{x+4}} \cdot \dfrac{x(x+4)}{x(x+4)}$

$= \dfrac{3x + 12 + x}{x + 4 + 3x} = \dfrac{4x + 12}{4x + 4}$

$= \dfrac{4(x+3)}{4(x+1)} = \dfrac{x+3}{x+1}$

17. $(x+5)^2 + 9 = 0$

$(x+5)^2 = -9$

$\sqrt{(x+5)^2} = \sqrt{-9}$

$x + 5 = \pm 3i$

$x = -5 \pm 3i$

The solutions are $-5 + 3i$ and $-5 - 3i$.

18. $a_n = a_1 + (n-1)d$

$a_n - a_1 = (n-1)d$

$d = \dfrac{a_n - a_1}{n-1}$

19. $\left(\dfrac{4x^2 y^{-1}}{3x^{-1} y}\right)^{-2} \left(\dfrac{2x^{-1} y^2}{9x^{-2} y^2}\right)^3$

$\dfrac{4^{-2} x^{-4} y^2}{3^{-2} x^2 y^{-2}} \cdot \dfrac{2^3 x^{-3} y^6}{9^3 x^{-6} y^6}$

$= \dfrac{2^3 \cdot 3^2 y^4 x^3}{4^2 \cdot 9^3 x^6}$

$= \dfrac{y^4}{162 x^3}$

20. $\left(\dfrac{3x^{2/3} y^{1/2}}{6x^2 y^{4/3}}\right)^6 = \dfrac{3^6 x^4 y^3}{6^6 x^{12} y^8} = \dfrac{1}{64 x^8 y^5}$

21. $x\sqrt{18x^2 y^3} - y\sqrt{50x^4 y}$

$= x\sqrt{3^2 x^2 y^2 (2y)} - y\sqrt{5^2 x^4 (2y)}$

$= 3x^2 y\sqrt{2y} - 5x^2 y\sqrt{2y}$

$= -2x^2 y\sqrt{2y}$

22. $\dfrac{\sqrt{16x^5y^4}}{\sqrt{32xy^7}} = \sqrt{\dfrac{16x^5y^4}{32xy^7}} = \sqrt{\dfrac{x^4}{2y^3}}$

$= \sqrt{\dfrac{x^4}{y^2(2y)}} = \dfrac{x^2}{y}\sqrt{\dfrac{1}{2y}} \cdot \sqrt{\dfrac{2y}{2y}}$

$= \dfrac{x^2}{y}\sqrt{\dfrac{2y}{(2y)^2}}$

$= \dfrac{x^2\sqrt{2y}}{2y^2}$

23. $\dfrac{5-2i}{3+4i} \cdot \dfrac{3-4i}{3-4i} = \dfrac{15-26i+8i^2}{9-16i^2} = \dfrac{15-26i-8}{9+16}$

$= \dfrac{7-26i}{25} = \dfrac{7}{25} - \dfrac{26}{25}i$

24. $f(x) = -x^2 + 4x + 2$

$x = -\dfrac{b}{2a} = -\dfrac{4}{2(-1)} = -(-2) = 2$

$f(2) = -(2)^2 + 4(2) + 2$
$= -4 + 8 + 2$
$= 6$

The maximum value is 6.

25. $2x^2 - 3x - 1 = 0$

$a = 2, b = -3, c = -1$

$x = \dfrac{-b \pm \sqrt{b^2 - 4ac}}{2a}$

$x = \dfrac{-(-3) \pm \sqrt{(-3)^2 - 4(2)(-1)}}{2(2)}$

$x = \dfrac{3 \pm \sqrt{9+8}}{4}$

$x = \dfrac{3 \pm \sqrt{17}}{4}$

The solutions are $\dfrac{3+\sqrt{17}}{4}$ and $\dfrac{3-\sqrt{17}}{4}$.

26. $x^{2/3} - x^{1/3} - 6 = 0$

$\left(x^{1/3}\right)^2 - x^{1/3} - 6 = 0$

$u^2 - u - 6 = 0$

$(u-3)(u+2) = 0$

$u - 3 = 0 \quad u + 2 = 0$

$u = 3 \qquad u = -2$

Replace u with $x^{1/3}$.

$x^{1/3} = 3 \qquad\qquad x^{1/3} = -2$

$\left(x^{1/3}\right)^3 = (3)^3 \quad \left(x^{1/3}\right)^3 = (-2)^3$

$x = 27 \qquad\qquad x = -8$

The solutions are -8 and 27.

27. $\dfrac{2}{x} - \dfrac{2}{2x+3} = 1$

$x(2x+3)\left(\dfrac{2}{x} - \dfrac{2}{2x+3}\right) = 1x(2x+3)$

$2(2x+3) - 2x = 2x^2 + 3x$

$4x + 6 - 2x = 2x^2 + 3x$

$0 = 2x^2 + x - 6$

$(2x-3)(x+2) = 0$

$2x - 3 = 0 \quad x + 2 = 0$

$2x = 3 \qquad x = -2$

$x = \dfrac{3}{2}$

The solutions are -2 and $\dfrac{3}{2}$.

28. $x = y^2 - 6y + 6$

$$y = -\frac{b}{2a} = -\frac{-6}{2(1)} = 3$$

$$x = (3)^2 - 6(3) + 6 = 9 - 18 + 6 = -3$$

The axis of symmetry is $y = 3$.

The vertex is $(-3, 3)$.

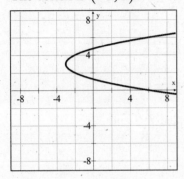

29. x-intercepts: $(4, 0)$ and $(-4, 0)$
y-intercepts: $(0, 2)$ and $(0, -2)$

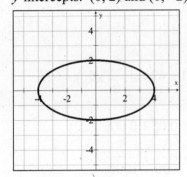

30. $f(x) = \frac{2}{3}x - 4$

$$y = \frac{2}{3}x - 4$$

$$x = \frac{2}{3}y - 4$$

$$x + 4 = \frac{2}{3}y$$

$$\frac{3}{2}(x + 4) = \frac{3}{2}\left(\frac{2}{3}y\right)$$

$$y = \frac{3}{2}x + 6$$

$$f^{-1}(x) = \frac{3}{2}x + 6$$

31. (1) $3x - 2y = 1$
(2) $5x - 3y = 3$
Eliminate y. Multiply equation (1) by -3
and equation (2) by 2. Add the two
equations.
$$-3(3x - 2y) = -3(1)$$
$$2(5x - 3y) = 2(3)$$

$$-9x + 6y = -3$$
$$10x - 6y = 6$$
$$x = 3$$

Substitute 3 for x in equation (1).
$$3(3) - 2y = 1$$
$$9 - 2y = 1$$
$$-2y = -8$$
$$y = 4$$
The solution is (3, 4).

32. $\begin{vmatrix} 3 & 4 \\ -1 & 2 \end{vmatrix} = 3(2) - (-1)(4) = 6 + 4 = 10$

33. (1) $x^2 - y^2 = 4$
(2) $x + y = 1$
Solve equation (2) for y and substitute into
equation (1) by 2.
$$x + y = 1$$
$$y = -x + 1$$
$$x^2 - (-x + 1)^2 = 4$$
$$x^2 - (x^2 - 2x + 1) = 4$$
$$x^2 - x^2 + 2x - 1 = 4$$
$$2x - 1 = 4$$
$$2x = 5$$
$$x = \frac{5}{2}$$

Substitute $\frac{5}{2}$ for x in equation (2).

$$\frac{5}{2} + y = 1$$

$$y = -\frac{3}{2}$$

The solutions is $\left(\frac{5}{2}, -\frac{3}{2}\right)$.

34. $2 - 3x < 6 \qquad 2x + 1 > 4$
$\qquad -3x < 4 \qquad\qquad 2x > 3$

$$x > -\frac{4}{3} \qquad x > \frac{3}{2}$$

The solution is $\left(\frac{3}{2}, \infty\right)$.

35. $|2x+5| < 3$

$-3 < 2x+5 < 3$

$-3-5 < 2x+5-5 < 3-5$

$-8 < 2x < -2$

$-4 < x < -1$

$\{x \mid -4 < x < -1\}$

36. $3x+2y > 6$

$2y > -3x+6$

$y > -\frac{3}{2}x+3$

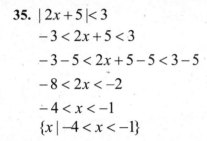

37. $f(x) = \log_2(x+1)$

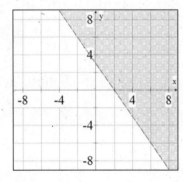

38. $2(\log_2 a - \log_2 b) = 2\log_2 \frac{a}{b} = \log_2 \frac{a^2}{b^2}$

39. $\log_3(x) - \log_3(x-3) = 2$

$\log_3 \frac{x}{x-3} = 2$

$\frac{x}{x-3} = 3^2$

$(x-3)\frac{x}{x-3} = 9(x-3)$

$x = 9x - 27$

$-8x = -27$

$x = \frac{-27}{-8} = \frac{27}{8}$

The solution is $\frac{27}{8}$.

40. $a_i = 2i - 1$

$a_1 = 2(1) - 1 = 1$

$a_{25} = 2(25) - 1 = 49$

$S_i = \frac{i}{2}(a_1 + a_i)$

$S_{25} = \frac{25}{2}(1 + 49) = \left(\frac{25}{2}\right)(50) = 625$

41. $0.\overline{5} = 0.5 + 0.05 + 0.005 + \ldots$

$0.\overline{5} = \frac{5}{10} + \frac{5}{100} + \frac{5}{1000} + \ldots$

$S = \frac{a_1}{1-r} = \frac{\frac{5}{10}}{1 - \frac{1}{10}} = \frac{\frac{5}{10}}{\frac{9}{10}} = \frac{5}{9}$

The equivalent fraction is $\frac{5}{9}$.

42. $n = 9, a = x, b = -2y, r = 3$

$\binom{9}{3-1}x^{9-3+1}(-2y)^{3-1} = \binom{9}{2}x^7(-2y)^2$

$= 36x^7(4y^2)$

$= 144x^7y^2$

43. Strategy: Let x represent the score on the last test.
To find the range of scores, solve the inequality.

Solution:
$$70 \le \frac{64 + 58 + 82 + 77 + x}{5} \le 79$$
$$70 \le \frac{281 + x}{5} \le 79$$
$$350 \le 281 + x \le 395$$
$$69 \le x \le 114$$

Since 100 is the maximum score, the range of scores is $69 \le x \le 100$.

44. Strategy: Let x represent the average speed of the jogger.
The average speed of the cyclist is $2.5x$.

	Rate	Time	Distance
Jogger	x	2	$2(x)$
Cyclist	$2.5x$	2	$2(2.5x)$

The distance traveled by the cyclist is 24 mi more than the distance traveled by the jogger.

Solution: $2x + 24 = 2(2.5x)$
$$2x + 24 = 5x$$
$$24 = 3x$$
$$x = 8$$
$2(2.5x) = 5(8) = 40$
The cyclist traveled 40 mi.

45. Strategy: Let x represent the amount invested at 8.5%.
The amount invested at 6.4% is $12000 - x$.

	Principal	Rate	Interest
8.5%	x	0.085	$0.085x$
6.4%	$12000 - x$	0.064	$0.064(12000 - x)$

The sum of the interest earned from the two investments is $936.

Solution:
$$0.085x + 0.064(12000 - x) = 936$$
$$0.085x + 768 - 0.064x = 936$$
$$0.021x + 768 = 936$$
$$0.021x = 168$$
$$x = 8000$$

$12000 - x = 12000 - 8000 = 4000$
$8,000 is invested at 8.5% and $4,000 is invested at 6.4%.

46. Strategy: Let x represent the width of the rectangle.
The length of the rectangle is $3x - 1$.
The area of the rectangle is 140 ft^2.

Solution: $A = LW$
$$140 = x(3x - 1)$$
$$140 = 3x^2 - x$$
$$0 = 3x^2 - x - 140$$
$$0 = (3x + 20)(x - 7)$$
$$3x + 20 = 0 \quad x - 7 = 0$$
$$3x = -20 \quad x = 7$$
$$x = -\frac{20}{3}$$

The width cannot be a negative number.
$3x - 1 = 3(7) - 1 = 20$
The width is 7 ft.
The length is 20 ft.

47. Strategy: Let x represent the number of additional shares. Write and solve a proportion.

Solution:
$$\frac{300}{486} = \frac{300 + x}{810}$$
$$(810 \cdot 486)\frac{300}{486} = \frac{300 + x}{810}(810 \cdot 486)$$
$$300(810) = 486(300 + x)$$
$$243{,}000 = 145{,}800 + 486x$$
$$97200 = 486x$$
$$x = 200$$

200 additional shares would need to be purchased.

48. Strategy: Let x represent the rate of the car. The rate of the plane is $7x$.

	Distance	Rate	Time
Car	45	x	$\dfrac{45}{x}$
Plane	1050	$7x$	$\dfrac{1050}{7x}$

The total time traveled is $3\dfrac{1}{4}$ h.

Solution: $\dfrac{45}{x} + \dfrac{1050}{7x} = 3\dfrac{1}{4}$

$$\dfrac{45}{x} + \dfrac{150}{x} = \dfrac{13}{4}$$

$$4x\left(\dfrac{45}{x} + \dfrac{150}{x}\right) = \left(\dfrac{13}{4}\right)4x$$

$$180 + 600 = 13x$$

$$780 = 13x$$

$$x = 60$$

$$7x = 7(60) = 420$$

The rate of the plane is 420 mph.

49. Strategy: To find the distance the object has fallen, substitute 75 ft/s for v and solve for d.

Solution: $v = \sqrt{64d}$

$$75 = \sqrt{64d}$$

$$75^2 = \left(\sqrt{64d}\right)^2$$

$$5625 = 64d$$

$$d = 87.89$$

The distance traveled is 88 ft.

50. Strategy: Let x represent the rate traveled during the first 360 mi.
The rate traveled during the next 300 mi is $x + 30$.

	Distance	Rate	Time
First part of the trip	360	x	$\dfrac{360}{x}$
Second part of the trip	300	$x + 30$	$\dfrac{300}{x+30}$

The total time traveled was 5 h.

Solution: $\dfrac{360}{x} + \dfrac{300}{x+30} = 5$

$$x(x+30)\left(\dfrac{360}{x} + \dfrac{300}{x+30}\right) = 5(x)(x+30)$$

$$360(x+30) + 300x = 5x^2 + 150x$$

$$360x + 10800 + 300x = 5x^2 + 150x$$

$$0 = 5x^2 - 510x - 10800$$

$$0 = 5(x^2 + 102x - 2160)$$

$$0 = 5(x+18)(x-120)$$

$$x + 18 = 0 \quad x - 120 = 0$$

$$x = -18 \quad\quad x = 120$$

The rate cannot be a negative number. The rate of the plane for the first 360 mi is 120 mph.

51. Strategy: Write the basic inverse variation equation, replacing the variable with the given values. Solve for k.
Write the inverse variation equation, replacing k with its value. Substitute 4 for d and solve for I.

Solution:

$$I = \dfrac{k}{d^2}$$

$$8 = \dfrac{k}{20^2}$$

$$8 = \dfrac{k}{400}$$

$$3200 = k$$

$$I = \dfrac{3200}{d^2} = \dfrac{3200}{4^2} = \dfrac{3200}{16} = 200$$

The intensity is 200 foot-candles.

52. Strategy: Let x represent the rate of the boat in calm water.
The rate of the current is y.

	Rate	Time	Distance
With current	$x+y$	2	$2(x+y)$
Against current	$x-y$	3	$3(x-y)$

The distance traveled with the current is 30 mi. The distance traveled against the current is 30 mi.
$2(x+y)=30$
$3(x-y)=30$

Solution:
$2(x+y)=30$
$3(x-y)=30$
$\frac{1}{2}\cdot 2(x+y)=\frac{1}{2}\cdot 30$
$\frac{1}{3}\cdot 3(x-y)=\frac{1}{3}\cdot 30$
$x+y=15$
$x-y=10$
$2x=25$
$x=12.5$
$x+y=15$
$12.5+y=15$
$y=2.5$
The rate of the boat in calm water is 12.5 mph. The rate of the current is 2.5 mph.

53. Strategy: To find the value of the investment after two years, use the compound interest formula.
$A=4000, n=24, i=\dfrac{9\%}{12}=\dfrac{0.09}{12}=0.0075$

Solution: $P=A(1+i)^n$
$P=4000(1+0.0075)^{24}$
$P=4000(1.0075)^{24}$
$P\approx 4785.65$
The value of the investment after 2 years is $4785.65.

54. Strategy: To find the height of the ball on the fifth bounce, use the formula for the nth Term of a Geometric Sequence.

Solution: $n=5, a_1=5$
$r=80\%=\dfrac{4}{5}$
$a_5=5\left(\dfrac{4}{5}\right)^5$
$a_5=5\left(\dfrac{1024}{3125}\right)\approx 1.64$

The ball bounces to a height of 1.64 ft on the fifth bounce.